Third Edition

Appleton & Lange's Review for the

RADIOGRAPHY EXAMINATION

Third Edition

Appleton & Lange's Review for the
RADIOGRAPHY EXAMINATION

D. A. Saia, BS, RT(R)
Director, Radiography Program
Department of Radiology
The Stamford Hospital
Stamford, Connecticut

APPLETON & LANGE
Stamford, Connecticut

Copyright © 1997 by Appleton & Lange
A Simon & Schuster Company

97 98 99 00 01 / 10 9 8 7 6 5 4 3 2

Prentice Hall International (UK) Limited, *London*
Prentice Hall of Australia Pty. Limited, *Sydney*
Prentice Hall Canada, Inc., *Toronto*
Prentice Hall Hispanoamericana, S.A., *Mexico*
Prentice Hall of India Private Limited, *New Delhi*
Prentice Hall of Japan, Inc., *Tokyo*
Simon & Schuster Asia Pte. Ltd., *Singapore*
Editora Prentice Hall do Brasil Ltda., *Rio de Janeiro*
Prentice Hall, *Englewood Cliffs, New Jersey*

Library of Congress Cataloging-in-Publication Data

Saia, D. A. (Dorothy A.)
    Appleton & Lange's review for the radiography examination / D.A. Saia.—3rd ed.
      p. cm.
    Includes bibliographical references.
    ISBN 0-8385-0280-6 (pbk. : alk. paper)
    1. Radiography, Medical—Examinations, questions, etc. I. Title.
    [DNLM: 1. Radiography—examination questions. 2. Technology, Radiologic—examination questions.   WN 18.2 S132a 1997]
    RC78.15.S25   1997
    616.07'57—dc20
    DNLM/DLC
    for Library of Congress   96-41411
                                             CIP

Acquisitions Editor: Marinita S. Timban
Production Editor: Jennifer Sinsavich
Production Assistant: Eileen Pendagast

ISBN 0-8385-0246-6

9 780838 502464

90000

PRINTED IN THE UNITED STATES OF AMERICA

*To my husband Tony, who makes every day
a joyous experience.*

Contents

Reviewers

Lynette Biglane, RT(R) (N)
Radiology Phase II Course Supervisor
Keesler Medical Center
Keesler AFB, Mississippi

Henry Y. Cashion, RT
Program Director,
    School of Radiologic Technology
Mineral Area Regional Medical Center
Farmington, Missouri

Cynthia A. Dennis, RT(R)
Program Director,
    Radiologic Technology Program
Louisiana State University Medical Center
Shreveport, Louisiana

Ted Hanson, RT
Nuclear Medicine Technologist
School of Radiologic Technology
Mineral Area Regional Medical Center
Farmington, Missouri

Jeff Pig, RT
Assistant Supervisor
Radiology Department
Mineral Area Regional Medical Center
Farmington, Missouri

Curt Serbus, RT(R), MEd
Program Director,
    Radiologic Technology Program
Assistant Professor of Radiologic Technology
University of Southern Indiana
Evansville, Indiana

Wanda E. Wesolowski, RT, MEd
Professor/Chairperson,
    Radiologic Technology Program
Community College of Philadelphia
Philadelphia, Pennsylvania

To the Student

Feedback on the first and second editions of this book has been encouraging. I sincerely appreciate all the comments and suggestions I have received. I hope that all who use this book, educators and students alike, will continue to provide me with their input in order that this book may continue to meet their needs.

This third edition contains at least 40 percent new material and 50 percent more radiographs and illustrations. It is based on the most recent American Registry of Radiologic Technologists (ARRT) Content Specifications for the Examination in Radiography, having an increased number of questions and explanations in the Patient Care and Management and Radiographic Procedures sections. This edition also uses the ARRT's recently revised *Conventions Specific to the Radiography Examination.*

I am sure you realize that a review book is not intended to be a "quick fix" preparation for the certification examination administered by the ARRT. It takes at least two years of didactic instruction and testing, and hours of clinical practice to prepare oneself as an entry-level radiographer.

As the student radiographer nears graduation, there is, understandably, an anxiety that begins to build. It is a time when you wonder if you are smart enough and if you are skillful enough. Although there will always be room for growth, these concerns arise from the realization that an important landmark has been reached. Formal education will soon be at an end—no more written examinations, no more competencies to complete. You will be on your own, proclaimed competent. How will you perform on the certification examination? How will you perform in the clinical arena? These are indeed sobering thoughts.

During about the last four months of training, the Registry ("R-word") becomes a new and more horrific reality. Confident, competent, even cavalier students suddenly become sober when the "R-word" is mentioned. They begin to question all they ever felt confident about. If you use this book the way it is designed to be used, and perhaps in conjunction with its companion book, *Radiography: Program Review and Exam Preparation,* you should be able to set aside any fears you may have.

I believe that proper use of the materials presented here will help you overcome your anxieties. First, read the introductory section carefully. This material was prepared by Robin Eubanks, Mark Palko, and Dr. Joseph Amato. Ms. Eubanks is a familiar face on the radiography circuit; she has lectured at many state and regional conferences. Her work at the University of Medicine and Dentistry of New Jersey, School of Health Related Professions involves her in motivational and academic workshops where she provides counseling and instruction on study skills and health careers. Mr. Palko is a mathematician and freelance writer whose work at the University of Arkansas includes teaching and developing educational media. Dr. Amato is a licensed psychologist on the staff at The Stamford Hospital in Stamford, Connecticut. He is also an adjunct faculty member in the Graduate School at New York University in the Department of Applied Psychology and maintains a private practice in Greenwich, Connecticut. They present proven, sensible suggestions to help improve test-taking performance. They elaborate on simple processes to help selection of the correct answer, and several methods and strategies that may be employed while taking "the" test. Probably the most important key to reducing apprehension is to reduce the unknowns to the fewest number. Ms. Eubanks, Mr. Palko, and Dr. Amato make an informative presentation and present several easy and effective suggestions for intelligent preparation and test taking.

Secondly, the format and content of the book has been specially designed in many ways to provide focus and direction for your review, and thus to help you do your very best on your certification examination. The ARRT has no secrets and springs no surprises on you. Just as your instructors have made known what is expected of you during your education, the ARRT has made known the content, question format, and terminology used on the certification examination. Each educational program receives the ARRT newsletter, *Educator Update,* and each program director receives the *ARRT Educator's Handbook.* The newsletter is published regularly by the ARRT and functions to keep educators and students current on the activities and policies of the ARRT. The *Educator's Handbook* is also published by the ARRT and serves as a source for general information, policies and procedures, and reprints of a number of documents including *Content Specifications for the Examination in Radiography, Conventions Specific to the Radiography Examination,* and *Standard Terminology for Positioning and Protection.* These documents are revised periodically and advise educators of terminology, categories, content, and approximate weight of content areas on the ARRT examination. Although the *Content Specifications* is by no means a curriculum, it does serve as a suitable guide for examination review and preparation.

It is important to note that the ARRT does not review, evaluate, or endorse publications. Permission to reproduce ARRT copyrighted materials within this publication should not be construed as an endorsement of the publication by the ARRT. It makes sense, however, to design a review book in which the content, question format, and terminology is similar to that which students can expect to find on their certification examination. The ARRT also publishes *Standard Terminology for Positioning and Protection* and advises educators to inform students of this standard terminology. You will find this information reprinted with permission at the beginnings of both Chapter 2, "Radiographic Procedures" and Chapter 6, "Practice Test" in this text. The number of questions found in each chapter is proportional to the number found in that category on the ARRT examination. The questions are designed to test your problem-solving skills and your ability to integrate facts that fit the situation.

Most important and practical, I believe, are the detailed explanations found at the end of each chapter. By themselves, the explanations are good reviews of essential material. Use them to confirm your correct answers and to better understand your weaker areas. You will see that most explanations will tell you not only why the correct answer is correct, but also why the other answer choices (distractors) are incorrect.

Once you have finished reviewing the first five chapters (in the manner suggested by Ms. Eubanks, Mr. Palko, and Dr. Amato), set aside special time for the practice test in Chapter 6. As they have suggested, try to simulate the actual examination environment as much as possible. Choose a quiet place free from distractions and interruptions, gather the necessary materials, and arrange to be uninterrupted for up to three hours.

In summary, use this book as recommended to help ease your pre-certification exam jitters. Excessive anxiety can impair clear thinking and lower your score. Avoiding excessive stress can improve your concentration and information retrieval process. Remember, you have been well prepared by your program director and instructors, and you have studied and worked hard for at least two years. So follow the advice found in the Introduction: prepare yourself sensibly, and keep a positive attitude. I wish you good luck and much happiness in your radiography career!

Acknowledgments

I would like to recognize and express my sincere appreciation to those who again have been so helpful and supportive during the preparation of this third edition.

Many of the radiographs are reproduced through the courtesy of The Stamford Hospital Department of Radiology. Many of the films in Chapter 4, "Image Production and Evaluation," have been reproduced courtesy of the American College of Radiology.

A special word of thanks goes to Mr. Joel R. Schenck, Marketing Product Manager of RMI, for permission to reproduce the mammographic phantom image found in Chapter 4. I wish to express appreciation again to the Dunlee Corporation, and especially to their Marketing Manager, Mr. Roger Flees, for his assistance in granting permission to reproduce various tube rating charts.

A most sincere thank you goes to the professional staff of Appleton & Lange, with special notes of appreciation to Marinita Timban, Amy Schermerhorn, Jennifer Sinsavich and Eileen Pendagast for their support and assistance.

I greatly appreciate the suggestions and expertise shared with me by the professionals who reviewed this project. Wanda Wesolowski, Curt Serbus, Henry Cashion, Jeff Pig, Ted Hanson, Cynthia Dennis, and Lynette Biglane were thorough and timely, and a pleasure to work with.

A very special thank you goes to Judy Vitanza for her substantial contribution to Chapter 1 of this edition; it was wonderful having the opportunity to work with her again. The contributions of Nancy Neligon, always on the lookout for useful radiographs as well, and Mike Sorbara are deeply appreciated. Nancy and Mike contributed to Chapter 3 of this edition. A special thank you again goes to David Perri for use of his expert illustrations.

Special acknowledgment also goes to Robin Eubanks of the University of Medicine and Dentistry of New Jersey, School of Health Related Professions, Mark Palko of the Department of Mathematical Science at the University of Arkansas, and Joseph J. Amato for their contributions to the Introduction of this edition.

Grateful acknowledgment also goes to Roya, Evelyn, Tina, Ray, Anthony, Barbara, Julia, Jacqueline, Kathy, and Melissa, to all former students, and to those who are still to come. Their questions, enthusiasm, and desire to learn make my job a most pleasant task and served as the original stimulation for the preparation of this text.

Most especially, a loving message of appreciation goes to my husband Tony for his love, understanding, and belief in me. Our plans frequently had to be altered as a result of deadline constraints. His encouragement, advice, and assistance were invaluable and deeply appreciated throughout the preparation of this third edition.

Introduction

Part 1: Getting Prepared

HOW THIS BOOK IS ARRANGED

There are three primary sections in this book: a topic by topic review with 1,000 exam-type questions and paragraph-length explanations; a 200-question practice test, also with paragraph-length explanations; and this Introduction, which includes information necessary to help you get the most out of the book and to do your best on the radiography examination.

The contents of the Introduction are arranged in the order they will be needed. Test strategies, study procedures, and guidelines for taking the practice test are covered in Part 1. Tests and test anxiety, suggestions for the day and week of the test, and checklists of what you need to remember before taking the test are covered in Part 2. We suggest reading Part 1 before you begin your review, and Part 2 when the testing date is nearing.

ABOUT THE CERTIFICATION EXAMINATION

The national certification examination for radiography is a standardized, multiple choice test given three times a year (in March, July, and October). It is administered by the American Registry of Radiologic Technologists, lasts three hours, and includes 200 questions. A score of at least 75 percent correct is required to pass.

The test is graded by a computerized device called an optiscanner. An optiscanner uses light sensors to determine which circles on the answer sheet have been filled in and which have been left blank. The computer marks a question incorrect if the wrong circle is filled in, if more than one circle is filled in, or if all the circles are left blank. Optiscanners are considerably more reliable than human graders, but, contrary to popular opinion, machines do make mistakes, particularly after extended use. The sensors may lose sensitivity. To guard against equipment failures, fill in each circle as completely as possible.

Correct .... Incorrect ....

It should take about two or three seconds to correctly mark an answer. This should give you enough time to fill in the circle without going outside the line. (If it takes you more than five seconds, you need to work faster.)

STRATEGIES FOR STUDYING AND TEST TAKING

The purpose of a test strategy is to make the most of your knowledge. No strategy, however elaborate, can help you if you don't know your subject. A good test strategy can do the following: (1) prevent you from making mistakes, (2) help you to use your time efficiently, and (3) improve your odds of getting the right answer.

The single most important trait of a good test strategy is simplicity. There are two ways to make and keep a procedure simple. The first way is to design it to be simple. The second is to practice the procedure as it is designed. The second part is up to you. If you use the following test strategies (particularly the elimination strategy) on all 1200 questions in this book, the strategies will become second nature to you, and you can then concentrate all your attention on passing the test.

Designing a Study Schedule

It is important to establish a routine study schedule. This schedule should allow you to study at a time when you are at your optimum. Some students are more alert in the morning for this kind of work, while others have better success in the afternoon. It would not be a good plan to try and study late at night after a full day unless this is an optimum time for you.

There are several advantages to designing a schedule. The first is that it forces you to face the reality of your study load. Many students underplay the amount of time it will take to complete a thorough study and this can adversely affect performance. If you write out a schedule that includes both your daily responsibilities and the time you need to study, you will have a sense of the pace needed to complete your review.

The second advantage to designing a schedule is that it will allow you to increase your concentration, because the schedule defines the allotted amount of time for each topic you need to cover. Otherwise a lot of time can be wasted in determining what to study during each session.

Time Management

Keeping track of your time and progress is harder than it might first appear. Most of us have been surprised while taking a test by how little time was left. This experience is even more upsetting in the middle of a certification exam. Knowing when there is a problem and knowing what to do about it are the objectives of time management.

Even with your eye on the clock, calculating the time you have left is not always easy. On the radiography examination you have three hours to finish 200 questions. That gives you 56 seconds (0.9 minute) per question. In other words, you have to answer approximately 66.67 questions per hour.

Another way to look at this is by breaking the time into two blocks. If, when you are half way through the allotted examination time, you have finished a minimum of 100 questions, you are working on time. However, there is one additional complication. Not all questions require equal time

to work. It is quite possible to run across a string of difficult questions early in the test and fall behind, then make up the time with easy questions later in the test. For this reason, being a few questions short at the half way mark is not a cause for concern. However, if you have finished significantly less than 100 questions after 90 minutes, you may be starting to fall behind.

If you do fall behind, what can you do to catch up? Sometimes simply seeing that you are behind and trying to work faster will be enough to motivate you to catch up. If not, you have other options. Try to read through the questions and answers faster. If you have checked only one answer as likely to be right, put that choice down immediately; do not reconsider your answer. Never skip a question if you are running seriously behind. Always mark your best choice and move forward. (Skipping is covered in detail in another section.)

As a rule of thumb, if a fact question (one requiring you to recall a fact) takes more than a minute or two, make a mark beside it in the question book and put down your best answer, or leave it for later, and go on. For a calculation problem (one requiring you to calculate some quantity), give yourself an extra minute or two.

Elimination

Good test performance is sometimes determined by the ability to recognize the incorrect answers as well as the correct ones. Eliminating incorrect answers not only improves your score, it actually makes the test a more accurate measure of your knowledge.

Eliminating a distractor reduces the possible wrong choices. If your knowledge allows you to eliminate two incorrect responses, your odds of a correct response would be increased from one out of four to one out of two. If you can eliminate three distractors, you would have a 100 percent probability of getting the right answer. Every distractor you eliminate increases your odds of picking the right answer.

Multiple choice questions usually have one distractor that is obviously wrong, one distractor that is closer to the correct response, and one distractor that is very close to the correct answer. If you know the subject, you can eliminate the distractors that are most incorrect and improve your chances. If you prepare thoroughly, you will be able to eliminate the others. The more you know, the better you will do.

Many books on test taking suggest complicated systems to eliminate bad answers and rank good ones, but in order to use elimination effectively, you need a procedure that is both quick and simple. One possible way is to mark each choice as you read it with a large X if you think it is wrong, or a large check if you think it may be right. If you don't have an opinion, don't mark the choice. What you do next depends on how much time you have. Unless you are falling behind, you may want to re-read the choices you have checked. Pick the one that best addresses the question. If you did not put a check by any choice but have X-ed out some responses, pick the best choice from the unmarked answers.

Skipping Questions

Though you should never leave a question blank when you hand in your test, there are situations when you might want to leave a question blank temporarily and come back to it when you have more time. Skipping *can* allow you to make the best possible use of your time during the test when used cautiously. *However*, it can also distract, slow you down, and cost you a good score if done without discretion.

Deliberately skipping questions has two important advantages. First, it allows you to decide in what order you will answer the questions. This can improve overall performance if you are running short of time because you can strategically miss one or two 3-minute questions instead of five or ten simpler questions that will take only 30 seconds each. (An easy question counts just as much as a hard one toward your final grade.) This is particularly useful for questions involving calculations. Second, skipping gives you a second chance at difficult questions. Everyone has had the experience of trying to remember the answer to a question without success and then finding that piece of information further along in the test. A problem that you stare at for an hour without progress might seem simple if you go on to other problems and come back 5 or 10 minutes later. Very often another question will jog your memory.

One drawback to using the skipping method is that answering the question actually takes longer. When you leave a question and come back to it, you have to reread it completely. Add in the time

spent deciding whether or not to skip the question, and you can end up using as much time as you were trying to save. Another argument against this method falls under the heading of common sense. The secret of successful test taking is in focusing the greatest possible amount of your attention and energy on answering the questions. Anything that distracts you can hurt your overall performance. When you skip questions, it is hard to keep track of which ones remain to be answered, and the possibility of entering an answer in the wrong place greatly increases. This can lead to a string of numbers one line above their proper spaces. If you do decide to use this method for answering some questions, the following rules may help you decide whether or not to skip:

1. If one of the answers seems better than the rest, put it down and mark the question for future reference. Come back and check the question at the end if you have time, but do not skip the question.
2. If you can eliminate two of the possible answers, it may be in your best interest to make an educated guess between the two remaining possibilities. Then mark the question for future reference.
3. Ask yourself, "Will more time really help me answer this question?" If your answer is no, do the best you can with what you know, using the process of elimination and making an educated guess. Again, mark the question for future reference so you can reread it if there is time at the end.
4. Never skip a question during the last 30 minutes.

Whenever you want to take a second look at a question (regardless of whether you left it blank), mark it clearly with a large slash, X, or arrow next to the problem and with a similar mark in the upper corner of the page. The marks have to be easy to see and quick to make (remember, you are doing this to save time).

Guessing

You've probably been given a great deal of information and advice about guessing on tests. Most of what you've been told may be confusing or contradictory. It may make the problem easier to think in terms of rolling a die. Imagine a game where you get a point every time the number 1, for example, comes up. How could you improve your score in this game? One way would be to roll the die as many times as you could. Another way would be to reduce the number of sides on the die so the "right" side would be more likely to come up; this way is called *the process of elimination,* and it plays a good part in test taking when you are unsure of the correct response.

You can improve your overall score on the radiography exam by making sure you attempt to answer every question you would have otherwise left blank. The radiography examination does not have penalties for wrong answers. An incorrect answer receives the same score as a blank, so it is in your best interest to make an educated guess whenever you are unsure.

Although we do **not** suggest guessing as an effective method of test taking, we do recognize that there will be times when it can be effective for you. Keep in mind the following things if you need to use this method:

1. The process of elimination will help you significantly in determining the right answer. Use this technique to narrow down the possible choices.
2. Mark the question so that, if you have time, you can come back to it. It is possible that the correct response may reveal itself through a question further ahead on the test.
3. Remember that guessing really doesn't work as an effective strategy by itself. You will need to study hard and use guessing in conjunction with other methods for it to be effective.

The Check-on-Five Rule

Probably the most aggravating experience any test-taker has ever had was looking down at the answer sheet and realizing that the last 20 or 30 answers were all one line off. Because the question book and the answer sheet of a standardized test are separate, this mistake is easy to make, particularly when you are leaving questions blank. One way of preventing this mistake is to pause before answering each question, read the number of the question, read the number of the answer, and then write down your answer. However, this could add 5 minutes to your time if you check all 200 questions. Checking after every fifth question is a convenient way that will not take too much time and

could save you from making costly errors in mismarked answers. In other words, check the number on the answer sheet whenever you get to a question number ending in "5" or "0." Do not check on the other four unless you have just skipped a question. More than any other technique described in this book, the Check-on-Five rule should be pure, undiluted habit.

SETTING UP A STUDY PLAN

After completing the best of radiography programs, even the best of students will have gaps in his or her knowledge, subjects that were somehow missed or forgotten, or that will not come to mind when needed. These gaps in your knowledge are often small; but since one piece of information often builds upon other pieces, a small gap in your knowledge can sometimes lead to a large drop in your test score. The best way to get around this problem is to use a well defined study plan. Listed below are two alternative plans for you to consider. The first is diagnosis and remediation, and the second is "SQ3R."

Diagnosis and Remediation

This is a two step approach: *diagnosis* (finding out what you do not know), and *remediation* (learning the material).

Diagnosis. Many students graduate from their programs without a good idea of what they do or do not know. Fortunately, this book has been designed to make diagnosis simple. By following the steps listed below, you will know what you need to learn before you take the certifying examination.

Step 1. Begin with Chapter 1, Patient Care and Management, or any of the other first five chapters. Go through the questions in one sitting, making the experience as similar to the actual exam as possible. Write your answers on a copy of the answer sheet in the back of the book. (Remember to practice test strategies while answering the questions. This will produce a more valid diagnosis.)

While taking the test, you should note or highlight words and phrases from the questions that you do not understand. After you have finished the questions (but before you have graded your work) make a list of the terms you noted and the number of the question that contained it.

Step 2. Analyze your results. Read the answers and make a list of the questions you missed. Compare this list with the subspecialty list at the end of the chapter. This will tell you if you are weak in a particular area. Once you have defined an area of weakness, pay special attention to the explanations provided. If the answer is still unclear, use the exact page references to your textbook for further study.

Anytime you go through your work, picking out and correcting your mistakes, you will gain a greater understanding of your strengths and weaknesses. However, by approaching the analysis systematically, the improvement can be dramatic. Concentrate on your areas of weakness, but be sure to read all the explanations at least once. This will allow you to compare your reasoning on right and wrong answers and to check for the possibility that you put down the right answer for the wrong reason.

Step 3. Repeat the process. The purpose of this study plan is to get important information into your long-term memory. The best way to ensure this is to begin your study plan early enough to allow yourself time to repeat your chapter study one more time before the examination. Keep and compare your results from each review and focus on any weaknesses still apparent from the comparison.

Remediation

Step 1. Read and cross-read. Starting with the subspecialty that you missed most often, make a reading list. For those areas in which you missed three or four questions, a single reference will probably be enough, but if you missed more than four, you should cross-read to cover the same information in more than one text. (You might also want to review these topics in your old class notes.)

When you study from texts, use the index and the table of contents to find the section you need. If you are using more than one text, compare and look for common ideas. Sometimes writing a summary of your reading helps to clarify the information. This technique has been proven to improve

retention and understanding, but it can be time consuming.

Step 2. Once you have finished your reading, go back to the questions that you missed. If they still aren't clear, consult an expert. Most students are reluctant to approach an instructor with a question that doesn't relate directly to a class. However, most instructors are glad to answer questions that will improve the chances for their students to obtain a high passing score on the certification examination. Instructors appreciate questions that are specific, well thought out, and which show that the student has done some independent work.

SQ3R

The second method for study is best suited for reviewing your textbooks for further study once you have identified a weakness. It is called the SQ3R and is presented by Frances P. Robinson in his book *Effective Study*. It makes study reading more efficient and long-term remembering more probable. SQ3R stands for Survey, Question, Read, Recite, and Review. The steps are as follows:

Survey. First, skim through an entire chapter.

1. Think about the title of the chapter. What do you already know about the subject? Write ideas in the margins. Read the conclusion. What better way is there to discover the main ideas of a chapter?
2. Read the headings. These are the main topics that have been developed by the author.
3. Read the captions under the diagrams, charts, and graphs.

Allow approximately 10 to 12 minutes for the survey step. Surveying will help increase your focus and interest in the material.

Question. Write out two or three questions relating to each heading. These should be questions that you believe will be answered within each section. Use the "who, what, when, where, why, and how" application when generating these questions.

Read. Now read the first section. Keep in mind the questions you have created and read, with a purpose, as quickly as possible.

Recite. At the end of the section, look away from the book for a few seconds. Recite and think about what you've just learned. It is best to recite aloud because hearing the information will help increase your memorization.

Review. Reviewing is a key step if you want to retain the material you've read. Reviewing as you study results in less time needed for test preparation.

Learning to utilize the SQ3R method is a skill that takes practice. Often times students feel that it takes too long and is too complicated, but its use results in an increase in comprehension, interest, and memorization. The SQ3R method allows you to study at the same time that you are doing your course reading.

Summary

Everyone does not have the same learning style, and, as a result, effective study techniques are not the same for everyone. It is important to choose the method that is best for you, and this will take some experimentation. For instance, some students become frustrated when they cannot comprehend the textbook material while reading when seated at a desk. Sometimes just getting up and pacing while memorizing can facilitate the learning process if you are having trouble at your desk. Other students learn faster with audio aids. If these are available to you and you are having trouble with learning just from your books, it may be worth the experiment to see if hearing the material will enhance your learning.

STUDY GROUPS

While preparing for an examination, properly organizing or attending a study group can be extremely helpful. However, a study group needs to be very focused with a specific agenda for each session. Otherwise it can be a time waster. Listed below are some important points to keep in mind when organizing and conducting a study group:

1. Limit the group size to four or five people.
2. Select classmates who share your academic goals.
3. Meet the first time to discuss the meeting times, meeting place, and group goals.

4. Select a group leader and time keeper. The group should meet for 2 to 3 hours for each session to ensure a thorough review.
5. Establish an agenda for each meeting that specifies the topics of discussion. This will save time and lend focus to the group. It ensures that the group reviews all pertinent topics by slotting time for all areas.
6. Establish group norms that define how the group will act. This would include things such as getting there on time, being prepared, and ending on time. It is important to emphasize that all members must do their fair share of the work for the group to gain the maximum benefit.

Study groups are used to review and compare both lecture and reading notes, to review textbook information together, and to review examination topics. Group sessions are a good time to review the question types used on the certification examination, to discuss test-taking strategies, to help each other design study plans, and to drill or review together all material expected to be on the examination.

The support of a study group is extremely helpful in building self-confidence and in overall preparation for examinations. They are not intended to replace individual study time, but serve as a supplement. If properly utilized, study groups are an enormous asset for test preparation. Finally, when working within a group, many students are more likely to exert their best effort because they are accountable to the other members of the group.

THE PRACTICE TEST

The practice test (Chapter 6) can be used in one of the following two ways: (1) as a way of determining strengths and weaknesses before you go through the review book, or (2) as a final preparation for the test after you have done your chapter by chapter review.

The practice test has been designed with an integrated format in an effort to duplicate the experience of taking the certification examination. Taking the practice test will make the process more familiar, so you won't be as nervous when you face the real test. The practice test will help you to determine whether or not you are answering the questions quickly enough, and whether your score is high enough to pass. In summary, the practice test simply gives you a chance to practice.

Taking the Practice Test

In order to use the practice test to determine how long the test will take you to complete or how high you will score, you must take it under conditions matching, as close as possible, the actual test conditions. If you try to eat supper while taking the test, take a 5 hour break in the middle of the test, or stop after every question to look up the answer, you will not get a clear picture of your current standing or potential to pass the exam. Following are some suggestions on how you can get the most out of the practice test:

1. Keep your schedule completely free. Find a time and a place that will guarantee that you won't be disturbed for the duration of the test. Most libraries work well for this purpose, as do unoccupied classrooms if you can get access to them. If you have to take the test at home, make sure that you won't be bothered by friends or family. Minimize your distractions, don't take phone calls and don't try to watch TV or concentrate on anything else other than the test.

2. Start at a predetermined time. The test is given on the afternoon of the same day at all the testing centers across the country. You may choose to take the practice test at the same time of day your test will be given.

3. Bring everything you will need to take the test. You will need a watch, two calculators, and five or six no. 2 pencils (sharpened). Make sure you don't need to stop the test to get anything once you have started.

4. Approach the practice test with the same strategies and attitudes that you plan on using with the actual examination. (Reviewing the section on test strategies would be a good idea.)

5. Note time-consuming questions. While taking the test, mark the questions that take longer than 2 or 3 minutes, including any question that you skip and answer later.

Don't spend too much time on any one question.

6. Note how far you get. You should be able to finish the whole test in the allotted time, but if you do run out of time, draw a line across the test book to show how far you got and then finish the rest of the test.

Checking Your Results

You should be able to finish all of the questions with enough time left over to go back and make sure that you answered all of the questions and to check your answers on those problems you marked as difficult. Your score should be at least 160 correct answers (80 percent). If you fail at either of these two goals, you need to go over the test carefully and try to analyze your problem.

Two questions you can address while analyzing a problem are "Was there a common factor in the questions that gave me trouble?" and "What were the subspecialties of the questions that I missed?" If you keep missing the same type of question, the problem could be easy to fix. Try going back and reworking the section of the book that corresponds to that topic.

Review your test-taking techniques. Did you spend too much time on a few questions? Did you skip too many questions and spend most of your time going back and forth in the test? Did you spend too much time rereading answers that you had already crossed out? If the answer is yes to any of these questions, you might want to review the earlier section on test strategies.

TEST STRATEGY CHECKLIST

1. CROSS AND CHECK
   Cross out bad answers and put checks next to good answers.
2. DO 50 EVERY 45
   Try to do at least 50 problems every 45 minutes.
3. NEVER SKIP IN THE LAST 30 MINUTES
   When nearing the end of the test, answer all questions.
4. MARK UNCERTAIN ANSWERS
   Mark questions you guessed on so you can come back and double-check them.
5. CHECK-ON-FIVE
   Check the problem number on every fifth problem.

Part 2: Optimizing Your Results

PREPARING FOR THE EXAM

When you take the radiography examination, two factors determine your score: your knowledge of the subject and your performance on the exam. Of the two, knowledge is the key element needed for success. Performance is harder to guarantee. For some people, it seems to come naturally. These people apply test-taking strategies almost unconsciously, concentrating all their attention on the test. For the rest of the population (which includes most people), standardized tests are some of the most stressful and unpleasant experiences that they will ever have to face.

Fortunately, it is possible to significantly improve your performance on tests, even if you've been taking tests all of your life with no apparent improvement. By mastering the following three areas, you can have better results and greatly reduce the trauma of test taking:

1. Know your test strategies.
2. Learn to manage your stress.
3. Avoid surprises on the day of the test.

The important points for each of these areas are explained in detail in the following sections and are summarized in checklists included at the end of each part of the Introduction. These checklists are designed so that you can read them on the day of the test to reassure yourself that you haven't forgotten anything.

TEST STRATEGIES

The strategies recommended in Part 1 of this Introduction serve two purposes: first, to give you a simple, systematic way of eliminating bad choices and improving your odds of getting a correct response; and second, to help you manage your time most efficiently during the test. A test strategies checklist is provided at the end of Part 1 and summarizes the various strategies described in the Introduction.

MANAGING STRESS

Over the past twenty or so years, educators have become increasingly concerned about the problems of test anxiety and excessive stress that prevent students from doing their best on examinations. In the following sections, you will learn some basics about the nature of stress, the difference between good stress and bad stress, and some management techniques. All of the items listed will be helpful in reducing test anxiety, but the most important point for you to remember is that you are well prepared for the test you are about to take and the odds of your doing well are very good.

Where You Stand

The best way to reduce test anxiety is to address the following points: (1) know the subject, (2) master a test-taking routine, (3) avoid surprises, (4) understand the role of stress in test taking, and (5) practice relaxation techniques.

The Stress Curve

Recently, stress has received national attention. Magazines discuss it, doctors warn against it, commercials promise to reduce it, and seminars claim to eliminate it. Stress is often treated as a psychological cancer.

There is, however, another aspect of stress that receives less attention. In demanding situations requiring optimum performance, moderate stress is not only natural, it is actually helpful. The relationship between stress and performance is called the stress curve, and is illustrated in Figure 1. The most important aspect of the curve is the location of the maxima. The maxima is the point of optimum performance, which occurs somewhere between too much stress and no stress at all.

There are plenty of familiar examples of stress improving performance. Athletes often set personal records during pressure situations such as playoff games or international events. Actors give their best performances and musicians play or sing best before an audience. You can probably think of personal examples as well. Most of us have surprised ourselves at one time or another by doing better than we expected under pressure.

How to Recognize Good Stress

If everyone dealt with stress equally well and experienced the same level of stress in the same situation, setting up guidelines for optimal stress levels would be easy. Unfortunately, everyone handles stress differently, and determining what level of stress is best has to be judged on an individual basis. Given the importance of the radiography exam and the amount of time you have spent preparing for it, there is little chance of your stress level being too low when you sit for the test.

How do you know if you have too much stress? Feeling nervous does not indicate excessive stress. You are just as likely to feel nervous when you are at your optimal stress level. Stress is excessive when it interferes with the test-taking process. If you have trouble reading the questions, if you lose your place because you're worrying about the test, or if you're too distracted to follow the test strategies you've been practicing, these are all signs of test anxiety.

Relaxation Techniques

Relaxation techniques should become part of your overall examination preparation. It is simple logic that a skill that is practiced regularly will be available when you need it most. This principle applies also to relaxation techniques. If you practiced inducing a relaxed state while preparing for the examination, confronting the particular material and questions, you will be able to call on your memory for relaxation more easily. Simply put, you can create a "mini" relaxed state by a few deep breaths. However, this ability will be determined by your previous practice of these exercises.

It is a good idea to take a couple of minutes to relax before the test. Stretch your muscles; take deep, slow breaths; and try to think about something unrelated to the test. If, during the test, you have trouble working effectively because of stress, stop, close your eyes, and count to five while taking some deep, slow breaths. Remind yourself that you are extremely well prepared for this test. Not many people realize that breathing and anxiety are related, or that a deep breathing relaxation exercise can be helpful in reducing anxiety.

When you feel anxious, you tend to tighten your chest muscles. "This results in breathing changes with movement predominately in your upper chest rather than your lower abdomen. Breathing this way often leads to undesirable conditions.

1. A reduction of oxygen in your blood, which can affect the way you think and lead to anxiety or fatigue;

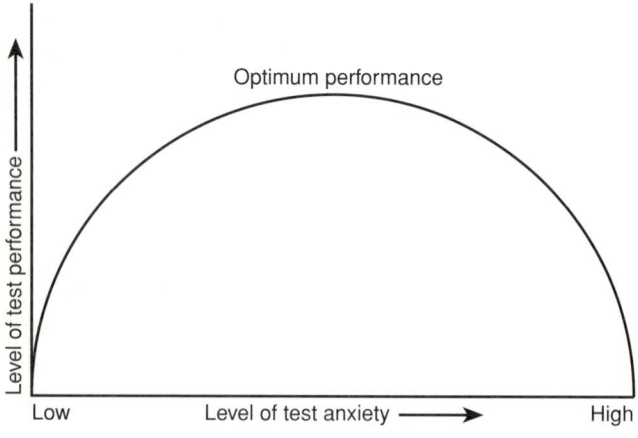

Figure 1. The Stress Curve.

2. Too much oxygen in your blood, resulting in an uncomfortable condition called hyperventilation." [1]

It is important, therefore, to learn how to control your breathing, which, in turn, will result in relaxation. When attempting the following breathing exercise, don't try too hard to relax—it can work against you. Instead, try to be as peaceful as possible.

First, take a deep full breath and exhale fully and completely. Next, inhale again, mentally counting from 1 to 4 while breathing in. Hold this breath in while again counting from 1 to 4. Then, begin fully exhaling, while slowly counting from 1 to 8. Repeat this sequence four times. If you run out of breath before reaching number 8, take deeper breaths and exhale more slowly. If you can learn this technique ahead of time, you can use it while in the exam room without anyone else knowing.

If the stress continues, try to think less about what you are doing. After practicing on 1200 questions, you have developed a kind of "automatic pilot," which will allow you to answer questions almost by reflex. Of course, this isn't the best way to take the test, but if you are faced with serious stress problems, it is an option.

Self-Statements

Examinations begin early in life, in grade school if not before. These early experiences may have had a profound effect on the way we feel about ourselves in the present. The internal message we have learned to give ourselves in the face of examinations is critical to the way we will interpret the experience. The manner in which we interpret the examination experience will determine how stressful it may be and how well we may handle the test.

During the examination preparation period, it is important to assure yourself regularly that you are competent to master the material and to demonstrate your mastery. Simple exercises consisting of repeating the phrase "I can" or "I am competent" have a way of becoming a self-fulfilling prophecy. Unfortunately, opposite self-statements have a similar influence. Practicing positive self-statements should be considered throughout the preparation process.

[1] Nieves L. *Coping in College.* Princeton, NJ: Educational Testing Service; 1984, pp 53, 66.

AVOIDING SURPRISES

The Week Before the Test

You have spent the past 2 (or more) years studying radiography, and the past 2 to 4 months reviewing for this examination. You probably know a great deal more than you think you do. Your top priority now should be getting yourself up to your best testing performance. If you follow these suggestions, you should have a good start.

Take Care of Yourself. When you take the test, you want to be as healthy and well rested as possible. The time it takes to get enough sleep, take a walk, or prepare a balanced meal is better spent than hours of last minute cramming.

Reread This Chapter. This may be unnecessary advice, but it is worth mentioning. Pay close attention to the figures and checklists.

Gather Your Supplies. You want to be sure to get everything you need, but, almost as importantly, you want everything organized so you can avoid extra effort and worry. You need to start with two small pouches, preferably transparent. A self-sealing plastic bag would be ideal, but a couple of old bread bags will do fine. One bag is for the things that will go on your desk (or table top) when you take the test; the other bag goes under your chair. Both bags should contain five or six sharpened no. 2 pencils and a calculator. (Remember to make sure that both calculators work.)

The calculator you use during the test should be the one with which you are most comfortable and familiar. Do not, under any circumstances, bring a calculator that you don't know how to use. You are far better off with a simple calculator that you understand than a sophisticated calculator that confuses you.

You also need a watch, in case you don't have a good view of the clock in the exam room. You can wear the watch during the exam, but many test-takers prefer to place it on the desk next to the test, so they can see it at all times and keep both hands free.

Finally, you probably need to find a sweater or extremely light jacket. This may seem like a strange item, particularly if your test is given in the summer, but an uncomfortably cold room is extremely distracting and many public buildings have a wide variation in temperature from room to room. A sweater or windbreaker is a quick, easy solution.

Scout the Location. It is a good idea to visit the location of the exam if possible. Getting lost on the morning of a test can add unwanted stress. Keep in mind the following questions: (1) What is the best route to the exam? (2) Where is the parking? (3) Where are the doors to the building? If possible, go into the building and look around.

Treat Yourself. Go out and do something special. See a movie. Eat out. Go for a drive. Just let yourself unwind.

The Day of the Test

There are few things more irritating then being told not to worry when you feel like worrying, but not worrying is the best thing to do. To help you avoid worrying, two checklists have been included at the end of Part 2, one physical checklist (things you need to bring) and one mental checklist (things you need to remember). Check off each item and put it out of your mind. Knowing you have both mentally and physically prepared will help you to relax before the exam.

You should plan to arrive a few minutes early. The test itself is preceded by 15 to 30 minutes of instructions, so be prepared to wait.

You will probably want to eat a light lunch before taking the test. Digestion tends to slow down when a person is under stress, so a large meal is, in most cases, a bad idea. You will also want to avoid excessive stimulants (and caffeine is definitely considered a stimulant).

This brings up another point. Though you want to be well-rested and alert, you should be careful not to disrupt your normal routine any more than necessary. Getting extra rest, eating light, and avoiding stimulants are relative suggestions—relative to your habits and lifestyle. Don't make drastic changes on the day of the test. Get an extra half hour or hour of sleep. Eat a lighter lunch than you usually would. If you drink coffee, drink a little less than normal.

When you get to the testing center, take a few minutes to relax. Walk around. Stretch your muscles. Remind yourself that you've put a great deal of work into doing well on this test and that work is the main factor for determining success.

MENTAL CHECKLIST
(What to remember)

1. Remind yourself that you are well prepared.
2. Fill in all the ovals thoroughly.
3. Use your test strategies.
4. Focus on the test, not the surroundings.
5. After you have finished, go back and make sure you have answered every question.

PHYSICAL CHECKLIST
(What to bring)

1. Sweater or Windbreaker
2. First Bag:
   a. Watch
   b. Calculator
   c. No. 2 Pencils (4 to 6)
3. Second Bag:
   a. Calculator
   b. No. 2 Pencils

REFERENCES

Benson H. *The Relaxation Response.* New York: Avon Books; 1975

Bosworth S, Brisk M. *Learning Skills for the Science Student.* Florida: H&H Publishing Company; 1986

Bragstad B, Stumpf S. *A Guidebook for Teaching: Study Skills and Motivation.* Boston: Allyn and Bacon; 1982

Coffman S. *How to Survive at College.* Indiana: College Town Press; 1988

Dobbin J. *How to Take a Test: Doing Your Best.* Princeton, NJ: Educational Testing Service; 1984

Kesselman-Turkel J, Peterson F. *Test-Taking Strategies.* Chicago: Contemporary Books, Inc.; 1981

Krantz H, Kemmelman J. *Keys to Reading and Study Skills.* New York: Holt, Rinehart & Winston; 1985

Nieves L. *Coping in College.* Princeton, NJ: Educational Testing Service; 1984

Robinson F. *Effective Study.* New York: Harper and Row; 1970

Master Bibliography

Here is a list of reference books pertaining to the Answers and Explanations sections found at the end of each chapter in this book.

On the last line of each answer/explanation, there appears the last name of the author or editor of one of the publications listed here, along with a number or numbers indicating the correct page or range of pages where information relating to the correct answer may be found. For example, (*Bushong, p 45*) refers to the 45th page of Bushong's *Radiologic Science for Technologists*.

American Red Cross Community CPR. American Red Cross; 1988

Ballinger PW. *Merrill's Atlas of Radiographic Positions and Radiologic Procedures, Vols 1, 2, and 3,* 8th ed. St. Louis, MO: Mosby; 1995

Bontrager KL, Anthony BT. *Textbook of Radiographic Positioning and Related Anatomy,* 2nd ed. St. Louis, MO: Mosby; 1987

Burns EF. *Radiographic Imaging: A Guide for Producing Quality Radiographs.* Philadelphia: Saunders; 1992

Bushong SC. *Radiologic Science for Technologists,* 5th ed. St. Louis, MO: Mosby; 1993

Cahoon JB. *Formulating X-ray Techniques,* 8th ed. Durham, NC: Duke University Press; 1984

Carlton RR, Adler AM. *Principles of Radiographic Imaging.* 2nd ed. Albany, NY: Delmar; 1996

Carroll QB. *Fuchs's Radiographic Exposure, Processing and Quality Control,* 5th ed. Springfield, IL: Charles C Thomas; 1993

Chapman S, Nakielny S. *A Guide to Radiological Procedures,* 2nd ed. Philadelphia: Saunders, 1986

Cullinan AM. *Producing Quality Radiographs.* 2nd ed. Philadelphia: Lippincott; 1994

Curry TS, Dowdey JE, Murry RC. *Christensen's Introduction to the Physics of Diagnostic Radiology,* 4th ed. Philadelphia: Lea & Febiger; 1990

Dowd SB, Wilson BG. *Encyclopedia of Radiographic Positioning.* Philadelphia: Saunders, 1995

Ehrlich RA, McCloskey ED. *Patient Care in Radiography,* 3rd ed. St. Louis, MO: Mosby; 1989

Eisenberg RL, Dennis CA. *Comprehensive Radiographic Pathology.* St. Louis, MO: Mosby; 1990

Gurley LT, Callaway WJ. *Introduction to Radiologic Technology,* 3rd ed. St. Louis, MO: Mosby; 1992

Hiss SS. *Understanding Radiography,* 2nd ed. Springfield, IL: Charles C Thomas; 1988

Hopp JW, Rogers EA. *Aids and the Allied Health Professions.* Philadelphia: F. A. Davis; 1989

Laudicina P. *Applied Pathology for Radiographers.* Philadelphia: Saunders; 1989

Martini FH. *Fundamentals of Anatomy and Physiology,* 3rd ed. Englewood Cliffs, NJ: Prentice Hall; 1995

McKinney WEJ. *Radiographic Processing and Quality Control.* Philadelphia: Lippincott; 1988

Miller RD. *Problems in Hospital Law,* 5th ed. Rockville, MD: Aspen; 1986

NCRP Report no. 105. *Radiation Protection for Medical and Allied Health Personnel.* National Council on Radiation Protection and Measurements; October 30, 1989

Noz ME, Maguire GQ Jr. *Radiation Protection in the Radiologic and Health Sciences.* Philadelphia: Lea & Febiger; 1985

Pizzutiello RJ, Cullinan FE. *Introduction to Medical Radiographic Imaging.* Rochester NY: Eastman Kodak Co; 1993

Saia DA. *Radiography Program Review and Exam Preparation,* 1st ed. Stamford, CT; Appleton & Lange; 1996

Selman J. *The Fundamentals of X-ray and Radium Physics,* 8th ed. Springfield, IL: Charles C Thomas; 1994

Statkiewicz MA, Ritenour ER. *Radiation Protection for Student Radiographers.* 2nd ed. St. Louis, MO: Mosby; 1993

Sweeney RJ. *Radiographic Artifacts: Their Cause and Control.* Philadelphia: Lippincott; 1983

Taber's Cyclopedic Medical Dictionary, 17th ed. Thomas CL, ed. Philadelphia: F. A. Davis; 1993

Thompson MA et al. *Principles of Imaging Science and Protection,* 1st ed. Philadelphia, WB Saunders, 1994

Torres LS. *Basic Medical Techniques and Patient Care for Radiologic Technologists,* 3rd ed. Philadelphia: Lippincott; 1989

Tortora GJ, Anagnostakos NP. *Principles of Anatomy and Physiology,* 6th ed. New York: Harper and Row; 1990

Travis EL. *Primer of Medical Radiobiology,* 2nd ed. Chicago: Year Book; 1989

Wallace JE. *Radiographic Exposure Principles and Practice,* 1st ed. Philadelphia: F. A. Davis; 1995

Wolbarst AB. *Physics of Radiology,* 1st ed. Norwalk: Appleton & Lange; 1993

Third Edition

Appleton & Lange's Review for the
RADIOGRAPHY EXAMINATION

CHAPTER 1
Patient Care and Management
Questions

1. Which ethical principle is related to the theory that patients have the right to decide what will or will not be done to them?

   (A) autonomy
   (B) beneficence
   (C) fidelity
   (D) veracity

2. Blood pressure is measured in units of

   (A) mm Hg
   (B) beats per minute
   (C) degrees Fahrenheit
   (D) L/min

3. The type of shock associated with pooling of blood in the peripheral vessels is classified as

   (A) neurogenic
   (B) cardiogenic
   (C) hypovolemic
   (D) septic

4. Which of the following is a violation of correct sterile technique?

   (A) gowns are considered sterile in the front, down to the waist, including the arms
   (B) sterile gloves must be kept above the waist level
   (C) persons in sterile dress should pass each other face to face
   (D) a sterile field should not be left unattended

5. While performing mobile radiography on a patient you note the requisition is for a chest film to check placement of a Swan–Ganz catheter. A Swan–Ganz catheter is a(n)

   (A) pacemaker
   (B) chest tube
   (C) intravenous catheter
   (D) urinary catheter

6. If a radiographer performed a barium enema on a patient who was supposed to have an obstruction series, which of the following charges may be brought against the radiographer?

   (A) assault
   (B) battery
   (C) false imprisonment
   (D) defamation

7. Which of the following should be used to disinfect after a blood spill?

   (A) soap and water
   (B) betadine solution
   (C) one part bleach to ten parts water
   (D) one part alcohol to ten parts water

8. A patient who is diaphoretic has

  (A) pale, cool, clammy skin

  (B) hot, dry skin

  (C) dilated pupils

  (D) warm, moist skin

9. What instructions might a patient receive upon completing a BE examination?

  1. drink plenty of fluids

  2. take a mild laxative

  3. withhold fluids for 6 hours

  (A) 1 and 2 only

  (B) 2 and 3 only

  (C) 2 only

  (D) 3 only

10. The type of isolation practiced that prevents the spread of infectious agents in aerosol form is

  (A) respiratory isolation

  (B) protective isolation

  (C) contact isolation

  (D) strict isolation

11. Which of the following drugs is considered a bronchodilator?

  (A) epinephrine

  (B) lidocaine

  (C) nitroglycerin

  (D) verapamil

12. Symptoms of impending diabetic coma include

  1. increased urination

  2. sweet-smelling breath

  3. extreme thirst

  (A) 1 and 2 only

  (B) 1 and 3 only

  (C) 2 and 3 only

  (D) 1, 2, and 3

13. The condition of slow heart rate, below 60 bpm (beats per minute), is termed

  (A) hyperthermia

  (B) hypotension

  (C) hypoxia

  (D) bradycardia

14. In what order should the following exams be performed?

  1. upper GI

  2. IVP

  3. barium enema

  (A) 3, 1, 2

  (B) 1, 3, 2

  (C) 2, 1, 3

  (D) 2, 3, 1

15. Which of the following conditions must be met in order for patient consent to be valid?

  1. the patient must sign the consent before receiving sedation

  2. the physician named on the consent form must perform the procedure

  3. all the blanks on the consent form must be filled in before the patient signs the form

  (A) 1 and 2 only

  (B) 1 and 3 only

  (C) 2 and 3 only

  (D) 1, 2, and 3

16. The Heimlich maneuver is used if a patient is

  (A) in cardiac arrest

  (B) choking

  (C) having a seizure

  (D) suffering from hiccups

17. Rapid onset of severe respiratory or cardiovascular symptoms after ingestion or injection of a drug, vaccine, contrast agent, or food, or after an insect bite, best describes

  (A) asthma

  (B) anaphylaxis

  (C) myocardial infarct

  (D) rhinitis

18. While in your care for a radiologic procedure, a patient asks to see his chart. Which of the following is the appropriate response?

   (A) inform the patient that the chart is for health care providers to view and not patients
   (B) inform the patient that you do not know where the chart is
   (C) inform the patient that he has the right to see his chart, and that he should request to review it with his physician, so that it is properly interpreted
   (D) give the patient the chart and leave him alone for a few minutes to review it

19. According to the CDC (Centers for Disease Control), all of the following isolation guidelines are true, EXCEPT

   (A) respiratory isolation requires masks for those who come in contact with the patient
   (B) masks are indicated for patients on enteric precautions
   (C) the hands must be washed after working with a patient on drainage/secretion precautions
   (D) gloves are not indicated when working with a patient on respiratory isolation

20. Which statement(s) would be true regarding tracheostomy patients?

   1. tracheostomy patients have difficulty speaking
   2. a routine chest x-ray requires the tracheostomy tubing to be rotated out of view
   3. audible rattling sounds indicates a need for suction

   (A) 1 only
   (B) 1 and 2 only
   (C) 1 and 3 only
   (D) 1, 2, and 3

21. The most effective method of sterilization is

   (A) dry heat
   (B) moist heat
   (C) pasteurization
   (D) freezing

22. Log-rolling is a method of moving patients with suspected

   (A) head injury
   (B) spine injury
   (C) bowel obstruction
   (D) extremity fracture

23. A nosocomial infection is a(n)

   (A) infection acquired at a large gathering
   (B) upper respiratory infection
   (C) infection acquired in a hospital
   (D) type of rhinitis

24. A radiographer who discloses confidential information to unauthorized individuals may be found guilty of

   (A) invasion of privacy
   (B) slander
   (C) libel
   (D) defamation

25. Which of the following is a vasopressor and may be used for an anaphylactic reaction or cardiac arrest?

   (A) nitroglycerin
   (B) epinephrine
   (C) hydrocortisone
   (D) digitoxin

26. The condition where pulmonary alveoli lose their elasticity and become permanently inflated, causing the patient to consciously exhale, is

   (A) bronchial asthma
   (B) bronchitis
   (C) emphysema
   (D) tuberculosis

27. When a radiographer is obtaining patient history, both subjective and objective data should be obtained. An example of subjective data would be

    (A) the patient appears to have a productive cough
    (B) the patient has a blood pressure of 130/95
    (C) the patient states he experiences extreme pain in the upright position
    (D) the patient has a palpable mass in the right upper quadrant of the left breast

28. The legal doctrine *res ipsa locquitur* relates to which of the following?

    (A) let the master answer
    (B) the thing speaks for itself
    (C) a thing or matter settled by justice
    (D) a matter settled by precedent

29. You receive a patient and the request form simply asks for left arm. You should

    (A) x-ray the entire arm
    (B) ask the patient what to x-ray
    (C) examine the patient and decide what to x-ray
    (D) check with the referring physician

30. The patient is usually required to drink barium sulfate suspension in order to demonstrate which of the following structure(s)?

    1. pylorus
    2. sigmoid
    3. duodenum

    (A) 1 and 2 only
    (B) 1 and 3 only
    (C) 2 and 3 only
    (D) 3 only

31. Which of the following are appropriate when making a correction in a patient's medical record?

    1. date and initial the correction
    2. write the word "error" next to the correction
    3. put a line through the error

    (A) 1 and 2 only
    (B) 1 and 3 only
    (C) 2 and 3 only
    (D) 1, 2, and 3

32. Diseases whose mode of transmission is through the air include

    1. tuberculosis
    2. mumps
    3. strep throat

    (A) 1 only
    (B) 1 and 2 only
    (C) 1 and 3 only
    (D) 1, 2, and 3

33. Radiographs are the property of the

    (A) radiologist
    (B) patient
    (C) health care institution
    (D) referring physician

34. Which of the following is (are) symptoms of shock?

    1. pallor and weakness
    2. increased pulse rate
    3. fever

    (A) 1 only
    (B) 1 and 2 only
    (C) 1 and 3 only
    (D) 1, 2, and 3

35. The complete killing of all microorganisms is termed

    (A) surgical asepsis
    (B) medical asepsis
    (C) sterilization
    (D) disinfection

36. Which of the following medical equipment may be used to monitor blood pressure?

    1. pulse oximeter
    2. stethoscope
    3. sphygmomanometer

    (A) 1 and 2 only
    (B) 1 and 3 only
    (C) 2 and 3 only
    (D) 1, 2, and 3

37. Chest drainage systems should always be kept

    (A) above the patient's chest
    (B) at the level of the patient's chest
    (C) below the level of the patient's chest
    (D) none of the above; the position of the chest drainage system is not important

38. The medical term for congenital clubfoot is

    (A) coxa plana
    (B) osteochondritis
    (C) talipes
    (D) muscular dystrophy

39. The AIDS virus is transmitted

    1. by sharing contaminated needles
    2. mother to child during birth
    3. by intimate contact with body fluids

    (A) 1 only
    (B) 1 and 2 only
    (C) 1 and 3 only
    (D) 1, 2, and 3

40. The patient is placed in the lithotomy position for which of the following procedures?

    (A) myelography
    (B) venography
    (C) T-tube cholangiography
    (D) hysterosalpingography

41. Symptoms of shock include

    1. rise in blood pressure
    2. increase in pulse rate
    3. restlessness and apprehension

    (A) 1 only
    (B) 1 and 2 only
    (C) 2 and 3 only
    (D) 1, 2, and 3

42. An esophagram might be requested for patients with which of the following esophageal disorders/symptoms?

    1. varices
    2. achalasia
    3. dysphasia

    (A) 1 only
    (B) 1 and 2 only
    (C) 1 and 3 only
    (D) 1, 2, and 3

43. The usual patient preparation for an upper GI exam is

    (A) npo 8 hours before the exam
    (B) light breakfast only, the morning of the exam
    (C) clear fluids only, the morning of the exam
    (D) 2 ounces castor oil and enemas until clear

44. Which of the following patient rights is violated by discussing privileged patient information with a friend not involved with the patient's care?

    1. the right to considerate and respectful care
    2. the right to privacy
    3. the right to continuity of care

    (A) 1 only
    (B) 2 only
    (C) 1 and 3 only
    (D) 2 and 3 only

45. When a GI series has been requested on a patient with a suspected perforated ulcer, the type contrast medium that should be used is

(A) thin barium sulfate suspension
(B) thick barium sulfate suspension
(C) water-soluble iodinated media
(D) oil-based iodinated media

46. Which of the following should be considered to determine the contrast medium dosage for pediatric intravenous urography?

1. patient age
2. patient weight
3. patient sex

(A) 1 only
(B) 1 and 2 only
(C) 2 and 3 only
(D) 1, 2, and 3

47. What is the appropriate action if a patient has signed consent for a procedure, but once on the radiographic table, refuses the procedure?

(A) proceed, the consent form is signed
(B) send the patient back to their room and take an extended lunch break
(C) honor the patient's request and proceed with the next patient
(D) immediately stop the procedure and inform the radiologist and the referring physician of the patient's request

48. An ambulatory patient is one who

(A) is able to walk
(B) is unable to walk
(C) has difficulty breathing
(D) arrives by ambulance

49. A compound fracture is one in which the

(A) fractured ends of bone are forced together
(B) fractured ends of bone penetrate the skin
(C) bone is broken in more than one place
(D) bone has been forced out of alignment

50. Increased pain threshold, breakdown of skin, and atrophy of fat pads and sweat glands are all important considerations when working with which group of patients?

(A) infants
(B) children
(C) adolescents
(D) geriatric patients

51. Abnormal accumulation of air in pulmonary tissues, resulting in overdistention of the alveolar spaces, is

(A) emphysema
(B) empyema
(C) pneumothorax
(D) pneumoconiosis

52. A life-threatening reaction to iodinated contrast media, anaphylactic shock, manifests early symptoms that include

1. dysphagia
2. itching of palms and soles
3. constriction of the throat

(A) 1 only
(B) 2 only
(C) 2 and 3 only
(D) 1, 2, and 3

53. What is the BEST way to be sure you have the correct patient for an x-ray examination?

(A) ask the receptionist to identify the patient
(B) call the patient's name
(C) check the patient's hospital identification
(D) ask the radiologist to identify the patient

54. An inanimate object that has been in contact with an infectious microorganism is termed a

(A) vector
(B) fomite
(C) host
(D) reservoir

55. An informed consent is required before performing which of the following exams?

   (A) upper GI
   (B) lower GI
   (C) sialogram
   (D) renal arteriogram

56. Anaphylaxis is the term used to describe

   (A) an inflammatory reaction
   (B) bronchial asthma
   (C) acute chest pain
   (D) allergic shock

57. The advantages of using non-ionic, water-soluble contrast media include

   1. cost containment benefits
   2. low toxicity
   3. less patient discomfort

   (A) 1 only
   (B) 1 and 2 only
   (C) 2 and 3 only
   (D) 1, 2, and 3

58. A diuretic is used to

   (A) induce vomiting
   (B) stimulate defecation
   (C) promote elimination of urine
   (D) inhibit coughing

59. The medical word for nosebleed is

   (A) vertigo
   (B) epistaxis
   (C) urticaria
   (D) aura

60. When caring for a patient with an IV, you should keep the bottle

   (A) 18 to 20 inches above the level of the vein
   (B) 18 to 20 inches below the level of the vein
   (C) 28 to 30 inches above the level of the vein
   (D) 28 to 30 inches below the level of the vein

61. The pain experienced by an individual whose coronary arteries are not conveying sufficient blood to the heart muscle is called

   (A) tachycardia
   (B) bradycardia
   (C) angina pectoris
   (D) syncope

62. Hypochlorite bleach (Clorox) and Lysol are examples of

   (A) antiseptics
   (B) bacteriostatics
   (C) antifungal agents
   (D) disinfectants

63. Medication may be administered by which of the following routes?

   1. orally
   2. intravenously
   3. intramuscularly

   (A) 1 and 2 only
   (B) 1 and 3 only
   (C) 2 and 3 only
   (D) 1, 2, and 3

64. Symptoms associated with a respiratory reaction to contrast media include

   1. sneezing
   2. dyspnea
   3. asthma attack

   (A) 1 and 2 only
   (B) 1 and 3 only
   (C) 2 and 3 only
   (D) 1, 2, and 3

65. Which of the following are TRUE regarding the proper care of a patient with a tracheostomy?

1. employ sterile technique if you must touch a tracheostomy for any reason
2. before suctioning a tracheostomy, the patient should be well aerated
3. never suction for longer than 15 seconds, permitting the patient to rest in between

(A) 1 and 2 only
(B) 1 and 3 only
(C) 2 and 3 only
(D) 1, 2, and 3

66. The condition allowing blood to shunt between right and left ventricles is called

(A) patent ductus arteriosus
(B) coarctation of the aorta
(C) atrial septal defect
(D) ventricular septal defect

67. Factors that are important to evaluate when selecting contrast media are

1. miscibility
2. potential toxicity
3. viscosity

(A) 1 and 2 only
(B) 2 and 3 only
(C) 1 and 3 only
(D) 1, 2, and 3

68. Which of the following effects does an analgesic have on the body?

(A) decreases pain
(B) helps delay clotting
(C) increases urine output
(D) combats bacterial growth

69. A vasodilator would most likely be used for

(A) angina
(B) cardiac arrest
(C) bradycardia
(D) antihistamine

70. When classifying intravenous contrast agents, the total number of particles in solution per kilogram of water defines

(A) osmolality
(B) toxicity
(C) viscosity
(D) miscibility

71. A patient in the recumbent position whose head is lower than the feet is said to be in which of the following positions?

(A) Trendelenburg
(B) Fowler's
(C) Sims
(D) Stenver's

72. Where is the "sterile corridor" located?

(A) just outside the operating rooms
(B) immediately inside each operating room door
(C) between the draped patient and the instrument table
(D) at the foot end of the draped patient

73. Which of the following statements are TRUE regarding the two-member team of radiographers performing mobile radiography on a patient under strict isolation for extensive burns?

1. one radiographer remains "clean"—that is, he or she have no physical contact with the patient
2. the radiographer who positions the mobile unit also makes the exposure
3. the radiographer who positions the cassette, retrieves the cassette and removes it from its plastic protective cover

(A) 1 and 2 only
(B) 1 and 3 only
(C) 2 and 3 only
(D) 1, 2, and 3

74. With which of the following conditions is protective or "reverse" isolation required?

   1. tuberculosis
   2. burns
   3. leukemia

   (A) 1 only
   (B) 1 and 2 only
   (C) 2 and 3 only
   (D) 1, 2, and 3

75. When radiographing the elderly, it is helpful to

   1. expedite the exam as swiftly as possible
   2. give simple, direct instructions
   3. address patients by their full name

   (A) 1 only
   (B) 1 and 2 only
   (C) 2 and 3 only
   (D) 1, 2, and 3

76. The medical abbreviation meaning "every hour" is

   (A) tid
   (B) qid
   (C) qh
   (D) pc

77. How should the radiographer proceed with an examination of a splinted extremity?

   (A) always remove the splint before the exam
   (B) never radiograph through the splint
   (C) ask the physician to remove the splint, if necessary
   (D) support the limb while another RT removes the splint

78. Which of the following may be used to effectively reduce the viscosity of contrast media?

   (A) warming
   (B) refrigeration
   (C) storage at normal room temperature
   (D) storage in a cool, dry place

79. You receive a patient complaining of pain in the area of the left fourth and fifth metatarsals; however, the requisition asks for a left ankle exam. What should you do?

   (A) a left foot exam
   (B) a left ankle exam
   (C) both left foot and left ankle
   (D) check with the referring physician

80. You receive an ambulatory patient for a GI series. As the patient is seated on the x-ray table he feels faint. You should

   1. lie the patient down on the x-ray table
   2. elevate the patient's legs or place the table slightly Trendelenburg
   3. leave quickly and call for help

   (A) 1 only
   (B) 1 and 2 only
   (C) 1 and 3 only
   (D) 1, 2, and 3

81. Nitroglycerin is used

   (A) to relieve pain from angina pectoris
   (B) to prevent a heart attack
   (C) as a vasoconstrictor
   (D) to increase blood pressure

82. Which of the following radiographic procedures requires an intrathecal injection?

   (A) intravenous pyelogram
   (B) myelogram
   (C) lymphangiogram
   (D) computed tomography

83. You and a fellow radiographer have received an unconscious patient from a motor vehicle accident. As you perform the examination, it is important to

   1. refer to the patient by name
   2. make only those statements you would make with a conscious patient
   3. reassure the patient about what you are doing

   (A) 1 only
   (B) 1 and 2 only
   (C) 2 and 3 only
   (D) 1, 2, and 3

84. For which of the following radiographic examinations is a consent form usually required?

   1. angiogram
   2. GI series
   3. skeletal survey

   (A) 1 only
   (B) 1 and 2 only
   (C) 1 and 3 only
   (D) 1, 2, and 3

85. The medical abbreviation meaning "three times a day" is

   (A) tid
   (B) qid
   (C) qh
   (D) pc

86. Which of the following conditions describes a patient who is unable to breath easily while in the recumbent position?

   (A) dyspnea
   (B) apnea
   (C) orthopnea
   (D) hyperpnea

87. When radiographing young children, it is helpful to

   1. let them bring in a toy
   2. tell them it will not hurt
   3. be cheerful and unhurried

   (A) 1 and 2 only
   (B) 1 and 3 only
   (C) 2 and 3 only
   (D) 1, 2, and 3

88. With a patient suffering abdominal pain, it is frequently helpful to

   1. elevate the head slightly with a pillow
   2. perform the exam in the Trendelenburg position
   3. place a support under the knees

   (A) 1 and 2 only
   (B) 1 and 3 only
   (C) 2 and 3 only
   (D) 1, 2, and 3

89. When disposing of contaminated needles, they are placed in special containers in which of the following manners?

   (A) recap the needle, remove the syringe, and dispose of
   (B) do not recap needle, remove from syringe, and dispose of
   (C) recap the needle, dispose of entire syringe
   (D) do not recap needle, dispose of entire syringe

90. Which of the following exams require(s) restriction of the patient's diet?

   1. GI series
   2. abdominal survey
   3. pyelogram

   (A) 1 only
   (B) 1 and 2 only
   (C) 1 and 3 only
   (D) 2 and 3 only

91. When reviewing patient blood chemistry levels, what is considered the normal creatinine range?

(A) 0.6 to 1.5 mg/100 mL
(B) 4.5 to 6 mg/100 mL
(C) 8 to 25 mg/100 mL
(D) up to 50 mg/100 mL

92. When an emergency trauma patient experiences hemorrhaging from a leg injury, you should

1. apply pressure to the bleeding site
2. call the ER for assistance
3. rebandage the site and complete the exam

(A) 1 and 2 only
(B) 1 and 3 only
(C) 2 and 3 only
(D) 1, 2, and 3

93. Forms of intentional misconduct include

1. slander
2. invasion of privacy
3. negligence

(A) 1 only
(B) 2 only
(C) 1 and 2 only
(D) 1, 2, and 3

94. When a patient with an arm injury needs help undressing, you should

(A) remove clothing from the injured arm first
(B) remove clothing from the uninjured arm first
(C) always remove clothing from the left arm first
(D) always cut clothing away from the injured extremity

95. A patient suffering from an episode of syncope should be placed in which of the following positions?

(A) dorsal recumbent with head elevated
(B) dorsal recumbent with feet elevated
(C) lateral recumbent
(D) seated with feet supported

96. The diameter of a needle is termed its

(A) hub
(B) gauge
(C) length
(D) bevel

97. All of the following rules regarding proper handwashing technique are correct, EXCEPT

(A) keep hands and forearms lower than elbows
(B) use paper towels to turn water on
(C) avoid using hand lotions whenever possible
(D) carefully wash all surfaces and between fingers

98. All drug packages must provide certain information required by the U.S. Food and Drug Administration. Some of the information that must be provided includes

1. generic name
2. contraindications
3. usual dose

(A) 1 only
(B) 1 and 2 only
(C) 1 and 3 only
(D) 1, 2, and 3

99. Which of the following must be contained in the patient's medical record or chart?

1. diagnostic and therapeutic orders
2. medical history
3. informed consent

(A) 1 and 2 only
(B) 1 and 3 only
(C) 2 and 3 only
(D) 1, 2, and3

100. The radiographer must perform the following procedures prior to entering a strict isolation room with a mobile x-ray unit

1. wear a gown and gloves
2. wear a gown, gloves, mask, and cap
3. clean the mobile x-ray unit

(A) 1 only
(B) 2 only
(C) 1 and 3 only
(D) 2 and 3 only

101. Ingestion of a gas-producing substance, such as powder or crystals, is usually preliminary to which of the following examinations?

1. oral cholecystogram
2. double-contrast GI
3. intravenous urogram

(A) 1 only
(B) 2 only
(C) 1 and 2 only
(D) 2 and 3 only

102. All of the following statements regarding osteoarthritis are true, EXCEPT

(A) osteoarthritis is a progressive disorder
(B) osteoarthritis is an inflammatory disorder
(C) osteoarthritis involves deterioration of the articular cartilage
(D) osteoarthritis involves the formation of bony spurs

103. During an arthrographic examination of the shoulder, it is normal for the patient to experience

(A) pain in the chest radiating down the arm
(B) a feeling of increased pressure within the joint
(C) temporary paralysis of the upper extremity
(D) numbness in the fingers

104. Blood pressure may be expressed as 120/95. What does 95 represent?

1. the phase of relaxation of the cardiac muscle tissue
2. the phase of contraction of the cardiac muscle tissue
3. a higher than average diastolic pressure

(A) 1 only
(B) 2 only
(C) 1 and 3 only
(D) 2 and 3 only

105. Local anesthetics may be used in all of the following radiographic examinations, EXCEPT

(A) lower extremity arteriography
(B) arthrography
(C) myelography
(D) postoperative or T-tube cholangiography

106. In which of the following conditions is a double-contrast BE preferable to a single-contrast BE?

1. polyps
2. colitis
3. diverticulosis

(A) 1 only
(B) 1 and 2 only
(C) 1 and 3 only
(D) 1, 2, and 3

107. If extravasation occurs during an intravenous injection of contrast media, correct treatment includes which of the following?

   1. remove needle and locate a sturdier vein immediately

   2. apply pressure to vein until bleeding stops

   3. apply warm moist heat

(A) 1 only

(B) 1 and 2 only

(C) 2 and 3 only

(D) 1, 2, and 3

108. Which of the following diastolic pressure readings may indicate shock?

(A) 50 mm Hg

(B) 70 mm Hg

(C) 90 mm Hg

(D) 110 mm Hg

109. To reduce the back strain associated with transferring patients from stretcher to x-ray table, you should

(A) pull the patient

(B) push the patient

(C) hold the patient away from your body and lift

(D) bend at the waist and pull

110. What is the most widely used method of vascular catheterization?

(A) Doppler

(B) Moniz

(C) Grandy

(D) Seldinger

111. Lyme disease is a condition caused by bacteria carried by deer ticks. The tick bite may cause fever, fatigue, and other associated symptoms. This is an example of transmission of an infection by

(A) droplet contact

(B) a vehicle

(C) the airborne route

(D) a vector

112. When the patient is in the Sims position for insertion of an enema tip, he or she is placed

(A) semi-Trendelenburg

(B) prone

(C) recumbent, on either side with the top knee flexed acutely and the body rotated forward slightly

(D) recumbent, on either side in a true lateral with the top knee flexed acutely

113. When radiographing a patient with a wet cast, the radiographer should be alert for signs of circulation impairment or nerve compression. This may be indicated by

   1. tingling of the fingers or toes

   2. bluish or pale color of the skin

   3. coldness in the extremity

(A) 1 only

(B) 1 and 2 only

(C) 1 and 3 only

(D) 1, 2, and 3

114. A radiologic technologist may be found guilty of a tort if which of the following types of behavior is demonstrated?

   1. failure to shield a patient of childbearing age from unnecessary radiation

   2. performing an examination on a patient who has refused the examination

   3. performing a patient's examination that was ordered by an emergency room physician

(A) 1 only

(B) 2 only

(C) 1 and 2 only

(D) 1, 2, and 3

115. The pulse can be detected only by the use of a stethoscope in which of the following locations?

(A) wrist

(B) apex of the heart

(C) groin

(D) neck

116. The following body fluids are potential carriers of HIV

    1. semen
    2. vaginal secretions
    3. blood

    (A) 1 only
    (B) 1 and 3 only
    (C) 2 and 3 only
    (D) 1, 2, and 3

117. Instruments needed to assess vital signs include

    1. thermometer
    2. tongue blade
    3. watch with second hand

    (A) 1 only
    (B) 1 and 2 only
    (C) 1 and 3 only
    (D) 1, 2, and 3

118. A patient suffering from orthopnea would experience the least discomfort in which body position?

    (A) Fowler's
    (B) Trendelenburg
    (C) recumbent
    (D) erect

119. What is the most common means of spreading infection?

    (A) improperly disposed of contaminated waste
    (B) instruments improperly sterilized
    (C) soiled linen
    (D) human hands

120. The normal average rate of respiration for a healthy adult patient is

    (A) 5 to 7 breaths per minute
    (B) 8 to 12 breaths per minute
    (C) 12 to 20 breaths per minute
    (D) 20 to 30 breaths per minute

121. A common cause of anaphylactic shock is administration of contrast agents. Which of the following is a symptom of anaphylactic shock?

    (A) weak, rapid pulse
    (B) blurred vision
    (C) tremors
    (D) convulsions

122. A diabetic patient who is prepared for a fasting radiographic exam is susceptible to a hypoglycemic reaction. This is characterized by

    1. shaking, nervousness
    2. cold, clammy skin
    3. cyanosis

    (A) 1 only
    (B) 2 only
    (C) 1 and 2 only
    (D) 1, 2, and 3

123. The least toxic contrast medium is

    (A) barium sulfate
    (B) metrizamide
    (C) ethiodized oil
    (D) meglumine diatrizoate

124. The mechanical device used to correct an ineffectual cardiac rhythm is a(n):

    (A) defibrillator
    (B) cardiac monitor
    (C) crash cart
    (D) resuscitation bag

125. Following a barium enema examination, the patient should be given which of the following instructions?

1. increase fluid and fiber intake for several days
2. changes in stool color will occur until all barium is evacuated
3. contact a physician if patient has not had a bowel movement in 24 hours

(A) 1 only
(B) 2 only
(C) 1 and 3 only
(D) 1, 2, and 3

126. Skin discoloration due to cyanosis may be observed in the

1. gums
2. nailbeds
3. lips

(A) 1 only
(B) 1 and 2 only
(C) 3 only
(D) 1, 2, and 3

127. The movement of synovial joints that involves an increase in the angle between articulating surfaces is called

(A) adduction
(B) protraction
(C) extension
(D) flexion

128. When a patient arrives in the radiology department with a urinary Foley catheter bag, it is important to

(A) place the drainage bag above the level of the bladder
(B) place the drainage bag at the same level as the bladder
(C) place the drainage bag below the level of the bladder
(D) clamp the Foley catheter

129. The abbreviation *pc* denotes

(A) every 4 hours
(B) after meals
(C) before meals
(D) whenever or as necessary

130. Which of the following parenteral routes is most often used for administration of contrast agents in the radiology department?

(A) subcutaneous
(B) intravenous
(C) intramuscular
(D) intradermal

131. Sterile technique is required when contrast agents are administered

(A) rectally
(B) orally
(C) intrathecally
(D) through a nasogastric tube

132. A patient whose systolic blood pressure is less than 90 mm Hg is usually considered

(A) hypertensive
(B) hypotensive
(C) average/normal
(D) baseline

133. Which of the following blood pressure readings would indicate shock?

(A) systolic pressure lower than 60 mm Hg
(B) systolic pressure higher than 140 mm Hg
(C) diastolic pressure higher than 140 mm Hg
(D) diastolic pressure lower than 90 mm Hg

134. Intrathecal contrast media administration may be performed in all of the following positions, EXCEPT

(A) prone with pillows under the abdomen

(B) supine

(C) seated and bent forward

(D) lying on the side with knees pulled up to chest and head tucked

135. Which of the following sites are commonly used for an intravenous injection?

1. antecubital vein
2. basilic vein
3. popliteal vein

(A) 1 and 2

(B) 1 and 3

(C) 2 and 3

(D) 1, 2, and 3

136. During a grand mal seizure, the patient should be

(A) protected from injury

(B) placed in a semi-upright position to prevent aspiration of vomitus

(C) allowed to thrash freely

(D) given a sedative to reduce jerky body movements and reduce the possibility of injury

137. A severe blow to the head resulting in injury on the side opposite the blow is termed

(A) concussion

(B) contrecoup injury

(C) seizure

(D) coma

138. An intravenous pyelogram requires the patient to remain in one position for an extended period of time. What can be done to make the patient as comfortable as possible?

1. place a pillow under the patient's head
2. place a support cushion under the patient's knees to relieve back strain
3. place a radiopaque pad on the entire table prior to the start of the examination

(A) 1 only

(B) 1 and 2 only

(C) 1 and 3 only

(D) 1, 2, and 3

139. Production or aggravation of decubitus ulcers may be avoided by

1. allowing the patient to change positions at least every 30 minutes when lying on an examination table
2. protecting the patient's heels during patient transfer
3. performing a modified exam while restricting patient movement

(A) 1 only

(B) 2 only

(C) 1 and 2 only

(D) 1, 2, and 3

140. In which of the following situations should a radiographer wear protective eye gear (goggles)?

1. when performing an upper gastrointestinal radiographic examination
2. when assisting the radiologist during an angiogram
3. when assisting the radiologist in a biopsy/aspiration procedure

(A) 1 and 2 only

(B) 1 and 3 only

(C) 2 and 3 only

(D) 1, 2, and 3

141. All of the following statements regarding handwashing and skin care are correct, ECEPT

   (A) hands should be washed after each patient examination

   (B) faucets should be opened and closed with paper towels

   (C) hands should be smooth and free from chapping

   (D) any cracks or abrasions should be left uncovered to facilitate healing

142. Upon review of a patient's blood chemistry, which of the following BUN ranges should be considered normal?

   (A) 0.6 to 1.5 mg/100 mL

   (B) 4.5 to 6 mg/100 mL

   (C) 8 to 25 mg/100 mL

   (D) up to 50 mg/100 mL

143. Of the four stages of infection, which is the stage during which the infection is MOST communicable?

   (A) latent period

   (B) incubation period

   (C) disease phase

   (D) convalescent phase

144. You have encountered a person who is apparently unconscious. Although you open his airway, there is no rise and fall of the chest and you can hear no breath sounds. You should

   (A) begin mouth-to-mouth rescue breathing, giving two full breaths

   (B) proceed with the Heimlich maneuver

   (C) begin external chest compressions at a rate of 80 to 100 per minute

   (D) begin external chest compressions at a rate of at least 100 per minute

145. Compared to an adult, when performing CPR on an infant it is required that the number of compressions per minute

   (A) remain the same

   (B) double

   (C) decrease

   (D) increase

146. To keep patient discomfort to a minimum, radiography of a traumatized shoulder should be performed in which position?

   (A) semi-erect position

   (B) erect position

   (C) recumbent position

   (D) lateral recumbent position

147. Which of the following statements is FALSE regarding the administration of a barium enema on a patient with a colostomy?

   (A) the dressing should be removed and disposed of

   (B) the drainage pouch should be retained unless a fresh one can be provided

   (C) the colostomy tip or catheter should be selected by the radiologist

   (D) never allow the patient to insert the colostomy tip

148. A patient developed hives several minutes following injection of an iodinated contrast medium. What drug should be readily available for this type of medical emergency?

   (A) analgesic

   (B) antihistamine

   (C) anti-inflammatory

   (D) antibiotic

149. The deer tick may transmit a disease known as

   (A) tinea nigra

   (B) tinea pedis

   (C) ringworm

   (D) Lyme disease

150. Due to medicolegal considerations, radiographs must include all of the following information, EXCEPT

   (A) patient's name and/or identification number

   (B) patient's birthdate

   (C) right or left side marker

   (D) date of examination

Answers and Explanations

1. **(A)** Autonomy is the ethical principle related to the theory that patients have the right to decide what will or will not be done to them. Beneficence is related to the idea of doing good and being kind. Fidelity is faithfulness and loyalty. Veracity is not only telling the truth, but also not practicing deception. *(Adler & Carlton, p 324)*

2. **(A)** Blood pressure is measured in units of mm Hg or millimeters of mercury. Heart rate, or pulse, is measured in units of beats per minute. Temperature is measured in degrees Fahrenheit. Oxygen delivery is measured in units of liters per minute or L/min. Table 1–1 outlines the normal ranges for vital signs in healthy adults. *(Torres, pp 73–80)*

TABLE 1–1. NORMAL RANGES FOR VITAL SIGNS IN ADULTS

Blood pressure	110–140 mm Hg
	60–90 mm Hg
Pulse	60–100 beats per minute
Temperature	97.7–99.5°F
Respiration	12–20 breaths per minute

3. **(A)** The type of shock associated with pooling of blood in the peripheral vessels is classified as neurogenic shock. This occurs in cases of trauma to the central nervous system, resulting in decreased arterial resistance and pooling of blood in peripheral vessels. Cardiogenic shock is related to cardiac failure, as a result of interference with heart function. It may occur in cases of cardiac tamponade, pulmonary embolus, or myocardial infarction. Hypovolemic shock is related to loss of large amounts of blood, either from internal bleeding or hemorrhage associated with trauma. Septic shock, as well as anaphylactic shock, is generally classified as vasogenic shock. *(Torres, p 93)*

4. **(C)** Persons in sterile dress should not pass each other face to face. Rather, they should pass each other back to back, to avoid contaminating each other. Gowns are considered sterile in the front, down to the waist, including the arms. Sterile gloves must be kept above the waist level. If the hands are accidentally lowered, or placed behind the back, they are no longer sterile. A sterile field should not be left unattended. Sterile fields should be set up immediately prior to a procedure, and covered with a sterile drape if a few moments are to elapse before the procedure can begin. A sterile field should be attended to. If it is not constantly monitored, the field can be contaminated. *(Adler & Carlton, p 231)*

5. **(C)** A Swan–Ganz catheter is a specific type of intravenous catheter used to measure the pumping ability of the heart, to obtain pressure readings, and to introduce medications and intravenous fluids. Pacemakers are inserted under the patient's skin and regulate heart rate. Pacemakers may be permanent or temporary. Chest tubes are used to remove fluid or air from the pleural cavity. Any of these items may be identified on a chest radiograph, provided the cassette is properly positioned and the correct exposure technique is employed. If a physician is interested in assessing the proper placement of a Swan–Ganz catheter, the lungs may have to be slightly overexposed to clearly delineate the

proper placement of the tip of the Swan–Ganz catheter, which will overlap with the denser cardiac silhouette. A urinary catheter would not appear on a chest radiograph. It should be noted that placement of any of these tubes, catheters, or pacemakers does require sterile technique. *(Adler & Carlton, pp 235–237)*

6. **(B)** A radiographer who performs the wrong examination on a patient may be charged with battery. Battery refers to the unlawful laying hands on a patient. Battery could also be charged if a patient is moved about roughly or touched in a manner that is inappropriate or without the patient's consent. *Assault* is the threat of touching or laying hands on. If a patient feels threatened by a health care provider, either because of the tone or pitch of voice, or because of words that are threatening, then an assault charge may be made. *False imprisonment* may be considered if a patient states that they no longer wish to continue with a procedure and they are ignored, or if restraining devices are improperly used or used without a physician order. *Defamation* may be slander, the spoken word, or libel, the written word. Defamation also occurs when patient confidentiality is not upheld, and as a result the patient suffers some embarrassment or mockery because of the breach of contract. *(Adler & Carlton, p 348)*

7. **(C)** The Centers for Disease Control considers all body substances as potential sources of infection. It lists several precautionary measures that should be taken when dealing with body substances. Gloves must be used when wiping up blood spills, and the area should be disinfected with a solution of one part bleach to ten parts water. *(Ehrlich & McCloskey, p 93)*

8. **(A)** Observation is an important part of the evaluation of acutely ill patients. The patient who is *diaphoretic* is in what is described as a "cold sweat," with pale, cool, moist skin. Hot and dry skin accompanies fever. Warm, moist skin may be a result of anxiety, or simply being in a warm room. The pupils dilate

in dimly illuminated places in order to allow more light into the eyes. *(Ehrlich, p 74)*

9. **(A)** Barium can dry and harden in the large bowel, causing symptoms from mild constipation to bowel obstruction. It is therefore essential that the radiographer provide *clear instruction of follow-up care,* along with its rationale, especially to outpatients. In order to avoid the possibility of fecal impaction, the patient should drink plenty of fluids for the next few days and take a mild laxative such as milk of magnesia. *(Torres, p 103)*

10. **(A)** The type of isolation practiced to prevent the spread of infectious agents in aerosol form is *respiratory isolation.* A mask is sufficient protection from aerosol transmission of pathogens. *Protective isolation,* also referred to as reverse isolation, is used to protect patients whose immune systems are compromised. Patients receiving chemotherapy, burn patients, or patients who are HIV positive may all have compromised immune systems. *Contact isolation* is used when there is a chance that infection may be spread by contact with body fluids. Gloves and a gown are used, and goggles and masks may be necessary if there is a chance of fluids spraying, such as in biopsy or drainage. *Strict isolation* is practiced with highly contagious disease or viruses that may be spread by air and/or contact. *(Adler & Carlton, p 211)*

11. **(A)** *Epinephrine* is a bronchodilator. Bronchodilators may be administered in a spray mister, such as for asthma, or injection form to relieve severe bronchospasm. *Lidocaine* (xylocaine) is an antiarrhythmic used to prevent or treat cardiac arrhythmias (dysrhythmia). *Nitroglycerin* and *verapamil* are vasodilators. Vasodilators permit increased blood flow by relaxing the walls of the blood vessels. *(Adler & Carlton, p 259)*

12. **(D)** When a diabetic patient misses an insulin injection, the body loses its ability to metabolize glucose, and ketoacidosis can occur. If not quickly corrected the patient may become comatose. Symptoms of impending

coma include increased urination, sweet (fruity) breath, and extreme thirst. Other symptoms are weakness and nausea. *(Torres, p 96)*

13. **(D)** The condition when a patient's heart rate slows below 60 bpm is *bradycardia. Hyperthermia* is the condition where the patient's temperature is well above the normal average range (97.7 to 99.5°F). *Hypotension* occurs if the blood pressure drops below the normal ranges (110 to 140/60 to 90 mm Hg). *Hypoxia* is a condition where there is a decrease of oxygen supplied to the tissues in the body. *(Adler & Carlton, p 181)*

14. **(D)** When scheduling patient examinations, it is important to avoid the possibility of residual contrast medium covering areas of interest on later examinations. The IVP should be scheduled first because the contrast medium used is excreted rapidly. The barium enema should be scheduled next. Lastly, the GI is scheduled. Any barium remaining from the previous BE should not be enough to interfere with the stomach or duodenum—although a preliminary scout film should be taken in each case. *(Ehrlich & McCloskey, p 192)*

15. **(D)** All of the statements in the question are true and are necessary in order for a consent form to be valid. The patient must *sign* the consent before receiving sedation. The *physician* named on the consent form must perform the procedure; no other physician should. Also, there should be *no blank* spaces on the consent form when the patient signs it. The consent form should be complete prior to the signing. In the case of a minor, a parent or guardian should sign the form. If a patient is not competent, then the legally appointed guardian should sign the consent. Remember that obtaining consent is a physician responsibility, and the explanation of the procedural risks should be performed by the physician, not the radiographer. *(Ehrlich & McCloskey, p 17)*

16. **(B)** The *Heimlich maneuver* is used if a patient is choking. If you suspect someone is chok-

ing, be certain the airway is indeed obstructed before attempting the Heimlich maneuver. A person with a completely obstructed airway will not be able to speak or cough. If the person cannot speak or cough, then the airway is obstructed, and the Heimlich maneuver should be performed. The proper method is to stand behind the choking victim with one hand in a fist, thumb side in, midway between the naval and the xyphoid tip. Place the other hand with the palm open over the closed fist and apply pressure, in and up. Repeat the thrust several times, until the object is dislodged. For an *infant*, the procedure is modified. Four back blows, midway between the scapula and using the heel of the hand are given. If the object is not dislodged, the baby is turned over and four chest thrusts are performed, using several fingers, just below the nipple line—being very careful to support the baby's head and spine. *(Adler & Carlton, p 262)*

17. **(B)** Anaphylaxis is an acute reaction characterized by sudden onset of urticaria, respiratory distress, vascular collapse, or systemic shock, and sometimes leads to death. It is caused by ingestion or injection of a sensitizing agent such as a drug, vaccine, contrast agent, food, or after an insect bite. Asthma and rhinitis are examples of allergic reactions. *(Ehrlich & McCloskey, p 176)*

18. **(C)** If a patient in your care asks to see her chart, the appropriate response is to refer her to her physician. A patient *does* have the right to review her own medical record; however, it should be *in the presence of her physician* so that the information is not misinterpreted and so that the physician can address concerns or answer questions. It is not appropriate to hand over the chart to a patient, nor is it appropriate to deceive the patient into believing that the chart is not available for her viewing or that she has no right to view the chart. *(Adler & Carlton, p 342)*

19. **(B)** Respiratory isolation does require the use of a mask for those who come in contact with the patient, but gloves are not necessary. Masks are not necessary for patients on

enteric precautions. Gloves should be worn for patients on enteric or drainage/secretion precautions, and the hands should be washed after handling the patient or the patient's linens and before contact with anyone else. *(Ehrlich & McCloskey, Appendix G, pp 287–288)*

20. **(C)** The tracheostomy patient will have difficulty speaking due to redirection of the air past the vocal cords. Gurgling or rattling sounds coming from the trachea indicate an excess accumulation of secretion, requiring suction with sterile catheters. Any rotation or movement of the tracheostomy tube may cause the tube to become dislodged and an obstructed airway would result. *(Torres, pp 123–124)*

21. **(B)** The most effective method of sterilization is *moist heat,* using steam under pressure. This is known as autoclaving. Sterilization by *dry heat* requires higher temperatures for longer periods of time then moist heat. *Pasteurization* is moderate heating with rapid cooling, and is frequently used in commercial preparation of milk, or alcoholic beverages such as wine and beer. It is not a form of sterilization. *Freezing* can also kill some microbes, but is not a form of sterilization. *(Adler & Carlton, p 208)*

22. **(B)** Patients arriving from the emergency department (ED) with suspected *spinal injury* should not be moved. A simple AP and horizontal lateral of the suspected area should be evaluated and a decision made about the advisability of further films. For a lateral projection, a patient should be moved along one plane, that is, rolled like a log. It is imperative that twisting motions be avoided. *(Ehrlich & McCloskey, p 179)*

23. **(C)** Nosocomial diseases are those acquired in hospitals, especially by patients whose resistance to infection has been diminished by their illness. Cleanliness is essential to decrease the number of nosocomial infections. X-ray tables must be cleaned and the pillowcase changed between patients. Probably the most common nosocomial infection is the

urinary tract infection (UTI), from contaminated urinary catheters. *(Ehrlich & McCloskey, p 89)*

24. **(A)** A radiographer who discloses confidential information to unauthorized individuals may be found guilty of *invasion of privacy.* If the disclosure is in some way detrimental or otherwise harmful to the patient, the radiographer may be accused of *defamation.* Spoken defamation is *slander;* written defamation is *libel. Assault* is to threaten harm; *battery* is to carry out the threat. *(Saia, Program Review and Exam Preparation: Radiography, or PREP, p 10)*

25. **(B)** Epinephrine (trade name Adrenalin) is the vasopressor used to treat an anaphylactic reaction or cardiac arrest. Nitroglycerin is a vasodilator. Hydrocortisone is a steroid that may be used to treat bronchial asthma, allergic reactions, and inflammatory reactions. Digitoxin is used to treat cardiac fibrillation. *(Ehrlich & McCloskey, p 135)*

26. **(C)** Emphysema is a progressive disorder caused by long-term irritation of the bronchial passages, such as by air pollution or cigarette smoking. Emphysema patients are unable to exhale normally due to the loss of elasticity of alveolar walls. If emphysema patients receive oxygen, it is usually administered at a very slow rate, because their respirations are controlled by the level of carbon dioxide in the blood. *(Tortora, p 723)*

27. **(C)** Obtaining a complete and accurate history from the patient for the radiologist is one of the most important aspects of a radiographer's job. Both subjective and objective data should be collected. *Objective data* include signs and symptoms that can be observed, such as a cough, a lump, or elevated blood pressure. *Subjective data* relate to what the patient feels, and to what extent. A patient may experience pain, but is it mild or severe? Is it localized or general? Does the pain increase or decrease under different circumstances? A radiographer should explore this with a patient and document any additional information on the requisition for the radiologist. *(Adler & Carlton, p 137)*

28. **(B)** The legal doctrine *res ipsa locquitur* relates to a matter that speaks for itself. For instance, if a patient went into the hospital to have a kidney stone removed and ended up with an appendectomy, that speaks for itself and negligence could be proven. *Respondeat superior* is the phrase meaning let the master answer, or the one ruling is responsible. If a radiographer were negligent, there may be an attempt to prove the radiologist was responsible, because the radiologist oversees the radiographer. *Res judicata* means a thing or matter settled by justice. *Stare decisis* refers to a matter settled by precedent. *(Miller, p 10; Adler & Carlton, pp 349–350)*

29. **(D)** *Only a physician is authorized to order an x-ray examination.* If the x-ray requisition is not specific enough, contact the referring physician for further clarification. This is in compliance with the Radiologic Technologist's Code of Ethics. *(Ehrlich & McCloskey, p 5)*

30. **(B)** Oral administration of barium sulfate is used to demonstrate the upper digestive system, esophagus, fundus, body and pylorus of the stomach, and barium progression through the small bowel. The large bowel, including sigmoid colon, is usually demonstrated via rectal administration of barium. *(Gurley, p 113)*

31. **(D)** All of the listed choices listed are appropriate when making a correction in a patient's medical record. A line should be put through the error and the word "error" may be written next to the crossed out line. The correction should then be initialed and dated. Radiographers should not skip pages in the patient chart or use "white out" when making corrections. *(Adler & Carlton, p 340)*

32. **(D)** Most pathogens found in the respiratory tract can be transmitted through the air. Some examples of bacterial diseases transmitted from one individual to another via airborne transmission are tuberculosis, strep throat, pneumonia, diptheria, and whooping cough. Mumps is a *viral* disease that is contagious via airborne transmission. *(Ehrlich & McCloskey, p 105)*

33. **(C)** Radiographs are the property of the health care institution. Radiographs are a part of the permanent medical record. They should be retained on file for 7 years, and in the case of pediatric patients, until the patient reaches maturity. Mammograms are also kept on file for extended time periods. The radiographs are not the personal property of the radiologist or the referring physician. If a patient decides to change doctors or to get a second opinion, they may request a copy of the radiographs. In this case, the patient may borrow the originals, which must be returned, or they may pay for a copy. If the courts subpoena a copy of the films, they must be provided at a reasonable fee. *(Ehrlich & McCloskey, p 51)*

34. **(B)** A patient going into shock may experience pallor and weakness, a significant drop in blood pressure, and an increase in pulse rate. The patient may also experience apprehension and restlessness, and have cool, clammy skin. A radiographer recognizing these symptoms should call them to a physician's attention immediately. Fever is not associated with shock. *(Ehrlich & McCloskey, pp 176, 177)*

35. **(C)** The complete killing of all microorganisms is termed *sterilization. Surgical asepsis* refers to the technique used when performing procedures to prevent contamination. *Medical asepsis* refers to practices that reduce the spread of microbes, and therefore the chance of spreading disease or infection. Washing your hands is an example of medical asepsis. It reduces the spread of infection, but does not eliminate all microorganisms. *Disinfection* involves the use of chemicals to either inactivate or inhibit the growth of microbes. *(Adler & Carlton, p 207)*

36. **(C)** A *pulse oximeter* is used to measure a patient's pulse rate and oxygen saturation level. A *stethoscope* and a *sphygmomanometer* are used together to monitor blood pressure. The first sound heard is the systolic pressure and the normal range is 110 to 140 mm Hg. When no more sound is heard, the diastolic pressure is recorded. The normal diastolic range is 60 to 90 mm Hg. Elevated blood pressure is *hypertension. Hypotension,* or a blood pressure below normal, is not of concern unless it is caused by injury or disease and the patient may go into shock. *(Adler & Carlton, pp 185–187)*

37. **(C)** Chest drainage systems should always be kept **below** the patient's chest. Chest tubes are used to remove air, blood, or fluid from the pleural cavity. By draining fluid from the pleural cavity, a collapsed lung, or *atelectasis,* may be relieved. By relieving pressure from air in the pleural cavity, a *pneumothorax* may be reduced. Radiographers must take care that the tubes of the chest drainage unit do not kink and do not get caught on intravenous poles or radiographic equipment. It is imperative that the unit remain *below the level of the chest.* The chest drainage system is composed of several components. One chamber collects the draining fluid. Another component is the suction control chamber. A third component is the water seal chamber, which prevents air from the atmosphere from entering the system. The last chamber is the water seal venting component, which functions to allow air out of the system, thus preventing pressure buildup. In order for the unit to work properly, it must remain below the chest. *(Adler & Carlton, pp 231–232)*

38. **(C)** *Talipes* is the term used to describe congenital clubfoot. There are several types of talipes generally characterized by a deformed talus and shortened Achilles tendon, giving the foot a club-like appearance. *Osteochondritis* (Osgood-Schlatter disease) is a painful incomplete separation of tibial tuberosity from the tibial shaft. It is fairly common in active adolescent boys. *Coxa plana* (Legg-Calvé-Perthes disease) is ischemic necrosis leading to flattening of the femoral head. *Muscular dystrophy* is a congenital disorder characterized by wasting of skeletal muscles. *(Taber's, p 1691)*

39. **(D)** Epidemiologic studies indicate that AIDS can be transmitted only by intimate contact with body fluids of an infected individual. This can occur through the sharing of contaminated needles, through sexual con-

tact, and from mother to baby at childbirth. AIDS can also be transmitted by transfusion of contaminated blood. *(Hopp, p 66)*

40. **(D)** The lithotomy position is generally employed for hysterosalpingography. The lithotomy position requires that the patient lie on her back with buttocks at the edge of the table. The hips are flexed, the knees are flexed over leg rests, and the feet are in stirrups. *(Ballinger, vol 2, p 200)*

41. **(C)** Shock occurs when blood pressure is unable to provide sufficient circulation of oxygenated blood to all body tissues. There are several types of shock. Symptoms common to all types of shock include a drop in blood pressure (therefore, inability to provide oxygenated blood to body tissues), increased pulse rate, and restlessness and apprehension. *(Ehrlich & McCloskey, pp 176, 177)*

42. **(B)** Dilated, twisted veins, or *varices,* of the esophagus are frequently associated with obstructive liver disease or cirrhosis of the liver. These esophageal veins enlarge and can rupture, causing serious hemorrhage. *Achalasia* is dilatation of the esophagus as a result of the cardiac sphincter's failure to relax in order to allow food to pass into the stomach. *Dysphasia* is a speech impairment resulting from a brain lesion; it is unrelated to the esophagus. *Dysphagia* refers to difficulty swallowing and is the most common esophageal complaint. Hiatal hernia is another common esophageal problem characterized by protrusion of a portion of the stomach through the cardiac sphincter. It is a common condition, and many individuals with it are asymptomatic. Each of these conditions of the esophagus may be evaluated with an esophagram. Positions usually include the PA, RAO, and right lateral. *(Linn-Watson, pp 102, 107)*

43. **(A)** In order to obtain a diagnostic examination of the stomach, it must first be empty. The usual preparation is *npo* (nothing by mouth) after midnight (approximately 8 hours before the exam). Any material in the stomach can simulate the appearance of disease. *(Torres, p 113)*

44. **(B)** The patient's right to privacy indicates that the patient's modesty and dignity will be respected. It also refers to the professional health care worker's obligation to respect the confidentiality of privileged information. Inappropriate communication of privileged information to anyone but health care workers involved with the patient's care is inexcusable. *(Ehrlich & McCloskey, p 17)*

45. **(C)** Whenever a perforation of the GI tract is suspected, a water-soluble contrast agent (such as Gastrografin or Oral Hypaque) should be used, because it is easily absorbable from within the peritoneal cavity. Barium sulfate leakage into the peritoneal cavity can have serious consequences. Water-soluble contrast agents may also be used in place of barium sulfate when the possibility of barium impaction exists. *(Ballinger, vol 2, p 83)*

46. **(B)** The contrast medium dosage administered for pediatric intravenous urography is dependent upon the age and weight of the infant or child. Much smaller doses, of course, are required for younger and smaller children compared to the dose delivered to the average adult. *(Ballinger, vol 2, p 168)*

47. **(D)** According to patient's bill of rights, the patient's verbal request supersedes any prior written consent. It is not appropriate to dismiss the patient without notifying the referring physician and the radiologist. The patient may very well need a particular radiographic examination to make a proper diagnosis, or for presurgical planning, and the radiographer must inform the physician of the patient's decision immediately. *(Ehrlich & McCloskey, pp 16–17)*

48. **(A)** An ambulatory patient is one who is able to walk with minimal or no assistance. Outpatients are usually ambulatory, and many inpatients are ambulatory. Patients who are not ambulatory are usually transported to the radiology department via stretcher. *(Taber's, p 66)*

49. **(B)** A compound fracture is one where the fractured ends of bone have penetrated the skin. A bone may be fractured in more than one place, or the fractured ends may be out of alignment, and yet not penetrating the skin. *(Ehrlich & McCloskey, p 179)*

50. **(D)** Increased pain threshold, breakdown of skin, and atrophy of fat pads and sweat glands are all important considerations when working with geriatric patients. As our bodies age, many changes in the organs take place. Although muscle is replaced with fat, the amount of subcutaneous fat is decreased, and the skin atrophies. Because of this, it is very important to treat the geriatric patient gently. A mattress or pad should always be placed on the radiographic table to help prevent skin abrasions. Additionally, paper tape should be used in place of adhesive tape. Geriatric patients are also more sensitive to hypothermia, because of breakdown of the sweat glands, and should always be kept covered, not just to preserve modesty, but for warmth as well. Loss of sensation in the skin increases the pain tolerance but may not alert a geriatric patient to undo stress on bony prominences like the elbow, wrist, coccyx, and ankles. The radiographer must be sensitive to this and treat geriatric patients delicately. *(Dowd & Wilson, vol 2, p 1026)*

51. **(A)** Overdistention of the alveoli with air is termed *emphysema*. The condition is frequently a result of many years of smoking and is characterized by dyspnea, especially when recumbent. *Empyema* is pus in the thoracic cavity; *pneumothorax* is air or gas in the pleural cavity. *Pneumoconiosis* is a condition of the lungs characterized by particulate matter having been deposited in lung tissue, sometimes resulting in emphysema. *(Taber's, p 537)*

52. **(D)** Adverse reactions to the intravascular administration of iodinated contrast media are not uncommon, and although the risk of a life-threatening reaction is relatively rare, the radiographer must be alert to recognize and deal effectively should a serious reaction occur. Minor reaction is characterized by flushed appearance and nausea, occasionally vomiting, and a few hives. Early symptoms

of a possible anaphylactic reaction include constriction of the throat, possibly due to laryngeal edema, dysphagia (difficulty swallowing), and itching of the palms and soles. The radiographer must maintain the patient's airway, summon the radiologist, and call a "code." (*Ehrlich & McCloskey, p 176*)

53. **(C)** The best way to be sure you have the correct patient is to check the name on the patient's wristband (hospital identification). Calling out the patient's name in the waiting room can have less accurate results. If patients have been anxiously awaiting their x-ray exam, they may expect to hear their name next, unintentionally responding to another name. It is always wise to doublecheck the patient's hospital ID. (*Torres, p 41*)

54. **(B)** A *fomite* is an inanimate object that has been in contact with an infectious microorganism. A *reservoir* is the site where an infectious organism can remain alive and from which transmission can occur. Although an inanimate object can be a reservoir for infection, so too can living objects (such as humans) be reservoirs. For infection to spread there must be a *host* environment. Although an inanimate object may serve as a temporary host for microbes to grow, microbes flourish on and in the human host, where plenty of body fluids and tissue nourish and feed the microbes. A *vector* is an animal host of an infectious organism that transmits the infection via a bite or sting. (*Adler & Carlton, p 203*)

55. **(D)** Informed consent is required before any examination that involves greater than usual risk. Routine procedures such as sialography and upper and lower GI series are examples of lower-risk procedures, for which the consent given on admission to the hospital is sufficient. (*Ehrlich, pp 15, 17*)

56. **(D)** Severe allergic reaction affecting several tissue functions is referred to as anaphylaxis or anaphylactic shock. It is characterized by dyspnea (difficulty breathing) caused by rapid swelling of the respiratory tract and a sharp drop in blood pressure. Individuals

sensitive to bee stings and certain medications, including iodinated contrast agents, are candidates for this reaction. (*Ehrlich & McCloskey, p 158*)

57. **(C)** The relatively new low-osmolality and non-ionic, water-soluble contrast media available to radiology departments have outstanding advantages, especially for patients with a history of allergic reaction. They were originally used for intrathecal injections (myelography), but were quickly accepted for intravascular injections. Side effects and allergic reactions are less likely and less severe with these media. Their one very significant disadvantage is their huge cost compared to ionic contrast media. Many institutions are choosing to use non-ionic contrast media in select procedures and with patients at greater risk of allergic reaction. (*Ehrlich & McCloskey, p 218*)

58. **(C)** Diuretics are used to promote urine elimination in individuals whose tissues are retaining excessive fluid. Emetics induce vomiting, and cathartics stimulate defecation. Antitussives are used to inhibit coughing. (*Ehrlich & McCloskey, p 121*)

59. **(B)** The medical word for nosebleed is epistaxis. Vertigo refers to the feeling of "whirling" or the sensation that the room is spinning. Some possible causes of vertigo include inner ear infection or an acoustic neuroma. Urticaria is a vascular reaction resulting in dilated capillaries and edema that cause the patient to break out in hives. An aura may be classified as either a feeling or a motor response (such as flashing lights, tasting metal, smelling coffee) that precedes an episode such as a seizure or a migraine headache. (*Ehrlich & McCloskey, p 186*)

60. **(A)** It is generally recommended that the IV bottle be kept 18 to 20 inches above the level of the vein. If the bottle is too high, the pressure of the IV fluid can cause it to pass through the vein into surrounding tissues, causing a painful and potentially harmful condition. If the IV bottle is too low, blood may return through the needle into the tub-

ing, form a clot, and cause obstruction to the flow of IV fluid. *(Ehrlich & McCloskey, p 159)*

61. **(C)** An individual whose coronary arteries are not carrying enough blood to the heart muscle (myocardium), due to partial or complete blockage of a cardiac vessel, experiences crushing pain in the chest, frequently radiating to the left jaw and arm. This is termed *angina pectoris* and may be relieved by the drug nitroglycerin, which dilates the coronary arteries, thus facilitating circulation. Tachycardia refers to rapid heart rate, and bradycardia to slow heart rate. Syncope is fainting. *(Ehrlich & McCloskey, p 142)*

62. **(D)** Hypochlorite bleach (Clorox) and Lysol are examples of *disinfectants.* Disinfectants are used in radiology departments to clean equipment and remove microorganisms from areas such as radiographic tables. *Antiseptics* are also used to stop the growth of microorganisms, but are often applied to the skin, not to radiographic equipment. *Antifungal* medications may be administered systemically or topically to treat or prevent fungal infections. *Antibacterial* medications may also be administered systemically or externally. Tetracycline is a systemic antibacterial medication. An example of an externally applied antibacterial medication is pHisoHex (hexachlorophene). *(Ehrlich & McCloskey, p 111)*

63. **(D)** Medications are commonly administered orally (by mouth). They may also be administered directly into a vein (intravenously), into a muscle (intramuscularly), or under the skin (subcutaneously). *(Ehrlich & McCloskey, pp 142, 143)*

64. **(D)** All of these symptoms are related to a respiratory reaction. There may also be hoarseness, wheezing, or cyanosis. The patient who has received contrast media should be watched closely, and if any symptoms arise, the radiologist should be notified immediately. *(Torres, p 202)*

65. **(D)** All of the statements in the question are true regarding the proper care of a patient with a tracheostomy. If you have to touch a

tracheostomy for any reason, employ sterile technique to avoid the possibility of infection. Patients with tracheostomies require frequent suction, but this is usually not performed by the technologist. Radiographers are often called upon to assist the nurse in suctioning, especially for those patients who are required to be in the radiology department for lengthy procedures. Patients who are to be suctioned should be aerated beforehand (that is, oxygen should be administered prior to suctioning). It is also important to note that patients should be permitted to rest during suctioning. Never suction for longer than 15 seconds. The nurse may check the breath sounds with a stethoscope to assure that the airway is clear. It is the radiographer's responsibility to check the work area and ensure that the suction is working and that ample ancillary supplies (suction kit, catheters, tubing) are available. *(Adler & Carlton, p 231)*

66. **(D)** Ventricular septal defect is a congenital heart defect characterized by a hole in the interventricular septum, which allows oxygenated and unoxygenated blood to mix. Some defects are small and close spontaneously; others require surgery. Coarctation of the aorta is a narrowing or constriction of the aorta. Atrial septal defect is a small hole (the remnant of the fetal foramen ovale) in the interatrial septum that usually closes spontaneously in the first months of life. If it persists, or is unusually large, surgical repair is necessary. The ductus arteriosus is a short fetal blood vessel connecting the aorta and pulmonary artery that usually closes within 10 to 15 hours after birth. A patent ductus arteriosus is one that persists and requires surgical closure. *(Taber's, p 1844)*

67. **(D)** All three factors are important considerations. *Miscibility* describes the ability of a contrast medium to mix with blood as it is injected into the bloodstream. Potential *toxicity* has always been an important consideration, as side effects to contrast media can be life threatening. The new non-ionic contrast media have considerably lower toxicity than the ionic compounds. *Viscosity* is a term that de-

notes the degree of stickiness or gumminess of the contrast medium. A contrast medium with increased viscosity can offer considerable resistance upon injection. *(Torres, p 201)*

68. (A) An *analgesic* is any drug that functions to relieve pain. An *anticoagulent* is used to prevent clotting of blood. A *diuretic* is used to increase urine output, and an *antibiotic* fights growth of bacterial microorganisms. *(Taber's, p 79)*

69. (A) Anginal pain, caused by constriction of blood vessels, may be relieved with administration of a vasodilator such as nitroglycerin. Bradycardia (abnormally slow heartbeat) and cardiac arrest are treated with vasoconstrictors such as dopamine or adrenalin to increase blood pressure. Antihistamines such as Benadryl are used to treat allergic reactions and anaphylactic shock. *(Ehrlich & McCloskey, p 142)*

70. (A) When classifying contrast agents, the total number of particles in solution per kilogram of water defines the *osmolality* of the contrast agent. The *toxicity* defines how noxious or harmful a contrast agent is. Low osmolar contrast agents have been found to cause less tissue toxicity than the ionic intravenous contrast agents. The *viscosity* defines the thickness or concentration of the contrast agent. Viscosity of a contrast agent can effect the injection rate. A thicker, or more viscous contrast agent will be more difficult to inject (more pressure is needed to push the contrast agent through the syringe and needle or angiocatheter). The *miscibility* of a contrast agent refers to the ability to mix with body fluids, such as blood. Miscibility is an important consideration in preventing thrombosis formation. It is generally preferable to use a contrast agent with low osmolality and low toxicity because they are safer for the patient and are less likely to cause any untoward reactions. When comparing ionic and non-ionic contrast, a non-ionic contrast agent has a lower osmolality. To further understand osmolality, remember that whenever intravenous contrast media is introduced, there is a notable shift in fluid and ions. This shift is

caused by an inflow of water from interstitial region into the vascular compartment, which increases the blood volume and cardiac output. Consequently, there will be a decrease in systemic arterial pressure and peripheral vascular resistance with peripheral vasodilatation. Additionally, the pulmonary pressure and heart rate increase. By understanding the effects of osmolality on the patient, it becomes clear that an elderly patient or one with cardiac disease or impaired circulation would greatly benefit from the use of a low osmolar agent rather than a high osmolar agent. *(Chapman & Nakielny, pp 9–21)*

71. (A) The patient is said to be in the Trendelenburg position when the head is positioned lower than the feet. This position is helpful in several radiographic procedures, such as separating redundant bowel loops and demonstration of hiatal hernias. It is also used in treating shock. Fowler's position is with the head higher than the feet. The Sims position is the LPO position with the right leg flexed up for insertion of an enema tip. *(Taber's, p 1353)*

72. (C) When radiographs in the surgical suite are requested, every precaution must be taken to maintain required surgical asepsis in the OR itself. This requires proper dress, cleanliness of equipment, and restricted access to certain areas. An example of a restricted area is the "sterile corridor," located between the draped patient and instrument table, and occupied only by the surgeon and instrument nurse. *(Ehrlich & McCloskey, pp 367, 369)*

73. (A) When a two-member team of radiographers is performing mobile radiography on a patient under strict isolation, such as a burn patient, one radiographer remains "clean," having no physical contact with the patient. The "clean" radiographer will position the mobile unit and make the exposure. The other member of the team will position the cassette and retrieve the cassette. As they fold down its protective plastic cover, the "clean" radiographer will remove the cassette from the plastic. Both radiographers should be protected in gowns, gloves, and masks if the patient is on strict isolation. Additionally, after completing the exam, the mobile unit should be disinfected. Patients on strict isolation may have highly contagious and easily spread viral diseases, such as Varicella (chickenpox). *(Ehrlich & McCloskey, pp 117–118)*

74. **(C)** Protective or "reverse" isolation is used to keep the susceptible patient from becoming infected. Patients who have suffered burns have lost a very important means of protection, their skin, and therefore have increased susceptibility to bacterial invasion. Patients whose immune systems are depressed have lost the ability to combat infection, and hence are more susceptible to infection. Active tuberculosis requires AFB (acid-fast bacilli) isolation. *(Gurley, p 153)*

75. **(C)** Elderly patients dislike being pushed or hurried along. They appreciate the radiographer who is caring enough to take the extra few minutes necessary for comfort. Some elderly patients are easily confused, and it is best to address them by the full name and keep instructions simple and direct. The elderly require the same respectful, dignified care as all other patients. *(Ehrlich & McCloskey, pp 42, 43)*

76. **(C)** Every hour is indicated by the abbreviation *qh*. The abbreviation *tid* means three times a day, and *qid* means four times a day. After meals is abbreviated *pc*. *(Torres, p 192)*

77. **(C)** A splint should never be removed from an extremity except by or under the direct supervision of a physician. Some splinting devices are not radiolucent and might need to be removed before the radiographic exam. In such a case, the physician must be present. *(Ehrlich & McCloskey, p 180)*

78. **(A)** Iodinated contrast material can become somewhat viscous (thick and sticky) at normal room temperatures. This makes injection much more difficult. Warming the contrast medium to body temperature serves to reduce viscosity. This may be achieved by placing the vial in warm water or into a special warming oven. *(Ehrlich & McCloskey, p 214)*

79. **(D)** Although it is never the responsibility of the radiographer to diagnose, it is the responsibility of every radiographer to be alert. The patient should not be subjected to unnecessary radiation from an unwarranted exam. Rather, it is the radiographer's responsibility to check with the referring physician and report the patient's complaint. *(Ehrlich & McCloskey, p 14)*

80. **(B)** A patient who has been npo since midnight, or one who is anxious, frightened, or in pain may suffer an episode of syncope (fainting) on exertion. The patient should be helped to a recumbent position with feet elevated, in order to increase blood flow to the head. *A patient who feels faint should never be left alone. (Ehrlich & McCloskey, p 184)*

81. **(A)** Angina pectoris is a crushing chest pain caused by circulatory disturbance of the coronary arteries. Nitroglycerin is used to dilate blood vessels (vasodilator) and decrease blood pressure in the treatment of pain from angina pectoris. Nitroglycerin is usually given sublingually, and thus absorbed directly into the bloodstream. *(Ehrlich & McCloskey, p 142)*

82. **(B)** A myelogram, or radiographic examination of the spinal canal, requires an intrathecal (intraspinal) injection. Intrathecal administration of contrast media is usually at the level of L2/3 or L3/4. An intravenous pyelogram is performed with an injection of contrast media into the venous system. A lymphangiogram requires contrast media be delivered into the lymphatic vessels. A computed tomography (CT) scan may or may not require the use of an intravenous injection. *(Torres, p 240)*

83. **(D)** *An unconscious patient is frequently able to hear and understand all that is going on, even though he or she is unable to respond.* The radiographer should go about performing the exam, always referring to the patient by

name and taking care to continually explain and reassure. *(Ehrlich & McCloskey, p 40)*

84. **(A)** Although patient consent for all routine procedures is implied on admission to the hospital, specific informed consent forms are required for procedures that involve significant risk. Routine procedures such as GI series and skeletal surveys do not involve significant patient risk, but invasive procedures such as angiography do, and therefore require that a consent form be signed. A family member may sign for a patient who is incompetent or too ill to sign. *(Ehrlich & McCloskey, pp 15, 17)*

85. **(A)** Three times a day is indicated by the abbreviation *tid*. The abbreviation *qid* means four times a day. Every hour is represented by *qh* and *pc* means after meals. *(Ehrlich & McCloskey, p 277)*

86. **(C)** A patient with *orthopnea* is unable to breath lying down. When the body is recumbent, abdominal organs move upward and push the abdominal organs up. It is therefore more difficult to breath deeply. Patients with orthopnea must be examined in the erect or semierect position. *Dyspnea* refers to difficulty breathing (any body position). *Apnea* describes cessation of breathing for short intervals. *Hyperpnea* is rapid breathing. *(Ehrlich & McCloskey, p 71)*

87. **(B)** Children are usually fearful of leaving familiar surroundings, and to be able to take along a familiar toy is helpful. A calm and cheerful radiographer can be reassuring to the anxious child. Honesty is essential and false reassurances, such as telling the child that it will not hurt, not only do more damage than good, but also focus the child's attention on pain. *(Ehrlich & McCloskey, p 40)*

88. **(B)** Strain on the abdominal muscles may be minimized by placing a pillow under the head and a support under the knees. The pillow also relieves neck strain, reduces the chance of aspiration in the nauseated patient, and permits the patient to observe surroundings. The Trendelenburg position causes the diaphragm to assume a higher position, and can cause a patient to become short of breath. *(Ehrlich & McCloskey, pp 71–73)*

89. **(D)** Most needle sticks occur while attempting to recap a needle. Proper disposal of contaminated needles and syringes is becoming more vital as HIV and AIDS reach epidemic proportions. To prevent the spread of any possible infection, handle contaminated materials as little as possible. Therefore do not attempt to recap a needle, but rather, dispose of the entire syringe with needle attached in the special container available. *(Ehrlich & McCloskey, p 113)*

90. **(C)** A patient having a GI series is required to be npo for at least 8 hours prior to the exam; food or drink in the stomach can simulate disease. A patient scheduled for a pyelogram must have the preceding meal withheld to avoid the possibility of aspirating vomitus in case of an allergic reaction. An abdominal survey does not require the use of contrast media, and no patient preparation is necessary. *(Ehrlich & McCloskey, p 206)*

91. **(A)** Creatinine is a normal alkaline constituent of urine and blood, but increased quantities of creatinine are present in advanced stages of renal disease. Creatinine and BUN (blood urea nitrogen) blood chemistry levels should be checked prior to beginning an IVP. Increased levels may forecast increased possibility of contrast media induced renal effects and poor visualization of the renal collecting systems. Normal creatinine range is 0.6 to 1.5 mg/100 mL. Normal BUN range is 8 to 25 mg/100 mL. *(Ballinger, vol 2, p 168)*

92. **(A)** It is unlikely that the radiographer would be faced with a wound hemorrhage, because bleeding from wounds is controlled before the patient is seen for x-ray examination. However, when radiographing a patient who does experience hemorrhaging from a wound, you should apply pressure to the bleeding site and call for assistance. Delay can lead to serious blood loss. *(Ehrlich & McCloskey, p 182)*

93. **(C)** Verbal defamation of another, or slander, is a type of intentional misconduct. Invasion of privacy, that is, public discussion of privileged and confidential information, is intentional misconduct. However, if a radiographer left a weak patient standing alone in order to check films or get supplies, and that patient fell and sustained injury, that would be considered unintentional misconduct or negligence. *(Gurley, p 21)*

94. **(B)** When assisting the patient with changing, *first remove clothing from the unaffected side.* In that way, removing clothing from the affected side will require less movement and effort. Patient clothing should be cut away only as a last resort in cases of extreme emergency and with the patient's consent. *(Torres, p 50)*

95. **(B)** Syncope, or fainting, is a result of a drop in blood pressure caused by insufficient blood (oxygen) to the brain. The patient should be helped into a dorsal recumbent position with feet elevated in order to facilitate blood flow to the brain. *(Ehrlich & McCloskey, pp 184, 185)*

96. **(B)** The diameter of a needle is the needle gauge. The higher the gauge number, the thinner the diameter. For example, a very tiny gauge needle of 25 gauge may be used on a pediatric patient for intravenous injection, whereas a large-gauge needle of 16 gauge may be used for donating blood. The *hub* of the needle is the portion of a needle that attaches to a syringe. The *length* of the needle varies depending on its use. A longer needle is needed for intramuscular injections, a shorter needle for a subcutaneous injection. The *bevel* of the needle is the slanted tip of the needle. For intravenous injections, the bevel should always face up. *(Adler & Carlton, p 287)*

97. **(C)** Frequent and correct handwashing is an essential part of medical asepsis; it is the best method to avoid the spread of microorganisms. If the faucet cannot be operated with the knee, it should be opened and closed using paper towels. Care should be taken to thoroughly wash all surfaces and between fingers. The hands and forearms should always be kept below the elbows. Hand lotions should be used frequently to keep hands from chapping. Remember, unbroken skin prevents entry of microorganisms; dry, cracked skin breaks down that defense and permits entry of microorganisms. *(Torres, pp 24, 25)*

98. **(D)** The U.S. Food and Drug Administration mandates that certain information be included in every drug package. Some of the information drug companies are required to provide are trade and generic names, indications and contraindications, usual dose, chemical composition and strength, and any reported side effects. *(Ehrlich & McCloskey, p 134)*

99. **(D)** The Joint Commission on the Accreditation of Healthcare Organizations (JCAHO) is the organization that accredits health care organizations in the United States. The JCAHO sets forth certain standards for medical records. In keeping with those standards, all diagnostic and therapeutic orders, the patient's medical history, and signed informed consent forms must all be contained in the patient's medical record or chart. Additionally, patient identification information, and any diagnostic and therapeutic reports should also be part of the patient's permanent record. The patient chart is a means of communication between various health care providers. *(Adler & Carlton, p 335)*

100. **(B)** When performing bedside radiography in an isolation room, the radiographer should wear a gown, gloves, mask, and cap. The cassettes are prepared for the examination by placing a pillowcase over them to protect them from contamination. Whenever possible, one person should manipulate the portable unit and remain "clean," while the other handles the patient. The portable unit should be cleaned with a disinfectant before exiting the patient's room. *(Ballinger, vol 1, p 7)*

101. **(B)** A double-contrast GI requires that the patient ingest gas-producing powder, crystals, pills, or beverage followed by a small

amount of high-density barium. The patient may then be asked to roll while in the recumbent position in order to coat the gastric mucosa, while the carbon dioxide expands. This procedure provides optimum visualization of the gastric walls. An oral cholecystogram may be performed approximately 3 hours after ingestion of special ipodate calcium granules. *(Ballinger, vol 2, pp 60, 101)*

102. **(B)** Osteoarthritis is a progressive degenerative joint disorder characterized by deterioration of articular cartilage. Once the subchondral bone is exposed, friction between adjacent bones occurs, and new bone formation begins. This bone tissue forms spurs, which get progressively larger and function to decrease joint space and restrict movement. Osteoarthritis is a noninflammatory disorder. Rheumatoid arthritis is an inflammatory disorder. *(Tortora, p 225)*

103. **(B)** For demonstration of the joint cavity, arthrography requires an injection of iodinated contrast medium and/or air, which distends the cavity and causes a sense of fullness and discomfort within the joint. Pain in the chest radiating down the arm may be a symptom of a heart attack. Temporary paralysis and numbness in the fingers are signs of nerve compression that may be caused by a traumatic injury or an improperly performed arthrographic examination. *(Ballinger, vol 1, pp 444, 445)*

104. **(C)** The normal blood pressure range for men and women is 110 to 140 mm Hg (systolic reading; top number) and 60 to 80 mm Hg (diastolic reading; bottom number). Systolic pressure is the contraction phase of the left ventricle and diastolic pressure is the relaxation phase in the heart cycle. *(Torres, p 66)*

105. **(D)** Local anesthetics are used to alleviate the pain caused by the insertion of a large-caliber needle for injection of contrast media, as in arteriography, arthrography, and myelography. A T-tube cholangiogram is a postoperative examination of the biliary tract. This examination is painless and permits administration of the contrast media via the T-tube. *(Ballinger, vol 2, p 105)*

106. **(B)** Double-contrast studies of the large bowel are particularly useful for demonstration of the bowel wall. Polyps are projections of bowel wall mucous membrane into the bowel lumen. Colitis is inflammation of the large bowel often associated with ulcerations of the mucosal wall. A single-contrast study would most likely obliterate these conditions, but *coating* of the bowel mucosa with barium and subsequent filling with air (double contrast) provides optimum delineation. *(Ballinger, vol 3, p 116)*

107. **(C)** Extravasation of contrast media into surrounding tissue is potentially very painful. If it does occur, the needle should be removed and the extravasation cared for immediately (before looking for another vein). Pressure should be applied to the vein until bleeding stops. Application of warm, moist heat to the affected area helps relieve pain. *(Adler & Carlton, p 292)*

108. **(A)** A diastolic pressure reading that may indicate the patient is in shock would be 50 mm Hg. Remember that the *diastolic* number is the *bottom* number in a blood pressure reading. The *normal range* for diastolic pressure is 60 to 90 mm Hg. A diastolic pressure reading of 110 mm Hg may indicate hypertension. The *systolic number* is the *top* number in a blood pressure reading. The *normal range* is 110 to 140 mm Hg. *(Ehrlich & McCloskey, p 94)*

109. **(A)** When transferring patients from stretcher to x-ray table, several rules apply that will reduce back strain. Pull, do not push the patient; pushing increases friction and makes the transfer more difficult. Do not bend at the waist and pull; use your biceps for pulling an object. Draw the object as closely to you as possible and then lift if necessary. *(Torres, p 41)*

110. **(D)** With the Seldinger technique, a needle with inner cannula is used to pierce an artery (usually femoral). The inner cannula is then

removed, a flexible guide wire inserted, the needle removed, and a catheter slipped over the guide wire into the artery. The guide wire is then removed, leaving the catheter in the artery. The Seldinger technique reduces the risk of extravasation (compared to direct needle stick). The patient may be positioned as required, and the radiographs may be inspected while the catheter remains safely in place. The axial or brachial arteries may also be used, but the femoral is the most common approach. Doppler is an ultrasonography term referring to the detection of movement (eg, blood flow through blood vessels). Grandy method describes the routine lateral projection of the cervical spine. Egaz Moniz introduced cerebral angiography in 1927. *(Ballinger, vol 3, p 130)*

111. **(D)** Lyme disease is a condition that results from transmission of an infection by a vector (deer tick). *Vectors* are insects and animals carrying disease. *Droplet contact* involves contact with secretions (from the nose, mouth, or eye) that travel via a sneeze or cough. *Airborne route* involves evaporated droplets in the air that transfer disease. *(Torres, p 22)*

112. **(C)** The Sims position requires the patient to lie on either side with the lower arm extended behind the body and the upper arm flexed. The body is rotated slightly anteriorly with the top knee flexed sharply and the bottom knee relaxed. This position facilitates insertion of the enema tip with the least discomfort to the patient. *(Torres, p 48)*

113. **(D)** Nerve compression and circulatory impairment may be caused by a cast that is too tight. Circulatory impairment or nerve compression may cause pain, immobility, numbness, burning or tingling of toes or fingers, coldness, and changes in color (blue or pale). The radiographer must be aware of these changes and notify a physician immediately. *(Torres, pp 52–53)*

114. **(C)** A *tort* is an *intentional or unintentional act that involves personal injury or damage to a patient.* Allowing a patient to be exposed to unnecessary radiation, either by neglecting to

shield the patient or by performing an unwanted, unnecessary examination, would be considered a tort and the radiographer would be held legally accountable. An emergency room physician is authorized to order x-ray examinations. *(Torres, p 4)*

115. **(B)** As blood pulsates through the arteries, a throb can be detected. This throb or pulse may be palpated where the arteries are superficial (examples are wrist, groin, neck, and posterior surface of the knee). The apical pulse is detected with the use of a stethoscope. *(Torres, p 63)*

116. **(D)** Blood and all other body fluids are carriers of HIV in the infected individual. Universal precautions are taken to avoid contact with any blood or body fluids. However, HIV cannot be transmitted via inanimate objects such as drinking fountains and glassware. *(Hopp, p 71)*

117. **(C)** The four *vital signs* are temperature, pulse, respirations, and blood pressure. Because radiographers may be required to take vital signs in an emergency, they should practice the more seldom-used skills. A thermometer is used to take patient's temperature. A watch with second hand is required to time the patient's pulse rate and respirations. To measure blood pressure, a blood pressure cuff, sphygmomanometer, and stethoscope are required. This is the skill the radiographer should practice frequently, as it is the one most likely needed in an emergency situation. A tongue blade is used to depress the tongue for inspection of the throat and is not part of vital sign assessment. *(Ehrlich & McCloskey, pp 86–98)*

118. **(D)** Orthopnea is a respiratory condition in which the patient has difficulty breathing (dyspnea) in any position other than erect. The patient is usually comfortable in the erect, standing, or seated positions. Trendelenburg position places the patient's head lower than the rest of the body. Fowler's position is a semi-erect position, and recumbent position is lying down. *(Taber's, p 1179)*

119. (D) Microorganisms are most commonly spread from one person to another by human hands and can be prevented from spreading by handwashing. Contaminated waste products, soiled linen, and improperly sterilized instruments are all ways that microorganisms can travel. Not every patient will come in contact with these items. However, the health care worker is in constant contact with the patient and is therefore a constant threat for the spread of infection. *(Torres, pp 23–24)*

120. (C) The *normal average* rate of respiration for a healthy adult patient is between 12 and 20 breaths per minute. For *children*, the rate is higher, averaging between 20 and 30 breaths per minute. Besides monitoring respiratory rate, it is also important to monitor the *depth* (shallow or labored) and pattern (regularity) of respiration. Respiratory rates greater than 20 breaths per minute in an adult would be considered *tachypnea*, and rates less than 12 would be *bradypnea*. *(Adler & Carlton, pp 182–183)*

121. (A) Anaphylactic shock is a result of a severe allergic reaction. The early signs and symptoms of shock are sneezing, coughing, nausea, vomiting, and itching at the injection site. The symptoms that follow may be edema of the hands, face, and other parts of the body; wheezing; weak and rapid pulse; and dilated pupils. Blurred vision, tremors, and convulsions are symptoms of hypoglycemia. *(Torres, pp 79–80)*

122. (C) Hypoglycemic reactions can be very severe and should be treated with an immediate dose of sugar. Early symptoms of an insulin reaction are shaking, nervousness, dizziness, cold and clammy skin, blurred vision, and slurred speech. Convulsions and coma may result if the patient is untreated. Cyanosis is the lack of oxygenated blood, which is a symptom of shock. *(Torres, pp 79–80)*

123. (A) The inert characteristics of barium sulfate renders it the least toxic contrast medium. Iodinated contrast media are absorbed by the body, while barium sulfate is not. Metrizamide, ethiodized oil, and meglu-

mine diatrizoate are iodinated contrast media that could cause anaphylactic shock and death in the allergic patient. *(Torres, pp 103–104)*

124. (A) The mechanical device used to correct an ineffectual cardiac rhythm is a *defibrillator*. The two paddles attached to the unit are placed on a patient's chest and used to introduce an electric current in an effort to correct the dysrhythmia. A *cardiac monitor* is used to display, and sometime record, ECG (electrocardiogram) readings and some pressure readings. The *crash cart* is a supply cart with various medications and equipment necessary for treating a patient who is suffering from a myocardial infarction or other serious medical emergencies. It is periodically checked and restocked, usually by nursing, although radiographers may be responsible for a daily check of the plastic throwaway locks. These locks are used to ensure that the cart has not been tampered with or supplies inadvertently used in non-emergency situations. A *resuscitation bag* is used for ventilation, as during cardiopulmonary resuscitation. *(Ehrlich & McCloskey, p 173)*

125. (D) Physicians often prescribe a mild laxative to aid in elimination of barium sulfate. If a laxative is not given, the patient should be instructed to increase dietary fluid and fiber and to monitor bowel movement (the patient should have at least one within 24 hours). Patients should also be aware of the white-colored appearance of the stool. It will be present until all barium is expelled. *(Torres, p 110)*

126. (D) Cyanosis is a condition resulting from a deficiency of oxygen circulating in the blood. It is characterized by bluish discoloration of the gums, nailbeds, earlobes, and around the mouth. Cyanosis may accompany labored breathing or other types of respiratory distress. *(Torres, p 66)*

127. (C) The body's synovial joints are afforded many types of motions and movements. *Adduction* is a type of angular movement defined as movement toward the midline. *Protraction* is the forward movement of a part, on a plane parallel to the floor (such as the mandible thrust forward). As *extension* occurs (as in the elbow), the angle between the articulating surfaces of the forearm and elbow increases. In *flexion*, the angle decreases. *(Tortora, p 215)*

128. **(C)** When caring for a patient with an indwelling Foley catheter, place the drainage bag and tubing below the level of the bladder to maintain the gravity flow of urine. Placement of the tubing or bag above or level with the bladder will allow backflow of urine into the bladder. This reflux of urine may increase the chance of urinary tract infection (UTI). *(Torres, pp 160–161)*

129. **(B)** The abbreviation *pc* denotes "after meals" (post cibum). The other abbreviations are as follows: every 4 hours, *q4h* (quaque 4 hora); before meals, *ac* (ante cibum); whenever or as necessary, *prn* (pro re nata). *(Torres, p 192)*

130. **(B)** A parenteral route of drug administration is one that bypasses the digestive system. In radiography, the intravenous method is commonly used to administer contrast agents. The four parenteral routes require varied needle placements: subcutaneous (under the skin), intramuscular (through the skin and into the muscle), intradermal (between the layers of the skin), and intravenous (into the vein). *(Torres, p 178)*

131. **(C)** Sterile technique is required for administration of contrast media by the intravenous and intrathecal (intraspinal) methods. Sterile technique is also required for injection of contrast media during arthrography. Aseptic technique is used for administration of contrast media by means of the oral and rectal routes, as well as through the nasogastric tube. *(Torres, pp 178–179)*

132. **(B)** Systolic blood pressure describes the pressure during contraction of the heart. It is expressed as the top number when recording blood pressure. Diastolic blood pressure is the reading during relaxation of the heart and is placed on the bottom when recording blood pressure. A patient is considered hypertensive when systolic pressure is consistently above 140 mm Hg, and hypotensive

when the systolic pressure is lower than 90 mm Hg. *(Torres, p 66)*

133. **(A)** *Shock* is indicated by extremely low blood pressure, that is, a systolic blood pressure reading lower than 60 mm Hg (below 90 mm Hg is considered low blood pressure). Normal blood pressure is 110 to 140 mm Hg systolic and 60 to 80 mm Hg diastolic. High blood pressure is indicated by systolic pressure higher than 140 mm Hg and diastolic pressure higher than 90 mm Hg. *(Torres, pp 76–77)*

134. **(B)** The intrathecal (intraspinal) method of contrast medium administration is used in myelography and may be performed in the seated, lateral, and prone positions (Figs. 1–1 and 1–2). Adding curvature to the spine, by bending forward and curling the spine, allows greater access to the intervertebral spaces. This makes it easier to insert the spinal needle into the subdural space for injection of contrast medium. *(Torres, p 206)*

135. **(A)** Either the antecubital vein or the basilic vein, both found in the elbow region, may be used for intravenous injection. Other veins in the area include the cephalic and accessory cephalic veins. The popliteal vein, found in the area of the knee, is not commonly used for an intravenous injection. *(Adler & Carlton, p 294)*

136. **(A)** When a patient is experiencing a seizure, he or she should be protected from hitting any hard surfaces or falling off the radiographic table. The patient will exhibit uncontrollable body movements. Any attempt to place the patient in a semi-erect position or to administer a sedative would prove to be futile. After the seizure, it is important to place the patient on his or her side to prevent aspiration of any vomitus or oral secretions. *(Torres, p 86)*

137. **(B)** A severe blow to the head that causes the brain to bounce from side to side, resulting in injury on the side opposite the blow, is a *contrecoup* injury. A *concussion* is a traumatic injury that may be characterized by "seeing

Level of iliac crests

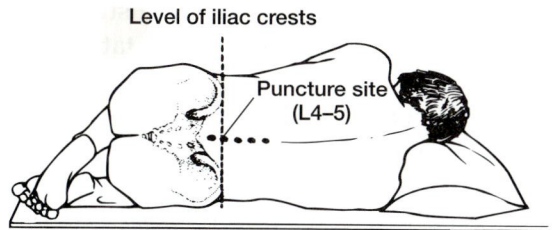

Puncture site
(L4–5)

Figure 1–1. Reprinted with permission from Krupp MA, et al: *Physician's Handbook.* 21 ed. Lange, 1985.

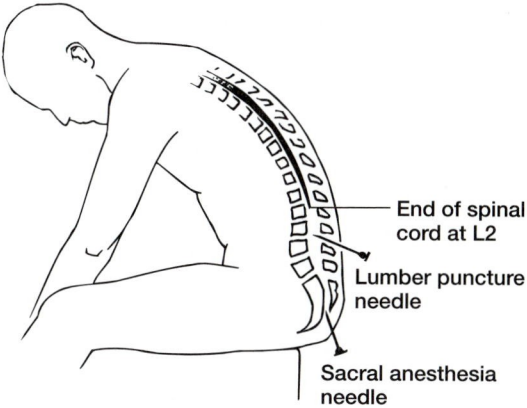

End of spinal
cord at L2

Lumber puncture
needle

Sacral anesthesia
needle

Figure 1–2. Reprinted with permission from de Groot J: *Correlative Neuroanatomy.* 21 ed. East Norwalk, CT: Appleton & Lange, 1991.

stars" or a brief loss of consciousness. A *seizure* is a disorder of the cerebral function characterized by a sudden brief attack of altered consciousness, motor activity, and sensory phenomena caused by an abnormal excessive discharge of cerebral neurons. Recurrent seizures are termed epilepsy. A patient who is in a coma is unarousable and unresponsive. The *Glasgow* coma scale rates the level of the coma to determine the level of the coma, based on motor response, verbal response, and to what degree the eyes open, if at all. *(Ehrlich & McCloskey , p 178)*

138. **(B)** It is important to make the patient as comfortable as possible during lengthy radiographic examinations; as comfort increases so does cooperation. Place a pillow under the patient's head and knees (this reduces back strain). A radio*lucent* pad will increase the patient's tolerance of the uncomfortable radiographic table and will not interfere with the examination. *(Ballinger, vol 2, pp 148, 152)*

139. **(C)** Disabled and elderly patients are occasionally bedridden for extended periods of time. Lying in one position reduces circulation and eventually leads to tissue breakdown. Careful position changes help alleviate pressure on stressed regions. Areas of the body that are more susceptible to "bed sores" or decubitus ulcers are the scapulae, sacrum, trochanters, knees, and heels of the feet. *(Torres, p 52)*

140. **(C)** It is recommended a radiographer wear protective eye gear (goggles) during any procedure in which there may be splattering of body fluids. This includes both angiography and any biopsy/aspiration procedures in which there may be splattering of blood or body fluids. This would not be expected during a routine upper gastrointestinal examinations. *(Torres, p 36)*

141. **(D)** In the practice of aseptic technique, *handwashing* is the most important precaution. The radiographer's hands should be thoroughly washed with warm, soapy running water after each patient examination. To avoid contamination of, or contamination by, the faucets, they should be opened and closed using paper towels. Care should be taken to avoid chapped hands by the use of handcream. Skin functions as a major factor in protecting bodies from invasion by bacteria and infection. Any cuts, abrasions, or other breaks in the continuity of this protective barrier should be protected from bacterial invasion with a bandage. *(Ballinger, vol 1, p 7)*

142. **(C)** BUN (blood urea nitrogen) level indicates the quantity of nitrogen in the blood in the form of urea. The normal concentration is 8 to 25 mg/100 mL. BUN and creatinine blood chemistry levels should be checked prior to beginning an IVP. An increase in the BUN level often indicates decreased renal function. Increased BUN and/or creatinine levels may forecast increased possibility of contrast media induced renal effects and poor visualization of the renal collecting systems. Normal creatinine range is 0.6 to 1.5 mg/100 mL. *(Ballinger, vol 2, p 168)*

143. **(C)** Of the four stages of infection, the stage during which the infection is most communicable is the *disease* phase. The initial phase, the *latent* period, is when the infection is introduced and lies dormant. As soon as the microbes begin to shed, the infection becomes communicable. The microbes reproduce and during the actual disease period, signs and symptoms of the infection may begin. The infection is most active and communicable at this point. As the patient fights off the infections, and the symptoms digress, the *convalescent* (recovery) stage occurs. *(Torres, p 29)*

144. **(A)** The *airway* of the victim should first be opened. This is accomplished by tilting back the head and lifting the chin. However, if the victim may have suffered a spinal cord injury, the spine should not be moved and the airway should be opened using the jaw-thrust method. The rescuer next listens for *breathing* sounds and watches for rise and fall of the chest to indicate breathing. If there is no breathing, the rescuer pinches the victims nose and delivers two full breaths via mouth-to-mouth rescue breathing. If rise and fall of the chest is still not present, the Heimlich maneuver is instituted. If ventilation does take place during the two full breaths, the victim's *circulation* is checked next (using the carotid artery). If there is no pulse, external chest compressions are begun at a rate of 80 to 100/minute for the adult and at least 100/minute for infants. *(Taber's, p 315)*

145. **(D)** The respiratory rate of an infant is much faster than that of an adult; therefore, the number of compressions per minute is also greater. Infant CPR requires 5 compressions to 1 breath. There should be at least 100 compressions per minute. *(American Red Cross Community CPR, pp 255–256)*

146. **(B)** Fractures of the humerus and shoulder girdle are quite painful, partly due to the difficulty encountered in immobilizing the upper extremity. Radiography should be performed as quickly as possible and with the least number of changes in body position; this may be accomplished in the upright position. An AP and transthoracic lateral humerus are taken by rotating the entire body, while the fractured extremity remains stationary. *(Ballinger, vol 1, p 116)*

147. **(D)** When preparing a colostomy patient for a barium enema, the dressing should be removed and disposed of. The drainage pouch may be retained for reuse, unless the patient specifies that they have a fresh pouch; or the radiographer may offer to provide one, if the department keeps them in stock. The choice of colostomy tip or catheter should be discussed with the radiologist. If a catheter is selected, the inflation cuff must be large enough to prevent barium from leaking, yet small enough so that the stoma area is not damaged. It is always a good idea to ask the patient if they would prefer to insert the colostomy tip. They are usually very used to caring for their stoma, and may be more comfortable in having control of the situation. *(Adler & Carlton, p 254)*

148. **(B)** When a contrast medium is injected, histamines are produced to protect the body from the foreign substance. An antihistamine (Benadryl) blocks the action of the histamine and reduces the body's inflammatory response to the contrast medium. An analgesic (aspirin) relieves pain. An anti-inflammatory drug (Ibuprofin) suppresses the inflammation of tissue. Antibiotics (penicillin) help fight bacterial infections. *(Torres, p 170)*

149. **(D)** *Lyme disease* is not a fungal infection, as the other listed choices are. It is carried by the deer tick and transmitted via the vector. A fungal infection that results in black or brown discoloration of the palmar surface of the hands and planter surface of the feet is *tinea nigra. Tinea pedis,* or athlete's foot, is also a fungal infection. *Ringworm* is the name given to denote the growth pattern of the fungi. Ringworm of the scalp in called tinea capitis, and tinea cruris is the medical term for "jock itch." *(Adler & Carlton, p 199)*

150. **(B)** Every radiograph must include: (1) the patient's name or ID number; (2) the side marker, right or left; (3) the date of the examination; and (4) institutional identity. In addition, other pertinent information may be included: patient's birthdate or age, attending physician, and time of day. When multiple films are taken on a patient in the same day, it becomes crucial that the time the radiographs were taken be included on the film. This allows the physician to track the patient's progress. (Ballinger vol 1, p 13)

Subspecialty List

74. Prevention and control of infection
75. Patient education, safety, and comfort
76. Patient education, safety, and comfort
77. Patient monitoring
78. Contrast media
79. Patient education, safety, and comfort
80. Patient monitoring
81. Patient education, safety, and comfort
82. Patient monitoring
83. Patient monitoring
84. Legal and professional responsibilities
85. Patient education, safety, and comfort
86. Patient education, safety, and comfort
87. Patient education, safety, and comfort
88. Patient monitoring
89. Prevention and control of infection
90. Contrast media
91. Contrast media
92. Patient monitoring
93. Legal and professional responsibilities
94. Patient education, safety, and comfort
95. Patient monitoring
96. Contrast media
97. Prevention and control of infection
98. Patient education, safety, and comfort
99. Legal and professional responsibilities
100. Prevention and control of infection
101. Contrast media
102. Patient education, safety, and comfort
103. Contrast media
104. Patient monitoring
105. Contrast media
106. Contrast media
107. Patient monitoring
108. Patient monitoring
109. Patient education, safety, and comfort
110. Contrast media
111. Prevention and control of infection
112. Contrast media
113. Patient education, safety, and comfort
114. Legal and professional responsibilities
115. Patient monitoring
116. Prevention and control of infection
117. Patient monitoring
118. Patient monitoring
119. Prevention and control of infection
120. Patient monitoring
121. Patient monitoring
122. Patient monitoring
123. Contrast media
124. Patient monitoring
125. Contrast media
126. Patient monitoring
127. Patient monitoring
128. Patient monitoring
129. Patient education, safety, and comfort
130. Patient monitoring
131. Prevention and control of infection
132. Patient monitoring
133. Patient monitoring
134. Contrast media
135. Patient monitoring
136. Patient monitoring
137. Patient monitoring
138. Contrast media
139. Patient monitoring
140. Prevention and control of infection
141. Prevention and control of infection
142. Contrast media
143. Contrast media
144. Patient monitoring
145. Patient monitoring
146. Patient monitoring
147. Contrast media
148. Contrast media
149. Prevention and control of infection
150. Legal and professional responsibilities

CHAPTER 2

Radiographic Procedures
Questions

DIRECTIONS (Questions 1 through 300): Each of the numbered items or incomplete statements in this section is followed by answers or by completions of the statement. Select the ONE lettered answer or completion that is BEST in each case.

1. In order to demonstrate the first two cervical vertebrae in the AP position, the patient is positioned so that

   (A) the acanthomeatal line is vertical
   (B) the infraorbitomeatal line is vertical
   (C) a line between the mentum and mastoid tip is vertical
   (D) a line between the occlusal plane and mastoid tip is vertical

2. Which of the following is (are) associated with a Colles fracture?

   1. transverse fracture of the radial head
   2. chip fracture of the ulnar styloid
   3. posterior or backward displacement

   (A) 1 only
   (B) 1 and 3 only
   (C) 2 and 3 only
   (D) 1, 2, and 3

3. Which of the positions illustrated in Figure 2–1 should be used to demonstrate the cervical apophyseal joints?

   1. Figure A
   2. Figure B
   3. Figure C

   (A) 1 only
   (B) 2 only
   (C) 1 and 3 only
   (D) 2 and 3 only

4. In which of the following positions/projections will the subtalar joint be visualized?

   (A) dorsoplantar projection of the foot
   (B) plantodorsal projection of the os calsis
   (C) medial oblique position of the foot
   (D) lateral foot

5. With the patient positioned as illustrated in Figure 2–2, and with the CR directed parallel to the patella, which of the following will be obtained?

   1. intercondyloid fossa
   2. patellofemoral articulation
   3. tangential patella

   (A) 1 only
   (B) 1 and 2 only
   (C) 2 and 3 only
   (D) 1, 2, and 3

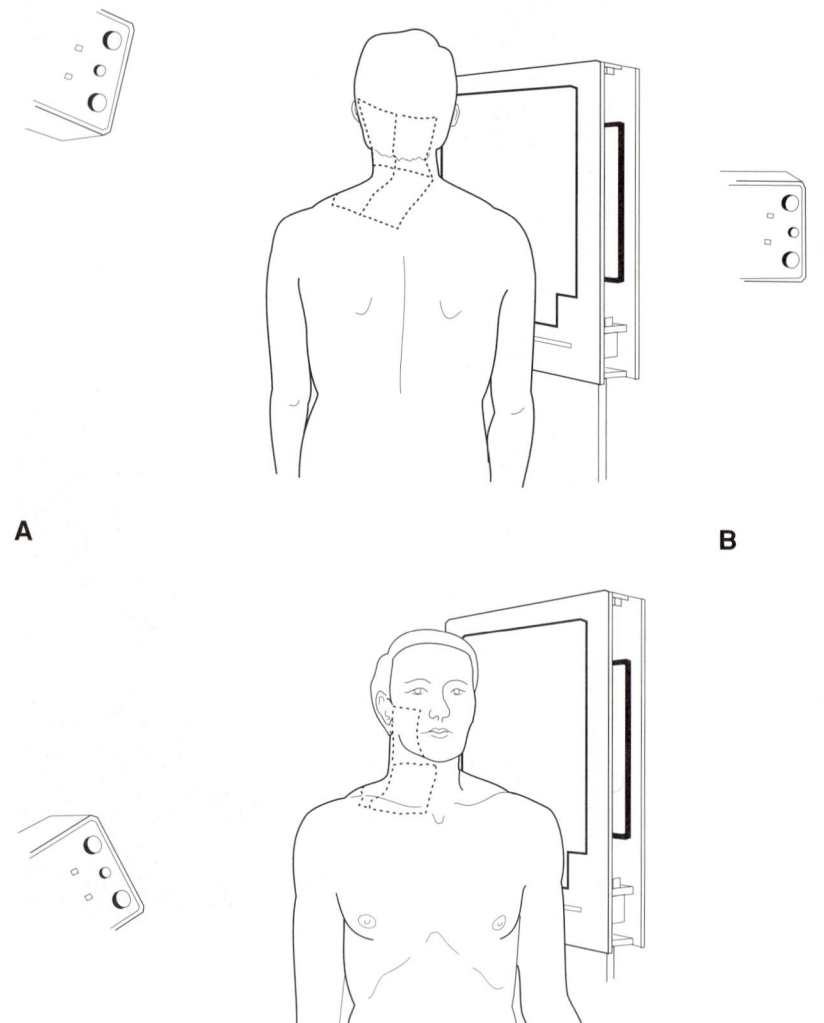

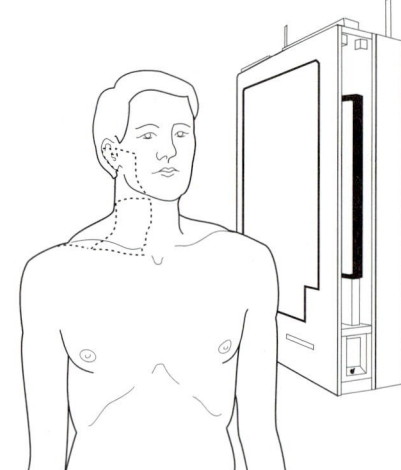

A

B

C

Figure 2–1.

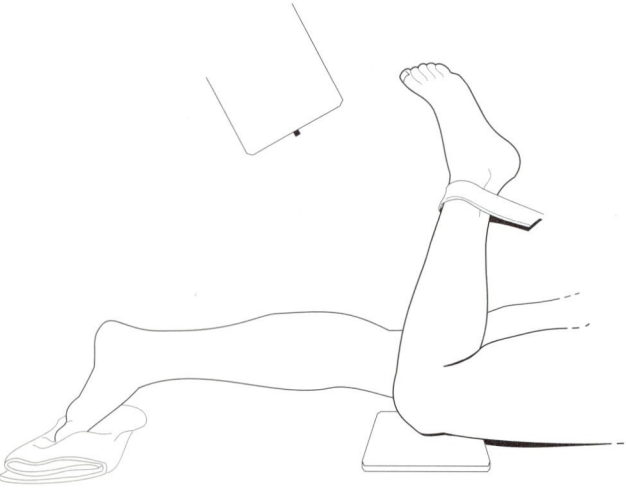

Figure 2–2.

6. Which of the following projection(s) require(s) that the shoulder be placed in external rotation?

1. AP humerus
2. lateral forearm
3. lateral humerus

(A) 1 only
(B) 1 and 2 only
(C) 2 and 3 only
(D) 1, 2, and 3

7. Which of the following is (are) part of the bony thorax?

1. 12 thoracic vertebrae
2. scapulae
3. 24 ribs

(A) 1 only
(B) 1 and 2 only
(C) 1 and 3 only
(D) 1, 2, and 3

8. Which of the following projections is most likely to demonstrate the carpal pisiform free of superimposition?

(A) radial flexion
(B) ulnar flexion
(C) AP oblique
(D) AP

9. All of the following statements regarding a PA projection of the skull, with central ray perpendicular to the film, are true EXCEPT

(A) orbitomeatal line is perpendicular to the film
(B) petrous pyramids fill the orbits
(C) midsagittal plane (MSP) is parallel to the film
(D) central ray exits at the nasion

10. All of the following statements regarding the radiograph in Figure 2–3 are true EXCEPT

(A) the tibial eminences are well visualized
(B) the intercondyloid fossa is demonstrated between the femoral condyles
(C) the femorotibial articulation is well demonstrated
(D) the radiograph was made with the knee extended

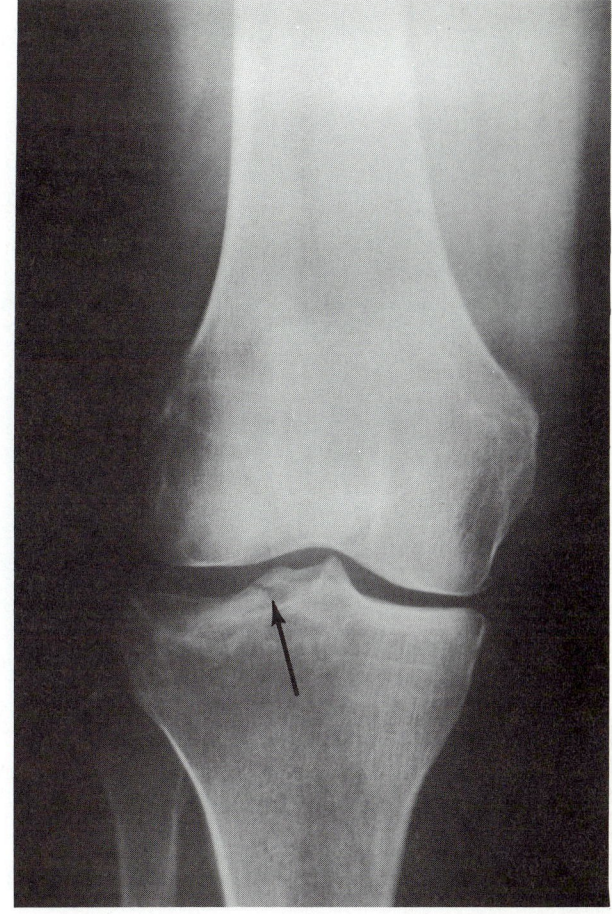

Figure 2–3. Reproduced with permission from Simon RR, Koenigsknecht SJ. *Emergency Orthopedics: The Extremities.* Stamford, CT: Appleton & Lange, 1996.

11. Involuntary motion can be caused by

1. peristalsis
2. severe pain
3. heart muscle contraction

(A) 1 only
(B) 2 only
(C) 1 and 2 only
(D) 1, 2, and 3

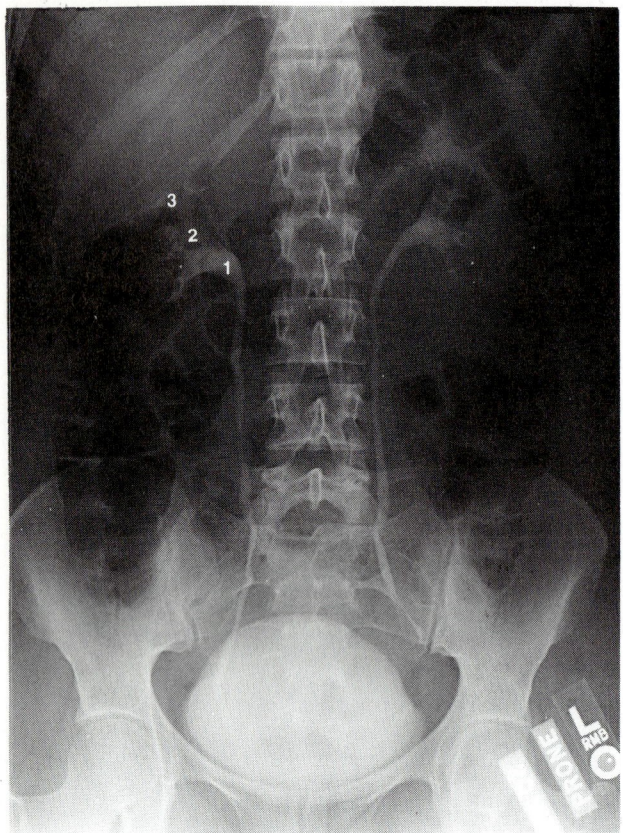

Figure 2–4. Courtesy of The Stamford Hospital, Department of Radiology.

12. An axial projection of the clavicle is often helpful in demonstrating a fracture not visualized using a perpendicular central ray. When examining the clavicle in the AP position, how is the central ray directed for the axial projection?

(A) cephalad

(B) caudad

(C) medially

(D) laterally

13. Which of the following is demonstrated in a 25° LPO position with the central ray entering 1 inch medial to the elevated anterior superior iliac spine (ASIS)?

(A) left sacroiliac joint

(B) right sacroiliac joint

(C) left ilium

(D) right ilium

14. The number *3* in the radiograph in Figure 2–4 represents which of the following renal structures?

(A) vesicoureteral junction

(B) renal pelvis

(C) minor calyx

(D) major calyx

15. During a gastrointestinal examination, the AP recumbent projection of a stomach of average shape will usually demonstrate

1. anterior and posterior aspects of the stomach
2. barium-filled fundus
3. double-contrast body and antral portions

(A) 1 only

(B) 1 and 2 only

(C) 2 and 3 only

(D) 1, 2, and 3

16. What are the positions most commonly employed for a radiographic examination of the sternum?

1. lateral
2. RAO
3. LAO

(A) 1 and 2 only

(B) 1 and 3 only

(C) 2 and 3 only

(D) 1, 2, and 3

17. For the AP projection of the scapula, the

1. patient's arm is abducted at right angles to the body
2. patient's elbow is flexed with hand supinated
3. exposure is made during quiet breathing

(A) 1 and 2 only

(B) 1 and 3 only

(C) 3 only

(D) 1, 2, and 3

18. The RAO position is used to project the sternum to the left of the thoracic vertebrae in order to take advantage of

    (A) pulmonary markings
    (B) heart shadow
    (C) posterior ribs
    (D) costal cartilages

19. In the AP axial position of the skull, with the central ray directed 30° caudad to the orbito-meatal line (OML) and passing midway between the external auditory meatuses, which of the following is best demonstrated?

    (A) occipital bone
    (B) frontal bone
    (C) facial bones
    (D) basal foramina

20. In order to demonstrate the entire circumference of the radial head, exposure(s) must be made with the

    1. epicondyles perpendicular to the cassette
    2. hand pronated and supinated as much as possible
    3. hand lateral and in internal rotation

    (A) 1 only
    (B) 1 and 2 only
    (C) 1 and 3 only
    (D) 1, 2, and 3

21. The best way to control voluntary motion is

    (A) immobilization of the part
    (B) careful explanation of the procedure
    (C) short exposure time
    (D) physical restraint

22. Figure 2–5 illustrates which of the following positions?

    (A) AP
    (B) medial oblique
    (C) lateral oblique
    (D) partial flexion

23. AP erect left and right bending films of the thoracic and lumbar vertebrae, to include 1 inch of the iliac crest, are performed to demonstrate

    (A) spondylolisthesis
    (B) subluxation
    (C) scoliosis
    (D) arthritis

24. Which of the following is (are) demonstrated in the lateral projection of the cervical spine?

    1. intervertebral joints
    2. apophyseal joints
    3. intervertebral foramina

    (A) 1 only
    (B) 1 and 2 only
    (C) 2 and 3 only
    (D) 1, 2, and 3

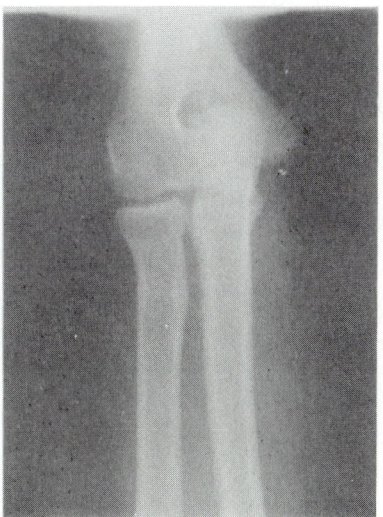

Figure 2–5. Courtesy of The Stamford Hospital, Department of Radiology.

25. The coronoid process should be visualized in profile in which of the following positions?

(A) scapular Y

(B) AP scapula

(C) medial oblique elbow

(D) lateral oblique elbow

26. Which of the following projections of the abdomen may be used to demonstrate air or fluid levels?

1. dorsal decubitus
2. lateral decubitus
3. AP Trendelenburg

(A) 1 only

(B) 1 and 2 only

(C) 1 and 3 only

(D) 1, 2, and 3

27. To best visualize the lower ribs, the exposure should be made

(A) on normal inspiration

(B) on inspiration, second breath

(C) on expiration

(D) during shallow breathing

28. All of the following are methods used to help reduce colonic spasms during radiographic examination of the barium-filled large bowel, EXCEPT

(A) lowering the enema bag a few inches

(B) placing the patient Trendelenburg

(C) administering glucagon prior to the exam

(D) slowing or stopping the flow of barium

29. Which of the following statements regarding the radiograph in Figure 2–6 is (are) true?

1. the position is used to demonstrate the frontal and ethmoidal sinuses
2. the sphenoid sinuses are seen near the medial aspect of the orbits
3. the chin should be elevated more to bring the petrous ridges below the maxillary sinuses

(A) 1 only

(B) 1 and 2 only

(C) 1 and 3 only

(D) 1, 2, and 3

30. All of the following are characteristics of the hypersthenic body type, EXCEPT

(A) short thoracic cavity

(B) short, wide, transverse heart

(C) diaphragm positioned low

(D) large bowel high and peripheral

31. All the following positions may be used to demonstrate the sternoclavicular articulations, EXCEPT

(A) weight-bearing

(B) RAO

(C) LAO

(D) PA

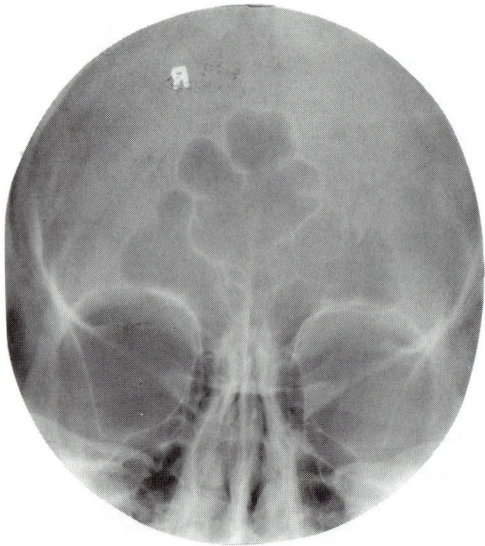

Figure 2–6. Courtesy of The Stamford Hospital.

32. Which of the following is (are) demonstrated in the AP projection of the cervical spine?

    1. intervertebral disc spaces

    2. C3–7 cervical bodies

    3. apophyseal joints

(A) 1 only

(B) 1 and 2 only

(C) 2 and 3 only

(D) 1, 2, and 3

33. When examining a patient whose elbow is in partial flexion, how should the AP projection be obtained?

    1. with humerus parallel to film, central ray perpendicular

    2. with forearm parallel to film, central ray perpendicular

    3. through the partially flexed elbow, resting on olecranon process, central ray perpendicular

(A) 1 only

(B) 1 and 2 only

(C) 2 and 3 only

(D) 1, 2, and 3

34. All of the following positions are likely to be employed for both single contrast and double contrast examinations of the large bowel, EXCEPT

(A) lateral rectum

(B) AP axial rectosigmoid

(C) right and left lateral decubitus abdomen

(D) RAO and LAO abdomen

35. Place the following anatomic structures in order from anterior to posterior

    1. trachea

    2. apex of heart

    3. esophagus

(A) trachea, esophagus, apex of heart

(B) esophagus, trachea, apex of heart

(C) apex of heart, trachea, esophagus

(D) apex of heart, esophagus, trachea

36. Which of the following skull positions will demonstrate the cranial base, sphenoid sinuses, atlas, and odontoid process?

(A) AP axial

(B) lateral

(C) parietoacanthial

(D) submentovertical (SMV)

37. In which of the following positions was the radiograph in Figure 2–7 made?

(A) AP with perpendicular plantar surface

(B) 45° lateral oblique

(C) 20° medial oblique

(D) 45° medial oblique

38. Which of the following anatomic structures is indicated by the number 3 in Figure 2–7?

(A) talus

(B) medial malleolus

(C) lateral malleolus

(D) lateral tibial condyle

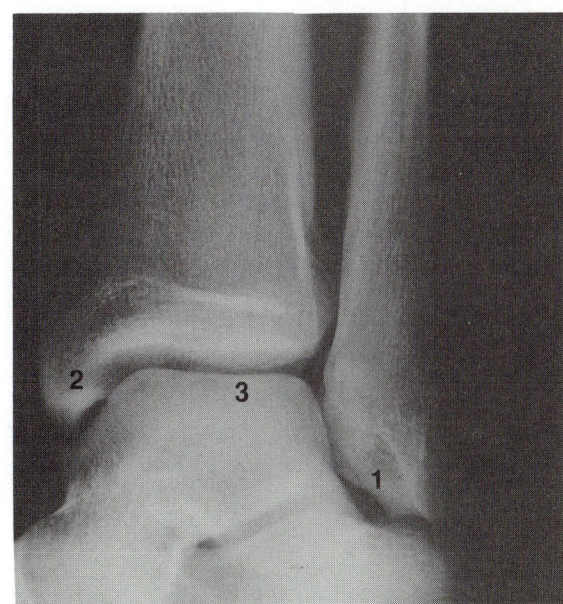

Figure 2–7. Courtesy of The Stamford Hospital, Department of Radiology.

39. Which of the following is (are) demonstrated in the lumbar spine pictured in Figure 2–8?

    1. intervertebral joints

    2. pedicles

    3. apophyseal joints

    (A) 1 only

    (B) 1 and 2 only

    (C) 1 and 3 only

    (D) 1, 2, and 3

40. Which of the following fracture classifications best describes a fragment of bone removed at the site of tendon or ligament attachment?

    (A) avulsion fracture

    (B) torus fracture

    (C) stress fracture

    (D) compound fracture

41. Which of the following should the patient be instructed to remove prior to x-ray examination of the chest?

    1. dentures

    2. earrings

    3. necklaces

    (A) 1 only

    (B) 1 and 2 only

    (C) 3 only

    (D) 1, 2, and 3

42. Which of the following is a functional study used to demonstrate the degree of AP motion present in the cervical spine?

    (A) open-mouth projection

    (B) moving mandible AP

    (C) flexion and extension laterals

    (D) right and left bending AP

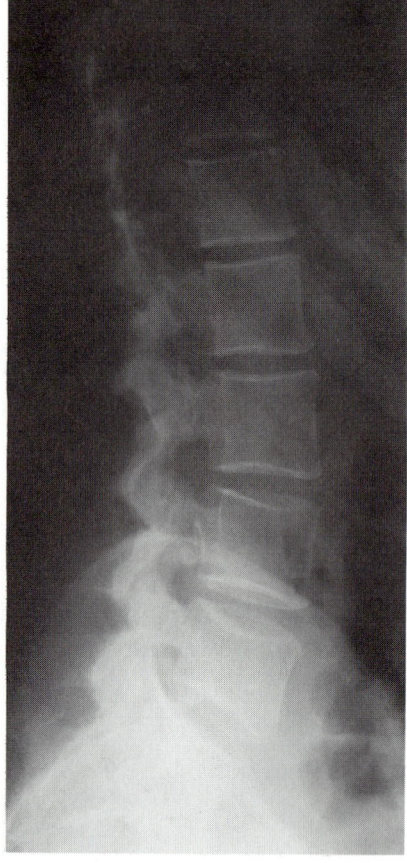

Figure 2–8. Courtesy of The Stamford Hospital, Department of Radiology.

43. The sigmoid colon is located in which of the following body areas?

    (A) left lower quadrant (LLQ)

    (B) right lower quadrant (RLQ)

    (C) hypogastrium

    (D) epigastrium

44. A lateral projection of the hand in extension is recommended to evaluate

    1. fracture

    2. foreign body

    3. soft tissue

    (A) 1 only

    (B) 2 only

    (C) 2 and 3 only

    (D) 1 and 3 only

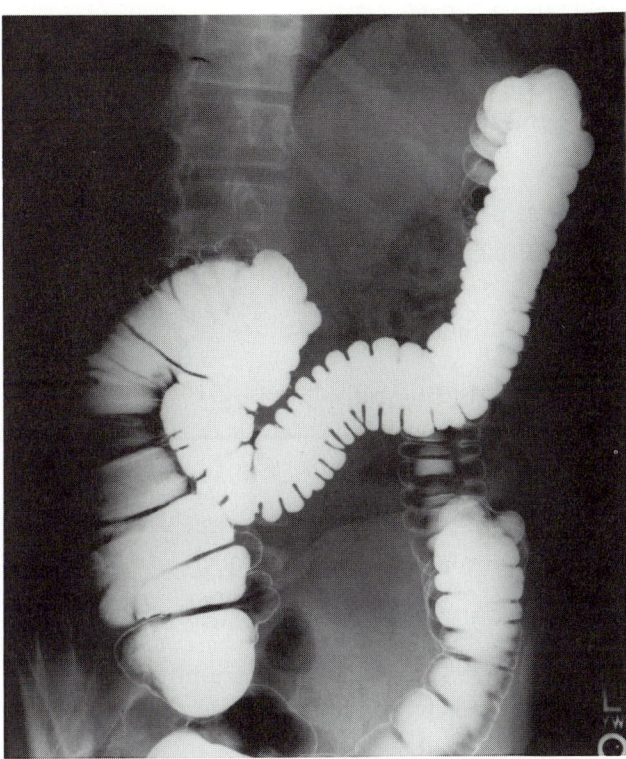

Figure 2–9. Courtesy of The Stamford Hospital, Department of Radiology.

45. In which of the following positions was the radiograph in Figure 2–9 taken?

(A) LPO

(B) RPO

(C) AP axial

(D) right lateral decubitus

46. Which of the following is (are) recommended when positioning the patient for a lateral projection of the chest?

1. patient should be examined upright
2. shoulders should be depressed
3. shoulders should be rolled forward

(A) 1 only

(B) 1 and 2 only

(C) 1 and 3 only

(D) 1, 2, and 3

47. Which of the following statements is (are) correct regarding the parietoacanthial projection (Water's method) of the skull?

1. head is rested on extended chin
2. OML is perpendicular to the film
3. the maxillary antra should be projected above the petrosa

(A) 1 only

(B) 1 and 2 only

(C) 1 and 3 only

(D) 1, 2, and 3

48. Which of the following will best demonstrate the size and shape of the liver and kidneys?

(A) lateral abdomen

(B) AP abdomen

(C) dorsal decubitus abdomen

(D) ventral decubitus abdomen

49. X-ray film identification markers should include

1. patient's name and/or ID number
2. name of hospital/office
3. right or left side marker

(A) 1 only

(B) 1 and 2 only

(C) 1 and 3 only

(D) 1, 2, and 3

50. In which projection of the foot are the sinus tarsi, cuboid, and tuberosity of the fifth metatarsal best demonstrated?

(A) lateral oblique foot

(B) medial oblique foot

(C) lateral foot

(D) weight-bearing foot

51. Which of the following is (are) well demonstrated in the oblique position of the cervical spine?

1. intervertebral foramina
2. disc spaces
3. apophyseal joints

(A) 1 only
(B) 1 and 2 only
(C) 1 and 3 only
(D) 1, 2, and 3

52. What is the position of the gallbladder in a hypersthenic patient?

(A) superior and medial
(B) superior and lateral
(C) inferior and medial
(D) inferior and lateral

53. In order to demonstrate esophageal varices, the patient must be examined in the

(A) recumbent position
(B) erect position
(C) anatomic position
(D) Fowler's position

54. Which of the following criteria are used to evaluate a PA projection of the chest?

1. ten posterior ribs should be visualized
2. sternoclavicular joints should be symmetrical
3. scapulae should be outside the lung fields

(A) 1 and 2 only
(B) 1 and 3 only
(C) 2 and 3 only
(D) 1, 2, and 3

55. Which of the following is an important consideration in order to avoid excessive metacarpophalangeal joint overlap in the oblique projection of the hand?

(A) oblique the hand no more than 45°
(B) use a support sponge for the phalanges
(C) clench the fist to bring the carpals closer to the film
(D) utilize ulnar flexion

56. Which of the following bones is NOT associated with a condyle?

(A) femur
(B) tibia
(C) fibula
(D) mandible

57. Proper and accurate film identification is essential for which of the following?

1. radiation protection considerations
2. follow-up examinations
3. medicolegal reasons

(A) 1 only
(B) 1 and 2 only
(C) 2 and 3 only
(D) 1, 2, and 3

58. In order to obtain an AP projection of the left ilium, the patient's

(A) left side is elevated 40°
(B) right side is elevated 40°
(C) left side is elevated 15°
(D) right side is elevated 15°

59. In the posterior oblique position of the cervical spine, the intervertebral foramina best seen are those

(A) nearest the film
(B) furthest from the film
(C) seen medially
(D) seen inferiorly

60. The usual patient preparation for an upper GI series is

   (A) clear fluids 8 hours prior to exam

   (B) npo after midnight

   (C) enemas until clear before exam

   (D) light breakfast day of the exam

61. Inspiration and expiration projections of the chest may be performed to demonstrate

   1. pneumothorax

   2. foreign body

   3. atelectasis

   (A) 1 only

   (B) 1 and 2 only

   (C) 1 and 3 only

   (D) 1, 2, and 3

62. Which of the following structures should be visualized through the foramen magnum in the AP axial projection (Grashey method) of the skull for occipital bone?

   1. posterior clinoid processes

   2. anterior clinoid processes

   3. dorsum sella

   (A) 1 only

   (B) 2 only

   (C) 1 and 3 only

   (D) 2 and 3 only

63. Which of the following criteria is (are) required for visualization of the greater tubercle in profile?

   1. epicondyles parallel to the film

   2. arm in external rotation

   3. humerus in AP position

   (A) 1 only

   (B) 1 and 3 only

   (C) 2 and 3 only

   (D) 1 2, and 3

64. What instructions might a patient be given after a GI examination?

   1. drink plenty of fluids

   2. take a mild laxative

   3. increase dietary fiber

   (A) 1 only

   (B) 1 and 2 only

   (C) 2 and 3 only

   (D) 1, 2, and 3

65. In the lateral projection of the foot, the

   1. plantar surface should be perpendicular to the film

   2. metatarsals are superimposed

   3. talofibular joint should be visualized

   (A) 1 only

   (B) 1 and 2 only

   (C) 2 and 3 only

   (D) 1, 2, and 3

66. Which of the following should be performed to rule out subluxation or fracture of the cervical spine?

   (A) oblique cervical spine, seated

   (B) AP cervical spine, recumbent

   (C) horizontal beam lateral

   (D) laterals in flexion and extension

67. The ulna articulates with what portion of the humerus to help form the elbow joint?

   (A) semilunar/trochlear notch

   (B) radial head

   (C) capitulum

   (D) trochlea

68. Which of the following describes correct centering for the lateral position of a barium-filled stomach?

   (A) midway between the vertebrae and left lateral margin of the abdomen

   (B) midway between the midcoronal plane and anterior surface of the abdomen

   (C) midway between the midsagittal plane and right lateral margin of the abdomen

   (D) midway between the midcoronal plane and posterior surface of the abdomen

69. During myelography, contrast medium is introduced into the

   (A) subdural space

   (B) subarachnoid space

   (C) epidural space

   (D) epidermal space

70. With the patient PA, MSP centered to the grid, the OML forming a 37° angle with the film, the CR perpendicular and exiting the acanthion, which of the following is best demonstrated?

   (A) occipital bone

   (B) frontal bone

   (C) facial bones

   (D) basal foramina

71. Which of the following statements is (are) true regarding the radiograph in Figure 2–10?

   1. the patient is in an RAO position

   2. the midcoronal plane is 60° to the film

   3. the acromion process is free of superimposition

   (A) 1 only

   (B) 1 and 2 only

   (C) 2 and 3 only

   (D) 1, 2, and 3

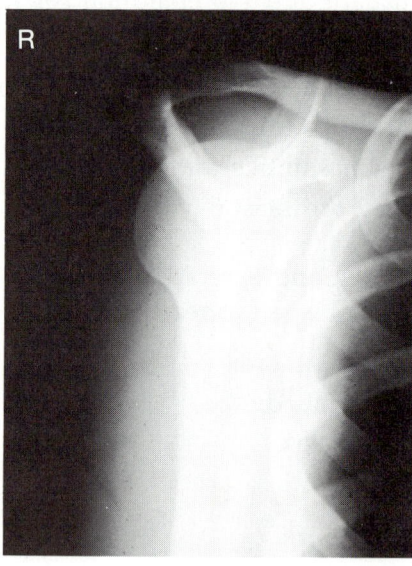

Figure 2–10. Courtesy of The Stamford Hospital, Department of Radiology.

72. Which of the following is the correct examination scheduling sequence?

   (A) upper GI, barium enema, intravenous pyelogram

   (B) barium enema, upper GI, intravenous pyelogram

   (C) intravenous pyelogram, barium enema, upper GI

   (D) intravenous pyelogram, upper GI, barium enema

73. The *pars interarticularis* is represented by what part of the "scotty dog" seen in a correctly positioned oblique lumbar spine?

   (A) eye

   (B) front foot

   (C) body

   (D) neck

74. Which of the following will separate the radial head, neck, and tuberosity from superimposition on the ulna?

   (A) AP

   (B) lateral

   (C) medial oblique

   (D) lateral oblique

75. Which of the following statements regarding myelography is (are) correct?

    1. spinal puncture may be performed in the prone or flexed lateral position
    2. contrast medium distribution is regulated through x-ray tube angulation
    3. the patient's neck must be in extension during Trendelenburg positions

    (A) 1 only
    (B) 1 and 2 only
    (C) 1 and 3 only
    (D) 1, 2, and 3

76. When the patient is unable to assume the upright body position, how should a lateral projection of the sinuses be obtained?

    (A) horizontal beam lateral
    (B) transthoracic lateral
    (C) recumbent RAO or LAO
    (D) recumbent RPO or LPO

77. With the patient's head in a PA position and the CR directed 20° cephalad, which part of the mandible will be best visualized?

    (A) symphysis
    (B) rami
    (C) body
    (D) angle

78. All of the following statements regarding respiratory structures are true, EXCEPT

    (A) the right lung has two lobes
    (B) the uppermost portion of the lung is the apex
    (C) each lung is enclosed in pleura
    (D) the trachea bifurcates into mainstem bronchi

79. The right anterior oblique position of the cervical spine requires which of the following combinations of tube angle and direction?

    (A) 15 to 20° caudad
    (B) 15 to 20° cephalad
    (C) 25 to 30° caudad
    (D) 25 to 30° cephalad

80. The PA chest radiograph seen in Figure 2–11 demonstrates

    1. rotation
    2. adequate inspiration
    3. air in the trachea

    (A) 1 only
    (B) 1 and 3 only
    (C) 2 and 3 only
    (D) 1, 2, and 3

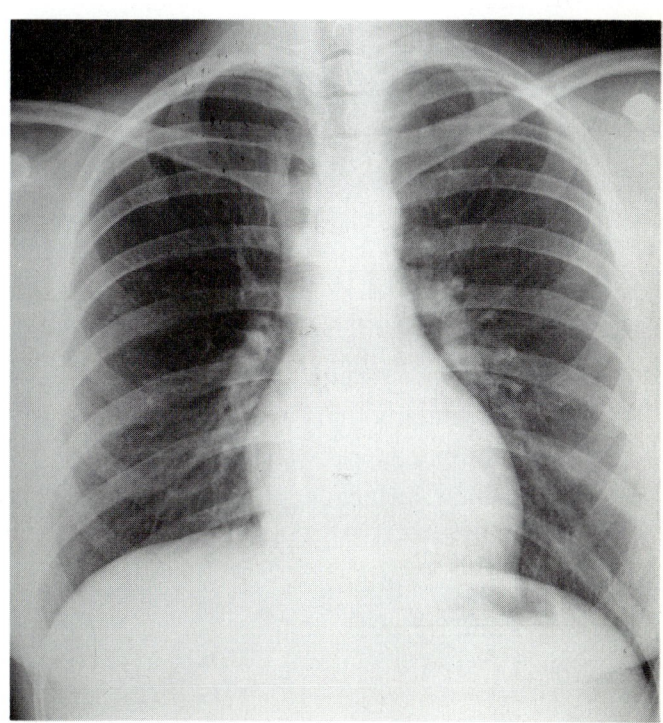

Figure 2–11. From the American College of Radiology Learning File. Courtesy of the ACR.

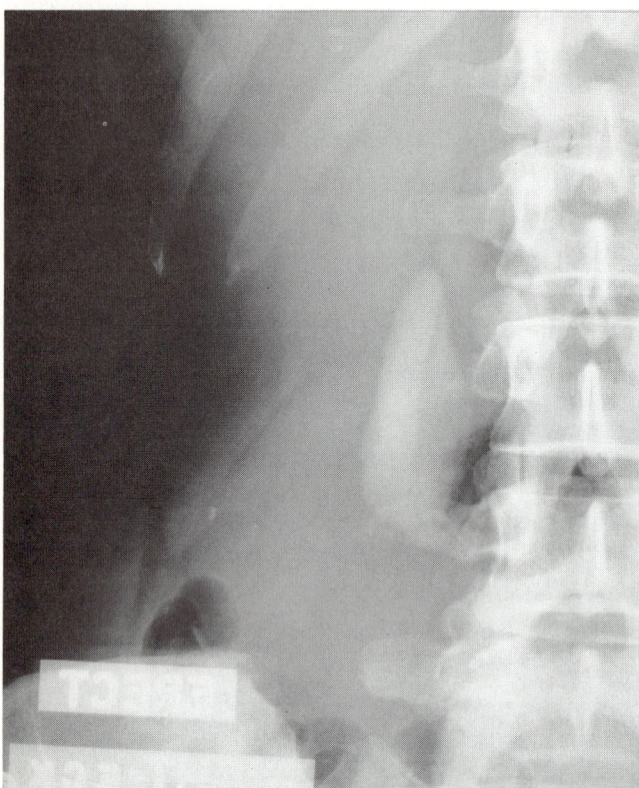

Figure 2–12. From the American College of Radiology Learning File. Courtesy of the ACR.

81. Which body type is most likely represented in Figure 2–12?

(A) asthenic
(B) hypersthenic
(C) sthenic
(D) decubitus

82. In the medial oblique projection of the ankle, the

1. talofibular joint is visualized
2. talotibial joint is visualized
3. plantar surface should be vertical

(A) 1 only
(B) 1 and 3 only
(C) 2 and 3 only
(D) 1, 2, and 3

83. A patient is usually required to drink barium sulfate suspension in order to demonstrate which of the following structures?

1. esophagus
2. pylorus
3. ilium

(A) 1 only
(B) 1 and 2 only
(C) 2 and 3 only
(D) 1, 2, and 3

84. Which of the following is (are) demonstrated in the lateral projection of the thoracic spine?

1. intervertebral spaces
2. apophyseal joints
3. intervertebral foramina

(A) 1 only
(B) 2 only
(C) 1 and 3
(D) 1, 2, and 3

85. The two palpable bony landmarks generally used for accurate localization of the hip are the

(A) ASIS and symphysis pubis
(B) iliac crest and greater trochanter
(C) symphysis pubis and greater trochanter
(D) iliac crest and symphysis pubis

86. In the AP projection of the knee joint space of an asthenic patient who measures less than 19 cm from ASIS to tabletop, the CR should be directed

(A) perpendicularly
(B) 5° medially
(C) 5° cephalad
(D) 5° caudad

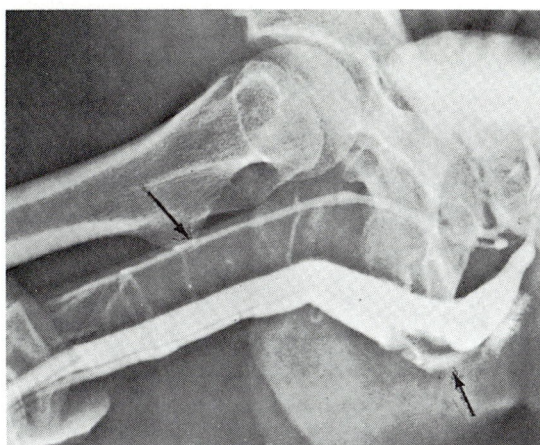

Figure 2–13. Reprinted with permission from Way, *Current Surgical Diagnosis & Treatment*, 10th ed. Norwalk, CT: Appleton & Lange, 1994.

87. Which of the following is used to evaluate the shoulder in its anatomic position?

    (A) external rotation position
    (B) internal rotation position
    (C) neutral rotation position
    (D) inferosuperior axial position

88. The contraction and expansion of arterial walls in accordance with forceful contraction and relaxation of the heart is called

    (A) hypertension
    (B) elasticity
    (C) pulse
    (D) pressure

89. The radiograph shown in Figure 2–13 illustrates which of the following studies?

    (A) cystogram
    (B) urethrogram
    (C) intravenous pyelogram
    (D) hysterosalpingogram

90. Which of the following correctly identifies the position illustrated in Figure 2–14?

    (A) AP axial mastoids (Towne/Grashey)
    (B) axiolateral TMJ (closed mouth)
    (C) axiolateral mastoids (Laws)
    (D) posterior profile mastoids (Stenvers)

91. All of the following statement completions regarding the oblique positions of the chest are correct, EXCEPT

    (A) for lungs, the patient is obliqued 45°
    (B) for heart size, the right oblique is 45° and the left oblique is 55 to 60°
    (C) the RAO demonstrates the right atrium
    (D) the RAO may be used with barium to demonstrate the esophagus

92. A "blowout" fracture is usually related to which of the following structures?

    (A) foot
    (B) elbow
    (C) orbit
    (D) pelvis

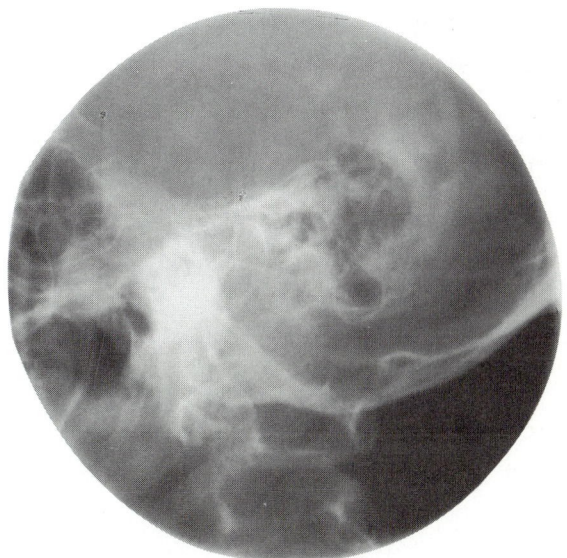

Figure 2–14. Courtesy of The Stamford Hospital, Department of Radiology.

93. Which of the following is the correct sequence of events when performing a double-contrast upper GI series?

    (A) patient given gas-producing substance, then small amount of high-density barium, then placed recumbent

    (B) patient recumbent, given a small amount of high-density barium, then a gas-producing substance

    (C) patient given a gas-producing substance, placed recumbent, then given a small amount of high-density barium

    (D) patient given a small amount of high-density barium, placed recumbent, then given a gas-producing substance

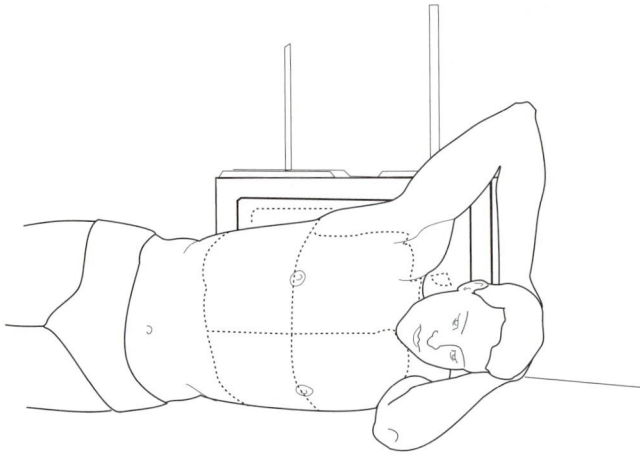

Figure 2–15.

94. An *intrathecal* injection is associated with which of the following examinations?

    (A) intravenous pyelogram

    (B) retrograde pyelogram

    (C) myelogram

    (D) arthrogram

95. The *pedicle* is represented by what part of the "scotty dog" seen in a correctly positioned oblique lumbar spine?

    (A) eye

    (B) front foot

    (C) body

    (D) neck

96. The position shown in Figure 2–15 is most often used to demonstrate

    1. free fluid on the right side

    2. free fluid on the left side

    3. free air on the right side

    (A) 1 only

    (B) 2 only

    (C) 1 and 3 only

    (D) 2 and 3 only

97. Which of the following projections will best demonstrate the tarsal navicular free of superimposition?

    (A) AP oblique, medial rotation

    (B) AP oblique, lateral rotation

    (C) mediolateral

    (D) lateral weight-bearing

98. Prior to bringing the patient into the radiographic room for a GI series, the radiographer should

    1. assemble the accessories needed for the exam

    2. be certain the x-ray room is clean and orderly

    3. check to see if the patient had a previous GI exam

    (A) 1 only

    (B) 1 and 2 only

    (C) 1 and 3 only

    (D) 1, 2, and 3

99. Which of the following are components of a trimalleolar fracture?

1. fractured lateral malleolus
2. fractured medial malleolus
3. fractured posterior tibia

(A) 1 only
(B) 1 and 3 only
(C) 2 and 3 only
(D) 1, 2, and 3

100. What is the structure indicated by the number 7 in Figure 2–16?

(A) common hepatic duct
(B) common bile duct
(C) cystic duct
(D) pancreatic duct

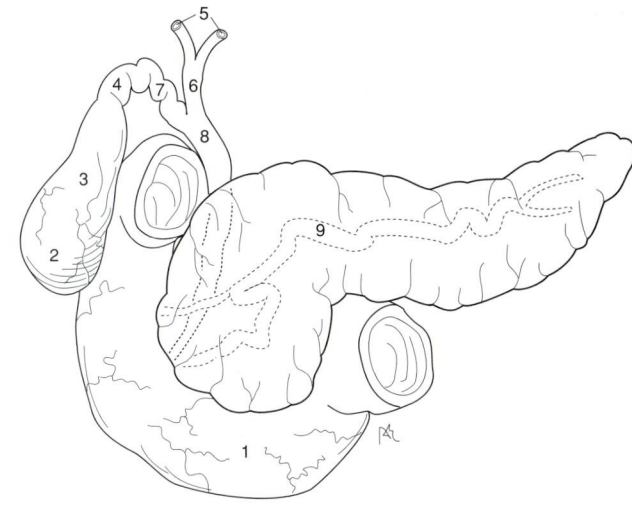

Figure 2–16.

101. What is the structure indicated by the number 8 in Figure 2–16?

(A) common hepatic duct
(B) common bile duct
(C) cystic duct
(D) pancreatic duct

answer missing in btek

102. In order to evaluate interphalangeal joints in the oblique and lateral positions, the fingers

(A) rest on the cassette for immobilization
(B) must be supported parallel to the film
(C) are radiographed in natural flexion
(D) are radiographed in palmar flexion

103. Which of the following examinations require(s) restriction of the patient's diet?

1. GI series
2. abdominal survey
3. pyelogram

(A) 1 only
(B) 1 and 2 only
(C) 1 and 3 only
(D) 2 and 3 only

104. All of the following statements regarding large bowel radiography are true, EXCEPT

(A) the large bowel must be completely empty prior to examination
(B) retained fecal material can simulate pathology
(C) single-contrast studies help demonstrate polyps
(D) double-contrast studies help demonstrate intraluminal lesions

105. In the lateral projection of the knee, the

1. femoral condyles are superimposed
2. patellofemoral joint is visualized
3. knee is flexed about 20 to 30°

(A) 1 only
(B) 2 only
(C) 1 and 3 only
(D) 1, 2, and 3

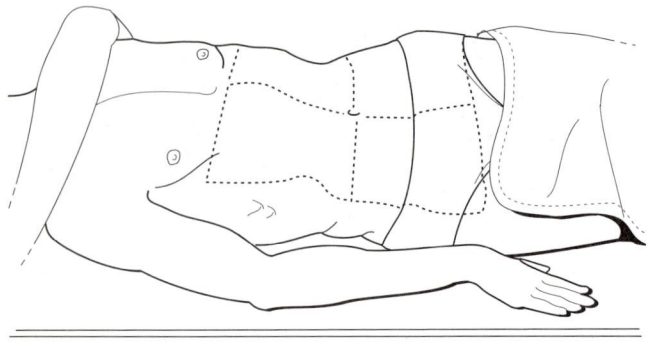

Figure 2–17.

106. In order to demonstrate the pulmonary apices with the patient in the AP position, the

(A) central ray is directed 15 to 20° cephalad

(B) central ray is directed 15 to 20° caudad

(C) exposure is made on full exhalation

(D) patient's shoulders are rolled forward

107. What angle is formed by the median sagittal plane and film in the parieto-orbital projection (Rhese method) of the optic canal?

(A) 90°

(B) 37°

(C) 53°

(D) 45°

108. Double-contrast examinations of the stomach or large bowel are performed to better visualize

(A) position of the organ

(B) size and shape of the organ

(C) diverticula

(D) gastric or bowel mucosa

109. Which of the following shoulder projections may be used to evaluate the lesser tubercle in profile?

(A) external rotation position

(B) internal rotation position

(C) neutral rotation position

(D) inferosuperior axial

110. Which of the following would be obtained by using the position illustrated in Figure 2–17?

1. splenic flexure and descending colon

2. hepatic flexure and ascending colon

3. hepatic flexure and descending colon

(A) 1 only

(B) 3 only

(C) 1 and 2 only

(D) 1 and 3 only

111. With the patient positioned as illustrated in Figure 2–18, which of the following structures is best demonstrated?

(A) patella

(B) patellofemoral articulation

(C) intercondyloid fossa

(D) tibial tuberosity

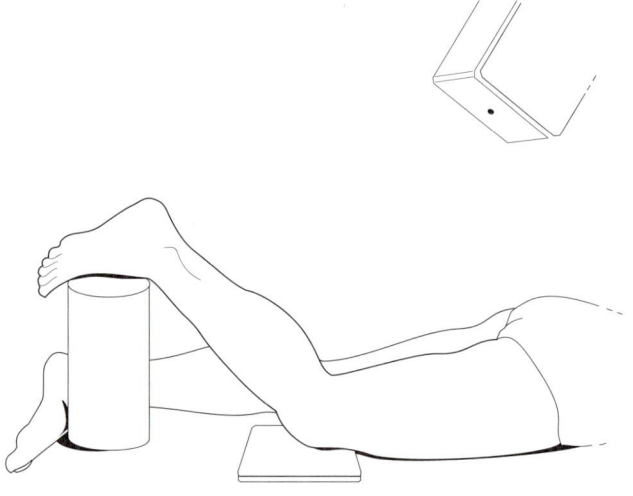

Figure 2–18.

112. Which of the positions listed below is used to demonstrate vertical patellar fractures and the femoropatellar articulation?

   (A) AP knee
   (B) lateral knee
   (C) tangential patella
   (D) "tunnel" view

113. Which of the following structures is illustrated by the 3 in Figure 2–19?

   (A) maxillary sinus
   (B) coracoid process
   (C) zygomatic arch
   (D) mandibular angle

114. Ingestion of barium sulfate is contraindicated in which of the following situation(s)?

   1. suspected perforation of a hollow viscus
   2. suspected large bowel obstruction
   3. presurgical patients

   (A) 1 only
   (B) 1 and 3 only
   (C) 2 and 3 only
   (D) 1, 2, and 3

115. Which of the following baselines is essential to the performance of the axial projection of the nasal bones?

   (A) infraorbitomeatal line
   (B) orbitomeatal line
   (C) acanthomeatal line
   (D) glabelloalveolar line

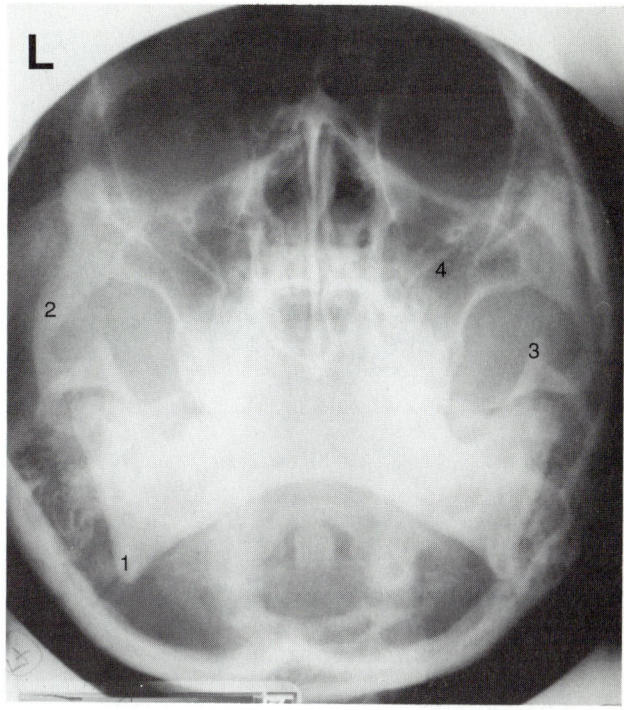

Figure 2–19. Courtesy of The Stamford Hospital, Department of Radiology.

116. The olecranon process is best demonstrated in which of the radiographs illustrated in Figure 2–20?

   (A) number 1
   (B) number 2
   (C) number 3
   (D) number 4

117. Which of the positions illustrated in Figure 2–21 will best demonstrate the lumbar intervertebral foramina?

   (A) number 1
   (B) number 2
   (C) number 3
   (D) number 4

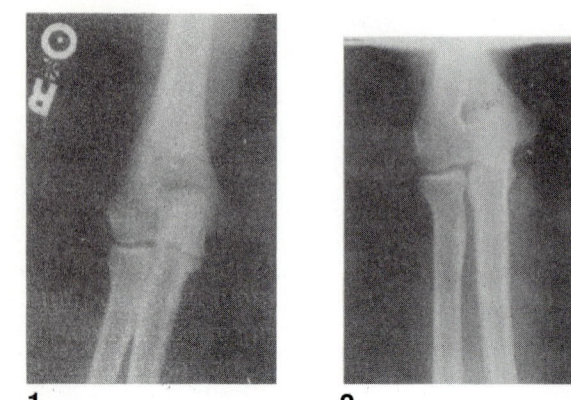

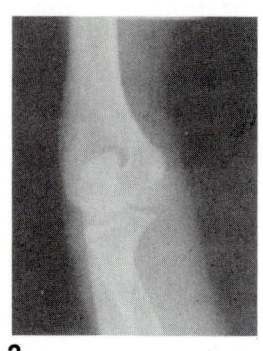

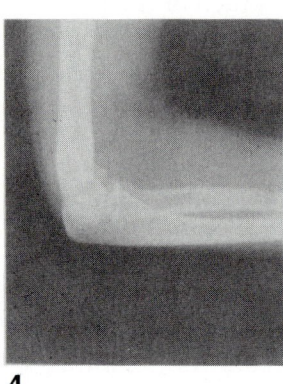

1   2   3   4

Figure 2–20. Courtesy of The Stamford Hospital, Department of Radiology.

118. Which of the positions illustrated in Figure 2–21 will best demonstrate the lumbar apophyseal joints *closest* to the film?

No Answer in Back

(A) number 1
(B) number 2
(C) number 3
(D) number 4

119. The apophyseal articulations of the thoracic spine are demonstrated with the

(A) coronal plane 45° to the film
(B) midsagittal plane 45° to the film
(C) coronal plane 70° to the film
(D) midsagittal plane 70° to the film

120. Which of the following may be used to evaluate the glenohumeral joint?

1. scapular Y projection
2. inferosuperior axial
3. transthoracic lateral

(A) 1 only
(B) 1 and 2 only
(C) 2 and 3 only
(D) 1, 2, and 3

121. Which of the following is (are) true regarding radiographic examination of the acromioclavicular joints?

1. the procedure is performed in the erect position
2. use of weights enhances demonstration of the joints
3. the procedure should be avoided if dislocation or separation is suspected

(A) 1 only
(B) 1 and 2
(C) 1 and 3 only
(D) 2 and 3 only

122. Which of the following women is most likely to have the most homogeneous, glandular breast tissue?

(A) a postpubertal adolescent
(B) a 20-year-old with one previous pregnancy
(C) menopausal women
(D) a postmenopausal 65-year-old

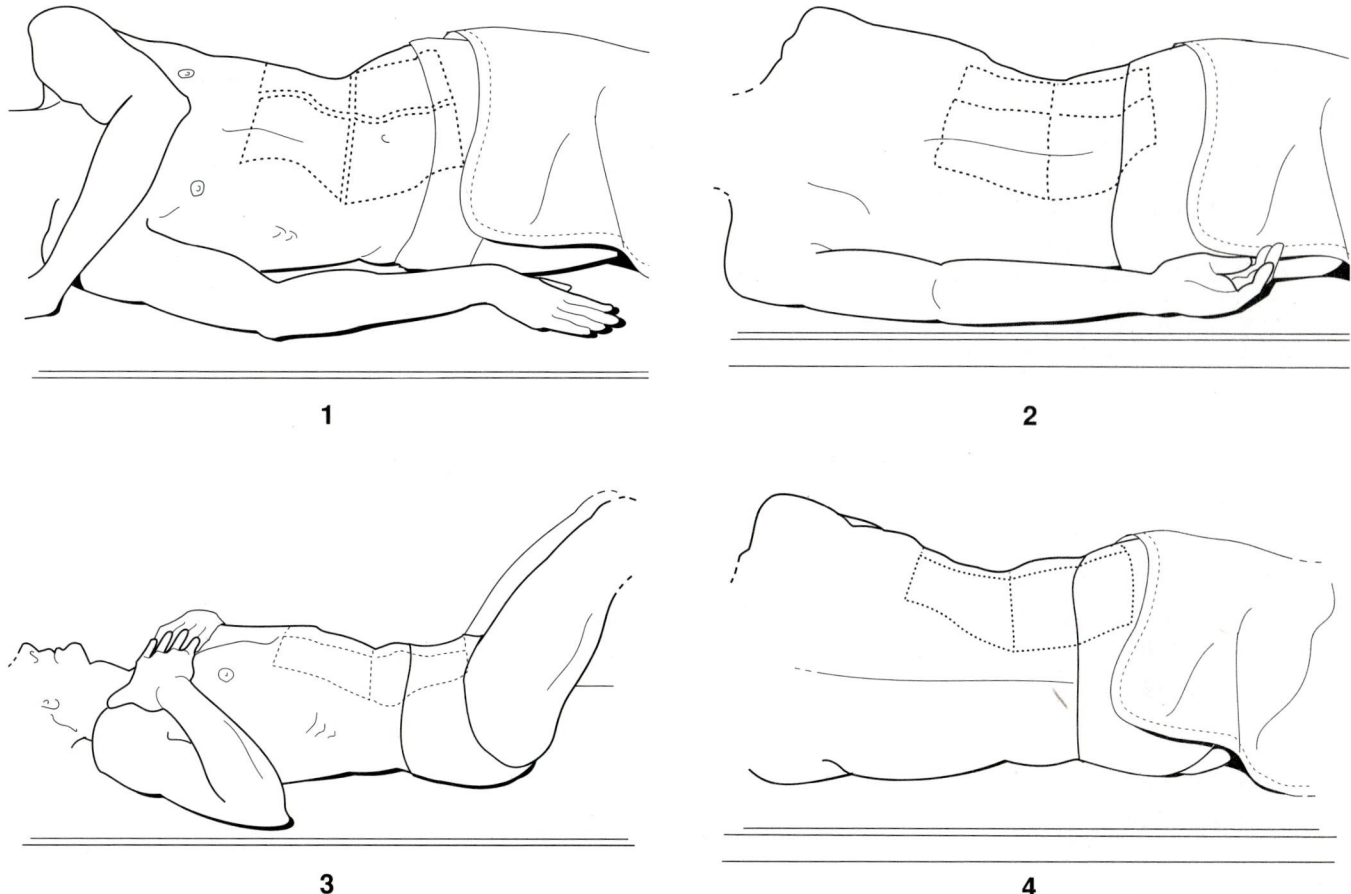

Figure 2–21.

123. A patient who is lying down is said to be

(A) prone
(B) supine
(C) recumbent
(D) decubitus

124. Which of the following examinations require(s) special identification markers in addition to the usual patient name and number, date, and side marker?

1. IVP
2. tomography
3. abdominal survey

(A) 1 only
(B) 1 and 2 only
(C) 2 and 3 only
(D) 1, 2, and 3

125. To make the patient as comfortable as possible during a single-contrast barium enema, the radiographer should

    1. instruct the patient to relax the abdominal muscles to prevent intra-abdominal pressure
    2. instruct the patient to concentrate on breathing deeply to reduce colonic spasm
    3. prepare a warm barium suspension (98 to 105°F) to aid in retention

    (A) 2 only
    (B) 1 and 2 only
    (C) 2 and 3 only
    (D) 1, 2, and 3

126. The uppermost portion of the iliac crest is approximately at the same level as the

    (A) costal margin
    (B) umbilicus
    (C) xiphoid tip
    (D) fourth lumbar vertebra

127. In order to better demonstrate the mandibular rami in the PA position, the

    (A) skull is obliqued toward the affected side
    (B) skull is obliqued away from the affected side
    (C) central ray is angled cephalad
    (D) central ray is angled caudad

128. Which of the following is *proximal* to the carpal bones?

    (A) distal interphalangeal joints
    (B) proximal interphalangeal joints
    (C) metacarpals
    (D) radial styloid process

129. The scapular Y projection of the shoulder demonstrates

    1. an oblique projection of the shoulder
    2. anterior or posterior dislocation
    3. a lateral projection of the shoulder

    (A) 1 only
    (B) 1 and 2 only
    (C) 1 and 3 only
    (D) 2 and 3 only

130. Angulation of the central ray may be required for the following reason(s)

    1. to avoid superimposition of overlying structures
    2. to avoid foreshortening or self-superimposition
    3. in order to project through certain articulations

    (A) 1 only
    (B) 2 only
    (C) 1 and 3 only
    (D) 1, 2, and 3

131. The fifth metacarpal is located on which aspect of the hand?

    (A) medial
    (B) lateral
    (C) ulnar
    (D) volar

132. In the AP projection of the ankle

    1. the plantar surface of the foot is vertical
    2. the fibula projects more distally than the tibia
    3. the calcaneus is well visualized

    (A) 1 only
    (B) 1 and 2 only
    (C) 2 and 3 only
    (D) 1, 2, and 3

133. During IV urography, the prone position is generally recommended to demonstrate

1. filling of obstructed ureters
2. the renal pelvis
3. the superior calyces

(A) 1 only
(B) 1 and 2 only
(C) 1 and 3 only
(D) 1, 2, and 3

134. The plane that passes vertically through the body dividing it into anterior and posterior halves is termed the

(A) median sagittal plane
(B) midcoronal plane
(C) sagittal plane
(D) transverse plane

135. In order to demonstrate a profile view of the glenoid fossa, the patient is AP recumbent and obliqued 45°

(A) toward the affected side
(B) away from the affected side
(C) with the arm at the side in the anatomical position
(D) with the arm in external rotation

136. Which of the following is (are) accurate criticism(s) of the open mouth projection of C1-C2 seen in Figure 2–22?

1. the MSP is not centered and perpendicular to the midline of the table
2. the neck should be flexed more
3. the neck should be extended more

(A) 1 only
(B) 1 and 2 only
(C) 3 only
(D) 1 and 3 only

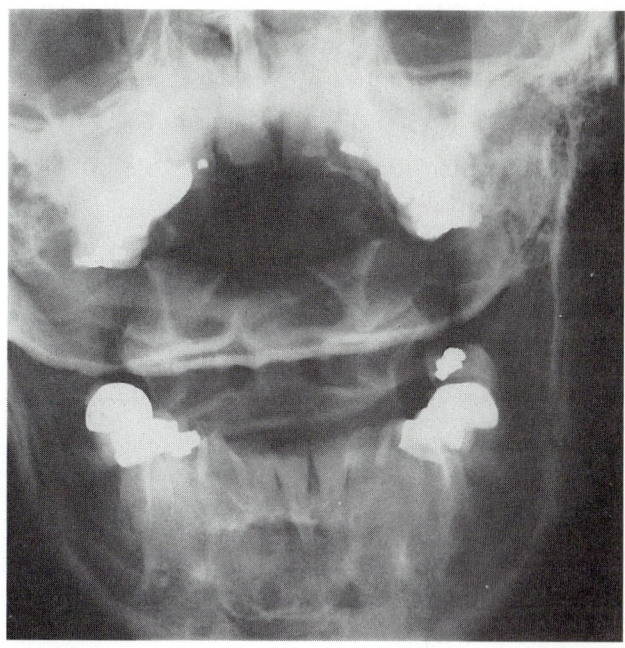

Figure 2–22. Courtesy of The Stamford Hospital, Department of Radiology.

137. Standard radiographic protocols may be reduced to include two views, at right angles to each other, in which of the following situations?

(A) barium examinations
(B) spine radiography
(C) skull radiography
(D) emergency and trauma radiography

138. AP stress studies of the ankle may be performed

1. to demonstrate fractures of the distal tibia and fibula
2. following inversion or eversion injuries
3. to demonstrate ligament tear

(A) 1 only
(B) 1 and 2 only
(C) 2 and 3 only
(D) 1, 2, and 3

139. Which of the following is recommended in order to better demonstrate the tarsometatarsal joints in the dorsoplantar projection of the foot?

(A) invert the foot

(B) evert the foot

(C) angle the central ray 10° posteriorly

(D) angle the central ray 10° anteriorly

140. The tissue that occupies the central cavity within the shaft of a long bone in an adult is termed

(A) red marrow

(B) yellow marrow

(C) cortical

(D) cancellous

141. Which of the following positions will provide an AP projection of the L5–S1 interspace?

(A) patient AP with 30 to 35° angle cephalad

(B) patient AP with 30 to 35° angle caudad

(C) patient AP with 0° angle

(D) patient lateral, coned to L5

142. Which of the following bony landmarks is in the same transverse plane as the symphysis pubis?

(A) ischial tuberosity

(B) most prominent part of the greater trochanter

(C) anterior superior iliac spine

(D) anterior inferior iliac spine

143. Which of the following structures is (are) located in the RLQ?

1. gallbladder
2. hepatic flexure
3. cecum

(A) 1 only

(B) 1 and 2 only

(C) 3 only

(D) 1, 2, and 3

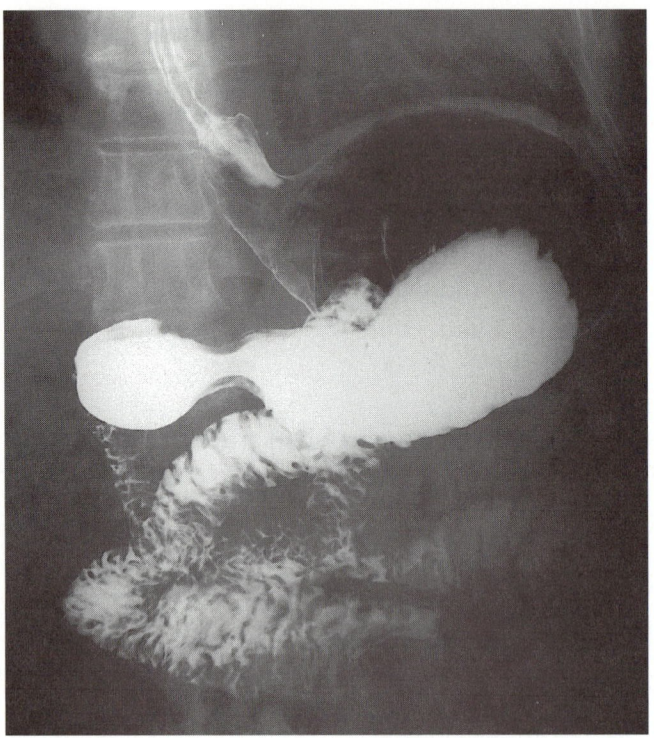

Figure 2–23. Courtesy of The Stamford Hospital, Department of Radiology.

144. Which of the following will BEST demonstrate acromioclavicular separation?

(A) AP recumbent, affected shoulder

(B) AP recumbent, both shoulders

(C) AP erect, affected shoulder

(D) AP erect, both shoulders

145. In which of the following positions was the radiograph in Figure 2–23 probably made?

(A) AP recumbent

(B) PA recumbent

(C) PA upright

(D) AP Trendelenburg

146. Which of the following vertebral groups form(s) lordotic curve(s)?

    1. cervical

    2. thoracic

    3. lumbar

(A) 1 only

(B) 2 only

(C) 1 and 2 only

(D) 1 and 3 only

147. Which of the following is (are) required for a lateral projection of the skull?

    1. infraorbitomeatal line (IOML) is parallel to the film

    2. midsagittal plane is perpendicular to the film

    3. central ray enters 2 inches superior to the EAM

(A) 1 only

(B) 1 and 3 only

(C) 2 and 3 only

(D) 1, 2, and 3

148. With which of the following does the trapezium articulate?

(A) fifth metacarpal

(B) first metacarpal

(C) distal radius

(D) distal ulna

149. Which of the following positions will most effectively move the gallbladder away from the vertebrae in the asthenic patient?

(A) LAO

(B) RAO

(C) LPO

(D) erect

150. The abdomen is divided into nine regions. The ileocecal valve is normally located in the lower-right region called the

(A) right iliac

(B) right lumbar

(C) right hypochondriac

(D) hypogastric

151. Which of the following tube angle and direction combinations is correct for an axial projection of the clavicle, with the patient in the AP recumbent position on the x-ray table?

(A) 10 to 15° caudad

(B) 10 to 15° cephalad

(C) 25 to 30° cephalad

(D) 25 to 30° caudad

152. Which of the following structures is (are) most likely to be demonstrated in a right lateral decubitus position of an double contrast barium enema?

    1. lateral wall of the descending colon

    2. medial wall of the ascending colon

    3. lateral wall of the ascending colon

(A) 1 only

(B) 3 only

(C) 1 and 2 only

(D) 2 and 3 only

153. Which of the following articulations participate in formation of the ankle mortise?

    1. talotibial

    2. talocalcaneal

    3. talofibular

(A) 1 only

(B) 1 and 3 only

(C) 2 and 3 only

(D) 3 only

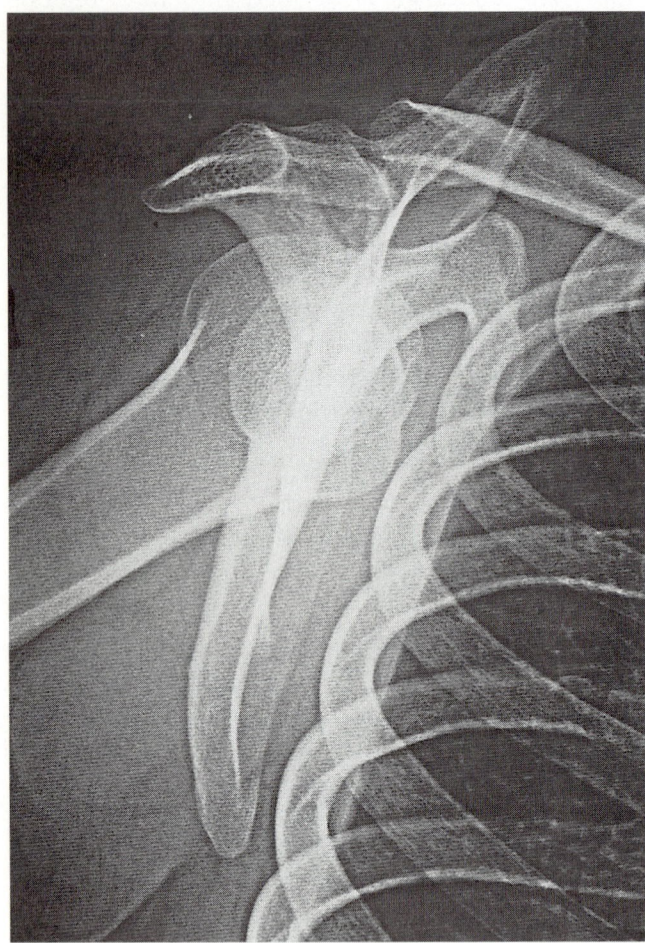

Figure 2–24. Courtesy of The Stamford Hospital, Department of Radiology.

154. Which of the following projections will BEST demonstrate the carpal navicular?

(A) lateral wrist
(B) ulnar flexion
(C) radial flexion
(D) carpal tunnel

155. Which of the following is (are) true with respect to the radiographs in Figure 2–24?

1. the coracoid process is seen partially superimposed on the third rib
2. this projection is performed to evaluate the acromioclavicular articulation
3. this projection is performed to evaluate possible shoulder dislocation

(A) 1 only
(B) 1 and 2 only
(C) 1 and 3 only
(D) 2 and 3 only

156. The manubrial notch is approximately at the same level as the

(A) fifth thoracic vertebra
(B) T2–T3 interspace
(C) T4–T5 interspace
(D) costal margin

157. Which of the following is most likely to be the correct routine for a radiographic examination of the forearm?

(A) PA and medial oblique
(B) AP and lateral oblique
(C) PA and lateral
(D) AP and lateral

158. In Figure 2–25, which of the following is represented by the number *1*?

(A) pedicle
(B) lamina
(C) spinous process
(D) superior articular process

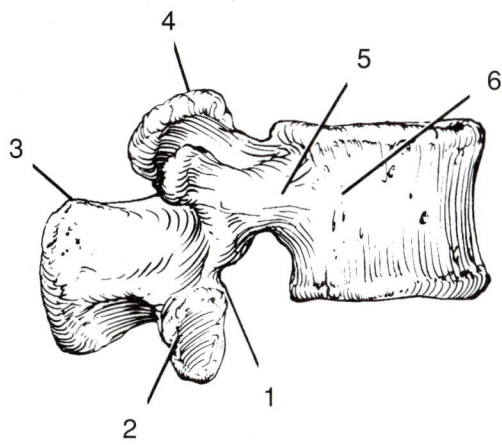

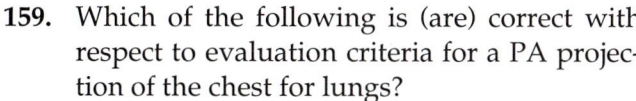

SECOND LUMBAR VERTEBRA
(RIGHT LATERAL VIEW)

Figure 2–25. Reprinted with permission from Chusid JG. *Correlative Neuroanatomy & Functional Neurology* (19th ed). East Norwalk, CT: Appleton & Lange, 1985.

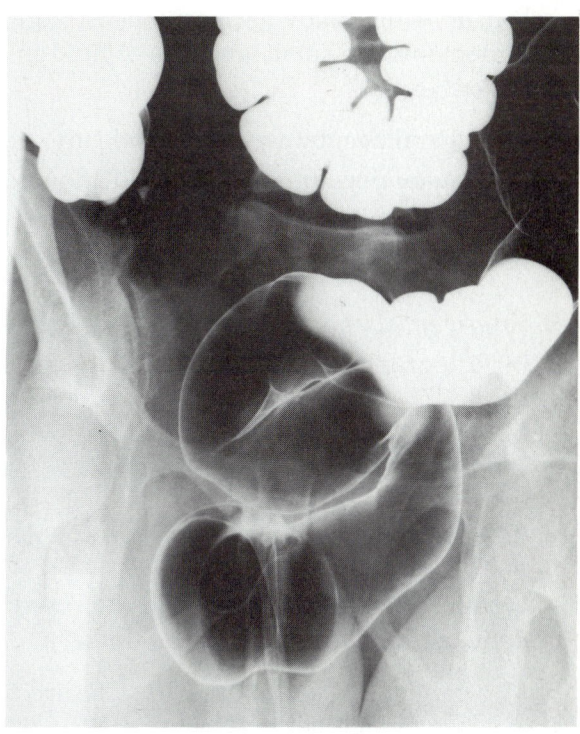

Figure 2–26. Courtesy of The Stamford Hospital, Department of Radiology.

159. Which of the following is (are) correct with respect to evaluation criteria for a PA projection of the chest for lungs?

   1. sternoclavicular joints should be symmetrical
   2. sternum is seen lateral without rotation
   3. 10 anterior ribs demonstrated above diaphragm

  (A) 1 only
  (B) 1 and 2 only
  (C) 1 and 3 only
  (D) 1, 2, and 3

160. Aspirated foreign bodies in older children and adults are most likely to lodge in the

  (A) right main bronchus
  (B) left main bronchus
  (C) esophagus
  (D) proximal stomach

161. In the lateral projection of the ankle, the

   1. talotibial joint is visualized
   2. talofibular joint is visualized
   3. tibia and fibula are superimposed

  (A) 1 only
  (B) 1 and 2 only
  (C) 1 and 3 only
  (D) 1, 2, and 3

162. The position illustrated in the radiograph in Figure 2–26 may be obtained with the patient

   1. supine and CR angled 30° caudad
   2. supine and CR angled 30° cephalad
   3. prone and CR angled 30° cephalad

  (A) 1 only
  (B) 2 only
  (C) 1 and 3 only
  (D) 2 and 3 only

163. Which of the following positions is required to demonstrate small amounts of fluid in the pleural cavity?

(A) lateral decubitus, affected side up

(B) lateral decubitus, affected side down

(C) AP Trendelenburg

(D) AP supine

164. Which of the following positions is essential in radiography of the paranasal sinuses?

(A) erect

(B) recumbent

(C) oblique

(D) Trendelenburg

165. Which of the following can be used to demonstrate the intercondyloid fossa?

1. patient PA, knee flexed 40°, CR directed caudad 40° to the popliteal fossa

2. patient AP, cassette under flexed knee, CR directed cephalad to knee, perpendicular to tibia

3. patient PA, patella parallel to film, heel rotated 5 to 10° lateral, CR perpendicular to knee joint

(A) 1 only

(B) 1 and 2 only

(C) 2 and 3 only

(D) 1, 2, and 3

166. The scapula pictured in Figure 2–27 demonstrates

1. its posterior aspect

2. its costal surface

3. its sternal articular surface

(A) 1 only

(B) 1 and 2 only

(C) 1 and 3 only

(D) 1, 2, and 3

167. In Figure 2–27, which of the following is represented by the number 2?

(A) acromion process

(B) scapular spine

(C) coracoid process

(D) acromioclavicular joint

168. In Figure 2–27, which of the following is represented by the number 9?

(A) vertebral border

(B) axillary border

(C) inferior angle

(D) superior angle

169. Which of the following techniques would provide a PA projection of the gastroduodenal surfaces of the barium-filled, high and transverse stomach?

(A) place the patient in a 35 to 40° RAO position

(B) place the patient in a lateral position

(C) angle the CR 35 to 45° cephalad

(D) angle the CR 35 to 45° caudad

170. Which of the following projections of the calcaneus is obtained with the leg extended, plantar surface of the foot vertical and perpendicular to the film, and CR directed 40° caudad?

(A) axial plantodorsal projection

(B) axial dorsoplantar projection

(C) lateral projection

(D) weight-bearing lateral

171. In the lateral projection of the knee, the central ray is angled 5° cephalad in order to prevent superimposition of which of the following structures on the joint space?

(A) lateral femoral condyle

(B) medial femoral condyle

(C) patella

(D) tibial eminence

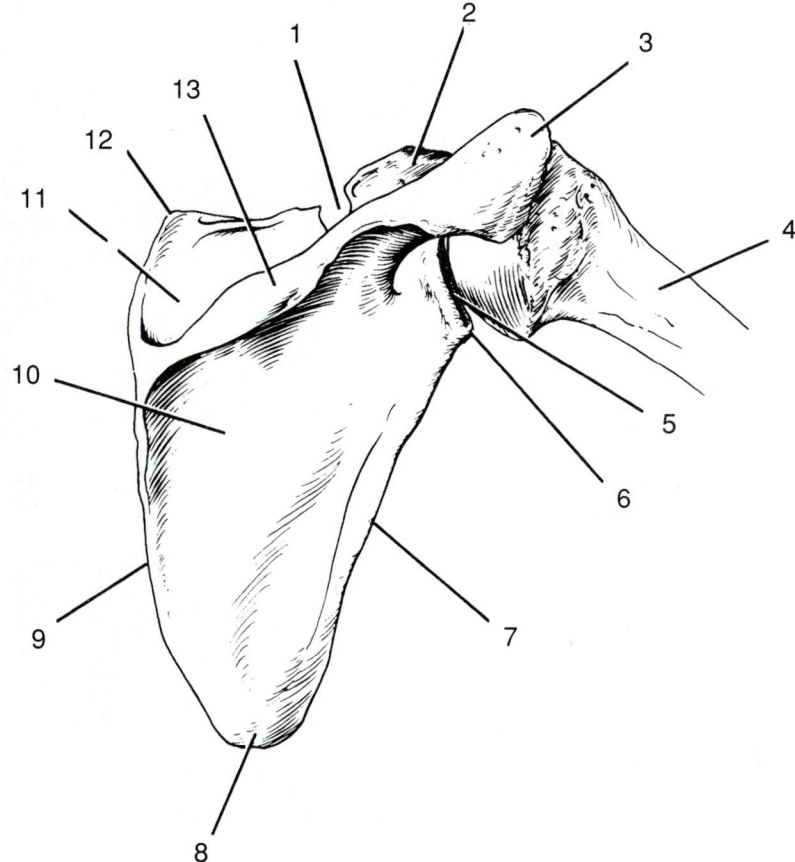

Figure 2–27. Reproduced with permission from Lindner HH. *Clinical Anatomy.* East Norwalk, CT: Appleton & Lange, 1989.

172. The body habitus characterized by a long and narrow thoracic cavity and low, midline stomach and gallbladder is the

(A) asthenic

(B) hyposthenic

(C) sthenic

(D) hypersthenic

173. Which of the following structures is (are) located in the LUQ?

1. stomach
2. spleen
3. cecum

(A) 1 only

(B) 2 only

(C) 1 and 2 only

(D) 1, 2, and 3

174. The advantages of digital subtraction angiography over film angiography include

1. greater contrast medium sensitivity
2. immediately available images
3. increased resolution

(A) 1 only

(B) 1 and 2 only

(C) 2 and 3 only

(D) 1, 2, and 3

175. Which of the following positions demonstrates all the paranasal sinuses?

(A) parietoacanthial

(B) PA axial

(C) lateral

(D) true PA

176. In the lateral projection of the scapula, the

1. vertebral and axillary borders are superimposed
2. acromion and coracoid processes are superimposed
3. patient may be examined in the erect position

(A) 1 only
(B) 1 and 2 only
(C) 1 and 3 only
(D) 1, 2, and 3

177. Which of the following statements are true regarding Figure 2–28?

1. the radiograph was made in the RAO position
2. the CR should be directed more inferiorly
3. the sternum should be projected onto the right side of the thorax

(A) 1 and 2 only
(B) 1 and 3 only
(C) 2 and 3 only
(D) 1, 2, and 3

178. In order to better visualize the joint space in the radiograph in Figure 2–29, the radiographer should

(A) flex the knee more acutely
(B) flex the knee less acutely
(C) angle the CR 5 to 7° cephalad
(D) angle the CR 5 to 7° caudad

179. Which of the following is a radiologic procedure that functions to dilate a stenotic vessel?

(A) percutaneous nephrolithotomy
(B) percutaneous angioplasty
(C) renal arteriography
(D) surgical nephrostomy

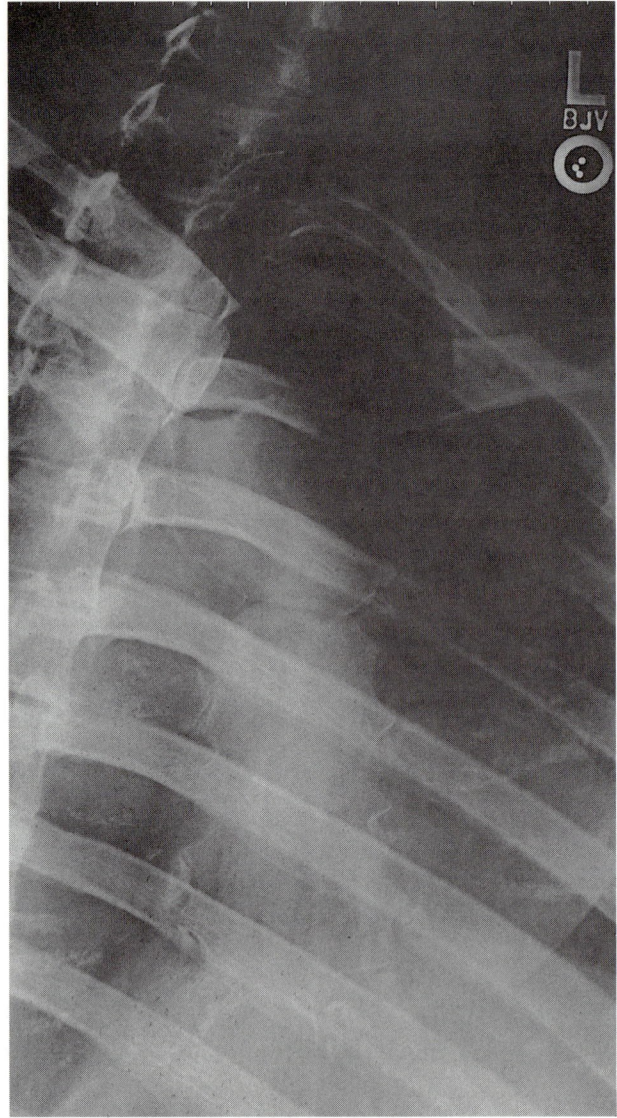

Figure 2–28. Courtesy of The Stamford Hospital, Department of Radiology.

180. That portion of a long bone where cartilage has been replaced by bone is known as the

(A) diaphysis
(B) epiphysis
(C) metaphysis
(D) apophysis

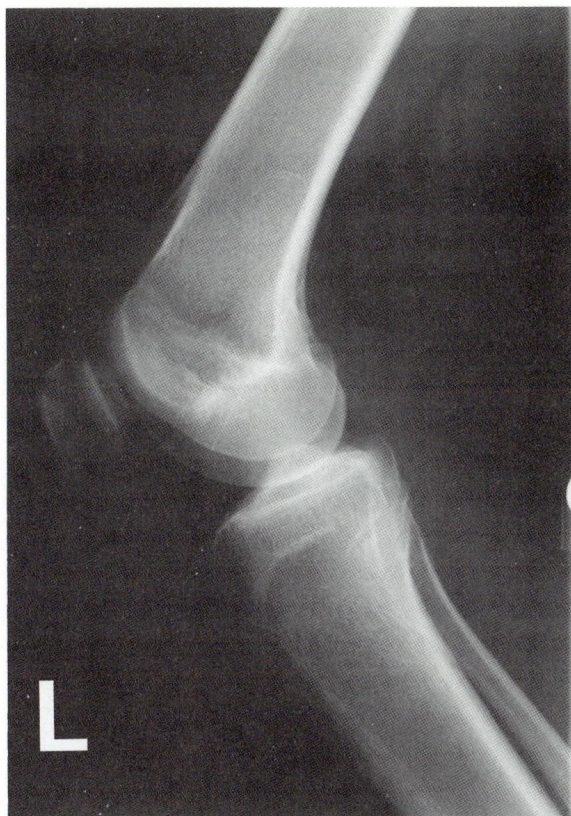

Figure 2–29. Courtesy of The Stamford Hospital, Department of Radiology.

181. Which of the following statements is (are) true regarding the PA axial projection of the paranasal sinuses?

   1. the central ray is directed caudally to the orbitomeatal line (OML)
   2. the petrous pyramids are projected into the lower third of the orbits
   3. the frontal sinuses are visualized

   (A) 1 only
   (B) 1 and 2 only
   (C) 1 and 3 only
   (D) 1, 2, and 3

182. Tracheostomy and intubation are effective techniques used to restore breathing when there is

   (A) respiratory pathway obstruction above the larynx
   (B) crushed tracheal rings due to trauma
   (C) lower respiratory pathway closure due to inflammation and swelling
   (D) respiratory pathway obstruction below the larynx

183. All of the following combining forms are related, EXCEPT

   (A) ile/o
   (B) gastr/o
   (C) bronch/o
   (D) col/o

184. Which of the following is recommended to demonstrate small amounts of air within the peritoneal cavity?

   (A) lateral decubitus, affected side up
   (B) lateral decubitus, affected side down
   (C) AP Trendelenburg
   (D) AP supine

185. For the average patient, the CR for a lateral projection of a barium-filled stomach should enter

   (A) midway between the midaxillary line and anterior abdominal surface
   (B) midway between the vertebral column and lateral border of abdomen
   (C) at the midaxillary line at the level of the iliac crest
   (D) perpendicular to the level of L2

186. Which of the following is (are) valid criteria for a lateral projection of the forearm?

    1. radius and ulna should be superimposed proximally and distally

    2. coronoid process and radial head should be superimposed

    3. the radial tuberosity should face anteriorly

    (A) 1 only

    (B) 1 and 2 only

    (C) 2 and 3 only

    (D) 1, 2, and 3

187. Which of the following articulations may be described as diarthrotic?

    1. knee

    2. intervertebral joints

    3. TMJ

    (A) 1 only

    (B) 2 only

    (C) 1 and 3 only

    (D) 1, 2, and 3

188. For which of the following diagnoses is operative cholangiography a useful tool?

    1. biliary tract calculi

    2. patency of the biliary ducts

    3. function of the sphincter of Oddi

    (A) 1 only

    (B) 2 only

    (C) 2 and 3 only

    (D) 1, 2, and 3

189. In what position was the radiograph in Figure 2–30 made?

    (A) flexion

    (B) extension

    (C) left bending

    (D) right bending

190. The structure labeled number *1* in Figure 2–30 is

    (A) intervertebral disc space

    (B) apophyseal joint

    (C) intervertebral foramen

    (D) spinous process

191. The erect position is occasionally requested as part of an IVP. This position is used to demonstrate

    (A) adrenal glands

    (B) renal surfaces

    (C) kidney mobility

    (D) the bladder neck

192. The ridge that marks the bifurcation of the trachea into right and left primary bronchi is the

    (A) root

    (B) hilus

    (C) carina

    (D) epiglottis

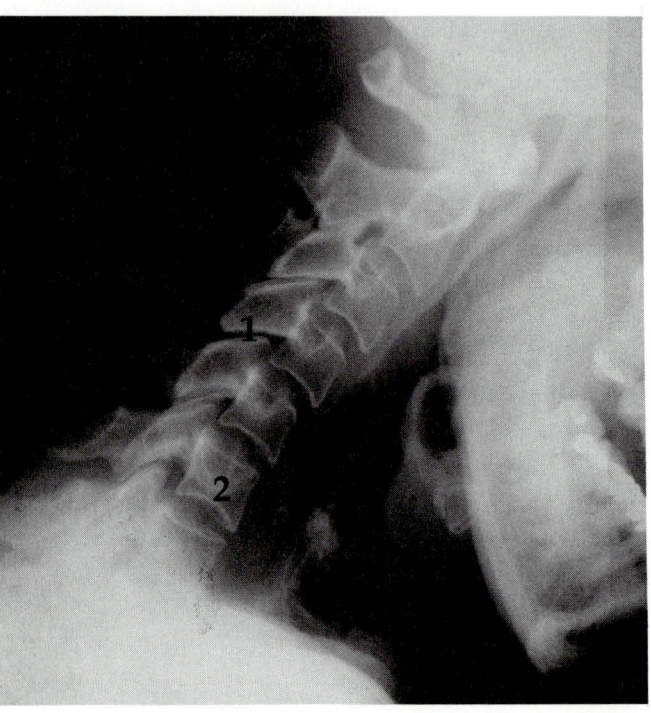

Figure 2–30. Courtesy of The Stamford Hospital, Department of Radiology.

193. In which of the following projections is the talofibular joint best demonstrated?

    (A) AP
    (B) lateral oblique
    (C) medial oblique
    (D) lateral

194. In the posterior profile projection (Stenvers method) of the petrous pyramids, the

    1. central ray is directed 12° cephalad
    2. midsagittal plane is 45° to the film
    3. head rests on zygoma, nose, and chin

    (A) 1 and 2 only
    (B) 1 and 3 only
    (C) 2 and 3 only
    (D) 1, 2, and 3

195. Which of the following sequences correctly describes the path of blood flow as it leaves the left ventricle?

    (A) arteries, arterioles, capillaries, venules, veins
    (B) arterioles, arteries, capillaries, veins, venules
    (C) veins, venules, capillaries, arteries, arterioles
    (D) venules, veins, capillaries, arterioles, arteries

196. Which of the following projections of the elbow should demonstrate the coronoid process free of superimposition and the olecranon process within the olecranon fossa?

    (A) AP
    (B) lateral
    (C) medial oblique
    (D) lateral oblique

197. Movement of a part away from the midline of the body is termed

    (A) eversion
    (B) inversion
    (C) abduction
    (D) adduction

198. To better demonstrate contrast-filled distal ureters during intravenous urography, it is helpful to

    1. use a 15° AP Trendelenburg position
    2. apply compression to the proximal ureters
    3. apply compression to the distal ureters

    (A) 1 only
    (B) 2 only
    (C) 1 and 2 only
    (D) 1 and 3 only

199. The term that refers to parts closer to the source or beginning is

    (A) cephalad
    (B) caudad
    (C) proximal
    (D) medial

200. Which of the following is a condition in which an occluded blood vessel stops blood flow to a portion of the lungs?

    (A) pneumothorax
    (B) atelectasis
    (C) pulmonary embolism
    (D) hypoxia

201. Which of the following positions will demonstrate the right axillary ribs?

    1. RAO
    2. LAO
    3. RPO

    (A) 1 only
    (B) 1 and 2 only
    (C) 2 and 3 only
    (D) 1, 2, and 3

202. Endoscopic retrograde cholangiopancreatography usually involves

    1. cannulation of the hepatopancreatic ampulla

    2. introduction of contrast medium into the common bile duct

    3. introduction of barium directly into the duodenum

    (A) 1 only

    (B) 1 and 2 only

    (C) 1 and 3 only

    (D) 1, 2, and 3

203. The location of the sacrum varies in the lateral position due to individual pelvic curves. The sacrum is located

    (A) 3 inches posterior to the midsagittal plane

    (B) 3 inches posterior to the midcoronal plane

    (C) 5 inches posterior to the midsagittal plane

    (D) 5 inches posterior to the midcoronal plane

204. Which of the following patient positions will BEST demonstrate the sternoclavicular joints?

    (A) supine

    (B) prone

    (C) posterior oblique

    (D) lordotic

205. Which of the following statements regarding the male pelvis is (are) true?

    1. the angle formed by the pubic arch is less than that of the female

    2. the pelvic outlet is wider than that of the female

    3. the ischial tuberosities are further apart

    (A) 1 only

    (B) 1 and 2 only

    (C) 2 and 3 only

    (D) 1, 2, and 3

206. Which of the following examinations involves the introduction of a radiopaque contrast medium through a uterine cannula?

    (A) retrograde pyelogram

    (B) voiding cystourethrogram

    (C) hysterosalpingogram

    (D) myelogram

207. The structure(s) BEST demonstrated on an AP axial projection of the skull with the central ray directed 40 to 60° caudally, are

    (A) entire foramen magnum and jugular foramina

    (B) petrous pyramids

    (C) occipital bone

    (D) rotundum foramina

208. Fluoroscopic imaging of the ileocecal valve is generally part of a(n)

    (A) esophagram

    (B) upper GI series

    (C) small bowel series

    (D) ERCP

209. During chest radiography, the act of inspiration

    1. elevates the diaphragm

    2. raises the ribs

    3. depresses the abdominal viscera

    (A) 1 only

    (B) 1 and 2 only

    (C) 2 and 3 only

    (D) 1, 2, and 3

210. The AP Trendelenburg position is often used during an upper GI examination to demonstrate

    (A) the duodenal loop

    (B) filling of the duodenal bulb

    (C) hiatal hernia

    (D) hypertrophic pyloric stenosis

211. The manubrial notch, a bony landmark used in radiography of the sternoclavicular joints, is located at the same level as the

    (A) vertebra prominens
    (B) first thoracic vertebra
    (C) third thoracic vertebra
    (D) ninth thoracic vertebra

212. The left sacroiliac joint is positioned perpendicular to the film when the patient is positioned in a

    (A) left lateral position
    (B) 25 to 30° LAO position
    (C) 25 to 30° LPO position
    (D) 30 to 40° LPO position

213. Which of the following bones participates in the formation of the acetabulum?

    1. ilium
    2. ischium
    3. pubis

    (A) 1 and 2 only
    (B) 1 and 3 only
    (C) 2 and 3 only
    (D) 1, 2, and 3

214. Which of the following radiologic procedures requires that contrast media be injected into the renal pelvis via a catheter placed within the ureter?

    (A) nephrotomography
    (B) retrograde urography
    (C) cystourethrography
    (D) IV urography

215. The AP projection of the sacrum requires that the central ray be directed

    1. 15° cephalad
    2. 2 inches superior to the pubic symphysis
    3. midline at the level of the lesser trochanter

    (A) 1 only
    (B) 2 only
    (C) 1 and 2 only
    (D) 1 and 3 only

216. Which cholangiographic procedure uses an indwelling drainage tube for contrast medium administration?

    (A) endoscopic retrograde cholangiographic pancreatography (ERCP)
    (B) operative cholangiography
    (C) T-tube cholangiography
    (D) percutaneous transhepatic cholangiography

217. Which of the following positions will demonstrate the lumbosacral apophyseal articulation?

    (A) AP
    (B) lateral
    (C) 30° RPO
    (D) 45° LPO

218. Which of the following is an accurate criticism of the PA chest illustrated in Figure 2–31?

    (A) The patient's shoulders are not rolled forward
    (B) the patient is rotated
    (C) the exposure was made on expiration
    (D) the patient breathed during the exposure

219. With the patient recumbent and the central ray directed horizontally, the patient is said to be in the

    (A) Trendelenburg position
    (B) Fowler's position
    (C) decubitus position
    (D) Sims position

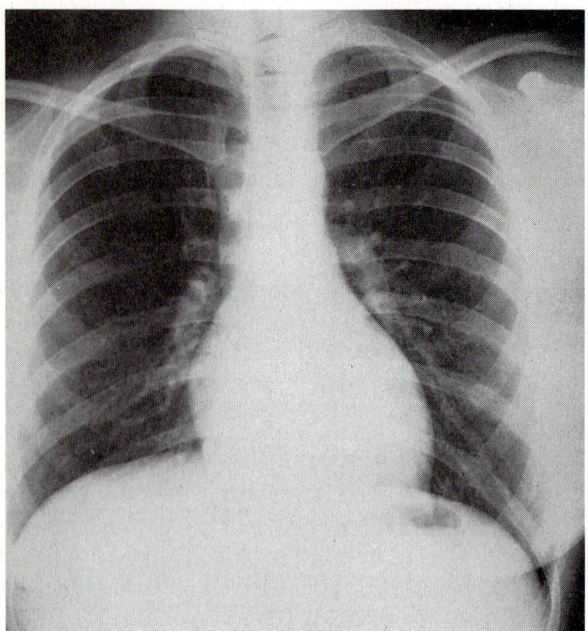

Figure 2–31. Courtesy of The Stamford Hospital, Department of Radiology.

220. Which of the following radiologic examinations can demonstrate ureteral reflux?

(A) intravenous urogram

(B) retrograde pyelogram

(C) voiding cystourethrogram

(D) nephrotomogram

221. The AP axial projection, or "frog leg" position, of the femoral neck places the patient in a supine position with the affected thigh

(A) adducted 25° from the horizontal

(B) abducted 25° from the vertical

(C) adducted 40° from the horizontal

(D) abducted 40° from the vertical

222. Which of the following precaution(s) should be observed when radiographing a patient who has sustained a traumatic injury to the hip?

1. when a fracture is suspected, manipulation of the affected extremity should be performed by a physician
2. the AP axiolateral projection should be avoided
3. to evaluate the entire region, the pelvis is typically included in the initial examination

(A) 1 only

(B) 1 and 2 only

(C) 2 and 3 only

(D) 1, 2, and 3

223. Which of the following positions would be obtained with the patient lying prone recumbent on the radiographic table, and the CR directed horizontally to the iliac crest?

(A) ventral decubitus position

(B) dorsal decubitus position

(C) left lateral decubitus position

(D) right lateral decubitus position

224. Prior to the start of an intravenous urogram, which of the following procedures should be carried out?

1. have patient empty the bladder
2. review the patient's allergy history
3. check the patient's creatinine level

(A) 1 only

(B) 2 only

(C) 1 and 2 only

(D) 1, 2 and 3

225. The cross-table or axiolateral projection of the hip requires the cassette to be placed in

1. contact with the lateral surface of the body, top edge slightly above the iliac crest
2. a vertical position and exactly perpendicular to the long axis of the femoral neck
3. a vertical position adjacent to the lateral surface of the affected hip

(A) 1 only
(B) 1 and 2 only
(C) 1 and 3 only
(D) 1, 2, and 3

226. The radiograph pictured in Figure 2–32 may be used to evaluate

1. polypoid lesions
2. the medial wall of the ascending colon
3. the posterior wall of the rectum

(A) 1 only
(B) 1 and 2 only
(C) 2 and 3 only
(D) 1, 2, and 3

227. In myelography, the contrast medium is generally injected into the

(A) cisterna magna
(B) individual intervertebral discs
(C) subarachnoid space between the first and second lumbar vertebrae
(D) subarachnoid space between the third and fourth lumbar vertebrae

228. Which of the following conditions is often the result of ureteral obstruction or stricture?

(A) pyelonephrosis
(B) nephroptosis
(C) hydronephrosis
(D) cystourethritis

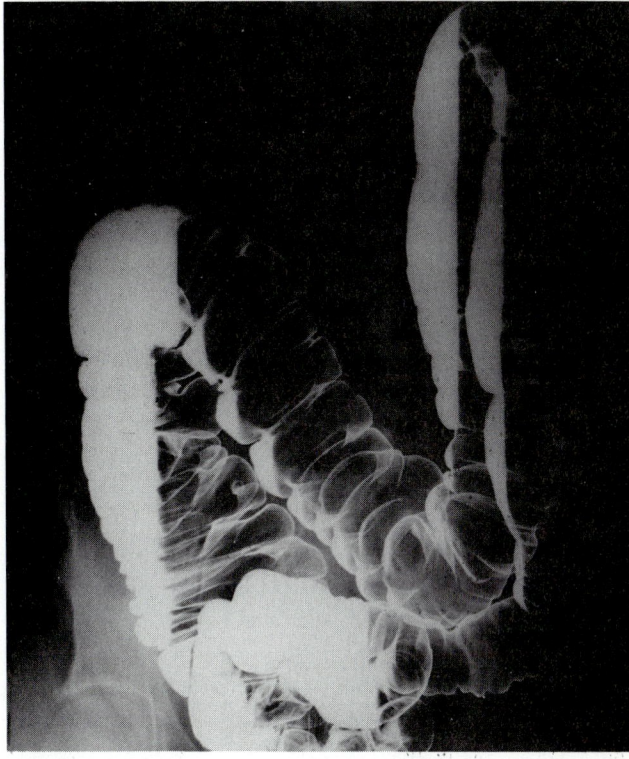

Figure 2–32. Courtesy of The Stamford Hospital, Department of Radiology.

229. Which of the following is (are) effective in reducing breast exposure during scoliosis examinations?

1. use of a high-speed imaging system
2. use of breast shields
3. use of compensating filtration

(A) 1 only
(B) 1 and 2 only
(C) 2 and 3 only
(D) 1, 2, and 3

230. Which type of articulation is evaluated in arthrography?

(A) synarthrodial
(B) diarthrodial
(C) amphiarthrodial
(D) cartilaginous

231. The true lateral position of the skull uses which of the following principles?

  1. interpupillary line perpendicular to the film
  2. MSP perpendicular to the film
  3. IOML parallel to the transverse axis of the film

(A) 1 only
(B) 1 and 2 only
(C) 1 and 3 only
(D) 1, 2, and 3

232. With the patient in the PA position and OML and CR perpendicular to the film, the resulting radiograph will demonstrate the petrous pyramids

(A) below the orbits
(B) in the lower ⅓ of the orbits
(C) completely within the orbits
(D) above the orbits

233. Which of the following describes the correct placement of mammographic markers?

  1. in the mediolateral projection they are placed along the upper border of the breast
  2. in the craniocaudal projection they are placed along the lateral aspect of the breast
  3. in the oblique projection they are placed along the lateral aspect of the breast

(A) 1 only
(B) 1 and 2 only
(C) 2 and 3 only
(D) 1, 2, and 3

234. When modifying the PA axial projection of the skull to demonstrate superior orbital fissures, the central ray is directed

(A) 20 to 25° caudad
(B) 20 to 25° cephalad
(C) 30 to 35° caudad
(D) 30 to 35° cephalad

235. Deoxygenated blood from the head and thorax is returned to the heart by the

(A) pulmonary artery
(B) pulmonary veins
(C) superior vena cava
(D) thoracic aorta

236. A flat and upright abdomen is requested on an acutely ill patient, to demonstrate the presence of air–fluid levels. Due to the patient's condition, the x-ray table can be tilted upright only 70° (rather than the desired 90°). How should the central ray be directed?

(A) perpendicular to the film
(B) parallel to the floor
(C) 20° caudad
(D) 20° cephalad

237. Myelography is a diagnostic examination used to demonstrate

  1. extrinsic spinal cord compression resulting from disc herniation
  2. post traumatic swelling of the spinal cord
  3. internal disc lesions

(A) 1 only
(B) 2 only
(C) 1 and 2 only
(D) 1 and 3 only

238. Which of the following positions will best demonstrate the left axillary portion of the ribs?

(A) left lateral
(B) PA
(C) LPO
(D) RPO

239. Following ingestion of a fatty meal, what hormone is secreted by the duodenal mucosa to stimulate contraction of the gallbladder?

(A) insulin
(B) cholecystokinin
(C) adrenocorticotrophic
(D) gastrin

240. The axiolateral position (Law method) of examining the mastoids uses which of the following?

1. orbitomeatal line
2. MSP parallel to tabletop
3. 15° caudad angulation

(A) 1 only
(B) 2 only
(C) 1 and 2 only
(D) 2 and 3 only

241. The stomach of an asthenic patient is most likely to be located

(A) high, transverse, and lateral
(B) low, transverse, and lateral
(C) high, vertical, and toward the midline
(D) low, vertical, and toward the midline

242. With which of the following is zonography associated?

1. thick tomographic cuts
2. long exposure amplitude
3. less blurring than pluridirectional tomography because a narrow exposure angle is used

(A) 1 only
(B) 2 only
(C) 1 and 3 only
(D) 2 and 3 only

243. During a double-contrast barium enema, which of the following positions would afford the BEST double-contrast visualization of the colic flexures?

(A) LAO and RPO
(B) lateral
(C) left lateral decubitus
(D) AP or PA erect

244. Mammography of the augmented breast is best accomplished using

(A) the Cleopatra method
(B) the Ecklund method
(C) magnification films
(D) the cleavage view

245. The four major arteries supplying the brain include the

1. brachiocephalic artery
2. common carotid arteries
3. vertebral arteries

(A) 1 and 2 only
(B) 1 and 3 only
(C) 2 and 3 only
(D) 1, 2, and 3

246. Which of the following articulates with the base of the first metatarsal?

(A) first cuneiform
(B) third cuneiform
(C) navicular
(D) cuboid

247. Which of the following is a major cause of bowel obstruction in children?

(A) appendicitis
(B) intussusception
(C) regional enteritis
(D) ulcerative colitis

248. Which of the following is (are) well demonstrated in the lumbar spine pictured in Figure 2–33?

    1. apophyseal articulations

    2. intervertebral foramina

    3. inferior articular processes

(A) 1 only

(B) 1 and 2 only

(C) 1 and 3 only

(D) 1, 2, and 3

249. Which of the following statements is (are) true regarding lower-extremity venography?

    1. the patient is often examined in the semi-erect position

    2. contrast medium is injected through a vein in the foot

    3. filming begins at the hip and proceeds inferiorly

(A) 1 only

(B) 1 and 2 only

(C) 1 and 3 only

(D) 1, 2, and 3

250. Which of the following is represented by the 3 in Figure 2–34?

(A) inferior vena cava

(B) aorta

(C) gallbladder

(D) psoas muscle

251. Which of the following bones participates in the formation of the knee joint?

    1. femur

    2. tibia

    3. patella

(A) 1 and 2 only

(B) 1 and 3 only

(C) 2 and 3 only

(D) 1, 2, and 3

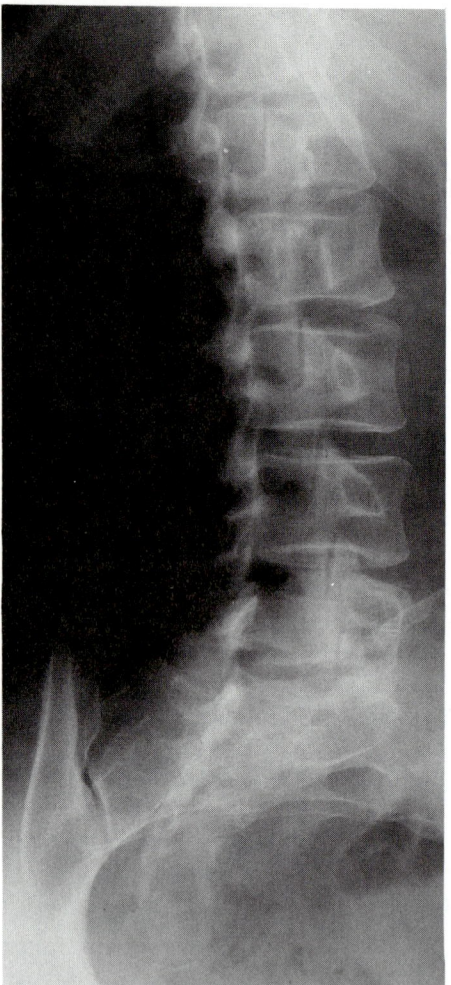

Figure 2–33. Courtesy of The Stamford Hospital, Department of Radiology.

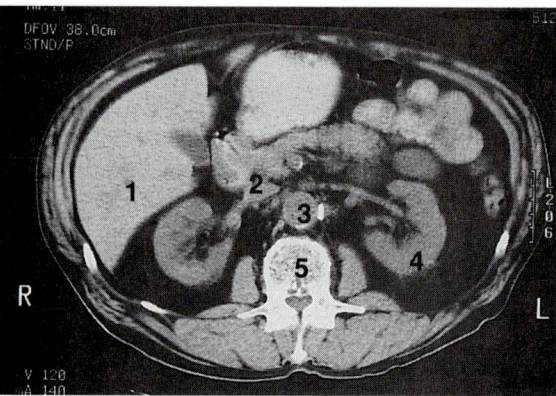

Figure 2–34. Courtesy of The Stamford Hospital, Department of Radiology.

252. All of the following are palpable bony landmarks used in radiography of the pelvis, EXCEPT

(A) femoral neck
(B) pubic symphysis
(C) greater trochanter
(D) iliac crest

253. Lateral deviation of the nasal septum may be best demonstrated in the

(A) lateral projection
(B) PA axial (Caldwell method) projection
(C) parietoacanthial (Waters method) projection
(D) AP axial (Grashey/Towne method) projection

254. The posterior oblique position of the lumbar spine will demonstrate the

(A) intervertebral foramina nearer to the film
(B) intervertebral foramina away from the film
(C) apophyseal joints nearer to the film
(D) apophyseal joints away from the film

255. All of the following may be determined by oral cholecystography, EXCEPT

(A) liver function
(B) concentrating ability of the gallbladder to concentrate bile
(C) emptying power of the gallbladder
(D) pancreatic function

256. The medial oblique projection of the elbow demonstrates the

1. olecranon process within the olecranon fossa
2. radial head free of superimposition
3. coronoid process free of superimposition

(A) 1 only
(B) 2 only
(C) 1 and 3 only
(D) 1, 2, and 3

257. In the posterior oblique position of the cervical spine, the CR should be directed

(A) parallel to C4
(B) perpendicular to C4
(C) 15° cephalad to C4
(D) 15° caudad to C4

258. What projection is required to determine bone age of the hand and wrist?

(A) AP
(B) PA
(C) lateral
(D) oblique

259. If the patient's zygomatic arch has been traumatically depressed or the patient has flat cheekbones, the arch may be demonstrated by modifying the SMV projection and rotating the patient's head

(A) 15° toward the side being examined
(B) 15° away from the side being examined
(C) 30° toward the side being examined
(D) 30° away from the side being examined

260. Which of the following barium-filled anatomic structures is best demonstrated in the LAO position?

(A) hepatic flexure
(B) splenic flexure
(C) sigmoid colon
(D) iliocecal valve

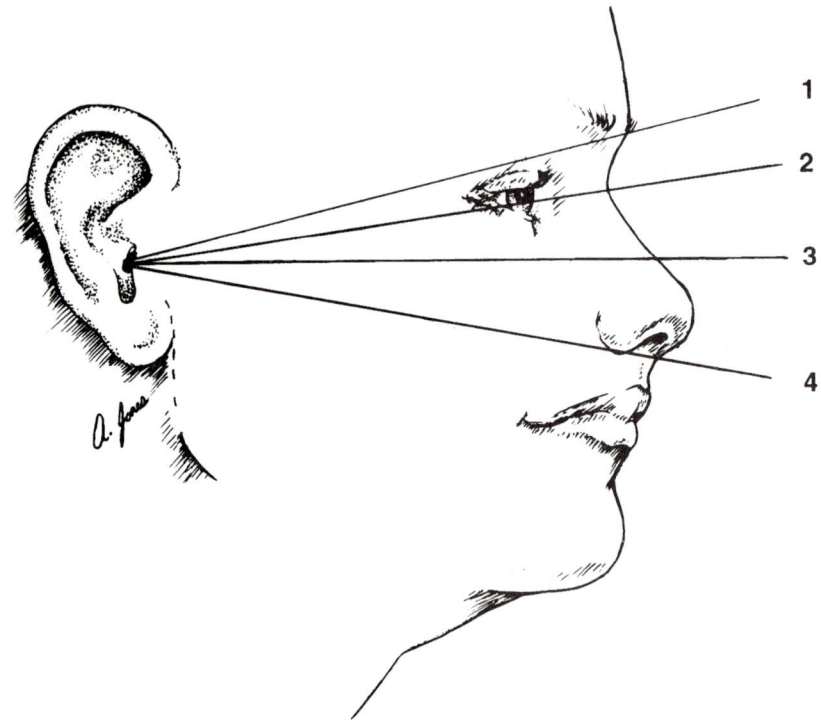

Figure 2–35. Reproduced with permission from Saia DA: *Radiography: Program Review and Exam Preparation.* Stamford, CT: Appleton & Lange, 1996.

261. In Figure 2–35, which localization line is used for the submentovertical (Schüller method) projection of the skull?

(A) line 1
(B) line 2
(C) line 3
(D) line 4

262. In Figure 2–35, which of the following localization lines is used for the lateral projection of the skull?

(A) line 1
(B) line 2
(C) line 3
(D) line 4

263. In Figure 2–35, which of the illustrated baselines are separated by 7°?

(A) lines 1 and 3
(B) lines 2 and 3
(C) lines 1 and 4
(D) lines 3 and 4

264. What is the name of the condition that results in the forward slipping of one vertebra on the one below it?

(A) spondylitis
(B) spondylolysis
(C) spondylolisthesis
(D) spondylosis

265. In a lateral projection of the nasal bones, the central ray is directed

(A) ½ inch posterior to anterior nasal spine
(B) ¾ inch posterior to glabella
(C) ¾ inch distal to nasion
(D) ½ inch anterior to EAM

266. Posterior displacement of a tibial fracture would be best demonstrated in the

(A) AP projection

(B) lateral projection

(C) medial oblique projection

(D) lateral oblique projection

267. Which of the following positions would best demonstrate the left apophyseal articulations of the lumbar vertebrae?

(A) LPO

(B) RPO

(C) left lateral

(D) PA

268. Blowout fractures of the orbit are BEST demonstrated using the

(A) lateral projection of the facial bones

(B) parietoacanthial projection (Water's method)

(C) posteroanterior projection with a 15° caudal angle

(D) Sweet's localization method

269. Which projection(s) of the abdomen would be used to demonstrate pneumoperitoneum?

1. right lateral decubitus

2. left lateral decubitus

3. upright

(A) 2 only

(B) 1 and 3 only

(C) 2 and 3 only

(D) 1, 2, and 3

270. Shoulder arthrography may be performed to evaluate

(A) humeral dislocation

(B) complete or incomplete rotator cuff tears

(C) osteoarthritis

(D) acromioclavicular joint separation

271. Examination of the pars petrosae in the posterior profile position (Stenvers method) requires

1. the use of the IOML

2. the MSP to be rotated 45°

3. that the head rest on forehead, nose, and chin

(A) 1 only

(B) 1 and 2 only

(C) 2 and 3 only

(D) 1, 2, and 3

272. Peripheral lymphatic vessels are located using

(A) ethiodized oil

(B) blue dye

(C) tiny scalpel

(D) water-soluble iodinated media

273. Which of the following is located at the interspace between the fourth and fifth thoracic vertebrae?

(A) manubrium

(B) jugular notch

(C) sternal angle

(D) xiphoid process

274. Which projection of the nasal bones will demonstrate medial or lateral displacement of fragments?

(A) oblique

(B) lateral

(C) AP axial

(D) tangential

275. The short, thick processes that project posteriorly from the vertebral body are the

(A) transverse processes
(B) vertebral arches
(C) laminae
(D) pedicles

276. All of the following are mediastinal structures, EXCEPT

(A) esophagus
(B) thymus
(C) heart
(D) terminal bronchiole

277. Which of the following sinus groups is demonstrated with the patient positioned as for a parietoacanthial projection (Water's method), and the CR directed through the patient's open mouth?

(A) frontal
(B) ethmoid
(C) maxillary
(D) sphenoid

278. To better demonstrate ribs below the diaphragm, the patient is instructed to

1. suspend respiration at the end of full exhalation
2. suspend respiration at the end of deep inhalation
3. lie in a recumbent position

(A) 1 only
(B) 2 only
(C) 1 and 3 only
(D) 2 and 3 only

279. To obtain an exact axial projection of the clavicle, place the patient

(A) supine and angle the central ray 30° caudally
(B) prone and angle the central ray 30° cephalad
(C) supine and angle the central ray 15° cephalad
(D) in a lordotic position and direct the CR at right angles to the coronal plane of the clavicle

280. When evaluating a PA axial projection of the skull with a 15° caudal angle, the radiographer should see

1. petrous pyramids in the lower third of the orbits
2. equal distance from the lateral border of the skull and the lateral rim of the orbit bilaterally
3. symmetrical petrous pyramids

(A) 1 and 2 only
(B) 1 and 3 only
(C) 2 and 3 only
(D) 1, 2, and 3

281. When performing gastrointestinal radiography, the position of the stomach may vary depending on

1. respiratory phase
2. body habitus
3. patient position

(A) 1 and 2 only
(B) 1 and 3 only
(C) 2 and 3 only
(D) 1, 2, and 3

282. With a patient in the PA position and the OML perpendicular to the table, a 15 to 20° caudal angulation would place the petrous ridges in the lower third of the orbit. In order to achieve the same result in a baby or a small child, it is necessary for the radiographer to modify the angulation to

(A) 10 to 15° caudal

(B) 25 to 30° caudal

(C) 15 to 20° cephalic

(D) no change in angulation is necessary

283. The patient positioned for an operative cholangiography is in a

(A) 15 to 20° left posterior oblique (LPO)

(B) 15 to 20° right posterior oblique (RPO)

(C) 45° left posterior oblique (LPO)

(D) 45° right posterior oblique (RPO)

284. When performing tomography, it is of paramount importance the radiographer:

1. properly apply immobilization
2. provide adequate radiation protection whenever possible
3. obtain and check a scout film

(A) 1 and 2 only

(B) 1 and 3 only

(C) 2 and 3 only

(D) 1, 2, and 3

285. Knee arthrography may be performed to demonstrate

1. torn meniscus
2. Baker's cyst
3. torn rotator cuff

(A) 1 and 2 only

(B) 1 and 3 only

(C) 2 and 3 only

(D) 1, 2, and 3

286. Indications for a myelographic cervical puncture include

1. demonstration of the upper level of a spinal block
2. suspected mass lesion in the upper cervical canal
3. failure of lumbar puncture

(A) 1 and 2 only

(B) 1 and 3 only

(C) 2 and 3 only

(D) 1, 2, and 3

287. During lower limb venography, tourniquets are applied above the knee and ankle to

1. suppress filling of the superficial veins
2. coerce filling of the deep veins
3. outline the anterior tibial vein

(A) 1 and 2 only

(B) 1 and 3 only

(C) 2 and 3 only

(D) 1, 2, and 3

288. Hysterosalpingography may be performed for demonstration of

1. uterine tubal patency
2. mass lesions in the uterine cavity
3. uterine position

(A) 1 and 2 only

(B) 1 and 3 only

(C) 2 and 3 only

(D) 1, 2, and 3

289. Which of the following equipment is necessary for endoscopic retrograde cholangiopancreatography?

1. a fluoroscopic unit with spot film and tilt table capabilities
2. a fiberoptic endoscope
3. polyethylene catheters

(A) 1 and 2 only

(B) 1 and 3 only

(C) 2 and 3 only

(D) 1, 2, and 3

290. Which of the following statements regarding pediatric positioning is FALSE?

(A) for radiography of the kidneys, the CR should be directed midway between the diaphragm and the symphysis pubis

(B) if a pediatric patient is in respiratory distress, a chest radiograph should be obtained in the AP projection rather than the standard PA projection

(C) chest radiography on a neonate should be performed in the supine position

(D) radiography of pediatric patients with a myelomeningocele defect should be performed in the supine position.

291. Operative cholangiography may be performed to

1. visualize biliary stones or a neoplasm
2. determine function of the hepatopancreatic ampulla
3. examine the patency of the biliary tract

(A) 1 and 2 only
(B) 1 and 3 only
(C) 2 and 3 only
(D) 1, 2, and 3

292. Linear, circular, elliptical, and hypocyclodial are all related to

(A) myelography
(B) arthrography
(C) venography
(D) tomography

293. T-tube cholangiography is performed

(A) preoperative
(B) during surgery
(C) postoperative
(D) with a chiba needle

294. Arthrography requires the use of

1. general anesthesia
2. sterile technique
3. fluoroscopy

(A) 1 and 2 only
(B) 1 and 3 only
(C) 2 and 3 only
(D) 1, 2, and 3

295. The contrast media of choice for use in myelography is

(A) ionic nonwater soluble
(B) ionic water soluble
(C) nonionic water soluble
(D) gas

296. Which of the following procedures will best demonstrate the cephalic, basilic, and subclavian veins?

(A) aortofemoral arteriogram
(B) upper limb venogram
(C) lower limb venogram
(D) renal venogram

297. Which of the following devices should NOT be removed before positioning for a radiograph?

1. a ring when performing hand radiography
2. an antishock garment
3. a pneumatic splint

(A) 1 and 2
(B) 1 and 3
(C) 2 and 3
(D) 1, 2, and 3

298. Patients are instructed to remove all jewelry, hair clips, metal prostheses, coins, and credit cards before entering the room for a examination in

(A) sonography
(B) computerized axial tomography
(C) magnetic resonance imaging
(D) nuclear medicine

299. Which of the following are appropriate techniques for imaging a patient with a possible traumatic spine injury?

1. instruct the patient to turn slowly and stop if anything hurts
2. maneuver the x-ray tube head instead of moving the patient
3. call for help and use the log-rolling method to turn the patient

(A) 1 and 2 only
(B) 1 and 3 only
(C) 2 and 3 only
(D) 1, 2, and 3

300. In which type of fracture are the splintered ends of bone forced through the skin?

(A) closed
(B) compound
(C) compression
(D) depressed

Answers and Explanations

1. **(D)** To clearly demonstrate the atlas and axis without superimposition of the teeth or base of the skull, a line between the occlusal plane (edge of upper teeth) and mastoid tip must be vertical. If the head is flexed too much, teeth will be superimposed. If the head is extended too much, the cranial base will be superimposed on the area of interest. A line between the mentum and mastoid tip is used to demonstrate the odontoid process only through the foramen magnum (Fuchs method). *(Ballinger, vol 1, p 292)*

2. **(C)** A Colles fracture is usually caused by a fall onto an outstretched (extended) hand, in order to "brake" a fall. The wrist then suffers an impacted transverse fracture of the distal inch of the *radius* with an accompanying chip fracture of the *ulnar* styloid process. Because of the hand position at the time of the fall, the fracture is usually *displaced backward* approximately 30°. *(Eisenberg, p 99)*

3. **(B)** Three views of the cervical spine are illustrated. Figure A is an RAO; figure C is an LPO. *Anterior oblique* positions (LAO, RAO) of the cervical spine demonstrate the intervertebral *foramina closer* to the film, while *posterior oblique* positions (LPO, RPO) demonstrate the intervertebral *foramina farther* from the film. Figure B is the left lateral position. Lateral projections of the cervical spine are done to demonstrate the intervertebral disc spaces, apophyseal joints, and spinous processes. *(Ballinger vol 1, p 337)*

4. **(B)** The subtalar, or talocalcaneal, joint is a three-faceted articulation formed by the talus and os calsis (calcaneus). The plantodorsal and dorsoplantar projections of the os calsis should exhibit density sufficient to visualize the subtalar joint (Fig. 2–36). This is the only "routine" position that will demonstrate the subtalar joint. If evaluation of the subtalar joint is desired, special views (such as the Broden and Isherwood methods) would be required. *(Ballinger, vol 1, p 215)*

5. **(C)** The relationship between the thigh, lower leg, patella, and the CR should be noted. The CR is directed parallel to the plane of the patella, thereby providing a tangential projection of the patella (patella in profile) and an unobstructed view of the patellofemoral articulation (Fig. 2–37). A

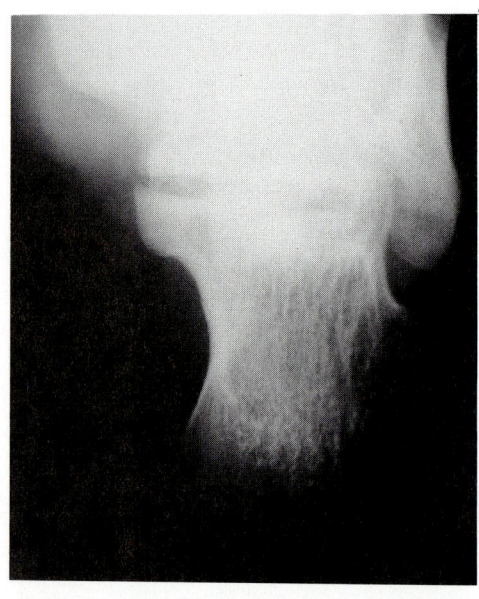

Figure 2–36. Courtesy of The Stamford Hospital, Department of Radiology.

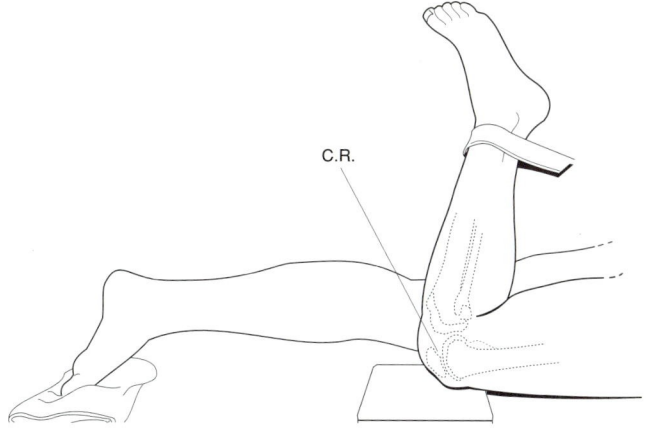

C.R.

Figure 2–37.

"tunnel view" is required to demonstrate the intercondyloid fossa and articulating surfaces of the tibia and femur. *(Ballinger, vol 1, p 264)*

6. **(A)** When the arm is placed in the *AP* position, the *epicondyles are parallel* to the plane of the cassette and the shoulder is placed in *external rotation*. In this position, an AP projection of the humerus, elbow, and forearm can be obtained. For the lateral projection of the humerus, elbow, or forearm, the epicondyles must be perpendicular to the plane of the cassette. *(Ballinger, vol 1, p 126)*

7. **(C)** The thorax consists of *12 pairs of ribs* and the structures to which they are attached anteriorly and posteriorly: the *sternum* and the *thoracic vertebrae* (Fig. 2–38). The scapulae, together with the clavicles, form the shoulder (pectoral) girdle. *(Martini, p 229)*

8. **(C)** In the direct PA projection of the wrist, the carpal pisiform is superimposed on the carpal triangular. The AP oblique projection (medial surface adjacent to the film) separates the pisiform and triangular and projects the pisiform as a separate structure. The pisiform is the smallest and only palpable carpal. *(Ballinger, vol 1, p 88)*

9. **(C)** In the "true" PA projection of the skull with perpendicular central ray exiting the na-

sion, the petrous pyramids should *fill* the orbits (Fig. 2–39). As the central ray (CR) is angled caudally, the petrous pyramids are projected lower in the orbits, and at about 25 to 30° they are below the orbits. The orbitomeatal line (OML) must be perpendicular to the film, or the petrous pyramids will not be projected into the expected location. The MSP (midsagittal plane) must be perpendicular to the film or the skull will be rotated. With the MSP parallel to the film, a lateral skull projection is obtained. *(Ballinger, vol 2, p 242)*

10. **(B)** The pictured radiograph is an AP projection of the knee with the knee in extension. The tibial intercondylar eminences are well demonstrated upon the tibial plateau and the femorotibial joint is well visualized. The intercondyloid fossa is not demonstrated here. A "tunnel" view of the knee is required to demonstrate the intercondyloid fossa. *(Ballinger, vol 1, p 240)*

11. **(D)** Patients who are able to cooperate are usually able to control *voluntary* motion. However, certain body functions and responses create *involuntary* motion not controllable by the patient. Severe pain, muscle spasm, and chills all cause involuntary movements. Peristaltic activity of the intestinal tract and motion caused by contraction of the heart muscle are other sources of involuntary motion. *(Ballinger, vol 1, p 12)*

12. **(A)** With the patient AP, the CR is directed cephalad 25 to 30°. This serves to project the clavicle away from the pulmonary apices and ribs. The reverse is true when the patient is examined in the PA position. *(Ballinger, vol 1, p 159)*

13. **(B)** The sacroiliac joints angle posteriorly and medially 25° to the MSP. Therefore, in order to demonstrate them with the patient in the *AP* position, the *affected* side must be elevated 25°. This places the joint space perpendicular to the film and parallel to the central ray. When performed with the patient *PA*, the *unaffected* side will be elevated 25°. *(Ballinger, vol 1, p 380)*

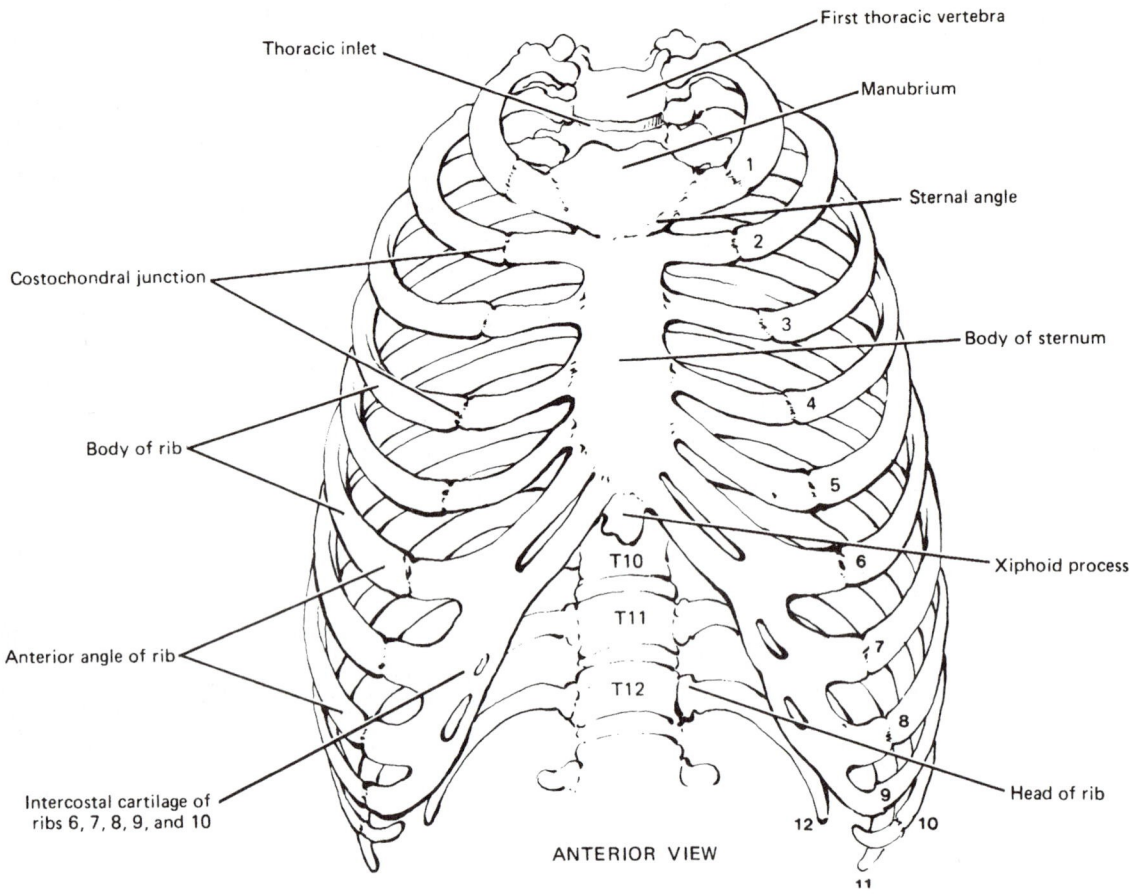

Figure 2-38. Reproduced with permission from Lindner HH. *Clinical Anatomy*. East Norwalk, CT: Appleton & Lange, 1989.

14. (C) The pictured radiograph is one of a series of IVP (IVU) films done prone at 20 minutes after injection of the contrast medium. The urinary collecting system is well demonstrated. The *renal pelvis* (number *1*) is the proximal expanded end of the ureter lying within the renal sinus. The *minor calyces* (number *3*) receive urine from the collecting tubules of the renal pyramids and convey it to the *major calyces* (number *2*), which empty into the renal pelvis. Urine is carried down the ureters by peristaltic waves. The *vesicoureteral junction* is located at the distal end of the ureter where it unites with the urinary bladder. (*Martini, pp 982–983*)

15. (C) With the body in the AP recumbent position, barium flows easily into the fundus of the stomach, displacing the stomach somewhat superiorly. The fundus, then, is filled solely with barium, while the air that had been in the fundus is displaced into the gastric body, pylorus, and duodenum, thus illustrating them in double-contrast fashion. Air contrast delineation of these structures allows us to see through the stomach, to retrogastric areas and structures. Anterior and posterior aspects of the stomach are visualized in the lateral position; medial and lateral aspects of the stomach are visualized in the AP projection. (*Ballinger, vol 2, p 110*)

16. (A) Because the sternum and vertebrae would be superimposed in a direct PA or AP projection, a slight oblique (just enough to separate the sternum from superimposition on the vertebrae) is used instead of a direct frontal projection. In the RAO position the heart superimposes a homogeneous density over the sternum, thereby providing clearer radiographic visualization of its bony struc-

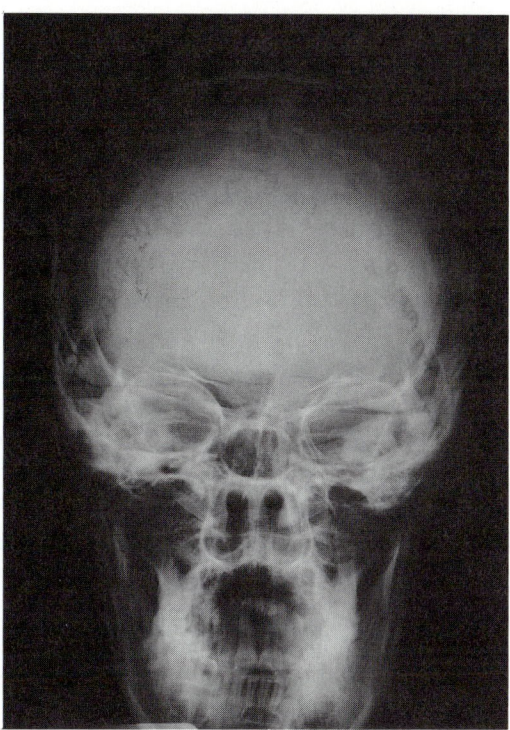

Figure 2–39. Courtesy of The Stamford Hospital, Department of Radiology.

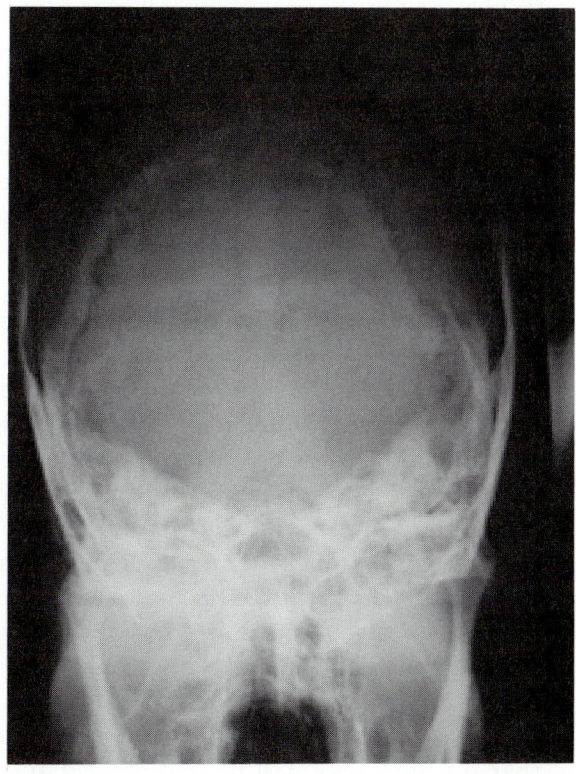

Figure 2–40. Courtesy of The Stamford Hospital, Department of Radiology.

ture. If the *LAO* position were used to project the sternum to the *right* of the thoracic vertebrae, posterior ribs and pulmonary markings would cast confusing shadows over the sternum because of their differing densities. The lateral projection requires that the shoulders be rolled back sufficiently to project the sternum completely anterior to the ribs. Prominent pulmonary vascular markings can be obliterated using a "breathing technique," that is, using an exposure time long enough (with appropriately low mA) to equal at least a few respirations. *(Ballinger, vol 1, pp 406–408)*

17. **(D)** With the patient in the AP position, the scapula and upper thorax are normally superimposed. With the arm abducted, the elbow flexed, and hand supinated, much of the scapula is drawn away from the ribs. The patient should not be rotated toward the affected side, as this causes superimposition of ribs on the scapula. The exposure is made during quiet breathing to obliterate pulmonary vascular markings. *(Ballinger, vol 1, p 164)*

18. **(B)** The heart superimposes a homogeneous density over the sternum in the RAO position, thus providing clearer radiographic visualization of its bony structure. If the LAO position were used to project the sternum to the right of the thoracic vertebrae, posterior ribs and pulmonary markings would cast confusing shadows over the sternum because of their differing densities. Prominent pulmonary markings can be obliterated using a "breathing" technique, that is, using an exposure time long enough (with appropriately low mA) to equal at least a few respirations. *(Ballinger, vol 1, p 408)*

19. **(A)** The AP axial position projects the anterior structures (frontal and facial bones) downward, thus permitting visualization of the occipital bone without superimposition (Grashey/Towne method). The dorsum sella and posterior clinoid processes of the sphenoid bone should be visualized within the foramen magnum. This projection may also

be obtained by angling the CR 30° caudad to the OML (Fig. 2–40). The frontal bone is best shown with the patient PA and a perpendicular central ray. The parietoacanthial projection is the single best position for facial bones. Basal foramina are well demonstrated in the submentovertical projection. *(Ballinger, vol 2, p 247)*

20. **(D)** Although routine elbow projections are essentially negative, conditions may exist (such as an elevated fat pad) that seem to indicate the presence of a small fracture of the radial head. In order to demonstrate the entire circumference of the radial head, four exposures are made with the elbow flexed 90° and with the humeral epicondyles superimposed and perpendicular to the cassette: one with the hand supinated as much as possible, one with the hand lateral, one with the hand pronated, and one with the hand in internal rotation, thumb down. Each maneuver changes the position of the radial head and a different surface is presented for inspection. *(Ballinger, vol 1, pp 102–103)*

21. **(B)** Patients who are able to cooperate are usually able to control *voluntary* motion if they are provided with adequate explanation of the procedure. Once patients understand what is needed, most will cooperate to the best of their ability (by suspending respiration and holding still for the exposure). Certain body functions and responses, such as heart action, peristalsis, pain, and muscle spasm, cause *involuntary* motion uncontrollable by the patient. The best and only way to control involuntary motion is by always selecting the shortest possible exposure time. Involuntary motion may also be minimized by careful explanation, immobilization, and (as a last resort and only in certain cases) restraint. *(Ballinger, vol 1, pp 12–13)*

22. **(C)** The radiograph is a lateral oblique (external rotation) projection of the elbow, removing the proximal radius from superimposition with the ulna and demonstrating its articulation with the ulna at the radial notch, that is, the proximal radioulnar articulation. An AP projection of the elbow would demonstrate partial overlap of the proximal radius and ulna. A medial oblique would demonstrate complete overlap of the proximal radius and ulna; this position is used to demonstrate the coronoid process in profile and the olecranon process within the olecranon fossa. *(Ballinger, vol 1, p 105)*

23. **(C)** Scoliosis is a lateral curvature of the spine and is typically noted in early adolescence. These young patients usually return for follow-up studies and it is imperative to limit their radiation dose as much as possible. Examining the patient in the PA position is frequently advisable, because the gonadal dose is significantly reduced and there is usually not appreciable loss of detail. A thyroid shield is also a valuable protection especially for the patient who requires follow-up examinations. Bending films would not be performed on a patient with suspected subluxation or spondylolisthesis, as further serious injury could result. *(Ballinger, vol 1, p 396)*

24. **(B)** Intervertebral joints are well visualized in the lateral projection of all the vertebral groups. Cervical articular facets (forming apophyseal joints) are 90° to the midsagittal plane and are therefore well demonstrated in the lateral projection. The cervical intervertebral foramina lie 45° to the midsagittal plane (and 15 to 20° to a transverse plane) and are therefore demonstrated in the oblique position. *(Ballinger, vol 1, p 336)*

25. **(C)** The coronoid process is located on the proximal anterior ulna. The medial oblique projection of the elbow demonstrates the coronoid process in profile, as well as the ulnar olecranon process within the humeral olecranon fossa. The lateral oblique elbow projects the proximal radius and ulna free of superimposition. The coracoid process is located on the scapula. *(Ballinger, vol 1, p 104)*

26. **(B)** Air or fluid levels will be clearly demonstrated only if the central ray is directed parallel to them. Therefore, to demonstrate air or fluid levels, erect or decubitus positions should be used. Small amounts of *fluid* are best

demonstrated in the lateral decubitus position, affected side *down*. Small amounts of *air* are best demonstrated in the lateral decubitus position, affected side *up*. Similarly, dorsal and ventral decubitus positions made with a horizontal x-ray beam can also be used to demonstrate air or fluid levels. *(Ballinger, vol 2, p 37)*

27. **(C)** *Full* or *forced expiration* is used to elevate the diaphragm and demonstrate ribs *below* the diaphragm to best advantage (with exposure adjustment). *Deep inspiration* is used to depress the diaphragm and demonstrate as many ribs *above the diaphragm* as possible. Shallow breathing technique is occasionally used to visualize above the diaphragm ribs, while obliterating pulmonary vascular markings. *(Ballinger, vol 1, p 426)*

28. **(B)** As the colon begins to fill with barium, filling often slows down in the rectosigmoid region. Unless preventative measures are taken, severe abdominal cramping and urge to defecate will occur. Stopping the flow of barium momentarily will usually be sufficient to relieve the patient's discomfort. Some patients have repeated cramping throughout the examination. These patients should be instructed to breath deeply through their mouth and the enema bag should be lowered a few inches to reduce the pressure of barium flow. Many patients benefit from administration of glucagon to relax the intestine. The Trendelenburg position is frequently used to separate redundant loops of bowel. *(Ballinger, vol 2, pp 124–125; Gurley, p 113)*

29. **(A)** The illustrated radiograph is a PA axial position (Caldwell method) of the frontal and anterior ethmoidal sinuses. The frontal sinuses are seen centrally in the vertical plate of the frontal bone and extending laterally over the superciliary arches. The patient is positioned with the OML perpendicular to the film; the CR should be angled 15° caudally to place the petrous pyramids in the lower third of the orbits (a little too much caudal angle was used here; the petrosae are projected to the inferior rim of the orbits). Projecting the petrous

pyramids below the orbits (no. 3) is the objective of the parietocanthial projection (Water's method). *(Ballinger, vol 2, pp 373, 381)*

30. **(C)** The hypersthenic body type is large and heavy. The thoracic cavity is short; the lungs are short with broad bases, and the heart is usually in an almost transverse position. The diaphragm is high; the stomach and gallbladder are high and transverse. The large bowel is positioned high and peripheral (and often requires that 14 × 17 cassettes be placed crosswise for filming a BE). *(Ballinger, vol 1, p 41)*

31. **(A)** Sternoclavicular articulations may be examined with the patient PA, either bilaterally with the patient's head resting on the chin, or unilaterally with the patient's head turned toward the side being examined. The sternoclavicular articulations may also be examined in the oblique position, with either the patient rotated slightly or the CR angled slightly medialward. Weight-bearing postions are frequently used for evaluation of acromioclavicular joints. *(Ballinger, vol 1, pp 414–416)*

32. **(B)** The AP projection of the cervical spine demonstrates the bodies and intervertebral spaces of the last five vertebra (C3–7). The cervical apophyseal joints are 90° to the midsagittal plane and are therefore demonstrated in the lateral projection. *(Ballinger, vol 1, p 334)*

33. **(B)** When a patient's elbow needs to be examined in partial flexion, the lateral projection offers little difficulty, but the AP projection requires special attention. If the AP is made with a perpendicular central ray and olecranon process resting on the tabletop, the articulating surfaces are obscured. With the elbow in partial flexion, two exposures are necessary. One is made with the forearm parallel to the film (humerus elevated)—this demonstrates the proximal forearm. The other with the humerus parallel to the film (forearm elevated)—this demonstrates the distal humerus. In both cases the central ray is perpendicular if the degree of flexion is not too great, or angled slightly into the joint space with greater degrees of flexion. *(Ballinger, vol 1, pp 106–107)*

34. (C) Radiographic examinations of the large bowel generally include the AP or PA axial position to "open" the S-shaped sigmoid colon, the lateral position especially for the rectum, and the LAO and RAO (or LPO and RPO) to "open" the colic flexures. Left and right decubitus positions are usually only employed in double-contrast barium enemas to better demonstrate the medial and lateral walls of the ascending and descending colon. *(Ballinger, vol 2, p 143)*

35. (C) The relationship of these three structures can be appreciated in a lateral projection of the chest. The heart is seen in the anterior half of the thoracic cavity, its apex extending inferior and anterior. The air-filled trachea can be seen in about the center of the chest, and the air-filled esophagus just posterior to the trachea (Fig. 2–41). *(Ballinger, vol 1, p 458)*

36. (D) The submentovertical (SMV) projection is made with the patient's head resting on the vertex and the central ray directed perpendicular to the infraorbitomeatal line (IOML). This position may be used as part of a sinus survey to demonstrate the sphenoid sinuses or as a view of the cranial base for basal foramina (especially foramina ovale and spinosum). It also demonstrates the bony part of the auditory (Eustachian) tubes. AP or PA axial projections are frequently used to demonstrate the occipital region or evaluate the sellar region: a lateral projection is usually part of a routine skull evaluation. The parietoacanthial projection is the single best position to demonstrate facial bones. *(Ballinger, vol 2, p 253)*

37. (C) The ankle mortise (the articulation formed by the talus, tibia, and fibula) in the radiograph is well demonstrated. This is accomplished with a 15 to 20° medial oblique position. The 45° medial and lateral obliques demonstrate the distal tibia and fibula considerably superimposed on the talus. In the AP ankle there is some superimposition of the fibula over the tibia and talus, thereby obscuring the medial aspect of the ankle mortise. *(Ballinger, vol 1, p 231)*

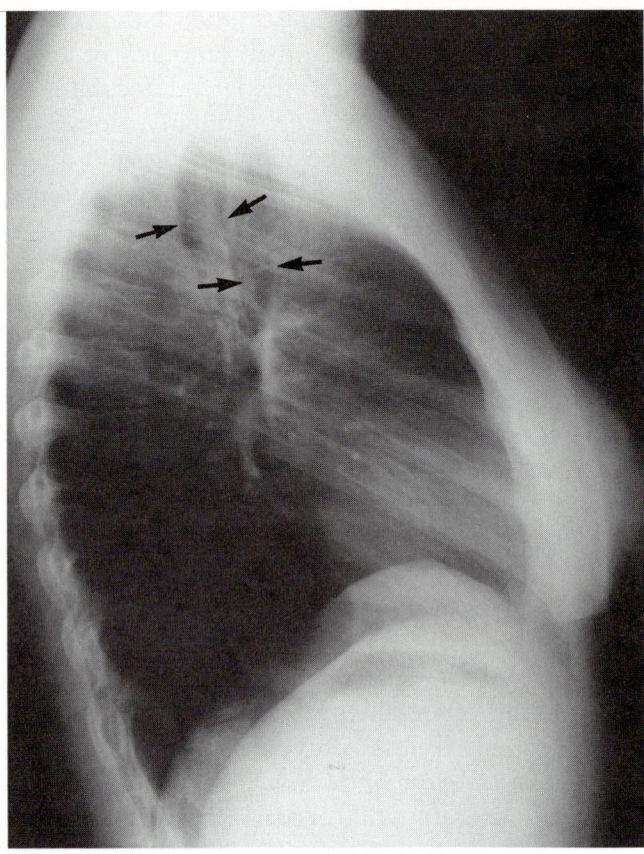

Figure 2–41. From the American College of Radiology Learning Files. Courtesy of the ACR.

38. (A) The ankle mortise is formed by the distal tibia and fibula and the talus. The distal tibia (the medial and larger bone) forms a club-shaped projection, the *medial malleolus* (no. 2). The distal fibula's projection is the *lateral malleolus* (no. 1). The distal articular surfaces of both the tibia and fibula articulate with the superior surface of the *talus* (no. 3) to form the ankle joint. *(Ballinger, vol 1, p 231)*

39. (B) A lateral projection of the lumbar spine is illustrated. The intervertebral joints (disc spaces) are well demonstrated. Because the intervertebral foramina are 90° to the midsagittal plane (MSP), they are also well demonstrated in the lateral projection. The articular facets, forming the apophyseal joints, lie 30 to 50° to the MSP and thus are visualized in the oblique position. *(Ballinger, vol 1, pp 368–369)*

40. **(A)** An *avulsion* fracture is a small fragment pulled from a bony process as a result of a forceful pull of the attached ligament or tendon. An example is an avulsion fracture of the medial malleolus as a result of eversion stress, causing pull on the deltoid ligament. A *torus* fracture is a greenstick fracture with one cortex buckled and the other intact. A *stress* or *fatigue* fracture is the result of a repeated force to a bony part. A *compound* fracture is an open fracture where the fractured ends have perforated the skin. *(Saia, Program Review & Exam Preparation or PREP, p 110)*

41. **(C)** The patient must remove any metallic foreign objects if they are within the area(s) of interest. Dentures, earrings, or necklaces can obscure bony details in a skull or cervical spine survey. However, only the necklace needs to be removed for a chest x-ray. The radiographer must be certain that the patient's belongings are cared for properly and returned following the examination. *(Ballinger, vol 1, p 14)*

42. **(C)** The degree of anterior and posterior motion is occasionally diminished with a "whiplash"-type injury. Anterior (forward, flexion) and posterior (backward, extension) motion is evaluated in the lateral position with the patient assuming flexion and extension as best as possible. Left- and right-bending films are frequently obtained of the thoracic and lumbar vertebrae when evaluating scoliosis. *(Saia, PREP, p 115)*

43. **(C)** The abdomen may be divided into four *quadrants:* the right and left upper quadrants (*RUQ, LUQ*) and right and left lower quadrants (*RLQ, LLQ*). The abdomen may also be divided into nine *regions:* the upper three regions are the right and left *hypochondrium* with the *epigastrium* between them, the middle three are the right and left *lumbar* regions with the *umbilical* region between them, and the lower three are the right and left *iliac* separated by the *hypogastric* region. The location of any of the abdominal viscera may be described in terms of the quadrants or regions. The sigmoid colon is the S-shaped distal portion of the large bowel, just proximal to the rectum and located in the hypogastric region.

Special angled projections are required to "open" its double curve. *(Saia, PREP, pp 62–63)*

44. **(C)** The lateral hand in extension, with appropriate technique adjustment, is recommended to evaluate foreign body location in soft tissue. A small lead marker is frequently taped to the spot thought to be the point of entry. The physician then uses this external marker and the radiograph to determine the exact foreign body location. Extension of the hand, in the presence of a fracture, would cause additional and unnecessary pain, and possibly additional injury. *(Ballinger, vol 1, p 79)*

45. **(A)** The pictured radiograph is an oblique position of the large bowel, illustrating an "open" view of the hepatic flexure, and the splenic flexure self-superimposed. Therefore, the radiograph must have been made in either an RAO (if the patient was prone) or an LPO (if done supine) position. The LAO and RPO positions are used to demonstrate the splenic flexure free of self-superimposition. Decubitus films demonstrate air–fluid levels, and the AP axial is generally used to visualize the sigmoid colon. *(Ballinger, vol 2, p 140)*

46. **(A)** The chest should be examined in the upright position whenever possible in order to demonstrate any air–fluid levels. The patient elevates the arms, flexes and grasps elbows. The midsagittal and midcoronal planes must remain vertical in order to avoid distortion of the heart. In the AP projection the shoulders should be relaxed and depressed to move the clavicles below the lung apices and the shoulders should be rolled forward in order to move the scapulae out of the lung fields. *(Ballinger, vol 1, p 456)*

47. **(C)** The parietoacanthial projection (Water's position) of the skull is valuable for the demonstration of facial bones or maxillary sinuses. The head is rested on extended chin so that the orbitomeatal line forms a 37° angle with the film. This projects the petrous pyramids below the floor of the maxillary sinuses and provides an oblique frontal view of the facial bones. *(Ballinger, vol 2, p 304)*

48. **(B)** The AP projection provides a general survey of the abdomen, showing the size and shape of the liver, spleen, and kidneys. When performed erect, it should demonstrate both hemidiaphragms. The lateral projection is sometimes requested and is useful for evaluating the prevertebral space occupied by the aorta. Ventral and dorsal decubitus positions provide a lateral view of the abdomen useful for demonstration of air–fluid levels. *(Ballinger, vol 1, p 38)*

49. **(D)** Complete and correct patient information on each radiograph is of paramount importance. Incorrect or delayed diagnosis can result from careless errors. The information that must appear on each film includes the patient's name, hospital or ID number, the date (occasionally the time), the name of the institution, and a right or left marker. *(Ballinger, vol 1, p 16)*

50. **(B)** To best demonstrate *most* of the tarsals and intertarsal spaces (including the cuboid, sinus tarsi, and tuberosity of the fifth metatarsal), a medial oblique is required (plantar surface and film form a 30° angle). The lateral oblique demonstrates the navicular and first and second cuneiforms. Weight-bearing lateral feet are used to demonstrate the longitudinal arches. *(Ballinger, vol 1, p 200)*

51. **(A)** The cervical intervertebral foramina form a 45° angle with the midsagittal plane and therefore are well visualized in a 45° oblique position. Apophyseal joints are formed by articulating surfaces of the inferior articular facet of one vertebra with the superior articular facet of the vertebra below, and are well demonstrated in the lateral position of the cervical spine. The intervertebral disc spaces are best demonstrated in the lateral position. *(Ballinger, vol 1, p 342)*

52. **(B)** The position, shape, and motility of various organs can differ greatly from one body habitus to another. The hypersthenic individual is large and heavy; the lungs and heart are high, the stomach is high and transverse, the gallbladder is high and lateral, and the colon is high and peripheral. In contrast, the other habitus extreme is the asthenic individual. This patient is slender and light, has a long narrow thorax, low and long stomach, low and medial gallbladder, and low medial and redundant colon. The radiographer must take these characteristic differences into consideration when radiographing individuals of various body types. *(Ballinger, vol 1, p 72)*

53. **(A)** Esophageal varices are tortuous dilatations of the esophageal veins. They are much less pronounced in the erect position and must always be examined with the patient recumbent. The recumbent position affords more complete filling of the veins, as blood flows against gravity. *(Ballinger, vol 2, p 88)*

54. **(D)** In order to evaluate sufficient inspiration and lung expansion, ten posterior ribs should be visualized. Sternoclavicular joints should be symmetrical; any loss of symmetry indicates rotation. In order to visualize maximum lung area, the shoulders are rolled forward to remove the scapulae from the lung fields. *(Ballinger, vol 1, p 413)*

55. **(B)** The oblique projection of the hand should demonstrate minimal overlap of the third, fourth, and fifth metacarpals. Excessive overlap of these metacarpals is caused by obliquing the hand *more than 45°*. The use of a 45° foam wedge ensures that the fingers will be extended and parallel to the film, thus permitting visualization of the interphalangeal joints and avoiding foreshortening of the phalanges. Clenching the fist and the use of ulnar flexion are maneuvers used to better demonstrate the carpal scaphoid. *(Ballinger, vol 1, p 76)*

56. **(C)** The distal femur is associated with two large condyles; the deep depression separating them is the intercondyloid fossa (Fig. 2–42). The proximal tibia has two condyles; their superior surfaces are smooth, forming the tibial plateau. The mandible has a condyle that articulates with the mandibular fossa of the temporal bone forming the temporomandibular joint. The fibula has a proximal styloid process and a distal malleolus, but no condyle. *(Martini, p 250)*

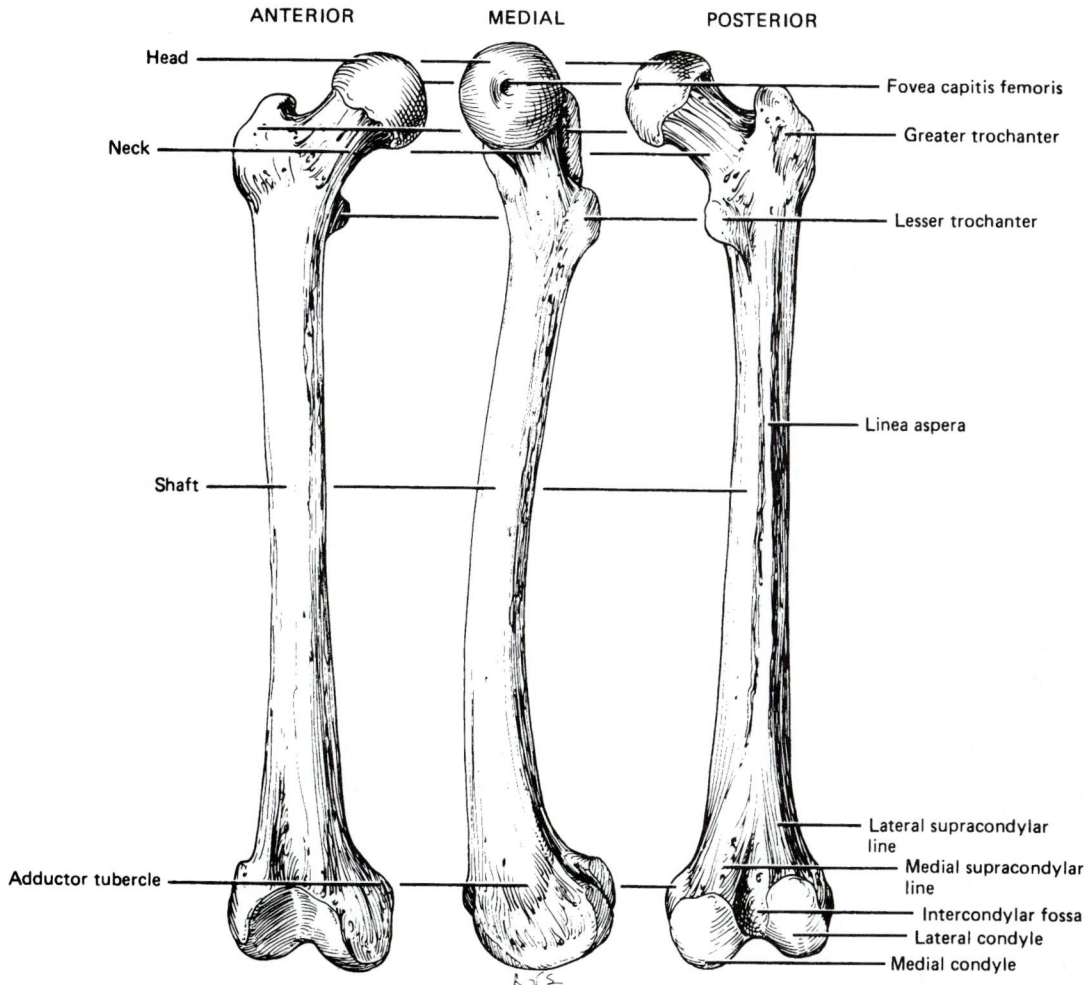

Figure 2–42. Reproduced with permission from Lindner HH. *Clinical Anatomy*. East Norwalk, CT: Appleton & Lange, 1989.

57. (D) Correct and complete patient information on every radiograph is of paramount importance. Correct ID is required on follow-up examinations that need to be accurately compared with previous films. Incorrect or delayed diagnosis can result from careless errors. Repeat exams may be required, resulting in needless additional radiation exposure. *(Ballinger, vol 1, p 16)*

58. (B) Observing the pelvis in the anatomic position, the ilia are seen to oblique forward, giving the pelvis a "basin-like" appearance. To view the *left* iliac bone, the radiographer must place it parallel to the film, by elevating the right side about 40°. The *right* iliac bone is radiographed in the LPO, 40° oblique, position (left AP oblique). *(Ballinger, vol 1, p 308)*

59. (B) The cervical intervertebral foramina lie 45° to the midsagittal plane and 15 to 20° to a transverse plane. When the *posterior oblique* position (LPO, RPO) is used, the cervical intervertebral foramina demonstrated are those *further* from the film. There is therefore some magnification of the foramina. In the anterior oblique position (LAO, RAO), the foramina disclosed are those *closer* to the film. *(Ballinger, vol 1, p 341)*

60. (B) The upper GI tract must be empty for best x-ray evaluation. Any food or liquid mixed with the barium sulfate suspension can simulate pathology. Preparation therefore is to withhold food and fluids for 8 to 9 hours before the exam, typically after midnight, as fasting exams are usually performed first thing in the morning. Enemas until clear prior to the exam is a part of the typical preparation for barium enema/air contrast. *(Ballinger, vol 2, p 98)*

61. **(D)** Phase of respiration is exceedingly important in thoracic radiography; lung expansion and the position of the diaphragm strongly influence the appearance of the finished radiograph. Inspiration and expiration radiographs of the chest are taken to demonstrate air in the pleural cavity (*pneumothorax*), to demonstrate *atelectasis* (partial or complete collapse of one or more pulmonary lobes) degree of *diaphragm excursion*, or to detect the presence of a *foreign body*. The expiration film will require a somewhat greater exposure (6 to 8 kV more) to compensate for the diminished quantity of air in the lungs. *(Ballinger, vol 1, p 444)*

62. **(C)** The AP axial projection (Grashey method) of the skull requires that the central ray be angled 30° caudad if the orbitomeatal line is vertical (37° caudad if the IOML is vertical). The frontal and facial bones are projected down and away from superimposition on the occipital bone. If positioned accurately, the dorsum sella and posterior clinoid processes will be demonstrated within the foramen magnum. If the CR is angled excessively, the posterior aspect of the arch of C1 will appear in the foramen magnum. *(Ballinger, vol 2, p 246)*

63. **(D)** The greater and lesser tubercles are prominences on the proximal humerus separated by the intertubercular (bicipital) groove. The AP projection of the humerus/shoulder places the epicondyles parallel to the film, the shoulder in external rotation, and demonstrates the greater tubercle in profile. The lateral projection of the humerus places the shoulder in extreme internal rotation with the epicondyles perpendicular to the film and demonstrates the lesser tubercle in profile. *(Ballinger, vol 1, p 126)*

64. **(D)** Barium can dry and harden in the large bowel, causing symptoms from mild constipation to bowel obstruction. It is therefore essential that the radiographer provide clear instructions of follow-up care, along with its rationale, especially to outpatients. In order to avoid the possibility of fecal impaction, patients should drink plenty of fluids for the next few days, increase their dietary fiber, and take a mild laxative such as milk of magnesia. *(Adler & Carlton, p 253)*

65. **(B)** When the foot is positioned for a lateral projection, the plantar surface should be perpendicular to the film, so as to superimpose the metatarsals. This may be accomplished with the patient lying on either the affected or unaffected side (usually affected), that is, mediolateral or lateromedial. The talofibular articulation is best demonstrated in the medial oblique projection of the ankle. *(Ballinger, vol 1, p 205)*

66. **(C)** When a cervical spine is requested to rule out subluxation or fracture, the patient will arrive in the Radiology area on a stretcher. The patient should *not* be moved before a subluxation is ruled out. Any movement of the head and neck could cause serious damage to the spinal cord. A horizontal beam lateral is performed and evaluated. The physician will then decide what further films are required. *(Ballinger, vol 1, p 350)*

67. **(D)** The distal humerus articulates with the proximal radius and ulna to form the elbow joint. Specifically, the semilunar/trochlear notch of the proximal ulna articulates with the trochlea of the distal medial humerus. The capitulum is lateral to the trochlea and articulates with the radial head (Fig. 2–43). *(Saia, PREP, p 76)*

68. **(B)** A *right* lateral position is generally performed *recumbent*; the *left* lateral is generally performed *upright*. In either case, a plane located midway between the midcoronal plane and the anterior surface of the abdomen is centered to the grid at the level of the pylorus (midway between the umbilicus and xiphoid process). Centering midway between the MSP/vertebrae and left lateral surface of the abdomen is correct for the AP or PA projection of the stomach. *(Ballinger, vol 2, p 109)*

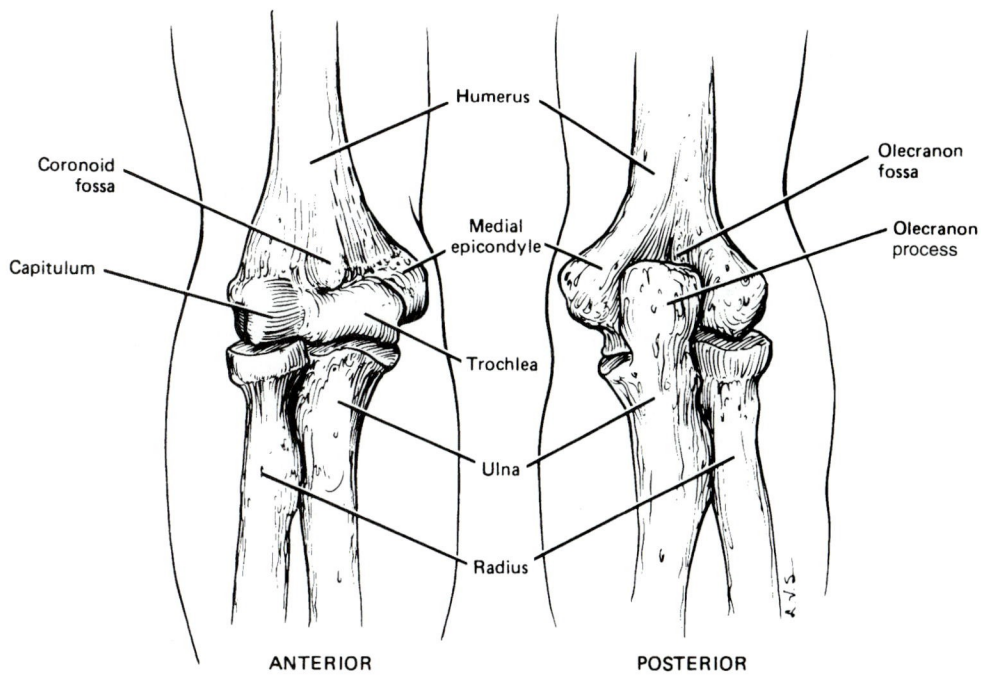

Figure 2–43. Reproduced with permission from Lindner HH. *Clinical Anatomy.* East Norwalk, CT: Appleton & Lange, 1989.

69. (B) The CNS (brain and spinal cord) is located within three protective membranes, the *meninges.* The inner membrane is the *pia* mater, the middle membrane is the *arachnoid,* and the outer membrane is the *dura* mater. The *subarachnoid space* is located between the pia and arachnoid mater and contains cerebrospinal fluid (CSF). During myelography, the needle is introduced into the subarachnoid space (between L3/L4 or L4/L5), a small amount of CSF is removed, and the contrast medium is introduced (Fig. 2–44). The subdural space is located between the arachnoid and dura mater. The epidural space is located between the two layers of the dura mater. *(Saia, PREP, p 179)*

70. (C) The parietoacanthial projection (Water's position) provides an oblique frontal projection of the facial bones. The maxilla (and antra), zygomatic arches, and orbits are well demonstrated. The patient is positioned PA with the head resting on extended chin so that the orbitomeatal line forms a 37° angle with the film. The position may be reversed if the patient is AP and the central ray is directed 30° cephalad to the IOML. This position is not preferred, however, because the facial bones are significantly magnified due to increased object-film distance. *(Ballinger, vol 2, p 304)*

71. (D) A right "scapular Y" is illustrated and refers to the characteristic Y formed by the humerus, acromion, and coracoid. The patient is positioned in a PA oblique position; in this case an RAO for the right side. The midcoronal plane is adjusted 60° to the film and the affected arm is left relaxed at the patient's side. The scapular Y position is employed to demonstrate anterior or posterior humeral dislocation. The humerus is normally superimposed on the scapula in this position; any deviation from this may indicate dislocation. *(Ballinger, vol 1, p 142)*

72. (C) When scheduling patient examinations it is important to avoid the possibility of residual contrast medium covering areas of interest on later exams. The IVP should be scheduled first because the contrast medium used is excreted rapidly. The BE should be scheduled next. The GI series is scheduled last, because any barium remaining from the previous BE should not be enough to interfere

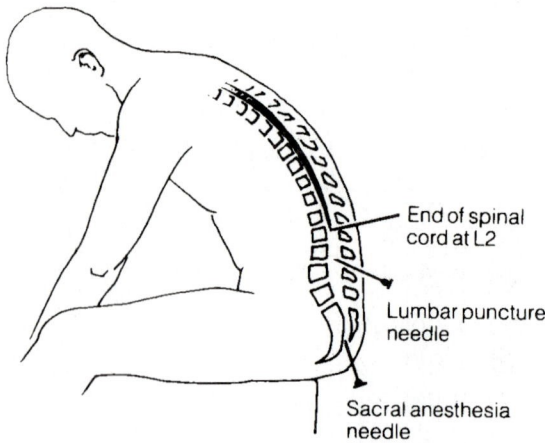

Figure 2–44. Reproduced with permission from deGroot, *Correlative Neuroanatomy*, 21st ed. Norwalk, CT: Appleton & Lange, 1996.

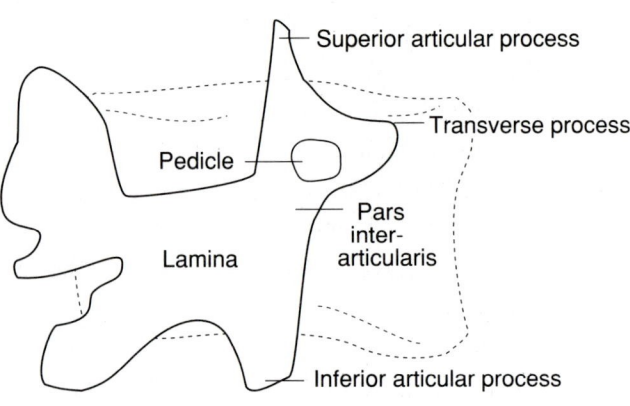

Figure 2–45.

with the stomach or duodenum, though a preliminary scout film should be taken in each case. *(Saia, PREP, pp 12–13)*

73. **(D)** The 45° oblique position of the lumbar spine is generally performed for demonstration of the apophyseal joints. In a correctly positioned oblique lumbar spine, "scotty dog" images are demonstrated (Figs. 2–45 and 2–46). The scotty's ear corresponds to the superior articular process, his nose to the transverse process, his eye is the pedicle, his neck is the pars interarticularis, his body is the lamina, and his front foot is the inferior articular process. *(Saia, PREP, p 119)*

74. **(D)** In the AP projection of the elbow, the proximal radius and ulna are partially superimposed. In the lateral position, the radial head is partially superimposed on the coronoid process, facing anteriorly. In the medial oblique position, there is even greater superimposition. The lateral oblique projection completely separates the proximal radius and ulna, projecting the radial head, neck, and tuberosity free of superimposition with the proximal ulna. *(Ballinger, vol 1, p 105)*

75. **(C)** Myelography is the radiologic examination of the structures within the spinal canal. Opaque contrast medium is usually used. Following injection, the contrast medium is distributed to the vertebral region of interest

by gravity; the *table* is angled Trendelenburg for visualization of the cervical region and in Fowler's position for visualization of the thoracic and lumbar regions. Although the table is Trendelenburg, care must be taken that the patient's neck be kept in acute *extension* in order to compress the cisterna magna and keep contrast medium from traveling into the ventricles of the brain. *(Ballinger, vol 2, pp 500–502)*

76. **(A)** Sinuses should be performed in the upright position in order to demonstrate fluid levels and to distinguish between fluid and other pathologies. When the patient cannot assume the upright position, the lateral projection can be obtained using a horizontal ("crosstable") x-ray beam. A transthoracic lateral position (distractor B) is used to obtain a lateral position of the upper one half to two thirds of the humerus when the arm cannot be abducted. *(Ballinger, vol 2, p 378)*

77. **(B)** With the patient in the PA position, the rami are well visualized with a perpendicular ray or with 20 to 25° cephalad angulation. A portion of the mandibular body is demonstrated but most is superimposed over the cervical spine. *(Ballinger, vol 2, pp 432–433)*

78. **(A)** The trachea ("windpipe") bifurcates into left and right *mainstem bronchi*, each entering its respective lung hilum. The *left* bronchus divides into *two* portions, one for each lobe of the left lung. The *right* bronchus divides into *three* portions, one for each lobe of the right lung

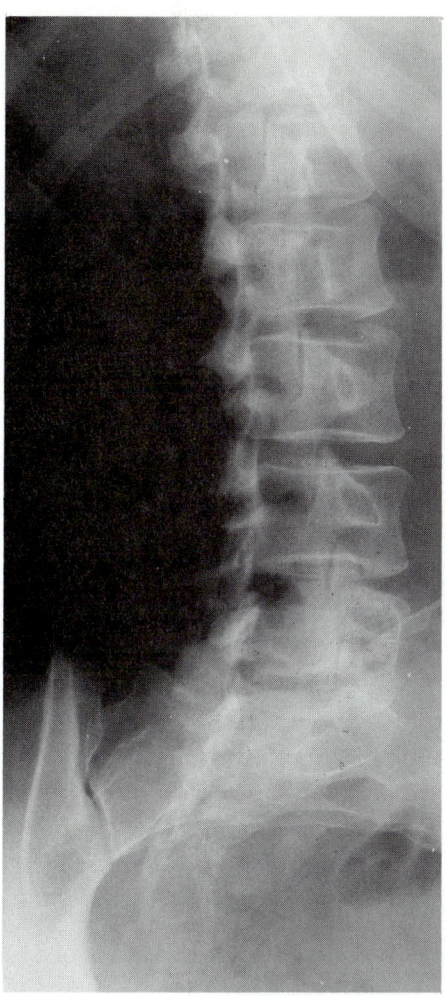

Figure 2–46. Courtesy of The Stamford Hospital, Department of Radiology.

(Fig. 2–47). The lungs are conical in shape, consisting of upper pointed portions, termed the *apices* (pleural for apex), and the broad lower portions (or *bases*). The lungs are enclosed in a double-walled serous membrane called the *pleura*. (*Ballinger, vol 1, pp 398–399*)

79. **(A)** The cervical intervertebral foramina lie 45° to the midsagittal plane and 15 to 20° to a transverse plane. When the *posterior oblique* position (LPO, RPO) is used, the CR is directed 15 to 20° caudad and the cervical intervertebral foramina demonstrated are those *further* from the film. There is therefore some magnification of the foramina. In the anterior oblique position (LAO, RAO), the CR is directed 15 to 20° cephalad and the foramina

disclosed are those *closer* to the film. (*Ballinger, vol 1, p 340*)

80. **(D)** A PA projection of the chest is pictured. Adequate *inspiration* is demonstrated by visualization of 10 pairs of ribs above the diaphragm. *Rotation* of the chest is demonstrated by unequal distance between the sternum and medial extremities of the clavicles. *Air-filled* trachea is seen in lower cervical and upper thoracic region, as a midline area of increased density. (*Saia, PREP, p 147*)

81. **(A)** There are four types of body habitus; listed from largest to slightest they are hypersthenic, sthenic, hyposthenic, and asthenic. The position, shape, and motility of various organs can differ greatly from one body type to another. In the illustrated radiograph the gallbladder is positioned low and medial—typical of the asthenic body habitus. The gallbladder of the hypersthenic individual occupies a high, lateral, and transverse position. (*Ballinger, vol 1, pp 41–42*)

82. **(D)** The medial oblique projection (15 to 20° mortise view) of the ankle (Fig. 2–48) is valuable because it demonstrates the talofibular joint as well as the talotibial joint, thereby visualizing all the major articulating surfaces of the ankle joint. In order to demonstrate maximum joint volume it is recommended that the plantar surface be vertical. (*Ballinger, vol 1, p 231*)

83. **(B)** Oral administration of barium sulfate is used to demonstrate the upper digestive tract: the esophagus, the fundus, body, and pylorus of the stomach; and the small bowel, consisting of duodenum, jejunum, and ileum. Consistent care must be taken to read and record patient information accurately and correctly. The large bowel is usually demonstrated via rectal administration of barium. (*Gurley, p 80*)

84. **(C)** The thoracic apophyseal joints are 70° to the midsagittal plane and are demonstrated in a steep (70°) oblique position. The thoracic intervertebral foramina, formed by the vertebral notches of the pedicles, are 90° to the

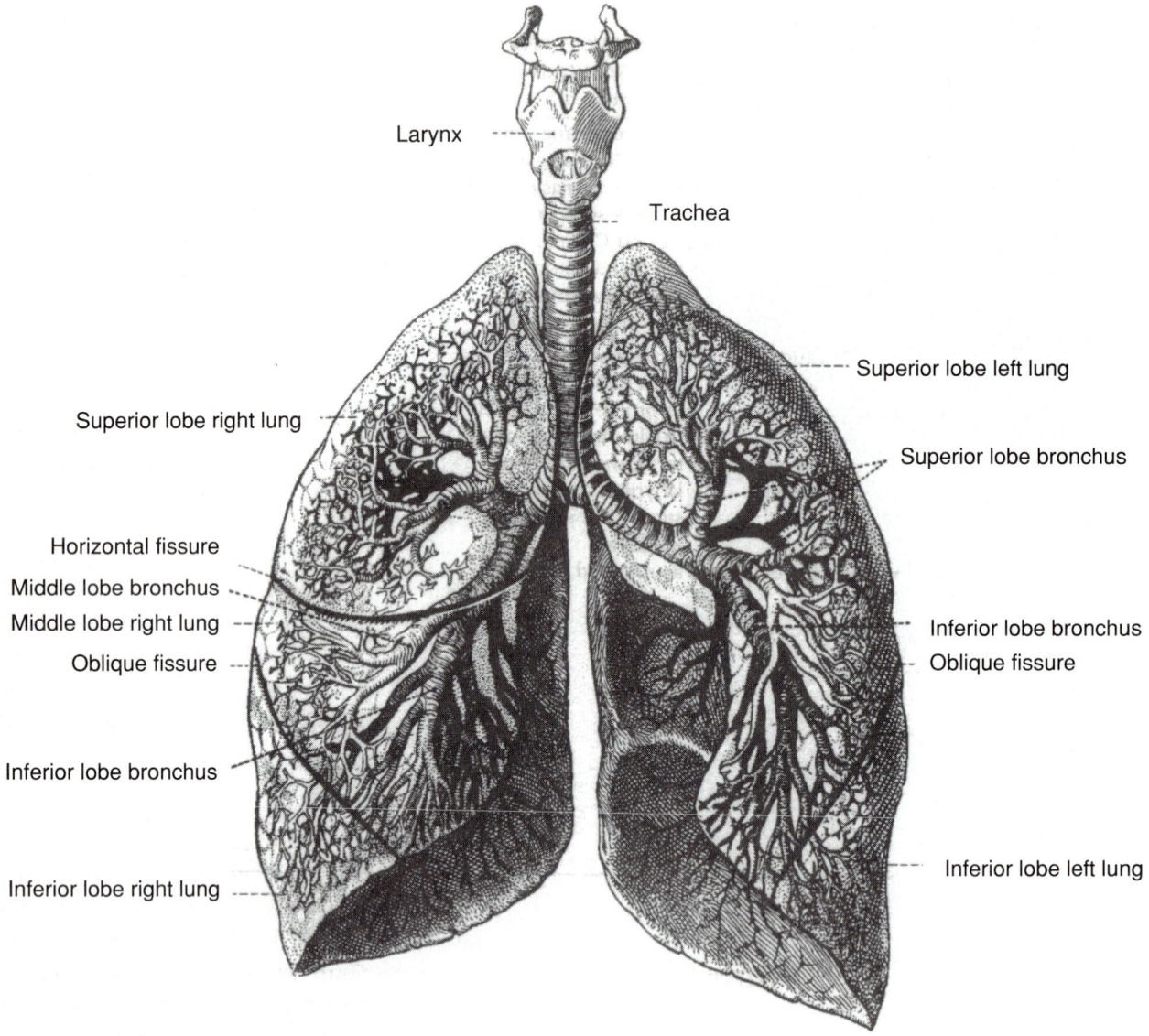

Larynx

Trachea

Superior lobe left lung

Superior lobe right lung

Superior lobe bronchus

Horizontal fissure
Middle lobe bronchus
Middle lobe right lung
Oblique fissure

Inferior lobe bronchus
Oblique fissure

Inferior lobe bronchus

Inferior lobe left lung

Inferior lobe right lung

Figure 2–47. Reproduced with permission from Montgomery RL. *Appleton & Lange's Review of Anatomy for the USMLE Step I*. East Norwalk, CT: Appleton & Lange, 1995.

midsagittal plane. They are therefore well demonstrated in the lateral position. The intervertebral foramina of the thoracic and lumbar vertebrae are demonstrated in the lateral position also. *(Ballinger, vol 1, p 361)*

85. **(A)** The dome of the acetabulum lies midway between the ASIS and symphysis pubis. On an adult of average size, a line perpendicular to this point will parallel the plane of the femoral neck. In an AP projection of the hip, the CR should be directed to a point approximately 2½ inches down that perpendicular

line, so as to enter the distal portion of the femoral head. *(Ballinger, vol 1, pp 274, 286)*

86. **(D)** In the AP projection of the knee, the position of the joint space is significantly affected by the patient's overall body habitus. It is recommended that the CR be directed to a point ½ inch distal to the patellar apex at an angle that depends on the patient's body habitus and the distance between the ASIS and tabletop. When the patient is of asthenic habitus with a distance of 19 to 24 cm between the ASIS and tabletop, the CR is di-

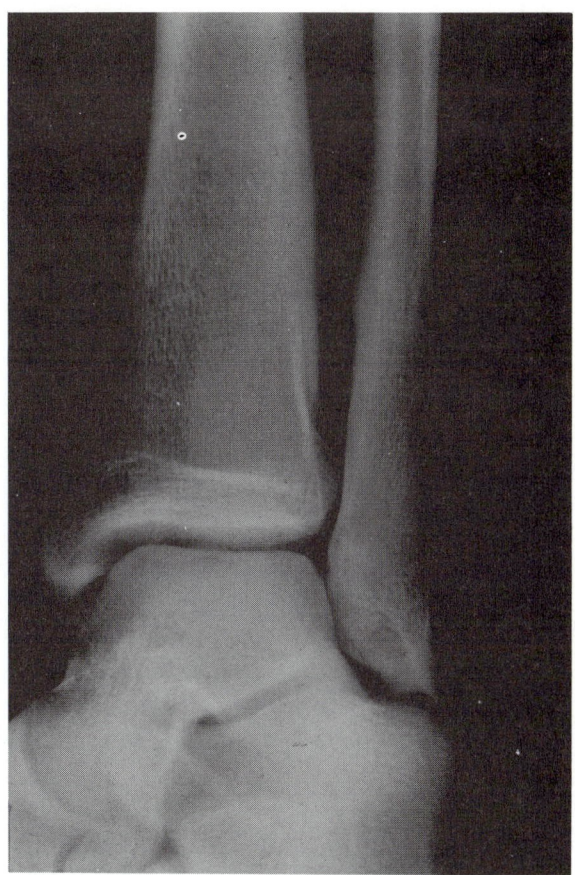

Figure 2–48. Courtesy of The Stamford Hospital, Department of Radiology.

rected perpendicularly. Whµen the patient is of asthenic habitus with an ASIS to table measurement less than 19 cm, the CR is directed 5° caudad. With a hypersthenic patient having an ASIS to table measurement greater than 24 cm, the CR is directed 5° cephalad. *(Ballinger, vol 1, p 241)*

87. **(A)** The external rotation position places the humeral epicondyles parallel to the film, the shoulder in a true AP position, and the greater tubercle in profile. The internal rotation position places the humeral epicondyles perpendicular to the film, the humerus in a lateral position, and the lesser tubercle in profile. The neutral position places the epicondyles about 45° to the film, and is frequently valuable in demonstrating calcific deposits in the shoulder joint. *(Ballinger, vol 1, p 126)*

88. **(C)** As the heart contracts and relaxes while functioning to pump blood from the heart, those arteries that are large and those in closest proximity to the heart will feel the effect of the heart's forceful contractions in their walls. The arterial walls pulsate in unison with the heart's contractions. This movement may be detected with the fingers in various parts of the body, and is referred to as the *pulse. (Ballinger, vol 2, p 517)*

89. **(B)** The illustrated radiograph is that of a voiding urethrogram, a radiologic examination of the contrast-filled urethra. Radiologic studies of the urethra are often requested to demonstrate *stricture* that can result from trauma or inflammation. A cystogram is a radiographic examination of the contrast-filled urinary bladder. An IVP demonstrates the structures of the urinary system (kidneys, ureters, and bladder). A hysterosalpingogram is a radiologic examination of the female reproductive system: the uterus, ovaries, and oviducts. *(Ballinger, vol 2, p 158)*

90. **(C)** The pictured radiograph shows an axiolateral projection (Laws method) of the right mastoid. The mastoid air cells are easily recognized in the temporal region just posterior to the auditory canal. In the posterior profile (Stenvers) position, the skull is seen more PA with the mastoid tip projected adjacent to the upper cervical spine. The AP axial (Towne/Grashey) position would demonstrate the petrous portions bilateral to the foramen magnum. *(Saia, PREP, p 142)*

91. **(C)** The RAO position demonstrates the left atrium of the heart. This position provides good visualization of the esophagus when it is full of barium, because it projects it between the vertebrae and heart. The 45° obliques are used to demonstrate the lungs: the LAO for the right lung and the RAO for the left lung. For heart size, the left oblique needs to be increased to 55 to 60° to separate superimposed aorta and vertebrae. *(Ballinger, vol 1, pp 461–463)*

92. **(C)** The orbits are formed by portions of seven bones: the frontal, lacrimal, ethmoid,

palatine, sphenoid, zygoma, and maxilla. The orbital walls are very thin and fragile and subject to "blowout" fractures. These fractures of the thin, delicate orbital wall may be demonstrated using the parietoacanthial (Waters) projection, radiographic tomography, and/or computed tomography. *(Ballinger, vol 2, p 270)*

93. **(A)** Many upper GI series are performed as double contrast studies today in order to better see the mucosal lining and small lesions within the stomach. For successful results, the examination preliminaries must be performed in the following sequence: the patient begins in the erect position, is given a gas-producing substance, followed by a small quantity of high-density barium. The small amount of barium coats the gastric mucosa and the air distends the stomach, making it possible to virtually "see through" the stomach (which would be impossible if the stomach was distended with opaque barium). *(Ballinger, vol 2, p 100)*

94. **(C)** An intrathecal injection is one made within the meninges. A myelogram requires an intrathecal injection in order to introduce contrast medium into the subarachnoid space. An IVP requires an intravenous injection; a retrograde pyelogram requires that contrast be introduced by way of cystoscopy, into the ureters. An arthrogram requires that contrast medium be introduced into a joint space. *(Saia, PREP, pp 32, 40)*

95. **(A)** The 45° oblique position of the lumbar spine is generally performed for demonstration of the apophyseal joints. In a correctly positioned oblique lumbar spine, "scotty dog" images are demonstrated. The scotty's ear corresponds to the superior articular process, his nose to the transverse process, his eye is the pedicle, his neck is the pars interarticularis, his body is the lamina, and his front foot is the inferior articular process (see Fig. 2–49). *(Saia, PREP, p 119)*

96. **(D)** The illustration shows the patient positioned on his left side, with the cassette behind his back. This is a left lateral decubitus

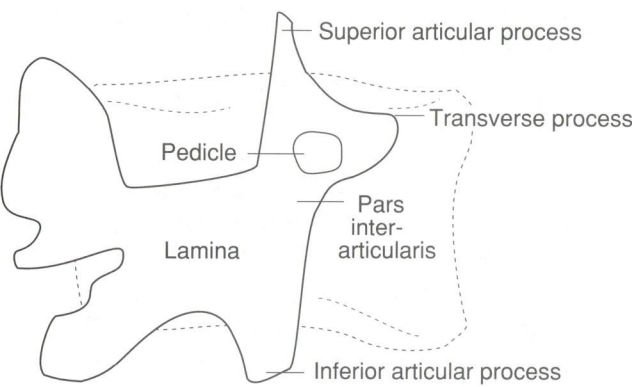

Figure 2–49.

position. The x-ray beam is directed horizontally in decubitus positions. Air or fluid levels will be clearly delineated only if the central ray is directed parallel to them. Therefore, to demonstrate air or fluid levels, erect or decubitus positions should be used. Small amounts of *fluid* within the pleural space are best demonstrated in the lateral decubitus position, *affected side down*. Small amounts of *air* within the pleural space are best demonstrated in the lateral decubitus position, *affected side up*. *(Ballinger, vol 1, p 476)*

97. **(A)** The medial oblique projection requires that the leg be rotated medially until the plantar surface of the foot forms a 30° angle with the cassette. This position demonstrates the navicular with minimal bony superimposition. The lateral oblique of the foot superimposes much of the navicular on the cuboid. The navicular is also superimposed on the cuboid in the lateral projections. *(Ballinger, vol 1, p 200)*

98. **(D)** A patient will feel more comfortable and confident if brought into a clean, orderly x-ray room that has been prepared appropriately for the examination to be performed. A disorderly, untidy room and a disorganized radiographer hardly inspire confidence. The radiographer should always check to see if the patient has had the same exam previously. It is useful to know if the patient has had previous ulcers or other conditions documented. In other cases it is useful to know if

you are looking for a healing fracture, the location of the fracture, any modifications of exposure that may be required, and so forth. *(Ballinger, vol 1, p 9)*

99. **(D)** A trimalleolar fracture involves three separate fractures. The lateral malleolus is fractured in the "typical" fashion, but the medial malleolus is fractured on both its medial and posterior aspects. The trimalleolar fracture is frequently associated with subluxation of the articular surfaces. *(Laudicina, p 184)*

100. **(C)** and 101. **(B)** Figure 2–16 illustrates the biliary system. Bile leaves the liver through the right and left hepatic ducts (no. 5) which join to form the common hepatic duct (no. 6). Bile enters the gallbladder through the cystic duct (7). The neck of the gallbladder is indicated by no. 4, its body is 3, and its fundus is 2. The gallbladder stores and concentrates bile, and when it contracts, bile flows out through the cystic duct and down the common bile duct (8). The common bile duct and pancreatic duct (9) unite to form the short hepatopancreatic ampulla (of Vater), which empties into the duodenum (1). *(Martini, p 913)*

102. **(B)** The fingers must be supported parallel to the film (for example, on a "finger sponge") in order that the joint spaces parallel the x-ray beam. When the fingers are flexed, resting on the cassette, the relationship between the joint spaces and film changes, and the joints appear "closed." *(Ballinger, vol 1, pp 76–77)*

103. **(C)** A patient having a GI series is required to be npo for at least 8 hours prior to the exam; food or drink in the stomach can simulate disease. A patient scheduled for a pyelogram must have the preceding meal withheld so as to avoid the possibility of aspirating vomitus in case of allergic reaction. An abdominal survey does not require the use of contrast media and no patient preparation is required. *(Saia, PREP, p 13)*

104. **(C)** Perhaps the most important prerequisite to a successful barium enema exam is a thoroughly clean large bowel. Any retained fecal material can simulate pathology. A single-

contrast examination demonstrates anatomy and contour of the large bowel, as well as anything that may project out from the bowel wall (eg, diverticula). In a double-contrast exam, the bowel wall is coated with barium and then the lumen filled with air. This enables visualization of any intraluminal lesions such as polyps and tumor masses. *(Ballinger, vol 2, pp 122–123)*

105. **(D)** To better visualize the joint space in the lateral projection of the knee, 20 to 30° flexion is recommended. The femoral condyles are superimposed so as to demonstrate the patellofemoral joint and the articulation between the femur and tibia. The correct degree of forward or backward body rotation is responsible for visualization of the patellofemoral joint. Cephalad tube angulation of 5 to 7° is responsible for demonstrating the articulation between the femur and tibia (by removing the magnified medial femoral condyle from superimposition on the joint space). *(Saia, PREP, p 102)*

106. **(A)** When the shoulders are relaxed, the clavicles are usually carried below the pulmonary apices. In order to examine the portions of the lungs lying behind the clavicles, the central ray is directed cephalad 15 to 20° to project the clavicles above the apices, when the patient is examined in the AP position. *(Ballinger, vol 1, p 472)*

107. **(C)** In the parieto-orbital projection the patient is PA with the acanthomeatal line perpendicular to the film. The head rests on the forehead, nose, and chin, and the midsagittal plane should form a 53° angle with the film (37° with the central ray). Radiographically, the optic canal should appear in the lower outer quadrant of the orbit. Incorrect rotation of the MSP results in lateral displacement and incorrect positioning of the baseline results in longitudinal displacement. *(Ballinger, vol 2, p 272)*

108. **(D)** Double-contrast studies of the stomach or large intestine involve coating the organ with a thin layer of barium sulfate, then introducing air. This permits seeing through

the organ to structures behind it and, most especially, allows visualization of the mucosal lining of the organ. Barium-filled stomach or large bowel demonstrates position, size, and shape of the organ and any lesion that projects *out* from its walls, such as diverticula. Polypoid lesions, which project *inward* from the wall of an organ, may go unnoticed unless a double-contrast exam is performed. *(Ehrlich & McCloskey, p 186)*

109. **(B)** The internal rotation position places the humeral epicondyles perpendicular to the film, the humerus in a true lateral position, and the lesser tubercle in profile. The external rotation position places the humeral epicondyles parallel to the film, the shoulder in a true AP position, and the greater tubercle in profile. The neutral position is often used for evaluation of calcium deposits in the shoulder joint. *(Ballinger, vol 1, p 129)*

110. **(A)** An RPO position is illustrated. The oblique projections in a barium enema are used to "open up" the flexures and adjacent colon. The RAO and LPO positions demonstrate the hepatic flexure and adjacent ascending colon. The LAO and RPO positions demonstrate the splenic flexure and descending colon. *(Ballinger, vol 2, p 141)*

111. **(C)** The PA axial projection (Camp-Coventry method) of the intercondyloid fossa ("tunnel view") is pictured. The knee is flexed about 40°, the CR directed caudally 40° and perpendicular to the tibia (Fig. 2–50). The patella and patellofemoral articulation are demonstrated in the axial/tangential view of the patella. *(Ballinger, vol 1, p 252)*

112. **(C)** In the tangential ("sunrise") projection of the patella, the central ray is directed parallel to the longitudinal plane of the patella, thereby demonstrating a vertical fracture and providing the best view of the patellofemoral articulation. The AP knee could demonstrate a vertical fracture through the superimposed femur but does not demonstrate the patellofemoral articulation. The "tunnel" view of the knee is used to demonstrate the intercondyloid fossa. *(Ballinger, vol 1, p 264)*

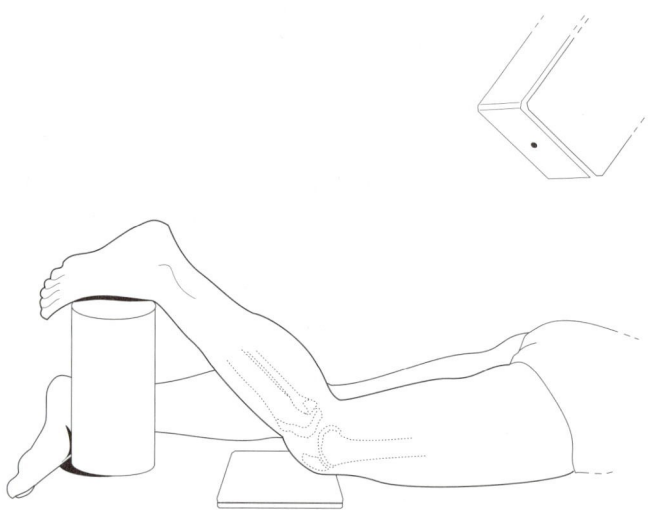

Figure 2–50.

113. **(B)** The parietoacanthial projection (Water's method) demonstrates a distorted view of the frontal and ethmoid sinuses. The maxillary sinuses (no. 4) are well demonstrated, projected free of the petrous pyramids. This is also the best single position for the demonstration of facial bones. The mandibular angle is illustrated by the *1*; the zygomatic arch is *2*, and the coronoid process is *3*. *(Ballinger, vol 2, p 305)*

114. **(D)** Barium sulfate suspension is the usual contrast medium of choice for investigation of the alimentary tract. There are, however, a few exceptions. Whenever there is the possibility of escape of contrast medium into the peritoneal cavity, barium sulfate is contraindicated and water-soluble iodinated medium is recommended, as it is easily aspirated before surgery. Rupture of a hollow viscus (eg, perforated ulcer) and patients who are scheduled for surgery are two examples. Patients suspected of large bowel obstruction should also ingest only water-soluble iodinated media. *(Bontrager, p 402)*

115. **(D)** The axial projection of the nasal bones is valuable for demonstrating medial or lateral displacement. Ideally, the film is placed into the mouth and the central ray directed horizontally, perpendicular to the film. Only that portion of the nasal bone that projects ante-

rior to a line between the glabella and alveolar ridge will be demonstrated. *(Ballinger, vol 2, p 317)*

116. **(D)** Four views of the elbow are shown. Position *1* is the AP, which demonstrates partial superimposition of the proximal radius and ulna. The lateral oblique position, *2*, demonstrates the proximal radius free of superimposition with the proximal ulna. The medial oblique position of the elbow, *3*, demonstrates the ulnar coronoid process free of superimposition and the olecranon process within the olecranon fossa of the humerus. Position *4* is the lateral position, elbow flexed 90° and epicondyles superimposed, providing best visualization of the olecranon process. *(Dowd, vol 1, p 161)*

117. **(D) and 118. (A)** Four positions for the lumbar spine are illustrated. Figure 1 is an RPO and figure 2 is an LAO. The *posterior oblique* positions (LPO, RPO) demonstrate the apophyseal joints *closer* to the film, while the *anterior oblique* positions (LAO, RAO) demonstrate the apophyseal joints *further* from the film. Figure 3 is the AP projection, which demonstrates the lumbar bodies and disc spaces, transverse and spinous processes. Figure 4 is the lateral position (Fig. 2–51), providing best demonstration of the lumbar bodies, intervertebral disc spaces, spinous processes, pedicles, and intervertebral foramina. *(Ballinger, vol 1, pp 364, 369, 372)*

118 (A) No Answer

119. **(C)** The thoracic apophyseal joints are demonstrated by placing the patient in an oblique position with the coronal plane 70° to the film (MSP 20° to the film). This may be accomplished by first placing the patient lateral, then obliquing the patient 20° "off lateral." The apophyseal joints closest to the film are demonstrated in the PA oblique and those remote from the film in the AP oblique. Comparable detail is obtained using either method, because OID is about the same. *(Ballinger, vol 1, p 327)*

120. **(D)** The scapular Y projection is an oblique projection of the shoulder and is used to demonstrate anterior or posterior shoulder dis-

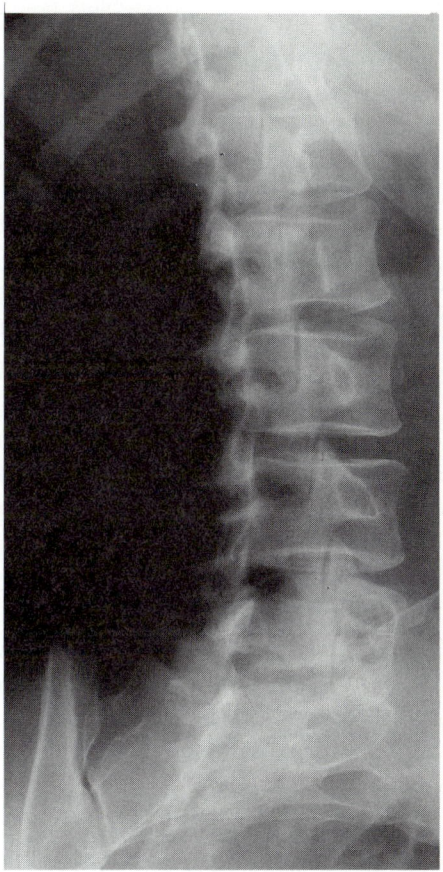

Figure 2–51. Courtesy of The Stamford Hospital, Department of Radiology.

location. The inferosuperior axial projection may be used to evaluate the glenohumeral joint when the patient is able to abduct the arm. The transthoracic lateral projection is used to evaluate the glenohumeral joint and upper humerus when the patient is unable to abduct the arm. *(Ballinger, vol 1, p 142)*

121. **(B)** Evaluation of the acromioclavicular joints requires bilateral AP or PA erect projections with and without the use of weights. Weights are used to emphasize the minute changes within a joint caused by separation or dislocation. Weights should be anchored from the patient's wrists rather than held in the patient's hands, as this encourages tightening of the shoulder muscles and obliteration of any small separation. *(Ballinger, vol 1, p 152)*

122. **(A)** Breast tissue is the most dense, glandular, and radiographically homogeneous in appearance in the postpubertal adolescent. Following pregnancy and lactation, changes occur within the breast that reduce the glandular tissue and replace it with fatty tissue (a process called fatty infiltration). Menopause causes further atrophy of glandular tissue. *(Ballinger, vol 2, p 461)*

123. **(C)** A person lying down in any position is said to be *recumbent*. The patient lying face down is said to be *prone* or *ventral recumbent*. Lying on the back, the patient is said to be *supine* or *dorsal recumbent*. The *decubitus* position refers to the use of a horizontal x-ray beam with respect to the patient's position. *(Ballinger, vol 1, p 52)*

124. **(D)** IVP films should indicate amount of time elapsed postinjection. Tomography films should indicate the fulcrum level. Abdominal survey films should be marked according to body position (such as erect or decubitus). *(Ballinger, vol 1, p 16)*

125. **(B)** To reduce anxiety prior to the examination, the radiographer should give the patient a full explanation of the enema procedure. This explanation should include keeping the anal sphincter tightly contracted, relaxing abdominal muscles, and deep breathing. The barium suspension should be either just below body temperature (at 85 to 90°F) to prevent injury and bowel irritation *or* cold (at 41°F) to produce less colonic irritation and to stimulate contraction of the anal sphincter. *(Ballinger, vol 2, p 124)*

126. **(D)** Surface landmarks, prominences, and depressions are very useful to the radiographer in locating anatomic structures not visible externally. The costal margin is about the same level as L3. The umbilicus is the same approximate level as the L3 to L4 interspace. The xiphoid tip is about at the same level as T10. The fourth lumbar vertebra is at the same approximate level as the iliac crest. *(Saia, PREP, p 64)*

127. **(C)** The straight PA (0°) projection effectively demonstrates the mandibular body, but the rami and condyles are superimposed on the

occipital bone and petrous portion of the temporal bone. To better visualize the rami and condyles, the central ray is directed cephalad 20 to 30°. This projects the temporal and occipital bones above the area of interest. *(Ballinger, vol 2, pp 343–345)*

128. **(D)** The term *proximal* refers to structures closer to the point of attachment. For example, the elbow is described as being *proximal* to the wrist, that is, the elbow is closer to the point of attachment (the shoulder) than the wrist is. Referring to the question, then, the interphalangeal joints (both proximal and distal) and the metacarpals are both *distal* to the carpal bones. The radial styloid process is *proximal* to the carpals. *(Saia, PREP, p 75)*

129. **(B)** The scapular Y projection requires that the coronal plane be about 60° to the film, thus resulting in an oblique projection of the shoulder. The vertebral and axillary borders of the scapula are superimposed on the humeral shaft and the resulting relationship between the glenoid fossa and humeral head will demonstrate anterior or posterior dislocation. Lateral or medial dislocation is evaluated on the AP projection. *(Ballinger, vol 1, pp 142–143)*

130. **(D)** If structures are overlying or underlying the area to be demonstrated (as with structures overlying the occipital bone in the AP skull), central ray angulation is employed (as in AP axial skull to visualize occipital bone). If structures would be foreshortened or self-superimposed, as in the navicular, central ray angulation may be employed to place the structure more closely parallel with the film. Some articulations, such as the knee, require angulation to better visualize the joint space. *(Ballinger, vol 1, p 17)*

131. **(A)** The fifth metacarpal is located on the *medial* aspect of the hand. Remember to always view a part in its *anatomical position*. With the arm in the anatomical position, the fifth metacarpal and ulna lie medially. *(Saia, PREP, pp 74–75)*

132. **(B)** In order to demonstrate the joint space to best advantage, the plantar surface of the foot

should be vertical in the AP projection of the ankle. Note that the fibula is the more distal of the two long bones of the lower leg, and forms the lateral malleolus. The calcaneus is not well visualized in this projection because of superimposition with other tarsals. *(Saia, PREP, p 100)*

133. (B) The kidneys lie obliquely in the posterior portion of the trunk, with their superior portion angled posteriorly and their inferior portion and ureters angled anteriorly. Therefore, in order to facilitate filling of the most anteriorly placed structures, the patient is examined in the prone position. Opacified urine then flows to the most dependent part of the kidney and ureter—the ureteropelvic region, inferior calyces, and ureters. *(Saia, PREP, p 170)*

134. (B) The *median sagittal,* or *midsagittal* plane (MSP), passes vertically through the midline of the body, dividing it into left and right halves. Any plane parallel to the MSP is termed a *sagittal* plane. The *midcoronal* plane is perpendicular to the MSP and divides the body into anterior and posterior halves. A *transverse* plane passes through the body at right angles to a sagittal plane. These planes, especially the MSP, are very important reference points in radiographic positioning (Fig. 2–52). *(Saia, PREP, p 61)*

135. (A) In the AP projection of the shoulder, there is superimposition of the humeral head and glenoid fossa. With the patient obliqued 45° toward the affected side, the glenohumeral joint is open and the glenoid fossa is seen in profile. The patient's arm is abducted somewhat and placed in internal rotation. *(Ballinger, vol 1, p 144)*

136. (B) The radiograph illustrated shows the odontoid process superimposed on the base of the skull. The maxillary teeth can be seen significantly superior to the base of the skull. A diagnostic image of C1–C2 depends upon adjusting the flexion of the neck *so that the maxillary occlusal plane and base of the skull are superimposed* (see dotted lines in Fig. 2–53). Accurate adjustment of these structures will usually allow good visualization of the odon-

toid process and the atlantoaxial articulation. Too much flexion superimposes teeth on the odontoid process; too much extension superimposes the base of the skull on the odontoid process. *(Saia, PREP, p 115)*

137. (D) Standard radiographic protocols may be reduced to include two views, at right angles to each other, in emergency and trauma radiography. Department policy and procedure manuals include protocols for radiographic examinations. In the best interest of the patient, and in order for the radiologist to make an accurate diagnosis, standard radiographic protocols should be followed. If the radiographer must deviate from the protocol, or believes that additional projections might be helpful, then this should be discussed with the radiologist. Emergency and trauma radiography is occasionally an exception to this rule. If the emergency room physician's request varies from the department protocol, the radiographer must respect this. A note should be added to the request so that the radiologist is informed of the reason for a change in protocol. For example, a patient that has been involved in a motor vehicle accident may need many radiographic studies, but the emergency room physician may order an AP chest and an AP and cross-table lateral C-spine only . Standard protocol may include a lateral chest and a cone down view of the atlas and axis as well as cervical oblique views. The emergency room physician has made a decision based on experience and expertise that overrules standard protocols. At a later time, when the patient has been stabilized, the patient may be sent back to radiology for additional views. *(Dowd & Wilson, vol 2, pp 1056–1057)*

138. (C) After forceful eversion or inversion injuries of the ankle, *AP stress studies* are valuable to confirm the presence of a ligament tear. Keeping the ankle in an AP position, the physician guides the ankle into inversion and eversion maneuvers. Characteristic changes in the relationship of the talus, tibia, and fibula will indicate ligament injury. Inversion stress demonstrates the lateral ligament, while eversion stress demonstrates the me-

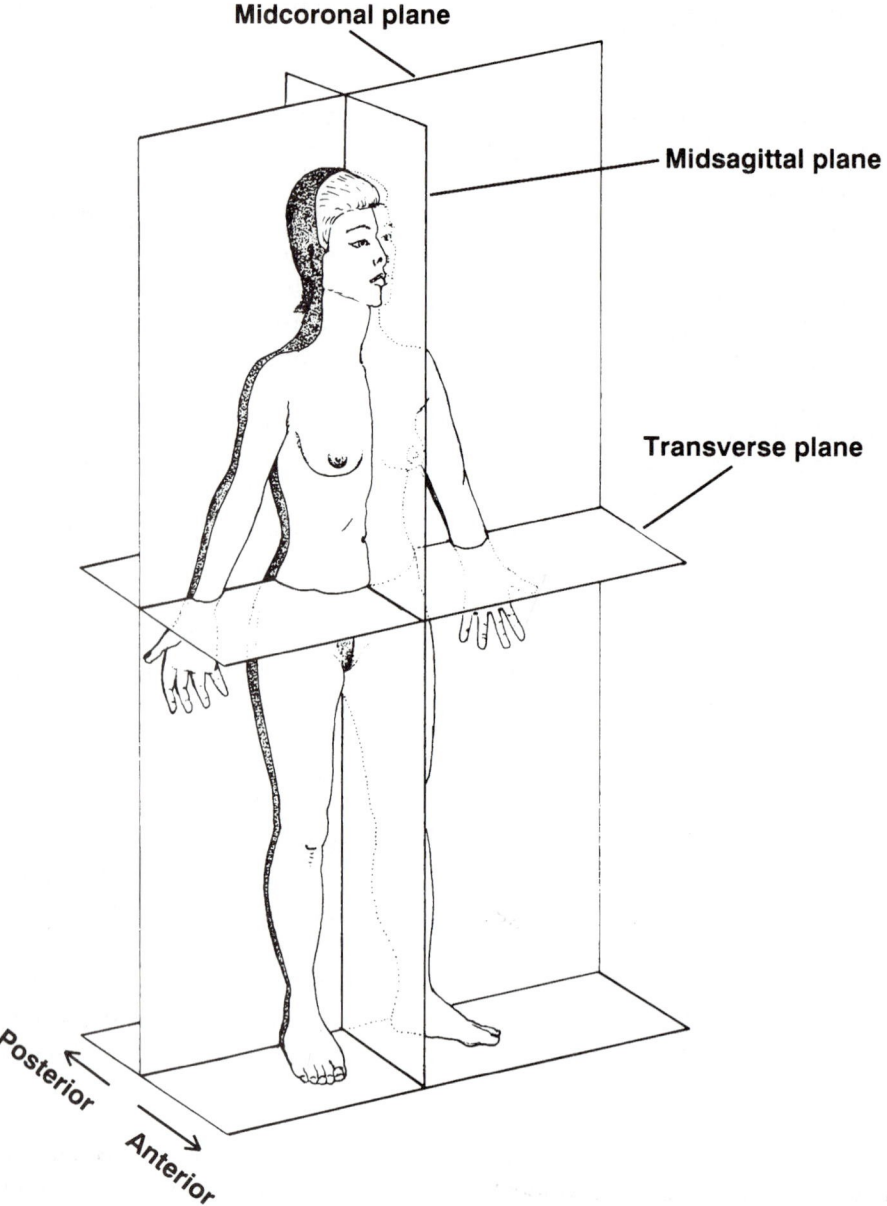

Midcoronal plane

Midsagittal plane

Transverse plane

Posterior ←

Anterior →

Figure 2–52 Reproduced with permission from Saia DA. *Radiography: Program Review and Exam Preparation*. Stamford, CT: Appleton & Lange, 1996.

dial ligament. A fractured ankle would not be manipulated in this manner. *(Ballinger, vol 1, p 233)*

139. **(C)** In the dorsoplantar projection of the foot, the central ray may be directed perpendicularly or angled 10° posteriorly. Angulation serves to "open" the tarsometatarsal joints not well visualized on the dorsoplantar projection with perpendicular ray. Inversion and eversion of the foot do not affect the tarsometatarsal joints. *(Ballinger, vol 1, p 198)*

140. **(B)** The central cavity of a long bone is the medullary canal. It contains yellow bone marrow, the most abundant type of marrow in the body. Red marrow is found within the cancellous tissue forming the extremities of long bones. *(Bontrager, p 8)*

141. **(A)** The routine AP projection of the lumbar spine demonstrates the intervertebral disc spaces between the first four lumbar vertebrae. The space between L5 to S1, however, is angled with respect to the other disc spaces.

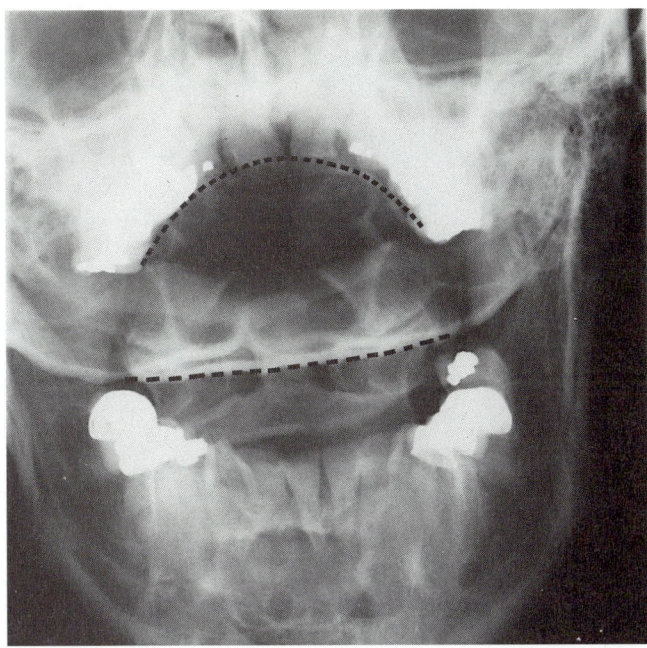

Figure 2–53. Courtesy of The Stamford Hospital, Department of Radiology.

Therefore, the central ray must be directed 30 to 35° cephalad to parallel the disc space, and thus project it open onto the film. (*Ballinger, vol 1, p 378*)

142. **(B)** The most prominent part of the greater trochanter is at the same level as the pubic symphysis—both are valuable positioning landmarks. The ASIS is in the same transverse plane as S2. The ASIS and pubic symphysis are the bony landmarks used to locate the hip joint, which is located midway between the two points. (*Ballinger, vol 1, p 245*)

143. **(C)** The gallbladder is located on the posterior surface of the liver in the RUQ. The hepatic flexure, named because of its close proximity to the liver, is also in the RUQ. The vermiform appendix projects from the first portion of the large bowel, the cecum, located in the RLQ. (*Saia, PREP, p 63*)

144. **(D)** Acromioclavicular joints are usually examined when separation or dislocation is suspected. They must be examined in the erect position, because when recumbent, a separation appears to reduce itself. Both AC

joints are examined simultaneously, for comparison because separations may be very minimal. (*Ballinger, vol 1, p 152*)

145. **(B)** The radiograph shown in Figure 2–53 is a *PA recumbent* projection. If the patient was *AP*, barium would be located in the fundus of the stomach (because the fundus is more posterior, and barium would flow down to fill the posterior structure). If the patient was in the *Trendelenburg* position, barium flow to the fundus would be even more facilitated. If the patient was *erect*, air–fluid levels would be clearly defined. Additionally, the barium-filled stomach tends to spread more horizontally in the PA position (as is seen in the radiograph). (*Ballinger, vol 2, p 102*)

146. **(D)** The *lordotic* curves are *secondary* curves, that is, they develop sometime after birth. The cervical and lumbar vertebrae form lordotic curves. The *thoracic* and *sacral* vertebrae exhibit the *primary kyphotic* curves, those which are present at birth. (*Saia, PREP, p 111*)

147. **(B)** In the lateral position of the skull, the midsagittal plane must be parallel to the film and the interpupillary line vertical. Flexion of the head is adjusted until the IOML is parallel to the film. The central ray should enter about 2 inches superior to the EAM. (*Ballinger, vol 2, pp 238–239*)

148. **(B)** The first metacarpal, on the lateral side of the hand, articulates with the most lateral carpal of the distal carpal row, the greater multangular. This articulation forms a rather unique and very versatile saddle joint, named for the shape of its articulating surfaces. (*Saia, PREP, p 75*)

149. **(A)** The position of the gallbladder varies with the body habitus of the patient. Hypersthenic patients are more likely to have their gallbladder located high and lateral. The asthenic patient's gallbladder is most likely to occupy a low and medial position, occasionally superimposed on the vertebrae or iliac fossa. The LAO position is most often used to move the gallbladder away from the spine. The erect position would make the gallblad-

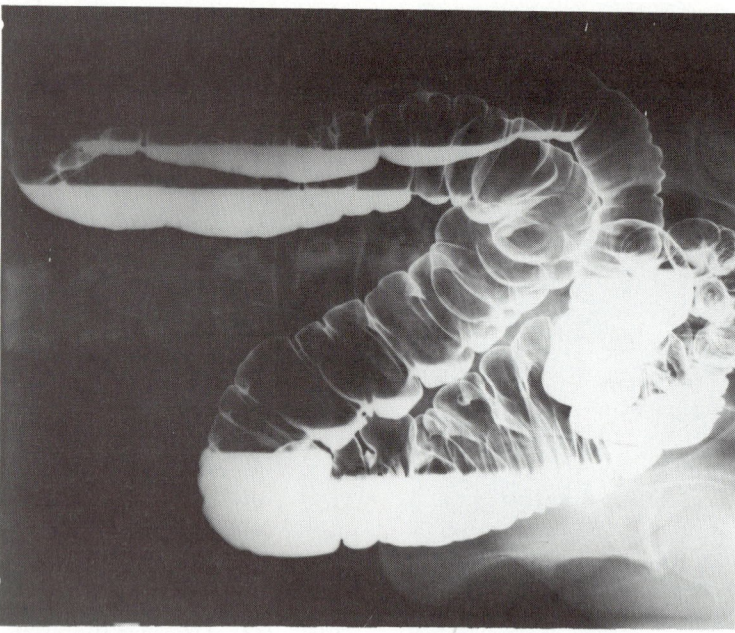

Figure 2–54. Courtesy of The Stamford Hospital, Department of Radiology.

der move even more inferior and medial. *(Ballinger, vol 2, p 62)*

150. **(A)** The abdomen is divided into nine regions. The upper lateral regions are the left and right hypochondriac, with the epigastric separating them. The middle lateral regions are the left and right lumbar, with the umbilical region between them. The lower lateral regions are the left and right iliac, with the hypogastric region between them. The ileocecal valve, cecum, and appendix (if present) are located in the lower right abdomen—therefore the right iliac region. *(Saia, PREP, p 63)*

151. **(C)** When the clavicle is examined in the AP recumbent position, the central ray must be directed 25 to 30° cephalad in order to project most of the clavicle's length above the ribs. The direction of the CR is reversed when examining the patient in the prone position. *(Ballinger, vol 1, p 159)*

152. **(C)** When performing decubitus positions of the double-contrast barium enema, the heavy barium moves to the side down, while the air rises to the side up. Therefore the *right* lateral

decubitus will best demonstrate the *up* or *left-sided* walls of the ascending and descending colon, that is, the lateral side of the descending colon and medial side of the ascending colon (Fig. 2–54). The *left* lateral decubitus, then, would demonstrate the lateral side of the ascending colon and medial side of the descending colon. *(Ballinger, vol 2, pp 143–144)*

153. **(B)** The ankle mortise, or ankle joint, is formed by the articulation of the tibia, fibula, and talus (Fig. 2–55). Two articulations form the ankle mortise, the talotibial and talofibular. The calcaneus is not associated with the formation of the ankle mortise. *(Saia, PREP, pp 100–101)*

154. **(B)** The carpal navicular (scaphoid) is somewhat curved, and consequently is foreshortened radiographically in the PA position. In order to better separate it from the adjacent carpals, the ulnar flexion maneuver is frequently employed. In addition to correcting foreshortening of the navicular, ulnar flexion opens the interspaces between adjacent lateral carpals. Radial flexion is used to better demonstrate medial carpals. *(Ballinger, vol 1, p 89)*

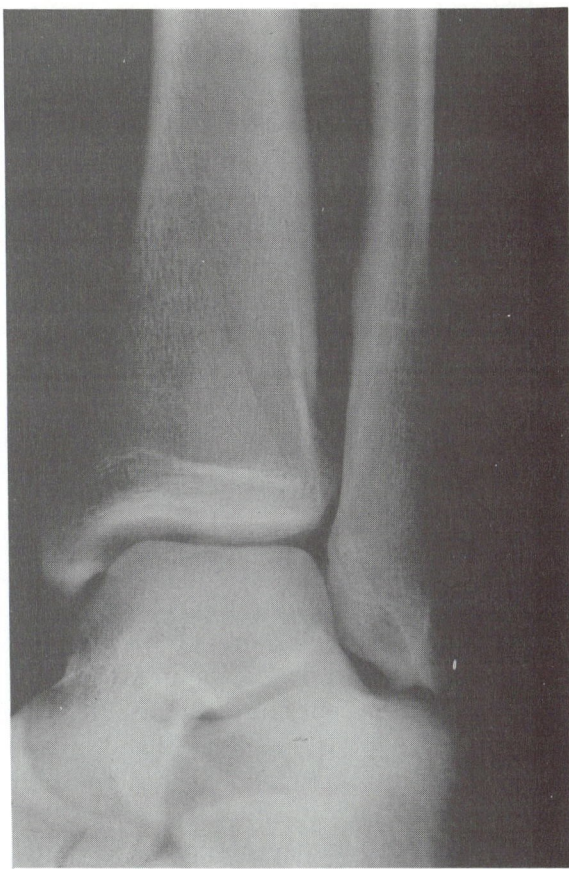

Figure 2–55. Courtesy of The Stamford Hospital, Department of Radiology.

155. (A) The radiograph in Figure 2–24 illustrates a *lateral* projection of the *scapula*. The axillary and vertebral borders are superimposed. The *acromion* and *coracoid* processes are visualized; the coracoid process is partially superimposed on the axillary portion of the third rib. A *scapular Y* projection is often performed to demonstrate shoulder dislocation, but the affected arm is left to rest at the patient's side; the arm in the illustrated radiograph is abducted somewhat to better view the body of the scapula. *(Ballinger, vol 1, p 166)*

156. (B) Surface landmarks, prominences, and depressions are very useful to the radiographer in locating anatomic structures not visible externally. The fifth thoracic vertebra is at the same approximate level as the sternal angle. The T2 to T3 interspace is about at the same

level as the manubrial (suprasternal) notch. The costal margin is about the same level as L3. *(Saia, PREP, p 64)*

157. (D) In order to demonstrate the radius and ulna free of superimposition, the forearm must be radiographed in the AP position, hand supinated. Pronation of the hand causes overlapping of the proximal radius and ulna. Two views, at right angles to each other, are generally required for each examination. Therefore, an AP and lateral is the usual routine for an exam of the forearm. *(Ballinger, vol 1, p 98)*

158. (B) The typical vertebra is divided into two portions, *body* (anteriorly) and *vertebral arch* (posteriorly). The vertebral arch supports seven processes: two transverse, one spinous (no. *3*), two superior articular processes (*4*), and two inferior articular processes (*2*). Superior articular processes and the superjacent inferior articular processes join to form apophyseal joints. Pedicles (*5*) project posteriorly from the vertebral body (*6*). Their upper and lower surfaces form vertebral notches. Superjacent vertebral notches form intervertebral foramina. The *lamina* is number 1. Transverse and spinous processes serve as attachment for muscles, or articulation for ribs in the thoracic region. The superior and inferior surfaces of the vertebral body are covered with articular cartilage, and between the vertebral bodies lie the intervertebral discs. *(Saia, PREP, p 117)*

159. (C) In the PA projection of the chest, there should be no rotation as evidenced by *symmetrical* sternoclavicular joints. The shoulders are rolled forward to remove the scapulae from the lung fields. Inspiration should be adequate to demonstrate 10 *posterior* ribs above the diaphragm (the anterior ribs angle downward; the 10th anterior rib is the last attached to the sternum and very unlikely to be imaged on inspiration). The sternum should be seen lateral without rotation in the *lateral* position of the chest. *(Ballinger, vol 1, pp 404, 455)*

160. (A) Because the right main bronchus is wider and more vertical, aspirated foreign bodies are more likely to enter it than the left main bronchus, which is narrower and angles more sharply from the trachea. An aspirated foreign body does not enter the esophagus or stomach, as they are not respiratory structures, but rather, digestive. *(Martini, p 834)*

161. (C) In the lateral projection of the ankle, the tibia and fibula are superimposed and the foot is somewhat dorsiflexed to better demonstrate the talotibial joint. The talofibular joint is not visualized because of superimposition with other bony structures. It may be well visualized in the medial oblique projection of the ankle. *(Ballinger, vol 1, p 230)*

162. (B) A double-contrast exam of the large bowel is performed in order to see through the bowel to its posterior wall and to visualize any intraluminal lesions or masses. Oblique projections are used to "open up" the flexures: the RAO for the hepatic flexure and the LAO for the splenic. In order to view the redundant S-shaped sigmoid in the AP position, the central ray is directed 30 to 40° cephalad. The central ray is reversed when the patient is PA, that is, CR is 30 to 40° caudal with the patient PA. *(Ballinger, vol 2, pp 134, 139)*

163. (B) Air or fluid levels will be clearly delineated only if the central ray is directed parallel to them. Therefore, to demonstrate air or fluid levels, erect or decubitus positions should be used. Small amounts of *fluid* within the pleural space are best demonstrated in the lateral decubitus position, *affected side down*. Small amounts of air within the pleural space are best demonstrated in the lateral decubitus position, *affected side up*. *(Ballinger, vol 1, p 476)*

164. (A) Because sinus exams are performed to evaluate the presence or absence of fluid, they must be performed in the *erect position with a horizontal x-ray beam*. The PA axial (Caldwell) projection demonstrates the frontal and ethmoid sinus groups, and the parietoacanthial projection (Water's) shows the maxillary sinuses. The lateral position demonstrates all the sinus groups and the SMV is frequently used to demonstrate the sphenoid sinuses. *(Ballinger, vol 2, p 376)*

165. (B) Statement 1 describes the PA axial (Camp-Coventry) projection, and statement 2 the AP axial (Beclere) projection, for demonstration of the intercondyloid fossa. The positions are actually the reverse of each other. Statement 3 describes the method of obtaining a PA projection of the patella. *(Ballinger, vol 1, pp 226–228)*

166. (A) The visualization of the scapular spine indicates that this is a view of the *posterior* aspect of the scapula. The scapula's *anterior*, or *costal*, surface is that which is adjacent to the ribs. The scapula has no sternal articulation. *(Saia, PREP, pp 77, 79, 81)*

167. (C) and 168. (A) Figure 2–27 depicts a posterior view of the right scapula and its articulation with the humerus (*4*). The scapula presents two borders, the lateral or axillary border (*7*) and the medial or vertebral border (*9*). It also presents three angles, the inferior angle (*8*), superior angle (*12*), and lateral angle (*6*). The processes of the scapula are the coracoid (*2*), acromion (*3*), and scapular spine (*13*). The scapula has a (supra)scapular notch (*1*), a supraspinatus fossa (*11*), and an infraspinatus fossa (*10*). *(Saia, PREP, p 81)*

169. (C) In the PA position, portions of the barium-filled hypersthenic stomach superimpose upon themselves. Thus, patients with a hypersthenic body habitus usually present a high transverse stomach, with poorly defined curvatures. If the PA stomach is projected with a 35 to 45° cephalad CR, the stomach "opens up." That is, the curvatures, the antral portion, and duodenal bulb all appear as a sthenic habitus stomach would appear. A 35 to 40° RAO position is used to demonstrate many of these structures in the average, or sthenic body habitus. A lateral position is used to demonstrate the anterior and posterior gastric surfaces and retrogastric space. *(Ballinger, vol 2, p 104)*

170. (B) An axial *dorsoplantar* projection is described; the central ray enters the dorsal surface of the foot and exits the plantar surface. The *plantodorsal* projection is done *supine* and requires cephalad angulation. The central ray enters the plantar surface and exits the dorsal surface. *(Ballinger, vol 1, pp 215–216)*

171. **(B)** For the lateral projection of the knee the patient is turned onto the affected side. This places the lateral femoral condyle closest to the film and the medial femoral condyle remote from the film. Consequently, there is significant magnification of the medial femoral condyle and, unless the central ray is angled slightly cephalad, subsequent obliteration of the joint space. *(Ballinger, vol 1, p 243)*

172. **(A)** The four types of body habitus describe differences in visceral shape, position, tone, and motility. One body type is *hypersthenic*, characterized by the very large individual with short, wide heart and lungs; high transverse stomach and gallbladder; and peripheral colon. The *asthenic individual* is the average, athletic, most predominant type. The *hyposthenic* patient is somewhat thinner and a little more frail, with organs positioned somewhat lower. The *asthenic* type is smaller in the extreme, with a long thorax; a very long, almost pelvic stomach; and low medial gallbladder. The colon is medial and redundant. Hypersthenic patients usually demonstrate the greatest motility. *(Ballinger, vol 1, p 42)*

173. **(C)** The stomach and spleen are both normally located in the LUQ. The cecum is the most distal end of the large bowel and is normally located in the RLQ. *(Ballinger, vol 1, p 39)*

174. **(B)** Superimposition of bony details frquently makes angiographic demonstration of blood vessels less than optimal. The method used to remove these superimposed bony details is called *subtraction*. *Digital subtraction angiography (DSA)* accomplishes this through the use of a computer. The advantages of DSA over film angiography include greater sensitivity to contrast medium, immediate availability of images, and lower total cost. Although DSA applications are increasing, film angiography is still preferred in cases where resolution is critical. *(Ballinger, vol 3, p 178)*

175. **(C)** The parietoacanthial (Water's) projection demonstrates the maxillary sinuses. The PA axial with a caudal CR (Caldwell) demonstrates the frontal and ethmoid sinus groups.

The lateral projection, with the central ray entering one inch posterior to the outer canthus, demonstrates all the paranasal sinuses. X-ray examinations of the sinuses should always be performed erect, to demonstrate leveling of any fluid present. *(Ballinger, vol 2, p 378)*

176. **(C)** A lateral projection of the scapula superimposes its medial and lateral borders (vertebral and axillary, respectively). The coracoid and acromion processes should be readily identified separately (not superimposed) in the lateral projection. The erect position is probably the most comfortable position for a patient with scapula pain. *(Ballinger, vol 1, p 166)*

177. **(A)** The pictured radiograph is an *RAO* position of the sternum. The sternum is projected to the *left* side of the thorax, over the heart and other mediastinal structures, in the RAO position, thus promoting more uniform density. Although the upper limits of the sternum are well demonstrated in the figure, not all of the xiphoid process is seen, because the CR was directed somewhat too inferiorly. The *CR* should be directed midway between the jugular (manubrial) notch and xiphoid process. *(Ballinger, vol 1, pp 407–408)*

178. **(C)** In the lateral projection of the knee, the joint space is obscured by the magnified medial femoral condyle unless the central ray is angled 5 to 7° cephalad. The degree of flexion of the knee is important when evaluating the knee for possible transverse patellar fracture. In such a case the knee should not be flexed more than 10°. The knee should normally be flexed 20 to 30° in the lateral position. *(Ballinger, vol 1, p 243)*

179. **(B)** Plaque, deposited on arterial walls in cases of atherosclerosis, causes arterial stenosis. Percutaneous transluminal angioplasty (PTA) is a procedure that uses a balloon catheter to permanently increase the size of the arterial lumen, thus reopening the vessel and restoring blood flow. A percutaneous nephrolithotomy is a procedure performed to remove a renal calculus from a kidney or proximal ureter. Renal arteriography is the radiologic investigation of the renal arteries. Nephrostomy is the surgical formation of an artificial opening into the kidney. *(Ballinger, vol 2, p 561)*

180. **(C)** Long bones are composed of a shaft, or diaphysis, and two extremities, or epiphyses. In the growing bone, the cartilaginous epiphyseal plate is gradually replaced by bone. The ossified growth area of long bones is the metaphysis. Apophysis refers to vertebral joints formed by articulation of superjacent articular facets. *(Martini, pp 185–186)*

181. **(D)** The PA axial (Caldwell) projection of the paranasal sinuses is used to demonstrate the frontal and ethmoid sinuses. The central ray is angled caudally 15° to the OML. This projects the petrous pyramids into the lower one third of the orbits, thus permitting optimum visualization of the frontal and ethmoid sinuses. *(Ballinger, vol 2, p 380)*

182. **(A)** The upper respiratory passageways include the nose, pharynx, and associated structures. The lower respiratory passageways include the larynx, trachea, bronchi, and lungs. If obstruction of the breathing passageways occurs in the upper respiratory tract, above the larynx (that is, in the nose or pharynx), tracheostomy or intubation may be performed in order to restore breathing. Tracheostomy or intubation cannot remedy obstructed passageways in the lower respiratory tract. *(Tortora, p 697)*

183. **(C)** The digestive system consists of several organs including the stomach (combining form: *gastr/o*), the ileum of the small bowel (*ile/o*), and the large bowel (*col/o*). The major organs of the respiratory system are the lungs. The combining form for their bronchial structures is *bronch/o*. *(Taber's, p 236)*

184. **(A)** Air or fluid levels will be clearly delineated only if the central ray is directed parallel to them. Therefore, the erect or decubitus position should be used. Small amounts of *fluid* within the peritoneal cavity are best demonstrated in the lateral decubitus position, *affected side down*. Small amounts of *air*

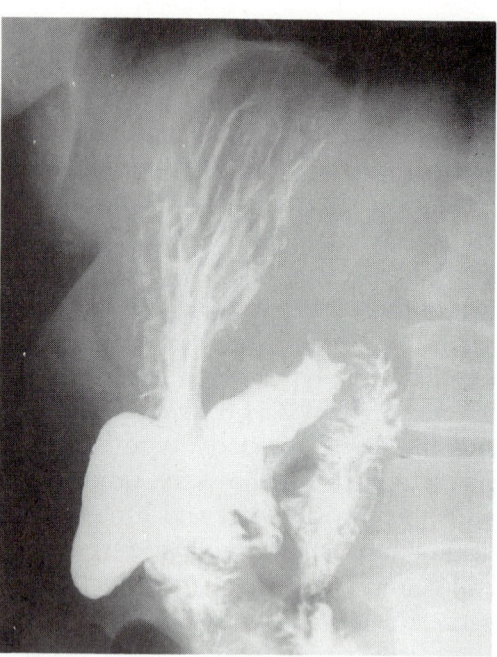

Figure 2–56. Courtesy of The Stamford Hospital, Department of Radiology.

within the peritoneal cavity are best demonstrated in the lateral decubitus position, *affected side up*. *(Ballinger, vol 2, pp 40–41)*

185. **(A)** The lateral projections of the barium-filled stomach (Fig. 2–56) may be performed recumbent or upright, for the demonstration of the retrogastric space. With the patient in the (usually right) lateral position, the central ray is directed to a point midway between the midaxillary line and anterior surface of the abdomen, at the level of L1. When the patient is in the LPO or RAO position, the CR should be directed midway between the vertebral column and lateral border of the abdomen. For the PA projection the CR is directed perpendicular to the film at the level of L2. *(Ballinger, vol 2, p 109)*

186. **(C)** To accurately position a lateral forearm, the elbow must form a 90° angle with the humeral epicondyles superimposed. The radius and ulna are only superimposed *distally*. Proximally, the coronoid process and radial head are superimposed, and the radial head faces anteriorly. Failure of the elbow to form a 90° angle, or the hand to be lateral, results

in a less than satisfactory lateral projection of the forearm. *(Ballinger, vol 1, p 100)*

187. **(C)** Diarthrotic, or synovial, joints such as the knee and TMJ, are freely movable. Most diarthrotic joints are associated with a joint capsule containing synovial fluid. Diarthrotic joints are the most numerous in the body and are subdivided according to type of movement. Amphiarthrotic joints are partially moveable joints whose articular surfaces are connected by cartilage—such as intervertebral joints. Synarthrotic joints, such as the cranial sutures, are immovable. *(Martini, p 263)*

188. **(D)** Operative cholangiography plays a vital role in biliary tract surgery. The contrast medium is injected and filming occurs *following* a cholecystectomy. This procedure is used to investigate the patency of the bile ducts, function of the hepatopancreatic sphincter (of Oddi), and previously undetected biliary tract calculi. *(Ballinger, vol 2, p 76)*

189. **(A) and 190. (B)** The radiograph shown is a lateral projection of the cervical spine taken in flexion. Flexion and extension views are useful in certain cervical injuries, such as whiplash, to indicate the degree of anterior and posterior motion. The structure labeled *1* is an apophyseal joint; because apophyseal joints are positioned 90° to the MSP they are well visualized in the lateral projection. The structure labeled *2* is a vertebral body. *(Saia, PREP, p 115)*

191. **(C)** The erect position in intravenous urography may be part of the departmental routine, but more often than not it is requested as a supplemental view to rule out nephroptosis. With the patient erect, the kidneys normally change position, dropping no more than 2 inches. More marked dropping of the kidney is termed nephroptosis, a condition actually due to loss of the surrounding perinephric fat. *(Ballinger, vol 2, p 170)*

192. **(C)** The *carina* is an internal ridge located at the bifurcation of the trachea into right and left primary, or mainstem, bronchi. The *epiglottis* is a flap of elastic cartilage that functions to prevent fluids and solids from entering the respiratory tract during swallowing. The *root* of the lung attaches the lung, via dense connective tissue, to the mediastinum. The root of the left lung is at the level of T6, and the right at T5. The *hilus* (hilum) is the slit-like opening on the medial aspect of the lung where arteries, veins, lymphatics, and so forth, enter and exit. *(Martini, p 834)*

193. **(C)** The AP projection demonstrates superimposition of the distal fibula on the talus; the joint space is not well seen. The 15 to 20° oblique position shows the entire mortise joint; the talofibular joint is well visualized, as well as the talotibial joint. There is considerable superimposition of the talus and fibula in the lateral rotation and lateral projections. *(Saia, PREP, p 100–101)*

194. **(A)** Stenvers method places the petrous pyramids parallel with the plane of the film. The mastoid tip is seen in profile next to the mandibular condyle and upper cervical spine. The internal and external auditory canals are demonstrated. The auditory ossicles and inner ear structures are frequently visualized. The patient is positioned so that the head rests on the *forehead*, *nose*, and *zygoma*, and the IOML is parallel to the film. The MSP is adjusted so as to be 45° to the film and the central ray is directed 12° cephalad to a point 1 inch anterior to the EAM closest to the film. The zygoma, nose, and chin form the "three-point landing" used in the parieto-orbital projection (Rhese method) of the optic foramen. *(Ballinger, vol 2, pp 408–409)*

195. **(A)** Blood is oxygenated in the lungs and carried to the left atrium by the four pulmonary veins. From the left atrium blood flows through the bicuspid (mitral) valve into the left ventricle. Blood leaving the left ventricle is bright red, oxygenated blood that travels through the systemic circulation, delivering oxygenated blood via arteries and returning deoxygenated blood to the lungs via veins. From the left ventricle blood first goes through the largest arteries, to progressively smaller arteries (arterioles), to the capillaries, to the smallest veins (venules), and on to progressively larger veins. *(Martini, p 702)*

196. (C) On the AP projection of the elbow, the radial head and ulna are normally somewhat superimposed. The lateral oblique demonstrates the radial head free of ulnar superimposition. The lateral projection demonstrates the olecranon process in profile. The medial oblique demonstrates considerable overlap of the proximal radius and ulna, but should clearly demonstrate the coronoid process free of superimposition and the olecranon process within the olecranon fossa. *(Ballinger, vol 1, p 104)*

197. (C) These are all terms used to describe particular body movements. Eversion refers to movement of the foot caused by turning the ankle outward. Inversion is foot motion caused by turning the ankle inward. Abduction is movement of a part away from the midline. Adduction is movement of a part toward the midline. *(Martini, pp 267–269)*

198. (A) A 15 to 20° AP Trendelenburg position during IV urography is often helpful in demonstrating filling of the distal ureters and the area of the vesicoureteral orifices. In this position, the contrast-filled urinary bladder moves superiorly, encouraging filling of the distal ureters and superior bladder, and provides better delineation of these areas. The CR should be directed perpendicular to the cassette. Compression of the *distal* ureters is used to prolong filling of the renal pelves and calyces. Compression of the proximal ureters is not advocated. *(Ballinger, vol 2, p 170)*

199. (C) There are many terms (with which the radiographer must be familiar) that are used to describe radiographic positioning techniques. *Cephalad* refers to that which is toward the head, and *caudad* to that which is toward the feet. Structures close to the source or beginning are said to be *proximal*, while those lying close to the midline are said to be *medial*. *(Martini, p 20)*

200. (C) Blood pressure in pulmonary circulation is relatively low, and therefore pulmonary vessels can easily become blocked by blood clots, air bubbles, or fatty masses—resulting in a *pulmonary embolism*. If the blockage stays in place, it results in an extra strain on the right ventricle, which is now unable to pump blood. This occurrence can result in congestive heart failure. *Pneumothorax* is air in the pleural cavity. *Atelectasis* is a collapsed lung or part of a lung. *Hypoxia* is a condition of low tissue oxygen. *(Martini, p 675)*

201. (C) The axillary portion of the ribs are best demonstrated in a 45° oblique position. The axillary ribs are demonstrated in the AP oblique projection with the affected side *adjacent* to the film, and in the PA oblique projection with the affected side *away from* the film. Therefore, the right axillary ribs would be demonstrated in the RPO (AP oblique with affected side *adjacent* to the film) or LAO (PA oblique with affected side *away* from the film) positions. *(Ballinger, vol 1, pp 428–431)*

202. (B) Endoscopic retrograde cholangiopancreatography (ERCP) may be performed to investigate abnormalities of the biliary system or pancreas. The patient's throat is treated with a local anesthetic in preparation for the passage of the endoscope. The hepatopancreatic ampulla (of Vater) is located and a cannula passed through it in order that contrast media may be introduced into the common bile duct. Spot films of the common bile duct and pancreatic duct are frequently taken in the oblique position. Direct injection of barium mixture into the duodenum occurs during enteroclysis procedure of the small bowel. *(Ballinger, vol 2, p 80)*

203. (B) Positioning the patient for a lateral projection of the sacrum and coccyx varies according to individual pelvic curvatures and male-female variations. The rule of thumb is to place the central ray 5 inches posterior to the *midcoronal plane* for a lateral coccyx and 3 inches posterior for a lateral sacrum. *(Ballinger, vol 1, pp 352–353)*

204. (B) The prone position is used to evaluate the sternoclavicular joints, unilaterally by turning the patient's face toward the affected side, and bilaterally with the patient's head resting on the chin. Anterior oblique positions and lateral projections are also used. Supine and posterior oblique positions place the sternoclavicular joints farther from the film. *(Ballinger, vol 1, pp 414–417)*

205. **(A)** The architectural features of the female pelvis are designed to accommodate childbearing. The female pelvis as a whole is broader and more shallow than its male counterpart, having a wider and more circular pelvic outlet. The ischial tuberosities and acetabula are further apart. The sacrum is wider and extends more sharply posteriorly. The pubic arch of the male is significantly narrower than that of the female. *(Saia, PREP, p 95)*

206. **(C)** Hysterosalpingography involves the introduction of a radiopaque contrast me-dium, through a uterine cannula, into the uterus and uterine (Fallopian) tubes. This examination is often performed to docu-ment patency of the uterine tubes in cases of infertility. A retrograde pyelogram requires cystoscopy and involves introduc-tion of contrast through the vesicoureteral orifices and into the renal collecting system. A voiding cystourethrogram also requires cystoscopy, and involves filling the bladder with contrast and documenting the voiding mechanism. A myelogram is peformed to investigate the spinal canal. *(Ballinger, vol 2, p 199)*

207. **(A)** The general survey AP axial projections (Grashey or Towne method) are performed with the central ray directed caudally at an angle of 30° with the OML perpendicular or 37° with the IOML perpendicular. These AP axial projections demonstrate the dorsum sellae and posterior clinoid processes within the foramen magnum (a portion of the foramen magnum is obscured by overlying bony shadows). A *40 to 60°* angulation shows the entire foramen magnum and the jugular foramina bilaterally. *(Ballinger, vol. 2, p 250)*

208. **(C)** The *ileocecal valve* is located at the terminal ileum where it meets with the first portion of the large bowel, the cecum. Most small bowel examinations are performed following oral administration of barium sulfate suspension. The first small bowel radiograph

is taken 15 minutes after the first swallow of barium, with subsequent radiographs made every 15 to 30 minutes depending on how quickly the barium is moving through the small bowel. Each film is shown to the radiologist and a decision is made regarding the time of the next film. When the barium reaches the terminal ileum, fluoroscopy may be performed and compression spot films taken of the *ileocecal valve*. *(Ballinger, vol 2, p 116)*

209. **(C)** With inspiration, the diaphragm is depressed, that is, moved into a lower position. The ribs and sternum are elevated. As the ribs are elevated, their angle is decreased. Radiographic density can vary considerably in appearance depending on which phase of respiration the exposure is made. *(Ballinger, vol 1, p 444)*

210. **(C)** Placing the patient in a 20 to 30° AP Trendelenburg position during an upper GI exam helps to demonstrate the presence of a *hiatal hernia*. A 10 to 15° Trendelenburg position with the patient rotated slightly to the right will also help demonstrate regurgitation and hiatal hernia. Filling of the *duodenal bulb* and demonstration of the *duodenal loop* is best seen in the RAO position. *Congenital hypertrophic pyloric stenosis* is caused by excessive thickening of the pyloric sphincter. It is noted in infancy and characterized by projectile vomiting. The pyloric valve will let very little pass through, and as a result the stomach becomes enlarged (hypertrophied). *(Ballinger, vol 2, p 99)*

211. **(C)** The manubrial or jugular notch is the depression on the superior border of the manubrium and is located at the level of third thoracic vertebra. The vertebra prominens is at the level of the seventh cervical vertebra. *(Saia, PREP, p 64)*

212. **(B)** Sacroiliac joints lie obliquely within the pelvis and open anteriorly at an angle of 25 to 30° to the midsagittal plane. A 25 to 30° oblique position places the joints perpendicular to the film. The left sacroiliac joint may be demonstrated in the LAO and RPO positions with little magnification variation. *(Ballinger, vol 1, p 382)*

213. (D) The acetabulum is the bony socket that receives the head of the femur to form the hip joint. The upper two fifths of the acetabulum is formed by the ilium, the lower anterior one fifth is formed by the pubis, and the lower posterior two fifths is formed by the ischium. Thus, the acetabulum is formed by all three bones that form the pelvis: the ilium, the ischium, and the pubis. *(Martini, pp 246–247)*

214. (B) Retrograde urography requires ureteral catheterization so that a contrast medium can be introduced directly into the pelvicalyceal system. This procedure provides excellent opacification and structural information but does not demonstrate function of these structures. Intravenous studies such as the intravenous urogram demonstrate function. Cystourethrography is an examination of the bladder and urethra, frequently performed during voiding. Nephrotomography is performed after intravenous administration of a contrast agent; it may be used to evaluate small intrarenal lesions and renal hypertension. *(Ballinger, vol 2, p 179)*

215. (A) The AP projection of the sacrum requires a 15° cephalad angle centered at a point midway between the pubic symphysis and the ASIS. The AP projection of the coccyx requires the central ray to be directed 10° caudally and centered 2 inches superior to the pubic symphysis. *(Ballinger, vol 1, p 387)*

216. (C) Contrast is administered in a variety of manners in cholangiography. These include:

1. An endoscope with a cannula placed in the hepatopancreatic ampulla (ampulla of Vater) for an ERCP
2. A needle or small catheter placed directly in the common bile duct for an operative cholangiogram
3. A very fine needle through the patient's side and into the liver for a percutaneous transhepatic cholangiogram
4. Via an indwelling T-tube for a postoperative, or T-tube, cholangiogram *(Ballinger, vol 2, p 78)*

217. (C) The articular facets (apophyseal joints) of the L5-S1 articulation form a 30° angle with the MSP; they are therefore well demonstrated in a 30° oblique position. The 45° oblique demonstrates the apophyseal joints of L1 through L4. *(Ballinger, vol 1, p 372)*

218. (B) The PA chest illustrated in Figure 2–31 is rotated, as evidenced by asymmetric distances between the sternal ends of the clavicles and the vertebrae. The patient's shoulders were rolled forward, as indicated by the majority of the scapulae being projected out of the lung fields. The exposure was made on adequate inspiration, as evidenced by visualization of 10 ribs above the diaphragm. The patient did not breathe during the exposure; there is no evidence of motion unsharpness. *(Ballinger, vol 1, p 455)*

219. (C) The decubitus position is used to describe the patient as recumbent (prone, supine, or lateral) with the central ray directed horizontally. When the patient is recumbent with the head lower than the feet, he or she is said to be in the Trendelenburg position. In the Fowler's position, the patient's head is positioned higher than the feet. The Sims position is the (LAO) position assumed for enema tip insertion. *(Bontrager, p 21)*

220. (C) Ureteral reflux is best demonstrated during voiding. It can occur even when the bladder is only partially filled with a contrast medium. The *vesicourethral* orifice, as well as other sphincter muscles, relaxes during urination; however the *vesicoureteral* orifices may also relax and cause reflux. *(Ballinger, vol 2, p 182)*

221. (D) The patient is supine with the leg *abducted* (drawn away from the midline) approximately 40°. This 40° abduction from the vertical places the long axis of the femoral neck parallel to the film. *Adduction* is drawing the extremity closer to the midline of the body. *(Ballinger, vol 1, p 282)*

222. (D) Typically, traumatic injury to the hip requires a cross-table lateral projection, and an AP projection of the entire pelvis. Both of these are performed using minimal manipulation of the affected extremity, reducing the possibility of further injury. A physician should perform any required manipulation of the traumatized hip. (*Ballinger, vol 1, pp 287, 290*)

223. **(A)** A decubitus projection is obtained using a horizontal x-ray beam. The type of decubitus projection is described depending on the patient's recumbent position. When the patient is lying AP recumbent, the patient is said to be in the dorsal decubitus position. When the patient is lying prone, he or she is in the ventral decubitus position. If the patient is lying in the left or right lateral recumbent positions with the x-ray beam directed horizontally, the patient is said to be in the left or right lateral decubitus positions, respectively. (*Ballinger, vol 1, p 52*)

224. **(D)** Prior to the start of an IVP, the patient should be instructed to empty the bladder. This is advised to avoid dilution of the contrast agent. Diluted contrast within the bladder will not affect diagnosis of *renal* abnormalities, but it may obscure bladder abnormalities. The patient's allergic history should be reviewed in order to avoid possibility of severe reaction to the contrast agent. The patient's creatinine level and BUN should be checked; significant elevation of these blood chemistry levels often suggests renal dysfunction. The normal BUN level is 8 to 25 mg/100 mL; normal creatinine range is 0.6 to 1.5 mg/100 mL. (*Ballinger, vol 2, p 168*)

225. **(C)** The cassette for a cross-table lateral projection of the hip is placed in a *vertical* position. The top of the cassette should be placed directly above the iliac crest, adjacent to the *lateral surface* of the affected hip. The cassette is positioned *parallel* to the femoral neck; the CR is *perpendicular* to the femoral neck and cassette. (*Ballinger, vol 2, p 290*)

226. **(B)** The pictured radiograph was made in the right lateral decubitus position. It is part of a series of radiographs made during an air contrast (double-contrast) barium enema examination. A double-contrast examination of

the large bowel is performed in order to see *through* the bowel to is posterior wall and to visualize any *intraluminal* (for example, polypoid) *lesions* or *masses*. Various body positions are used to redistribute the barium and air. In order to demonstrate the medial and lateral walls of the bowel, decubitus positions are performed. The radiograph presents a right lateral decubitus position, because the *barium has gravitated* to the right side (the side of the hepatic flexure). The *air rises* and delineates the medial side of the ascending colon and the lateral side of the descending colon. The posterior wall of the rectum could be visualized using the ventral decubitus position and a horizontal beam lateral of the rectum. (*Ballinger, vol 2, p 143*)

227. **(D)** Generally, contrast medium is injected into the subarachnoid space between the third and fourth lumbar vertebrae (Fig. 2–57). Because the spinal cord ends at the level of the first or second lumbar vertebrae, this is considered to be a relatively safe injection site. The cisterna magna can be used, but the risk of contrast entering the ventricles and causing side effects increases. Discography requires injection of contrast medium into the individual intervertebral discs. (*Ballinger, vol 2, p 500*)

228. **(C)** *Hydronephrosis* is a collection of urine in the renal pelvis due to obstructed outflow as from a stricture or obstruction. If the obstruction occurs at the level of the bladder or along the course of the ureter, it will be accompanied by the condition of hydroureter above the level of obstruction. These conditions may be demonstrated during intravenous urography. The term *pyelonephrosis* refers to some condition of the renal pelvis. *Nephroptosis* refers to drooping or downward displacement of the kidneys. This may be demonstrated using the erect position during IV urography. *Cystourethritis* is inflammation of the bladder and urethra. (*Taber's, p 925*)

229. **(D)** Spinal column studies are often required for evaluation of adolescent scoliosis, thus presenting a twofold problem: radiation exposure to youthful gonadal and breast tis-

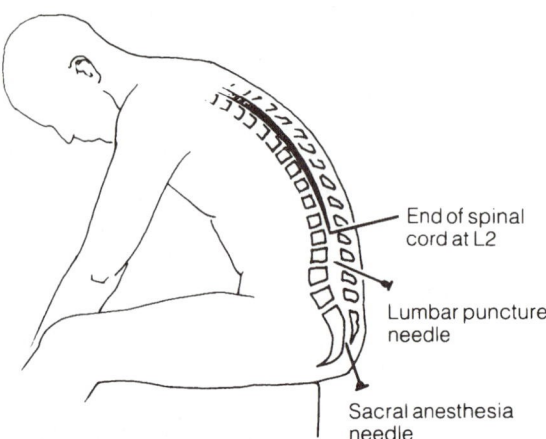

End of spinal
cord at L2

Lumbar puncture
needle

Sacral anesthesia
needle

Figure 2–57. Reproduced with permission from deGroot, *Correlative Neuroanatomy*, 21st ed. Norwalk, CT: Appleton & Lange, 1991.

sues, and significantly differing tissue densities/thicknesses. The use of high-speed film/screen combination helps reduce the exposure required for the examination. Exposure dose concerns can also be resolved with the use of a compensating filter (for uniform density) that incorporates lead shielding for the breasts and gonads (Fig. 2–58). *(Cullinan, pp 140–141)*

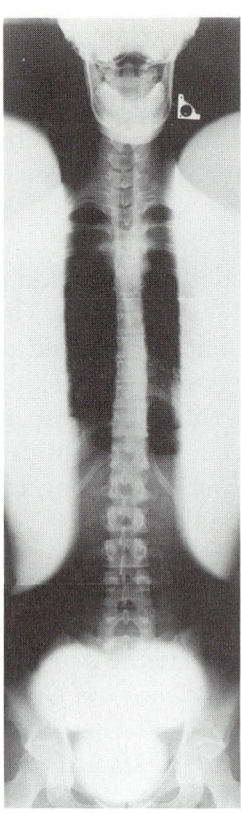

Figure 2–58. Courtesy of Nuclear Associates.

230. **(B)** Diarthrodial joints are freely movable joints that distinctively contain a joint capsule. Contrast is injected into this joint capsule to demonstrate menisci, articular cartilage, bursae, and ligaments of the joint under investigation. Synarthrodial joints are immovable joints, composed of either cartilage or fibrous connective tissue. Amphiarthrodial joints allow only slight movement. *(Saia, PREP, p 106)*

231. **(C)** A lateral projection is generally included in a routine skull series. The patient is placed in a PA oblique position. The *MSP* is positioned parallel to the film and the *IOML* adjusted so as to be parallel to the long axis of the cassette. The *interpupillary line* must be perpendicular to the film. In a routine lateral projection of the skull, the CR should enter approximately 2 inches superior to the EAM. *(Saia, PREP, p 134)*

232. **(C)** For the PA projection of the skull the OML is adjusted perpendicular to the film

and the MSP must be perpendicular to the film. The CR is directed so as to exit the nasion. In this position the petrous pyramids should completely fill the orbits. When caudal angulation is used with this position, the petrous pyramids are projected in the lower portion, or out of, the orbits. If cephalad angulation is employed with this position, the petrous pyramids are projected up toward the occipital region (as in the nuchofrontal projection). *(Ballinger, vol 2, p 242)*

233. **(B)** Mammographic identification markers are generally placed according to established standards. In the *mediolateral* and *oblique* projections they are placed *along the upper border* of the breast. In *the craniocaudad* projection they are placed along the *lateral side* of the breast. *(Ballinger, vol 2, p 470)*

234. **(A)** The PA projection can be easily modified by redirecting the central ray to demonstrate

a variety of structures. The central ray can be directed

1. 25 to 30° caudad for rotundum foramina
2. 20 to 25° caudad for the superior orbital fissures
3. 20 to 25° cephalad for the inferior orbital fissures *(Ballinger, vol 2, p 242)*

235. **(C)** Deoxygenated (venous) blood from the *upper* body (head, neck, thorax, upper extremities) empties into the *superior vena cava*. Deoxygenated (venous) blood from the *lower* body (abdomen, pelvis, lower extremities) empties into the *inferior vena cava*. The superior and inferior vena cava empty into the right atrium. The coronary sinus, returning venous blood from the heart, also empties into the right atrium. Deoxygenated blood passes from the right atrium, through the tricuspid valve, into the right ventricle. From the right ventricle blood is pumped (during ventricular systole) through the pulmonary semilunar valve into the pulmonary artery—the only artery that carries deoxygenated blood. From the pulmonary artery blood travels to the lungs, picks up oxygen, and is carried by the four pulmonary veins (the only veins carrying oxygenated blood) to the left atrium. The oxygenated blood passes through the mitral (or bicuspid) valve during atrial systole and into the left ventricle. During ventricular systole, oxygenated blood from the left ventricle passes through the aortic semilunar valve, into the aorta, and into systemic circulation. *(Martini, p 702)*

236. **(B)** Whenever a part is being radiographed for demonstration of air–fluid levels, the central ray *must* be directed parallel to the floor. In this example, the patient was unable to tolerate the 90° tilt of the x-ray table. If the radiographer compensated for this by directing the CR perpendicular to the film (angling 20° caudad), it is very possible that any air–fluid level would be blurred, indistinct, and go unrecognized. Remember that air or fluid always levels out parallel to the floor. Thus, if it's the air–fluid level that needs to be demonstrated, the CR must also be parallel to the floor. *(Ballinger, vol 2, p 376)*

237. **(C)** Myelography is used to demonstrate encroachment upon, and compression of, the spinal cord as a result of herniated disc, tumor growth, or posttraumatic swelling of the cord. This is accomplished by placing positive or negative contrast media into the subarachnoid space. Myelography will demonstrate posterior protrusion of herniated intervertebral discs or spinal cord tumors. Anterior protrusion of a herniated intervertebral discs does not impinge on the spinal cord, and is not demonstrated in myelography. Internal disc lesions can only be demonstrated by injecting contrast media into the individual discs (this procedure is termed *discography*). *(Ballinger, vol 2, p 500)*

238. **(C)** The axillary portions of the ribs are foreshortened and self-superimposed in the AP and PA positions. However, they are "opened" and placed more parallel to the film in the oblique positions. Thus, right axillary portions are best demonstrated in the RPO position. The LPO position demonstrates the left axillary ribs. The *affected* side should be placed closest to the film. *(Ballinger, vol 1, p 448)*

239. **(B)** About 30 minutes after ingestion of fatty foods, cholecystokinin is released from the duodenal mucosa and absorbed into the bloodstream. As a result, the gallbladder is stimulated to contract, releasing bile into the intestine. *(Ballinger, vol 2, p 64)*

240. **(D)** A direct lateral of the mastoids is contraindicated because of superimposed structures; an angled projection is necessary to separate structures of interest. The patient's head is placed in a true lateral position, placing the *IOML* parallel to the long axis of the film and the interpupillary line perpendicular to the film. The *MSP* is then rotated 15° toward the tabletop and the CR angled *15° caudad*. An alternate method is to angle the MSP 15° toward the tabletop *and* the interpupillary line 15° from the vertical. In either case, the CR enters approximately 2 inches above and posterior to the uppermost EAM. *(Ballinger, vol 2, p 400)*

241. (D) The four body types (from largest to smallest) are hypersthenic, sthenic, hyposthenic, and asthenic. The abdominal viscera of the asthenic person are generally located quite low, vertical, and toward the midline. The opposite is true of the hypersthenic individual: organs are located high, transverse, and laterally. *(Saia, PREP, p 62)*

242. (C) A zonogram is a thick tomographic section, or "cut," and appears more similar to conventional radiography. A thick tomographic slice is produced by using a short exposure amplitude (arc), resulting in limited blurring of the radiographic image. Pluridirectional tomography produces maximal blurring of the radiographic image and generally uses a long exposure amplitude, resulting in a thin tomographic section or "cut." *(Ballinger, vol 3, pp 46–47)*

243. (D) With the patient in the erect position, barium moves inferiorly and air rises to provide double-contrast visualization of the hepatic and splenic flexures. The LAO or RPO position is used to demonstrate especially the hepatic flexure; the splenic flexure generally appears self-superimposed in this position. A left lateral decubitus position will demonstrate a double-contrast visualization of right-sided bowel structures—that is, the right side of the ascending colon, right side of the sigmoid and rectum, and so on. The lateral position offers a singularly valuable view of the rectum. *(Ballinger, vol 2, p 146)*

244. (B) Over 2 million women in the United States have had breast *augmentation* (implants) for cosmetic or reconstructive purposes. The augmented breast presents a challenge to the radiographer, because breast implants can obscure up to 85% of breast tissue. The *Ecklund method* of mammography pushes the implant posteriorly against the chest wall, and the breast tissue is then pulled forward and compressed. The Cleopatra projection is performed in the seated position with the patient leaning backward and laterally over the cassette. This adjustment from the craniocaudad projection demonstrates the tail of the breast and often a por-

tion of the axilla. Magnification mammography utilizes a 0.1 mm focal spot and is used to evaluate calcifications and the margins of any lesions (to determine if they are likely to be benign or malignant). The cleavage view is a bilateral craniocaudal view of the medial aspect of the breasts. *(Ballinger, vol 2, p 479)*

245. (C) Major branches of the common carotid arteries (internal carotids) function to supply the anterior brain, while the posterior brain is supplied by the vertebral arteries (branches of the subclavian). The brachiocephalic (innominate) artery is unpaired and is one of the three branches of the aortic arch, from which the right common carotid artery is derived. The left common carotid artery comes directly off the aortic arch. *(Martini, pp 754–755)*

246. (A) The base of the first metatarsal articulates with the first (medial) cuneiform. The base of the second metatarsal articulates with the second (intermediate) cuneiform; the third metatarsal articulates with the third (lateral) cuneiform. The bases of the fourth and fifth metatarsals articulate with the cuboid. The navicular articulates with the first and second cuneiforms anteriorly and the talus posteriorly. *(Saia, PREP, p 92)*

247. (B) *Intussusception* is the telescoping of one part of the intestinal tract into another. It is a major cause of bowel obstruction in children, usually in the region of the ileocecal valve, and is much less common in the adult. Radiographically, intussusception appears as the classic "coil spring," with barium trapped between folds of the telescoped bowel. The diagnostic barium enema procedure can occasionally reduce the intussusception, though care must be taken to avoid perforation of the bowel. *Appendicitis* occurs when an obstructed appendix becomes inflamed. Distention of the appendix occurs and, if left untended, gangrene and perforation can result. *Regional enteritis*, Crohn's disease, is a chronic granulomatous inflammatory disorder that can affect any part of the GI tract but generally involves the area of the terminal ileum. Ulceration and formation of fistulous tracts often occur. *Ulcerative colitis* occurs most often in the young adult; its etiology is unknown, although psychogenic or autoimmune factors seem to be involved. *(Eisenberg, p 138)*

248. **(C)** An oblique projection of the lumbar spine is illustrated. This is a 45° LPO position demonstrating the apophyseal joints closest to the film. The apophyseal joints are formed by the articulation of the inferior articular facets of one vertebra with the superior articular facets of the vertebra below. Note the "scotty dog" images that appear in the oblique lumbar spine. Intervertebral foramina are best visualized in the lateral lumbar position. *(Ballinger, vol 1, p 373)*

249. **(B)** To increase the concentration of contrast media in the deep veins of the leg, a Fowler's position is used with the x-ray table angled at least 45°. Tourniquets can also be used to force the contrast into the deep veins of the leg, especially when the patient is examined in the recumbent position. Contrast medium is injected through a superficial vein in the foot. Filming may be performed with or without fluoroscopy, and may include AP, lateral, and 30° obliques of the lower leg in internal rotation. Filming begins at the ankle and proceeds superiorly, usually to include the inferior vena cava. *(Ballinger, vol 2, p 542)*

250. **(B)** A cross-sectional image of the abdomen is pictured in Figure 2–34. The large structure on the right, labeled *1*, is the liver. The gallbladder is seen as a somewhat darker density on the medial border of the liver. The left kidney is labeled *4*; the right kidney is clearly seen on the other side. The vertebra is labeled *5* and the psoas muscles are seen just posterior to the vertebra. Just anterior to the body of the vertebra is the circular aorta, labeled *3* (some calcification can be seen as brighter densities). The somewhat flattened inferior vena cava (labeled *2*) is seen to the left of, and slightly anterior to, the aorta. *(Von Hagens, p 92)*

251. **(A)** The knee (tibiofemoral joint) is the largest joint of the body, formed by the articulation of the femur and tibia. However, it actually consists of three articulations: the

patellofemoral joint, the lateral tibiofemoral joint (lateral femoral condyle with tibial plateau), and the medial tibiofemoral joint (medial femoral condyle with tibial plateau). Though the knee is classified as a synovial (diarthrotic) hinge-type joint, the patellofemoral joint is actually a gliding joint, and the medial and lateral tibiofemoral joints are hinge type. *(Martini, p 279)*

252. **(A)** Femoral necks are nonpalpable bony landmarks. The ASIS, pubic symphysis, and greater trochanter are palpable bony landmarks used in radiography of the pelvis and for localization of the femoral necks. *(Ballinger, vol 1, p 274)*

253. **(C)** The full length of the nasal septum is best demonstrated in the parietoacanthial (Waters method) projection. This is the single best view for nasal bones. The PA axial (Caldwell method) projection superimposes petrous structures over the nasal septum, while the lateral projection superimposes and obscures good visualization of the septum. *(Ballinger, vol 2, p 307)*

254. **(C)** The *posterior oblique* (AP oblique) positions demonstrate the apophyseal joints *nearer* the film. The anterior oblique positions demonstrate the apophyseal joints farther from the film. Lumbar intervertebral foramina are demonstrated in the lateral position. *(Saia, PREP, p 118)*

255. **(D)** A successful oral cholecystogram depends on the ability of the liver to remove contrast from the portal bloodstream and to excrete it with bile. A healthy gallbladder should concentrate and store bile as well as contrast medium. With a functioning gallbladder and liver, an opacified gallbladder should result. The pancreas plays an integral part in the digestive process, but is not in the biliary system. *(Ballinger, vol 2, p 58)*

256. **(C)** In the AP projection of the elbow, the radial head and ulna are normally somewhat superimposed. The lateral oblique demonstrates the radial head free of superimposition with the ulna. The *lateral* projection

demonstrates the olecranon process in profile. The *medial oblique* position demonstrates considerable overlap of the proximal ulna, but should clearly demonstrate the coronoid process free of superimposition and the olecranon process within the olecranon fossa. *(Ballinger, vol 1, p 104)*

257. **(C)** The *posterior oblique* positions of the cervical spine (LPO, RPO) require that the CR be directed 15° to C4. The posterior obliques demonstrate the intervertebral foramina *farther* away from the film. The *anterior* oblique positions require a 15° caudal angulation and demonstrate the intervertebral foramina *closest* to the film. *(Ballinger, vol 1, p 340)*

258. **(B)** Generally, bone age is evaluated in a single PA projection of the left hand. Another series of films may be used in determining bone age of children under age 3; this includes AP humerus to include shoulder, one half of the clavicle and elbow, PA hand and wrist, AP and lateral knee, and AP and lateral foot and ankle. *(Saia, PREP, p 82)*

259. **(A)** When one cheekbone is depressed a tangential projection is required to "open up" the zygomatic arch and draw it away from the overlying cranial bones. This is accomplished by placing the patient in the *SMV* position and rotating the head 15° toward the affected side and centering to the zygomatic arch. A 30° rotation places the mandibular shadow over the zygomatic arch. *(Ballinger, vol 2, p 230)*

260. **(B)** In the prone oblique positions (RAO, LAO), the flexure disclosed is the one closer to the film. Therefore, the LAO position will "open up" the splenic flexure; the RAO will demonstrate the hepatic flexure. The AP oblique positions (RPO, LPO) demonstrate the side further from the film. *(Ballinger, vol 2, p 136)*

261. **(C)** The SMV (Schüller method) projection of the skull requires that the patient's neck be extended, placing the vertex adjacent to the film holder/upright Bucky, so that the *IOML* is parallel with the film. This projection is

useful for demonstrating ethmoid and sphenoid sinuses, pars petrosae, mandible, foramina ovale, and spinosum. In the illustration, line *1* represents the *glabellomeatal* line (GML), line *2* is the *orbitomeatal* line (OML), *3* is the *infraorbitomeatal* line (IOML), and *4* is the *acanthomeatal* line. *(Saia, PREP, p 134)*

262. **(C)** The lateral projection of the skull requires that the patient be in the prone oblique position with the MSP parallel to the film and the interpupillary line perpendicular to the film. The *IOML* must be parallel to the long axis of the film. The supraorbital margins, anterior clinoid processes, and posterior clinoid processes should be superimposed. *(Saia, PREP, p 134)*

263. **(B)** Accurate positioning of the skull requires the use of several baselines. The *OML* (number 2) and *IOML* (number 3) are usually separated by 7°. The orbitomeatal line and glabellomeatal line are usually separated by 8° (therefore, 15° exists between the GML and IOML). It is useful to remember these differences, as CR angulation must be adjusted when using a baseline other than the one recommended for a particular position. For example, if it is recommended that the CR be angled *30° to the OML*, then the CR would be angled *37° to the IOML*. *(Saia, PREP, p 133)*

264. **(C)** The forward slipping of one vertebra upon the one below it is called *spondylolisthesis*. *Spondylolysis* is the breakdown of the pars interarticularis; it may be unilateral or bilateral, and results in forward slipping of the involved vertebra—the condition of spondylolisthesis. Inflammation of one or more vertebrae is called *spondylitis*. *Spondylosis* refers to degenerative changes occurring in the vertebra. *(Ballinger, vol 1, p 321)*

265. **(C)** The patient is placed in a true lateral position and the central ray is directed perpendicular to a point ¾ inch distal to the nasion. An 8 × 10 cassette divided in half or an occlusal film is used for this procedure. *(Ballinger, vol. 2, p 315)*

266. **(B)** A frontal projection, AP or PA, demonstrates medial and lateral relationship of structures. A lateral projection demonstrates anterior and posterior relationship of structures. Two views, at right angles to each other, are generally taken of most structures. *(Saia, PREP, p 67)*

267. **(A)** The posterior oblique positions (LPO, RPO) of the lumbar vertebrae demonstrate the apophyseal joints closer to the film. The left apophyseal joints are demonstrated in the LPO position, while the right apophyseal joints are demonstrated in the RPO position. The lateral position is useful to demonstrate the intervertebral disc spaces, intervertebral foramina, and spinous processes. *(Ballinger, vol 1, p 372)*

268. **(B)** Blowout fractures of the orbital floor are well demonstrated in the Water's method (parietoacanthial [PA] projection) and by using tomographic studies. A PA with the OML perpendicular and the central ray angled 30° caudad will demonstrate the orbital floor in profile. Sweet's localization method shows exact placement of foreign bodies within the eye. *(Ballinger, vol 2, p 270)*

269. **(C)** An erect abdomen or left lateral decubitus should be performed for demonstration of air–fluid levels in the abdomen. The right lateral decubitus position is used to demonstrate the layering of gallstones. It will not show free air within the peritoneum because of the overlying gastric bubble on the elevated left side of the body. *(Eisenberg, pp 161–162)*

270. **(B)** Shoulder arthrograms (Fig. 2–59) are used to evaluate rotator cuff tear, glenoid labrum (a ring of fibrocartilaginous tissue around the glenoid fossa), and frozen shoulder. Acromioclavicular joint separation is demonstrated on erect AP films with and without the use of weights. Routine radiographs demonstrate arthritis, and the addition of a transthoracic humerus or scapular Y would demonstrate dislocation. *(Ballinger, vol 1, p 496)*

271. **(B)** The posterior profile position (Stenvers method) demonstrates a profile image of the

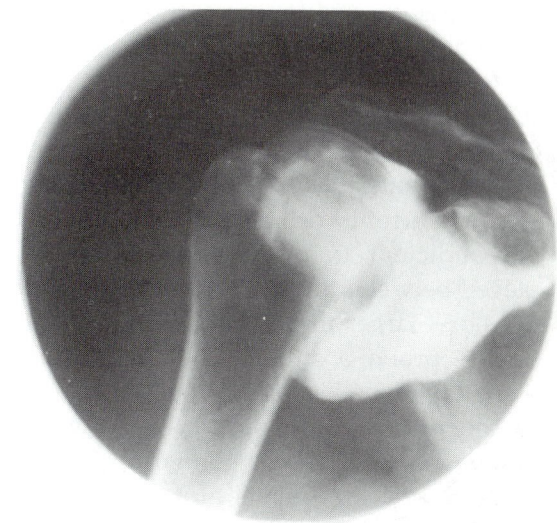

Figure 2–59.

pars petrosa, placing it *parallel to the film*. The patient is recumbent prone with the head resting on the *forehead, nose, and zygoma*. The *IOML* is placed parallel to the film. The *MSP* is rotated 45°. The CR is directed 12° cephalad and exits just anterior to the dependent EAM. For patients unable to assume the prone position, an anterior profile position (Arcelin method) may be performed using a 10° caudal angle. *(Ballinger, vol 2, p 408)*

272. **(B)** Peripheral lymphatic vessels are very difficult to identify because of their small size and colorless lymphatic fluid. In order to locate these vessels, a blue dye such as 11% patent blue violet or 4% sky blue is used. This dye is absorbed specifically by the lymphatic vessels following subcutaneous injection. Ethiodized oil is the contrast medium employed to radiographically demonstrate the lymphatic vessels and nodes. Water-soluble iodinated medium is too rapidly diluted with lymph to be useful. *(Ballinger, vol 2, p 574)*

273. **(C)** There are several surface landmarks and localization points that can help the radiographer in positioning of various body structures. The *jugular notch* is located approximately opposite the T2-3 interspace. The *sternal angle* is located opposite the T4-5 interspace. The *xiphoid* (or ensiform) *process* is located opposite T10. *(Saia, PREP, p 64)*

274. (D) The tangential projection may be used to demonstrate medial or lateral displacement of fragments. The patient is positioned supine or prone with the glabelloalveolar line perpendicular to the film. Occlusal film is placed between the teeth (in the supine position) or a larger film is placed under the chin (in the prone position). The CR is directed along the glabelloalveolar line, perpendicular to the film. Nasal bones will not be well demonstrated in the tangential projection on patients with a prominent forehead, recessed nose, or protruding upper teeth. *(Ballinger, vol 2, pp 316–317)*

275. (D) The typical vertebra has two parts, the body and the vertebral arch. The body is the dense, anterior bony mass. Posteriorly attached is the vertebral arch, a ringlike structure. The vertebral arch is formed by two pedicles (short, thick processes projecting posteriorly from the body) and two laminae (broad, flat processes projecting posteriorly and medially from the pedicles). *(Saia, PREP, p 117)*

276. (D) The mediastinum is the space between the lungs that contains the heart, great vessels, trachea, esophagus, and thymus gland. It is bound anteriorly by the sternum and posteriorly by the vertebral column, and extends from the upper thorax to the daphragm. *(Martini, p 24)*

277. (D) This is a modification of the parietoacanthial projection (Waters method) in which the patient is requested to first open the mouth, and then the skull is positioned so that the OML forms a 37° angle with the film. The CR is directed through the sphenoid sinuses and exits the open mouth. The routine parietoacanthial projection (with mouth closed) is used to demonstrate the maxillary sinuses projected above the petrous pyramids. The frontal and ethmoidal sinuses are best visualized in the PA axial position (modified Caldwell method). *(Ballinger, vol 2, p 382)*

278. (C) Ribs below the diaphragm are best demonstrated with the diaphragm elevated. This is accomplished by placing the patient in a recumbent position and by taking the exposure at the end of exhalation. Conversely, the ribs above the diaphragm are best demonstrated with the diaphragm depressed. Placing the patient in the erect position and taking the exposure at the end of deep inspiration accomplishes this. *(Ballinger, vol 1, p 428)*

279. (D) The AP axial projection is performed by placing the patient in a lordotic position, leaning against the vertical grid device. This places the clavicle at right angles, or nearly so, to the plane of the film. The central ray is directed to enter the inferior border of the clavicle, at right angles to its coronal plane. Other axial projections may include a prone position with a 25 to 30° caudal angle. However, none of these produces an exact axial projection of the clavicle. *(Ballinger, vol 1, p 159)*

280. (D) A PA axial projection of the skull with a 15° caudad angle will show the petrous pyramids in the lower third of the orbits. Using *no* angulation, the petrous pyramids will fill the orbits. Either PA projection should demonstrate symmetrical petrous pyramids and an equal distance from the lateral border of the skull to the lateral border of the orbit on both sides. This determines that the is no rotation of the skull. *(Ballinger, vol 2, p 244)*

281. (D) When performing gastrointestinal radiography the position of the stomach may vary depending on the respiratory phase, the body habitus, and the patient position. Inspiration causes the lungs to fill with air and the diaphragm to descend, thereby pushing the abdominal contents downward. On expiration, the diaphragm will rise, allowing the abdominal organs to ascend. Body habitus is an important factor in determining the size and shape of the stomach. An asthenic patient may have a long J-shaped stomach, while the stomach may be transverse on a hypersthenic patient. The body habitus in an important consideration in determining the positioning and placement of the cassette. The patient position can also alter the position of the stomach. If a patient turns from the right anterior oblique (RAO) position, into the anteroposterior (AP) position, the stomach will move into a more horizontal position. Although the cardiac sphincter and the pyloric sphincter are relatively fixed, the fundus is quite mobile, and will vary in position. *(Dowd & Wilson, vol 2, p 778)*

282. **(A)** With a patient in the PA position and the OML perpendicular to the table, a 15 to 20° caudal angulation would place the petrous ridges in the lower third of the orbit. In order to achieve the same result in a baby or a small child, it is necessary the radiographer to decrease the angulation or modify the angulation to 10 to 15° caudal. The reason for this is understood by examining the base lines for skull positioning. In the adult skull, the OML and IOML are about 7° apart. In a baby or small child, the difference is larger, about 15° apart. Remember, in the adult our heads make up about one-seventh the length of our bodies. In children, the head is about one-fourth the length of the body. These differences must be considered in radiographic examination of the skull on babies. *(Ballinger, vol 3, p 26)*

283. **(B)** The patient position for operative cholangiography is a 15 to 20° right posterior oblique (RPO). Remember that the gallbladder lies in upper right quadrant. Because the radiographs are obtained during surgery, they must be obtained with the patient in the supine rather than the prone position. A slight oblique (15 to 20°) will allow visualization of the biliary tract free of superimposition from the vertebrae. The LPO would place the biliary vessels over the spine. A 45° oblique is too steep for visualization of the gall bladder and biliary tree. *(Dowd & Wilson, vol 2, p 840)*

284. **(D)** When performing tomography, it is of paramount importance the radiographer properly apply immobilization, provide adequate radiation protection whenever possible, and obtain and check a scout film. Tomography differs from conventional radiography in several important aspects. Tomographic examinations are often lengthy and positioning is crucial. A patient must be made as comfortable as possible, and the exam should be thoroughly explained. The sudden, quick movement of a tomographic tube can frighten the patient if it is unexpected. Proper padding and support devices should be provided and the part under examination should be immobilized. Because tomographic examinations frequently use high-mAs techniques, a single exposure can result in a significant dose. Multiple exposures at high mAs techniques require proper shielding. Additionally, a scout film will not only ensure proper technique and positioning, but is used to decide the fulcrum levels necessary. *(Dowd & Wilson, p 926)*

285. **(A)** Knee arthrography may be performed to demonstrate torn meniscus (cartilage), Baker's cyst, loose bodies, and ligament damage. A torn rotator cuff would be demonstrated on a shoulder, not a knee arthrogram. *(Dowd & Wilson, vol 2, p 890)*

286. **(B)** Indications for a myelographic cervical puncture include demonstration of the upper level of a spinal block or when lumbar puncture has failed. If there is a suspected mass lesion of the upper cervical canal, cervical puncture is contraindicated. Cervical puncture is also contraindicated if the thoracic or lumbar region needs to be demonstrated. The contrast becomes too diluted if cervical puncture is performed. Lumbar puncture is indicated in these instances. *(Chapman & Nakielny, p 224)*

287. **(A)** During lower limb venography, tourniquets are applied above the knee and ankle to suppress filling of the more superficial veins and coerce filling of the deep veins. The anterior tibial vein may be blocked when tourniquets are used. The patient is positioned so that the table is tilted with the head up to slow the transit time of the contrast media, in order that films may be obtained of the entire lower limb and pelvic area. *(Ballinger, vol 2, p 542)*

288. **(D)** Hysterosalpingography may be peformed for demonstration of uterine tubal patency, mass lesions in the uterine cavity, and uterine position. Although hysterosalpingography is often performed to check tubal patency, the uterine anatomy, position, and morphology are exhibited. Additionally, polyps, fibroids, or space-occupying lesions within the uterus are well demonstrated. *(Ballinger, vol 2, p 199)*

289. **(D)** A fluoroscopic unit with spot film and tilt table should be used for endoscopic retrograde pancreatography. The Trendelenburg position is sometimes necessary to fill the interhepatic ducts and a semierect position to fill the lower end of the common bile duct. Also necessary are a fiber-optic endoscope for locating the hepatopancreatic ampulla and polyethylene catheters for the introduction of contrast media. *(Dowd & Wilson, vol 2, p 842)*

290. **(D)** Radiography of pediatric patients with a myelomeningocele defect should be performed in the prone position, rather than the routine supine position. The supine position would put unnecessary pressure on the protrusion of the meninges and spinal cord. All of the other statements in the question are true. Anatomic dimensions for children are different than for adults, and this must be kept in mind when performing pediatric radiography. The liver occupies a larger area of the abdominal cavity in a child than in the adult. This causes the kidneys to be in a lower position. Generally, the kidneys will be midway between the diaphragm and the symphysis pubis. Chest radiography for the pediatric patient varies depending on the age of the child. Neonates are routinely radiographed in the supine position. Although infants may also be positioned supine, it is preferable to examine them by placing the infant securely in a support device to obtain a good PA erect radiograph. Exceptions to this rule are made if the infant is in respiratory distress. To avoid aggravating the respiratory distress, an erect AP radiograph is usually obtained. *(Dowd & Wilson, vol 2, pp 1004–1005, 1013)*

291. **(D)** Operative cholangiography may be performed to visualize biliary stones or a neoplasm, determine the function of the hepatopancreatic ampulla, and examine the patency of the biliary tract. Any strictures or obstructions may be localized when contrast media is introduced into the catheter and films are obtained. It is important that no air bubbles are introduced into the biliary tract because they can imitate radiolucent stones. The radiographer can coordinate the time of exposure with the anesthesiologist to obtain the radiographs during suspended respiration. *(Dowd & Wilson, vol 2, p 840)*

292. **(D)** Linear, circular, elliptical, and hypocyclodial are all types of tomographic motions. The names refer to the motion of the x-ray tube and blurring pattern. The blurring patterns may be divided into two groups: unidirectional (linear) and pluridirectional (circular, elliptical, hypocyclodial, and spiral). There are advantages and disadvantages to each different blurring technique. Although linear tomography allows for short exposure times, variable exposure angles, and maximum blurring, linear streaking can occur, which can inadvertently mask pathology. On the other hand hypocyclodial tomography, which will allow for demonstration of structures as small as 1 mm, requires longer exposure times and many exposures to "cut" completely through a section of anatomy. In order to optimize tomographic examinations, the choice of blurring pattern should be determined by size and shape of the part under examination. For example, although the hips may be examined at 5 mm slices, the internal auditory canals may be examined at 1 or 2 mm intervals. Although computerized axial tomography (CAT scan) and magnetic resonance imaging (MRI) have reduced the number of tomograms performed today, many institutions still practice conventional tomography, particularly in conjunction with intravenous urography examinations. *(Ballinger, vol 3, pp 48–49)*

293. **(C)** T-tube cholangiography, also referred to as postoperative cholangiography involves the introduction of contrast media via the T-tube, or postoperative drainage tube. The exam is usually performed several days after surgery to check for residual stones and biliary tree function. A chiba needle is not used for T-tube cholangiography, but rather for percutaneous transhepatic cholangiography. *(Dowd & Wilson, vol 2, p 840)*

294. **(C)** Arthrography requires the use of a local, rather than general anesthesia. Sterile technique should be employed to avoid introducing infection into the joint. Other possible complications of arthrography include pain, trauma to nearby structures, and capsular rupture. It is recommended that contrast agents with meglumine salts be used, rather than sodium salts, as they have been found to be less painful when introduced into joint spaces. Fluoroscopy is used for proper placement of the needle and to obtain films immediately after the introduction of contrast media. *(Chapman & Nakielny, pp 204–207)*

295. **(C)** The contrast media of choice for use in myelography is nonionic water soluble. For years Pantopaque, an non-water-soluble (ethyl ester) contrast agent, was used for radiographic demonstration of the spinal canal. Because it was nonsoluble, it had to be removed after the procedure. Metrizamide was the first non-ionic contrast agent introduced for use in myelography, but it has been replaced with iohexol and iopamidol, which are cheaper, safer, and do not dissipate as quickly. Ionic contrast is not used for intrathecal injections, because it is too toxic and gas or air and does not provide adequate demonstration. *(Ballinger, vol 2, p 500)*

296. **(B)** The cephalic, basilic, and subclavian veins should be demonstrated on an upper limb venogram. Venography of the upper limb is usually performed to rule out venous obstruction or thrombosis. The injection site is usually in the hand or wrist, and films should be obtained up to the area of the superior vena cava. *(Ballinger, vol 2, p 540)*

297. **(C)** Neither an antishock garment nor a pneumatic splint should be removed by the radiographer prior to performing radiographic examination. A ring may certainly be removed whenever possible, before performing a hand radiograph. An antishock garment is used when a patient has suffered a traumatic incident and is suffering from internal bleeding; it functions to slow the rate of bleeding. An air cast may be used to temporarily support a fractured limb until surgery and/or a more permanent cast is in place. Both antishock garment and air splints are radiolucent; most rings are radiopaque. *(Adler & Carlton, p 170)*

298. **(C)** Patients are instructed to remove all jewelry, hair clips, metal prostheses, coins, and credit cards before entering the room for a *magnetic resonance imaging* (MRI) scan. MRI does not use radiation to produce images, but uses a very strong magnetic field. All patients must be screened prior to entering the magnetic field to be sure they do not have any metal on or within them. Proper screening included questioning the patient about any eye injury with metal, cardiac pacemakers, aneurysm clips, insulin pumps, heart valves, shrapnel, or any metal in the body. This is extremely important, and if there is any doubt, the patient should be rescheduled until a time when it is determined safe to enter the room. Patients who have done metal work or welding are frequently sent to diagnostic radiology for screening films of the orbits, to ensure that there are no metal fragments near the optic nerve. Any metallic objects externally, such as bobby pins, hair clips, or coins in the pocket, must be removed, or they will be pulled by the magnet, and can cause harm to the patient. Credit cards and any plastic card with a magnetic strip will be wiped clean if they come in contact with the magnetic field. *(Torres, p 244)*

299. **(C)** When imaging a patient with a possible traumatic spine injury, it is appropriate to either maneuver the x-ray tube head or, if the patient must be moved, to use the log-roll method. This cannot be done by one person; the radiographer must summon assistance. If

the patient is on a backboard and in a neck collar, as are most patients with suspected spine injury, it is never appropriate to ask the patient to turn, scoot, or slide over. The only movement that should be permitted is when the entire spine, body, and head move together, as in the log-roll. Any twisting could cause severe and permanent damage to the spinal cord, resulting in paralysis or even death. *(Ehrlich & McCloskey, p 180)*

300. **(B)** A type of fracture in which the splintered ends of bone are forced through the skin is a compound fracture. In a closed fracture, no bone would protrude through the skin. Compression fractures are seen in stressed areas, such as the vertebrae. A depressed fracture would not protrude, but rather be pushed in. *(Ehrlich & McCloskey, pp 180–181)*

Subspecialty List

74. Extremity imaging
75. Cardiovascular/neurological/miscellaneous imaging
76. Head and neck imaging
77. Head and neck imaging
78. Thorax imaging
79. Spine and pelvis imaging
80. Thorax imaging
81. General procedural considerations
82. Extremity imaging
83. Abdomen and GI studies
84. Spine and pelvis imaging
85. Extremity imaging
86. Extremity imaging
87. Extremity imaging
88. Cardiovascular/neurological/miscellaneous imaging
89. Urological studies
90. Head and neck imaging
91. Thorax imaging
92. Head and neck imaging
93. Abdomen and GI studies
94. Cardiovascular/neurological/miscellaneous imaging
95. Spine and pelvis imaging
96. Thorax imaging
97. Extremity imaging
98. General procedural considerations
99. Extremity imaging
100. Abdomen and GI studies
101. Abdomen and GI studies
102. Extremity imaging
103. General procedural considerations
104. Abdomen and GI studies
105. Extremity imaging
106. Thorax imaging
107. Head and neck imaging
108. Abdomen and GI studies
109. Extremity imaging
110. Abdomen and GI studies
111. Extremity imaging
112. Extremity imaging
113. Head and neck imaging
114. Abdomen and GI studies
115. Head and neck imaging
116. Extremity imaging
117. Spine and pelvis imaging
118. Spine and pelvis imaging
119. Spine and pelvis imaging
120. Extremity imaging
121. Extremity imaging
122. Cardiovascular/neurological/miscellaneous imaging
123. General procedural considerations
124. General procedural considerations
125. Abdomen and GI studies
126. General procedural considerations
127. Head and neck imaging
128. Extremity imaging
129. Extremity imaging
130. General procedural considerations
131. Extremity imaging
132. Extremity imaging
133. Urological studies
134. General procedural considerations
135. Extremity imaging
136. Spine and pelvis imaging
137. General procedural considerations
138. Extremity imaging
139. Extremity imaging
140. Extremity imaging
141. Spine and pelvis imaging
142. General procedural considerations
143. Abdomen and GI studies
144. Thorax imaging
145. Abdomen and GI studies
146. Spine and pelvis imaging
147. Head and neck imaging
148. Extremity imaging
149. Abdomen and GI studies
150. Abdomen and GI studies
151. Extremity imaging
152. Abdomen and GI studies
153. Extremity imaging
154. Extremity imaging
155. Extremity imaging
156. General procedural considerations
157. Extremity imaging
158. Spine and pelvis imaging
159. Thorax imaging
160. Thorax imaging
161. Extremity imaging
162. Abdomen and GI studies
163. Thorax imaging
164. Head and neck imaging
165. Extremity imaging
166. Extremity imaging
167. Extremity imaging
168. Extremity imaging
169. Abdomen and GI studies
170. Extremity imaging
171. Extremity imaging

172. General procedural considerations
173. Abdomen and GI studies
174. Cardiovascular/neurological/miscellaneous imaging
175. Head and neck imaging
176. Thorax imaging
177. Thorax imaging
178. Extremity imaging
179. Cardiovascular/neurological/miscellaneous imaging
180. Extremity imaging
181. Head and neck imaging
182. Thorax imaging
183. General procedural considerations
184. Abdomen and GI studies
185. Abdomen and GI studies
186. Extremity imaging
187. Extremity imaging
188. Abdomen and GI studies
189. Spine and pelvis imaging
190. Spine and pelvis imaging
191. Urological studies
192. Thorax imaging
193. Extremity imaging
194. Head and neck imaging
195. Cardiovascular/neurological/miscellaneous imaging
196. Extremity imaging
197. General procedural considerations
198. Urological studies
199. General procedural considerations
200. Thorax imaging
201. Thorax imaging
202. Abdomen and GI studies
203. Spine and pelvis imaging
204. Thorax imaging
205. Spine and pelvis imaging
206. Cardiovascular/neurological/miscellaneous imaging
207. Head and neck imaging
208. Abdomen and GI studies
209. Thorax imaging
210. Abdomen and GI studies
211. General procedural considerations
212. Spine and pelvis imaging
213. Spine and pelvis imaging
214. Urological studies
215. Spine and pelvis imaging
216. Abdomen and GI studies
217. Spine and pelvis imaging
218. Abdomen and GI studies
219. General procedural considerations
220. Urological studies
221. Spine and pelvis imaging
222. Spine and pelvis imaging
223. General procedural considerations
224. Urological studies
225. Spine and pelvis imaging
226. Abdomen and GI studies
227. Cardiovascular/neurological/miscellaneous imaging
228. Urological studies
229. Spine and pelvis imaging
230. Cardiovascular/neurological/miscellaneous imaging
231. Head and neck imaging
232. Head and neck imaging
233. Cardiovascular/neurological/miscellaneous imaging
234. Head and neck imaging
235. Cardiovascular/neurological/miscellaneous imaging
236. Abdomen and GI studies
237. Cardiovascular/neurological/miscellaneous imaging
238. Thorax imaging
239. Spine and pelvis imaging
240. Head and neck imaging
241. Abdomen and GI studies
242. General procedural considerations
243. Abdomen and GI studies
244. Cardiovascular/neurological/miscellaneous imaging
245. Cardiovascular/neurological/miscellaneous imaging
246. Extremity imaging
247. Abdomen and GI studies
248. Spine and pelvis imaging
249. Cardiovascular/neurological/miscellaneous imaging
250. Abdomen and GI studies
251. Extremity imaging
252. General procedural considerations
253. Head and neck imaging
254. Spine and pelvis imaging
255. Abdomen and GI studies
256. Extremity imaging
257. Spine and pelvis imaging
258. Extremity imaging
259. Head and neck imaging
260. Abdomen and GI studies
261. Head and neck imaging

262. Head and neck imaging
263. Head and neck imaging
264. Spine and pelvis imaging
265. Head and neck imaging
266. Extremity imaging
267. Spine and pelvis imaging
268. Head and neck imaging
269. Abdomen and GI studies
270. Cardiovascular/neurological/miscellaneous imaging
271. Head and neck imaging
272. Cardiovascular/neurological/miscellaneous imaging
273. General procedural considerations
274. Head and neck imaging
275. Spine and pelvis imaging
276. Thorax imaging
277. Head and neck imaging
278. Thorax imaging
279. Extremity imaging
280. Head and neck imaging
281. Abdomen and GI studies
282. Head and neck imaging
283. Abdomen and GI studies
284. General procedural considerations

285. Cardiovascular/neurological/miscellaneous imaging
286. Cardiovascular/neurological/miscellaneous imaging
287. Cardiovascular/neurological/miscellaneous imaging
288. Cardiovascular/neurological/miscellaneous imaging
289. Abdomen and GI studies
290. General procedural considerations
291. Abdomen and GI studies
292. Cardiovascular/neurological/miscellaneous imaging
293. Abdomen and GI studies
294. Cardiovascular/neurological/miscellaneous imaging
295. Cardiovascular/neurological/miscellaneous imaging
296. Cardiovascular/neurological/miscellaneous imaging
297. General procedural considerations
298. General procedural considerations
299. Spine and pelvis imaging
300. General procedural considerations

CHAPTER 3

Radiation Protection
Questions

1. How much protection is provided from a 75 kVp x-ray beam when using a 0.25-mm lead equivalent apron?

   (A) 51 percent
   (B) 66 percent
   (C) 88 percent
   (D) 99 percent

2. Gonadal shielding should be used on the male patient for which of the following exams?

   1. hip
   2. thoracic spine
   3. femur

   (A) 1 only
   (B) 1 and 2 only
   (C) 1 and 3 only
   (D) 1, 2, and 3

3. The operation of personnel radiation monitoring devices depends on which of the following?

   1. ionization
   2. thermoluminescence
   3. resonance

   (A) 1 only
   (B) 1 and 2 only
   (C) 2 and 3 only
   (D) 1, 2, and 3

4. Which of the following is (are) composed of nondividing, differentiated cells?

   1. neurons and neuroglia
   2. epithelial tissue
   3. lymphocytes

   (A) 1 only
   (B) 1 and 2 only
   (C) 1 and 3 only
   (D) 1, 2, and 3

5. Types of secondary radiation barriers include

   1. control booth wall
   2. lead aprons
   3. mobile x-ray barriers

   (A) 2 only
   (B) 1 and 2 only
   (C) 2 and 3 only
   (D) 1, 2, and 3

6. Which of the following factors is (are) important in determining thickness of protective barriers?

1. distance between x-ray source and barrier
2. time of occupancy factor
3. workload (mA-min/wk)

(A) 1 only
(B) 1 and 2 only
(C) 1 and 3 only
(D) 1, 2, and 3

7. Which of the following MOST effectively minimizes radiation exposure to the patient?

(A) small focal spot
(B) low-ratio grids
(C) long focal-film distance
(D) beam restriction

8. Which of the following radiation-induced conditions is MOST likely to have the longest latency period?

(A) leukemia
(B) temporary sterility
(C) erythema
(D) acute radiation lethality

9. Which of the following are radiation protection measures appropriate for mobile radiography?

1. the radiographer must be at least 6 feet from the patient and x-ray tube during the exposure
2. the radiographer must announce in a loud voice that an exposure is about to be made, and wait for personnel, visitors, and patients to temporarily leave the area
3. the radiographer must try to use the shortest practical SID

(A) 1 and 2 only
(B) 1 and 3 only
(C) 2 and 3 only
(D) 1, 2, and 3

10. A student radiographer who is under 18 years of age must not receive an annual occupational dose greater than:

(A) 0.1 rem (1 mSv)
(B) 0.5 rem (5 mSv)
(C) 5 rem (50 mSv)
(D) 10 rem (100 mSv)

11. Which of the following factors will affect both the quality and quantity of the primary beam?

1. half-value layer (HVL)
2. kV
3. mA

(A) 1 only
(B) 1 and 2 only
(C) 1 and 3 only
(D) 1, 2, and 3

12. The primary function of filtration is to

(A) reduce patient skin dose
(B) reduce operator dose
(C) reduce film noise
(D) reduce scattered radiation

13. Some patients, such as infants and the elderly, are frequently unable to maintain the necessary radiographic position without assistance. If mechanical restraining devices cannot be used for a child, for example, who of the following should be requested or permitted to hold this patient?

    (A) transporter
    (B) patient's father
    (C) patient's mother
    (D) student radiographer

14. A "controlled area" is defined as one

    1. that is occupied by people trained in radiation safety
    2. that is occupied by people who wear radiation monitors
    3. whose occupancy factor is one (1)

    (A) 1 and 2 only
    (B) 2 only
    (C) 1 and 3 only
    (D) 1, 2, and 3

15. Which of the following defines the gonadal dose that, if received by every member of the population, would be expected to produce the same total genetic effect on that population as the actual doses received by each of the individuals?

    (A) genetically significant dose
    (B) somatically significant dose
    (C) maximum permissible dose
    (D) lethal dose

16. Which of the following is (are) important for patient protection during fluoroscopic procedures?

    1. intermittent fluoroscopy
    2. fluoroscopic field size
    3. focus-to-table distance

    (A) 1 and 2 only
    (B) 1 and 3 only
    (C) 2 and 3 only
    (D) 1, 2, and 3

17. What is the annual dose-limit equivalent for the skin and hands of an occupationally exposed individual?

    (A) 5 rem
    (B) 25 rem
    (C) 50 rem
    (D) 100 rem

18. The x-ray interaction with matter that is responsible for the majority of scattered radiation reaching the film is

    (A) photoelectric effect
    (B) Compton scatter
    (C) classical scatter
    (D) Thompson scatter

19. One rule of radiation protection is that radiation must scatter at LEAST how many times before reaching the operator?

    (A) 1
    (B) 2
    (C) 3
    (D) 4

20. Which of the following formulas is a representation of the Inverse Square Law and may be used to determine x-ray intensity at different distances?

(A) $\dfrac{I_1}{I_2} = \dfrac{D_2^2}{D_1^2}$

(B) $\dfrac{I_1}{I_2} = \dfrac{D_1^2}{D_2^2}$

(C) $\dfrac{kVp_1}{kVp_2} = \dfrac{D_2^2}{D_1^2}$

(D) $\dfrac{kVp_1}{kVp_2} = \dfrac{D_1^2}{D_2^2}$

21. Sources of natural background radiation exposure include

  1. the food we eat
  2. air travel
  3. medical and dental x-rays

(A) 1 only
(B) 1 and 2 only
(C) 2 and 3 only
(D) 1, 2, and 3

22. Each time an x-ray beam scatters, its intensity at 1 meter from the scattering object is what fraction of its original intensity?

(A) 1/10
(B) 1/100
(C) 1/500
(D) 1/1000

23. Types of gonadal shielding include which of the following?

  1. flat contact
  2. shaped (contour) contact
  3. shadow

(A) 1 only
(B) 1 and 2 only
(C) 2 and 3 only
(D) 1, 2, and 3

24. Most radiation detectors operate on which of the following x-ray characteristics?

(A) luminescent effects
(B) ionization effects
(C) physiologic effects
(D) thermionic emission

25. Sources of secondary radiation include

  1. background radiation
  2. leakage radiation
  3. scattered radiation

(A) 1 only
(B) 1 and 2 only
(C) 2 and 3 only
(D) 1, 2, and 3

26. Filters used in radiographic x-ray tubes are generally composed of

(A) aluminum
(B) copper
(C) tin
(D) lead

27. The photoelectric process is an interaction between an x-ray photon and

(A) an inner-shell electron
(B) an outer-shell electron
(C) a nucleus
(D) another photon

28. The reduction in radiation intensity as it passes through material is termed

   (A) absorption
   (B) scattering
   (C) attenuation
   (D) divergence

29. Which of the following factors can affect the amount or nature of radiation damage to biologic tissue?

   1. radiation quality
   2. absorbed dose
   3. size of irradiated area

   (A) 1 only
   (B) 2 only
   (C) 1 and 2 only
   (D) 1, 2, and 3

30. According to NCRP regulations, leakage radiation from the x-ray tube must NOT exceed

   (A) 10 mR/hr
   (B) 100 mR/hr
   (C) 10 mR/min
   (D) 100 mR/min

31. The annual dose-limit to medical imaging personnel includes radiation from

   1. medical x-rays
   2. occupational exposure
   3. background radiation

   (A) 1 only
   (B) 2 only
   (C) 2 and 3 only
   (D) 1, 2, and 3

32. In which type of monitoring device do photons release electrons by their interaction with air?

   (A) film badge
   (B) thermoluminescent dosimeter
   (C) ion chamber
   (D) scintillation detector

33. The symbols $^{130}_{56}$Ba and $^{138}_{56}$Ba are examples of which of the following?

   (A) isotopes
   (B) isobars
   (C) isotones
   (D) isomers

34. Following exposure to 1 rad of each of the following ionizing radiations, which would result in the greatest dose to the individual?

   (A) external source of 1 MeV x-rays
   (B) external source of diagnostic x-rays
   (C) internal source of alpha particles
   (D) external source of beta particles

35. Which of the following is (are) possible long-term somatic effects of radiation exposure?

   1. nausea and vomiting
   2. carcinogenesis
   3. leukemia

   (A) 1 only
   (B) 1 and 2 only
   (C) 2 and 3 only
   (D) 1, 2, and 3

36. The purpose of inherent and added filtration in the x-ray tube is to

   (A) reduce patient skin dose
   (B) shorten the scale of contrast
   (C) reduce scattered radiation
   (D) soften the x-ray beam

37. The skin response to radiation exposure, that appears as reddening of the irradiated skin area, is known as

   (A) dry desquamation
   (B) moist desquamation
   (C) erythema
   (D) epilation

38. How does use of rare earth intensifying screens contribute to lowering patient dose?

1. permits the use of lower mAs
2. permits the use of lower kVp
3. eliminates the need for patient shielding

(A) 1 only
(B) 1 and 2 only
(C) 1 and 3 only
(D) 2 and 3 only

39. Immature cells are referred to as

1. undifferentiated
2. stem
3. genetic

(A) 1 only
(B) 1 and 2 only
(C) 1 and 3 only
(D) 1, 2, and 3

40. The unit of measure used to express occupational exposure is the

(A) Roentgen (C/kg)
(B) rad (Gray)
(C) rem (Seivert)
(D) RBE

41. Which of the following statements regarding the human gonadal cells is (are) true?

1. the female oogonia reproduce only during fetal life
2. the male spermatogonia reproduce continuously
3. both male and female stem cells reproduce only during fetal life

(A) 1 only
(B) 2 only
(C) 1 and 2 only
(D) 3 only

42. To be in compliance with radiation safety standards, the fluoroscopy switch must

(A) sound during fluoro-on time
(B) be on a 6-foot long cord
(C) terminate fluoro after 5 minutes
(D) be the dead-man type

43. Primary radiation barriers must be at LEAST how high?

(A) 5 feet
(B) 6 feet
(C) 7 feet
(D) 8 feet

44. The MPD for occupationally exposed individuals is valid for

(A) alpha, beta, and x-radiations
(B) x and gamma radiations only
(C) beta, x, and gamma radiations
(D) all ionizing radiations

45. The interaction between x-ray photons and matter pictured in Figure 3–1 is associated with

1. high-energy x-ray photons
2. ionization
3. characteristic radiation

(A) 1 only
(B) 1 and 2 only
(C) 1 and 3 only
(D) 2 and 3 only

46. The likelihood of adverse radiation effects to any radiographer whose dose is kept below the recommended guideline is

(A) very probable
(B) possible
(C) very remote
(D) zero

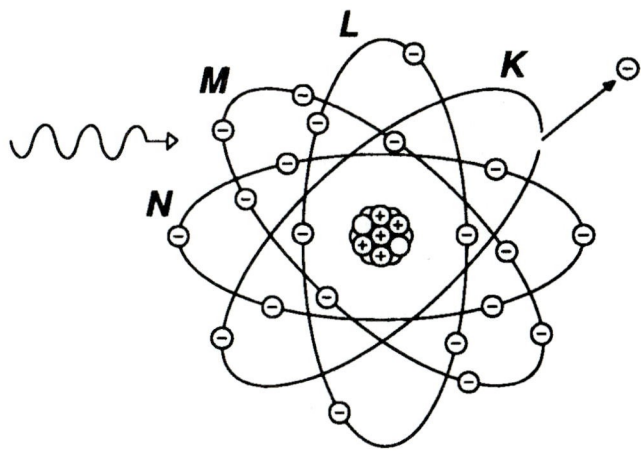

Figure 3–1. Courtesy of David Perri.

47. The purpose of filters in a film badge is

(A) to eliminate harmful rays
(B) to measure radiation quality
(C) to prevent exposure from alpha particles
(D) as a support for film contained within

48. The unit of absorbed dose is the

(A) Roentgen (C/kg)
(B) rad (Gy)
(C) rem (Sv)
(D) RBE

49. The Law of Bergonié and Tribondeau states that cells are more radiosensitive if they are

1. highly mitotic
2. undifferentiated
3. mature cells

(A) 1 only
(B) 1 and 2 only
(C) 2 and 3 only
(D) 1, 2, and 3

50. Medical and dental radiation accounts for what percentage of the general public's exposure to man-made radiation?

(A) 10 percent
(B) 50 percent
(C) 75 percent
(D) 90 percent

51. Skin response to radiation exposure, appearing as hair loss, is known as

(A) dry desquamation
(B) moist desquamation
(C) erythema
(D) epilation

52. Biologic material is MOST sensitive to irradiation under which of the following conditions?

(A) anoxic
(B) hypoxic
(C) oxygenated
(D) deoxygenated

53. Which of the following terms refers to the period between conception and birth?

(A) gestation
(B) congenital
(C) neonatal
(D) in vitro

54. Which of the following have an effect on the amount and type of radiation-induced tissue damage?

1. quality of radiation
2. type of tissue being irradiated
3. fractionation

(A) 1 only
(B) 1 and 2 only
(C) 1 and 3 only
(D) 1, 2, and 3

55. The presence of ionizing radiation may be detected in which of the following ways?

    1. ionizing effect on air
    2. physiologic effect on living tissue
    3. fluorescent effect on certain crystals

    (A) 1 only
    (B) 1 and 2 only
    (C) 1 and 3 only
    (D) 1, 2, and 3

56. What unit of measure expresses the amount of energy deposited in tissue?

    (A) Roentgen (C/kg)
    (B) rad (Gy)
    (C) rem (Sv)
    (D) RBE

57. An undifferentiated cell is a

    1. precursor cell
    2. stem cell
    3. mature cell

    (A) 1 only
    (B) 1 and 2 only
    (C) 2 and 3 only
    (D) 1, 2, and 3

58. Which stage of mitosis is considered the MOST radiosensitive?

    (A) prophase
    (B) metaphase
    (C) anaphase
    (D) telophase

59. The dose–response curve that appears to be valid for genetic and some somatic effects is the

    1. linear
    2. sigmoidal
    3. non-threshold

    (A) 1 only
    (B) 1 and 3 only
    (C) 2 and 3 only
    (D) 1, 2, and 3

60. Which of the following are precautions recommended for the pregnant radiographer?

    1. send for weekly dosimeter readings
    2. wear a second dosimeter under the lead apron
    3. wear two dosimeters and switch their position periodically

    (A) 1 only
    (B) 2 only
    (C) 1 and 3 only
    (D) 2 and 3 only

61. Under what circumstances might a radiographer be required to wear two dosimeters?

    1. during pregnancy
    2. during vascular procedures
    3. during mobile radiography

    (A) 1 and 2 only
    (B) 2 only
    (C) 2 and 3 only
    (D) 1, 2, and 3

62. Isotopes are atoms that have the same

    (A) mass number but a different atomic number
    (B) atomic number but a different mass number
    (C) atomic number but a different neutron number
    (D) atomic number and mass number

63. Gamma rays have

(A) mass and a charge of 1
(B) no mass and a charge of 1
(C) mass of 1 and a charge of zero
(D) no mass and no charge

64. Which of the following accounts for the x-ray beam's heterogeneity?

1. incident electrons interacting with several layers of tungsten target atoms
2. energy differences among incident electrons
3. electrons moving to fill different shell vacancies

(A) 1 only
(B) 1 and 2 only
(C) 1 and 3 only
(D) 1, 2, and 3

65. In the production of Bremsstrahlung radiation, the incident electron

(A) ejects an inner-shell tungsten electron
(B) ejects an outer-shell tungsten electron
(C) is deflected with resulting energy loss
(D) is deflected with resulting energy increase

66. What is the term used to describe x-ray photon interaction with matter and the transference of part of the photons energy to matter?

(A) absorption
(B) scattering
(C) attenuation
(D) divergence

67. Which of the following contributes MOST to patient dose?

(A) photoelectric effect
(B) Compton scatter
(C) classical scatter
(D) Thompson scatter

68. Which of the following illustrates the Inverse Square Law?

1. that distance is a most effective protection from radiation
2. that distance is a rather ineffective protection from radiation
3. as distance from the radiation source decreases, radiation decreases

(A) 1 only
(B) 1 and 2 only
(C) 1 and 3 only
(D) 2 and 3 only

69. Which of the following disorders is MOST likely to result from irradiation of the fetus in utero during the first trimester?

(A) leukemia
(B) sterility
(C) CNS abnormalities
(D) bone marrow syndrome

70. Stochastic effects of radiation are those that

1. are "late effects"
2. may be described as "all or nothing" effects
3. do not exhibit a threshold

(A) 1 only
(B) 1 and 2 only
(C) 2 and 3 only
(D) 1, 2, and 3

71. Which of the following is a measure of dose to biologic tissue?

(A) Roentgen (C/kg)
(B) rad (Gy)
(C) rem (Sv)
(D) RBE

72. A thermoluminescent dosimetry system would use which of the following crystals?

(A) silver bromide
(B) sodium sulfite
(C) lithium fluoride
(D) ferrous sulfate

73. What is the intensity of scattered radiation perpendicular to and 1 meter from the patient, compared to the useful beam at the patient's surface?

    (A) 0.01 percent
    (B) 0.1 percent
    (C) 1.0 percent
    (D) 10.0 percent

74. Which of the following are features of fluoroscopic equipment, designed especially to eliminate unnecessary radiation exposure to the patient and/or personnel?

    1. bucky slot cover
    2. exposure switch
    3. cumulative timer

    (A) 1 only
    (B) 1 and 2 only
    (C) 2 and 3 only
    (D) 1, 2, and 3

75. The biologic effect to an individual is dependent on which of the following?

    1. type of tissue interaction(s)
    2. amount of interactions
    3. biologic differences

    (A) 1 and 2 only
    (B) 1 and 3 only
    (C) 2 and 3 only
    (D) 1, 2, and 3

76. Protective devices such as lead aprons function to protect the user from

    1. scattered radiation
    2. the primary beam
    3. remnant radiation

    (A) 1 only
    (B) 1 and 2 only
    (C) 1 and 3 only
    (D) 1, 2, and 3

77. The amount of time that x-rays are being produced and directed toward a particular wall is referred to as

    (A) work load
    (B) use factor
    (C) occupancy factor
    (D) controlling factor

78. Radiographers use monitoring devices to record their monthly exposure to radiation. The type(s) of device(s) used would MOST likely be which of the following?

    1. film badge
    2. TLD
    3. cutie pie

    (A) 1 only
    (B) 1 and 2 only
    (C) 2 and 3 only
    (D) 1, 2, and 3

79. Which of the following is (are) a result of beam restriction due to collimation?

    1. less scatter radiation production
    2. less patient hazard
    3. less radiographic contrast

    (A) 1 only
    (B) 1 and 2 only
    (C) 2 and 3 only
    (D) 1, 2, and 3

80. What is the established fetal dose-limit guideline for pregnant radiographers during the entire gestation period?

    (A) 100 mrem
    (B) 250 mrem
    (C) 500 mrem
    (D) 1000 mrem

81. Early symptoms of acute radiation syndrome include

    1. leukemia

    2. nausea and vomiting

    3. cataracts

    (A) 1 and 2 only

    (B) 2 only

    (C) 1 and 3 only

    (D) 2 and 3 only

82. Referring to the nomogram in Figure 3–2, what is the approximate patient skin exposure from an AP projection of the cervical spine made at 105 cm using 60 kVp, 300 mA, 0.2 second, and 2.5 mm Al total filtration?

    (A) 3 mR

    (B) 10 mR

    (C) 140 mR

    (D) 180 mR

83. Diagnostic x-radiation may be correctly described as

    (A) low energy, low LET

    (B) low energy, high LET

    (C) high energy, low LET

    (D) high energy, high LET

84. If 100 R or more is received as whole-body dose in a short period of time, certain symptoms will occur and are referred to as

    (A) short-term effects

    (B) long-term effects

    (C) lethal dose

    (D) acute radiation syndrome

85. Which of the following contributes MOST to occupational exposure?

    (A) photoelectric effect

    (B) Compton scatter

    (C) classical scatter

    (D) Thompson scatter

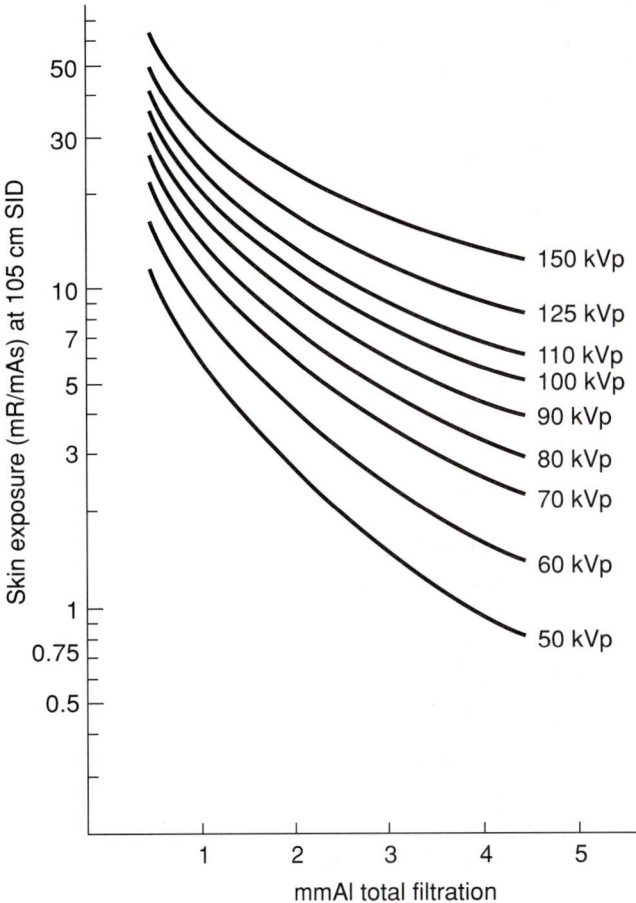

Figure 3–2.

86. In radiation protection, the product of absorbed dose and the correct modifying factor (rad × QF) is used to determine

    (A) Roentgen (C/kg)

    (B) rem (Sv)

    (C) rad (Gy)

    (D) radiation quality

87. Which of the following is (are) features of fluoroscopic equipment, designed especially to eliminate unnecessary radiation to patient and personnel?

    1. protective curtain

    2. filtration

    3. collimation

    (A) 1 only

    (B) 1 and 2 only

    (C) 1 and 3 only

    (D) 1, 2, and 3

88. The MOST efficient type of male gonadal shielding for use during fluoroscopy is

    (A) flat contact
    (B) shaped (contour) contact
    (C) shadow
    (D) cylindrical

89. If a patient received 3000 mrad during a 15-minute fluoroscopic examination, what was the dose rate?

    (A) 0.2 rad/min
    (B) 2.0 rad/min
    (C) 20 rad/min
    (D) 200 rad/min

90. Which of the following refers to a regular program of evaluation that ensures proper functioning of x-ray equipment, thereby protecting both radiation workers and patients?

    (A) sensitometry
    (B) densitometry
    (C) quality assurance
    (D) modulation transfer function

91. Methods of reducing radiation exposure to patients and/or personnel include

    1. beam restriction
    2. shielding
    3. high kV, low mAs factors

    (A) 1 only
    (B) 1 and 2 only
    (C) 2 and 3 only
    (D) 1, 2, and 3

92. Factors that determine the amount of scattered radiation produced include which of the following?

    1. radiation quality
    2. field size
    3. pathology

    (A) 1 only
    (B) 1 and 2 only
    (C) 2 and 3 only
    (D) 1, 2, and 3

93. The focal spot-to-table distance, in mobile fluoroscopy, must be

    (A) a minimum of 15 inches
    (B) a maximum of 15 inches
    (C) a minimum of 12 inches
    (D) a maximum of 12 inches

94. The effects of radiation to biologic material are dependent on several factors. If a quantity of radiation is delivered to a body over a long period of time, the effect

    (A) will be greater than if it were delivered all at one time
    (B) will be less than if it were delivered all at one time
    (C) has no relation to how it is delivered in time
    (D) is solely dependent on the radiation quality

95. Linear energy transfer (LET) is

    1. a method of expressing radiation quality
    2. a measure of the rate at which radiation energy is transferred to soft tissue
    3. absorption of polyenergetic radiation

    (A) 1 only
    (B) 1 and 2 only
    (C) 1 and 3 only
    (D) 1, 2, and 3

96. It is essential to question female patients of child-bearing age regarding

    1. the date of her last menstrual period
    2. the possibility of her being pregnant
    3. the number of children she presently has

    (A) 1 only
    (B) 1 and 2 only
    (C) 1 and 3 only
    (D) 2 and 3 only

97. In 1906, Bergonié and Tribondeau established their law, which states that cells are more radiosensitive if they

    1. are young
    2. are stem cells
    3. have a low proliferation rate

    (A) 1 only
    (B) 1 and 2 only
    (C) 2 and 3 only
    (D) 1, 2, and 3

98. What is the effect on relative biologic effectiveness (RBE) as linear energy transfer (LET) increases?

    (A) as LET increases, RBE increases
    (B) as LET increases, RBE decreases
    (C) as LET increases, RBE stabilizes
    (D) LET has no effect on RBE

99. Which of the following would MOST likely cause the greatest skin dose?

    (A) short SID
    (B) high kVp
    (C) increased filtration
    (D) increased mA

100. The NCRP recommends an annual effective occupational dose equivalent limit of

    (A) 2.5 rem (25 mSv)
    (B) 5 rem (50 mSv)
    (C) 10 rem (100 mSv)
    (D) 20 rem (200 mSv)

101. Which of the following groups of exposure factors will deliver the LEAST amount of exposure to the patient?

    (A) 100 mAs, 100 kVp
    (B) 200 mAs, 90 kVp
    (C) 400 mAs, 80 kVp
    (D) 800 mAs, 70 kVp

102. If an individual received 90 mR while standing 4 feet from a source of radiation for 2 minutes, which of the options listed below will MOST effectively reduce his or her radiation exposure?

    (A) standing 3 feet from the source for 2 minutes
    (B) standing 2 feet from the source for 1 minute
    (C) standing 5 feet from the source for 1 minute
    (D) standing 6 feet from the source for 2 minutes

103. Which of the following tissues or organs listed below is the MOST radiosensitive?

    (A) rectum
    (B) esophagus
    (C) small bowel
    (D) CNS

104. The dose of radiation that will cause a noticeable skin reaction is referred to as the

    (A) LET
    (B) SSD
    (C) SED
    (D) SID

105. The photoelectric effect is the interaction between x-ray photons and matter that is largely responsible for patient dose. The photoelectric effect is likely to occur under which of the following conditions?

    1. with absorbers of high atomic number
    2. with low-energy incident photons
    3. with use of positive contrast media

(A) 1 and 2 only
(B) 1 and 3 only
(C) 2 and 3 only
(D) 1, 2, and 3

106. All of the following affect patient dose, EXCEPT

(A) inherent filtration
(B) added filtration
(C) source-image distance
(D) focal spot size

107. The type of damage that can occur in a DNA molecule following irradiation includes which of the following?

    1. single-chain break
    2. cross-linking
    3. base damage

(A) 1 only
(B) 1 and 2 only
(C) 2 and 3 only
(D) 1, 2, and 3

108. Guidelines for the use of gonadal shielding state that gonadal shielding should be used

    1. if the patient has reproductive potential
    2. when the gonads are within 5 cm of a well-collimated field
    3. when tight collimation is not possible

(A) 1 only
(B) 1 and 2 only
(C) 1 and 3 only
(D) 2 and 3 only

109. To within what percent of the FFD must the collimator light and actual irradiated area be accurate?

(A) 2 percent
(B) 5 percent
(C) 10 percent
(D) 15 percent

110. Which of the following is considered the unit of exposure in air?

(A) Roentgen (C/kg)
(B) rad (Gy)
(C) rem (Sv)
(D) RBE

111. Which of the following are considered especially radiosensitive tissues?

    1. blood forming organs
    2. reproductive organs
    3. lymphocytes

(A) 1 only
(B) 1 and 2 only
(C) 2 and 3 only
(D) 1, 2, and 3

112. How many half-value layers (HVL) are required to reduce the intensity of a beam of polyenergetic photons to less than 10 percent of its original value?

(A) 2
(B) 3
(C) 4
(D) 5

113. The person responsible for ascertaining that all radiation guidelines are adhered to and that personnel understand and employ radiation safety measures is designated as the

(A) radiology department manager
(B) radiation safety officer
(C) chief radiologist
(D) CEO

114. Radiation dose to personnel is reduced by which of the following exposure control cord guidelines?

    1. exposure cords on fixed equipment must be very short
    2. exposure cords on mobile equipment should be fairly long
    3. exposure cords on fixed and mobile equipment should be the coiled expandable type

    (A) 1 only
    (B) 1 and 2 only
    (C) 2 and 3 only
    (D) 1, 2, and 3

115. When the exposure rate of an x-ray beam gradually decreases as it passes through matter, it is termed

    (A) attenuation
    (B) absorption
    (C) scatter radiation
    (D) secondary radiation

116. Which of the following body parts is (are) included in whole-body dose?

    1. gonads
    2. lens
    3. extremities

    (A) 1 only
    (B) 1 and 2 only
    (C) 1 and 3 only
    (D) 1, 2, and 3

117. If the exposure rate to a body standing 5 feet from a radiation source is 12 mR/min, what will be the dose to that body at a distance of 12 feet from the source?

    (A) 2 mR/min
    (B) 5 mR/min
    (C) 29 mR/min
    (D) 69 mR/min

118. Which of the following personnel monitoring devices used in diagnostic radiography is considered to be the MOST sensitive and accurate?

    (A) thermoluminescent dosimeter
    (B) film badge
    (C) Geiger counter
    (D) pocket dosimeter

119. That maximum amount of ionizing radiation that an individual may receive during a lifetime (or in a single exposure) that carries only negligible risk of somatic or genetic injury is termed

    (A) genetically significant dose
    (B) somatically significant dose
    (C) maximum permissible dose
    (D) lethal dose

120. Which of the following radiation situations is potentially the MOST harmful?

    (A) a large dose, to a specific area, all at once
    (B) a small dose, to the whole body, over a period of time
    (C) a large dose, to the whole body, all at one time
    (D) a small dose, to a specific area, over a period of time

121. The interaction between x-ray photons and tissue that is responsible for radiographic contrast but which contributes significantly to patient dose is

    (A) photoelectric effect
    (B) Compton scatter
    (C) coherent scatter
    (D) pair production

122. The correct way(s) to check for cracks in lead aprons is

    1. to fluoroscope them once a year
    2. to radiograph them at low kV twice a year
    3. by visual inspection

  (A) 1 only
  (B) 1 and 2 only
  (C) 2 and 3 only
  (D) 1, 2, and 3

123. Which of the following types of radiation is (are) considered electromagnetic?

    1. x-ray
    2. gamma
    3. beta

  (A) 1 only
  (B) 1 and 2 only
  (C) 2 and 3 only
  (D) 1, 2, and 3

124. Which of the following is (are) associated with Compton scattering?

    1. high-energy incident photons
    2. outer-shell electrons
    3. characteristic radiation

  (A) 1 only
  (B) 1 and 2 only
  (C) 2 and 3 only
  (D) 1, 2, and 3

125. Which of the following cells is the MOST radiosensitive?

  (A) myelocytes
  (B) erythroblasts
  (C) megakaryocytes
  (D) myocytes

126. Which of the following personnel radiation monitors will provide an immediate reading?

  (A) thermoluminescent dosimeter
  (B) film badge
  (C) lithium fluoride chips
  (D) pocket dosimeter

127. Irradiation of water molecules within the body, and their resulting breakdown, is termed

  (A) epilation
  (B) radiolysis
  (C) proliferation
  (D) repopulation

128. Radiation that passes through the tube housing in directions other than that of the useful beam is termed

  (A) scattered radiation
  (B) secondary radiation
  (C) leakage radiation
  (D) remnant radiation

129. The cell division of genetic cells is termed

  (A) mitosis
  (B) meiosis
  (C) synthesis
  (D) replication

130. The tabletop exposure rate during fluoroscopy shall NOT exceed

  (A) 5 mR/min
  (B) 10 mR/hr
  (C) 5 R/hr
  (D) 10 R/min

131. In the production of characteristic radiation at the tungsten target, the incident electron

  (A) ejects an inner-shell tungsten electron
  (B) ejects an outer-shell tungsten electron
  (C) is deflected with resulting energy loss
  (D) is deflected with resulting energy increase

132. What is (are) the major effect(s) of DNA irradiation?

    1. genetic damage
    2. malignant disease
    3. radiolysis

  (A) 1 only
  (B) 1 and 2 only
  (C) 2 and 3 only
  (D) 1, 2, and 3

133. The advantages of beam restriction include

    1. less scattered radiation is produced

    2. less biologic material is irradiated

    3. less total filtration will be necessary

(A) 1 only

(B) 1 and 2 only

(C) 2 and 3 only

(D) 1, 2, and 3

134. Which of the following may be used to express patient dose?

    1. skin dose

    2. gonadal dose

    3. midline dose

(A) 1 only

(B) 1 and 2 only

(C) 1 and 3 only

(D) 1, 2, and 3

135. If the exposure rate at 2.0 m from a source of radiation is 18 R/min, what will be the exposure rate at 5 m from the source?

(A) 2.8 R/min

(B) 7.2 R/min

(C) 45 R/min

(D) 113 R/min

136. Which one or more of the following choices are acceptable ways to monitor radiation exposure to those occupationally employed?

    1. film badge

    2. TLD

    3. quarterly blood count

(A) 1 only

(B) 1 and 2 only

(C) 1 and 3 only

(D) 1, 2, and 3

137. The exposure rate to a body 3 feet from a source of radiation is 22 R/hr. What distance from the source would be necessary to decrease the exposure to 8 R/hr?

(A) 1 foot

(B) 2 feet

(C) 5 feet

(D) 8 feet

138. Biologic material irradiated under hypoxic conditions is

(A) more sensitive than when irradiated under oxygenated conditions

(B) less sensitive than when irradiated under anoxic conditions

(C) less sensitive than when irradiated under oxygenated conditions

(D) unaffected by presence or absence of oxygen

139. Which of the following is (are) helpful in minimizing patient exposure?

    1. accurate positioning

    2. high kV, low mAs factors

    3. rare earth screens

(A) 1 only

(B) 1 and 2 only

(C) 1 and 3 only

(D) 1, 2, and 3

140. What is the single MOST important scattering object in both radiography and fluoroscopy?

(A) x-ray table

(B) x-ray tube

(C) patient

(D) film

141. All of the following statements regarding thermoluminescent dosimeters (TLDs) are true, EXCEPT

(A) TLDs are reusable

(B) TLDs store energy

(C) the TLD's response is proportional to the quantity of radiation received

(D) following x-ray exposure, TLDs are exposed to light and emit a quantity of heat in response

142. Elective abdominal radiographic examinations on women of reproductive age should be limited to the

(A) 10 days following the menses

(B) 10 days following the onset of menses

(C) 10 days before the onset of menses

(D) last 10 days of the menstrual cycle

143. Which of the following systems is the most radiosensitive?

(A) hematapoietic

(B) gastrointestinal

(C) CNS

(D) skeletal

144. Classify the following tissues according to increasing radiosensitivity

1. liver cells

2. intestinal crypt cells

3. muscle cells

(A) 1, 3, 2

(B) 2, 3, 1

(C) 2, 1, 3

(D) 3, 1, 2

145. What minimum total amount of filtration (inherent plus added) is required in equipment operated above 70 kVp?

(A) 2.5 mm Al equivalent

(B) 3.5 mm Al equivalent

(C) 2.5 mm Cu equivalent

(D) 3.5 mm Cu equivalent

146. How does filtration affect the primary beam?

(A) filtration increases the average energy of the primary beam

(B) filtration decreases the average energy of the primary beam

(C) filtration makes the primary beam more penetrating

(D) filtration increases the intensity of the primary beam

147. A test radiograph like the one pictured in Figure 3–3 would be made by the radiation safety officer (RSO) or equipment service person and is used to evaluate

(A) focal spot size

(B) linearity

(C) collimator alignment

(D) spatial resolution

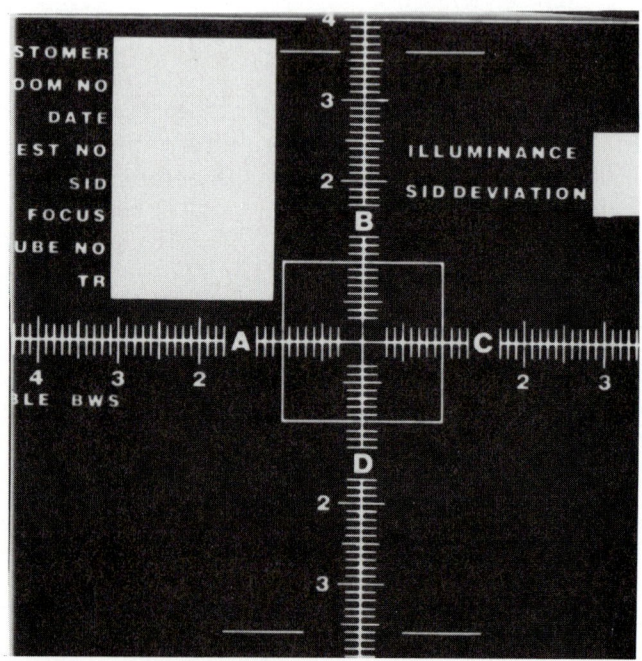

Figure 3–3. Courtesy of Manny DaSilva and The Stamford Hospital, Department of Radiology.

148. Which of the following can be an effective means of reducing radiation exposure?

    1. barriers
    2. distance
    3. time

    (A) 1 only
    (B) 2 only
    (C) 1 and 2 only
    (D) 1, 2, and 3

149. What should be the radiographer's main objective regarding personal radiation safety?

    (A) not to exceed his or her MPD
    (B) to keep personal exposure as far below the MPD as possible
    (C) to avoid whole-body exposure
    (D) to wear protective apparel when "holding" patients for films

150. Gamma rays and x-ray photons differ in their

    (A) velocity
    (B) charge
    (C) mass
    (D) origin

Answers and Explanations

1. **(B)** Lead aprons are worn by occupationally exposed individuals during fluoroscopic procedures. Lead aprons are available with various lead equivalents; 0.25, 0.5, and 1.0 mm are the most common. The *1.0-mm* lead equivalent apron will provide close to 100 percent protection at most kVp levels, but it is rarely used because it weighs anywhere from 12 to 24 pounds! A *0.25-mm* lead equivalent apron will attenuate about 97 percent of a 50-kVp x-ray beam, 66 percent of a 75-kVp beam, and 51 percent of a 100-kVp beam. A *0.5-mm* apron will attenuate about 99 percent of a 50-kVp beam, 88 percent of a 75-kVp beam, and 75 percent of a 100-kVp beam. *(Thompson, p 457)*

2. **(C)** Gonadal shielding should be used when the gonads lie within *5 cm of the collimated primary beam*, when the patient has *reasonable reproductive potential*, and when *clinical objectives permit*. Because their reproductive organs lie outside the body, male patients are more easily and effectively shielded than female patients. Therefore, it is recommended that *radiographic examinations of the male abdomen and pelvic structures should include evidence of gonadal shielding. (Statkiewicz, pp 173–174)*

3. **(B)** Ionization is the fundamental principle of operation of both the film badge and pocket dosimeter. In the film badge, the film's silver halide emulsion is ionized by x-ray photons. The pocket dosimeter contains an ionization chamber, and the number of ionizations taking place may be equated to exposure dose. Thermoluminescent dosimeters (TLD) contain lithium fluoride crystals that undergo characteristic changes upon irradiation. When the crystals are subsequently *heated*, they emit a quantity at visible *light* in proportion to the amount of radiation absorbed. Resonance refers to motion, and has no application to personnel radiation monitoring. *(Statkiewicz, pp 246–256)*

4. **(A)** Nondividing, differentiated cells are specialized, mature cells that *do not undergo mitosis*. Having these qualities, they are rendered radioresistant, according to the theory proposed by Bergonié and Tribondeau. The adult nervous system is composed of nondividing, differentiated cells, and thus is the most radioresistant system in the adult. Epithelial tissue and lymphocytes contain many precursor stem cells, and hence are among the most radiosensitive cells in the body. *(Travis, p 134)*

5. **(D)** *Secondary* radiation is comprised of *leakage and scattered* radiation. The control booth wall is a secondary barrier; the primary beam must therefore never be directed toward it. Lead aprons, lead gloves, mobile x-ray barriers, and so forth, are also designed to protect the user from exposure to *scattered* radiation, and will not protect from the primary beam. *(Statkiewicz p 217)*

6. **(D)** The *closer* the x-ray source is to the barrier (wall), the greater the thickness necessary. *Occupancy factor* refers to the degree of occupancy of the room adjacent to the barrier; a stairway would require less shielding than a busy work area. *Workload* is important in determining barrier thickness, and refers to the number of exams performed in the

x-ray room measured in mA-min/wk: the greater the number of exams per week, the greater the barrier thickness required. *Use factor* is also important in determining barrier thickness, and refers to the amount of time x-rays are directed to a particular wall: the greater the amount of time, the greater the thickness required. *(Statkiewicz, pp 219–222)*

7. **(D)** Focal spot size has no effect on patient dose. Low-ratio grids, although they require less mAs than high-ratio grids, are not a means of patient protection. Long focal-film distances usually require the use of higher mAs and so would not be an effective means of patient protection. *Limiting the size of the field being irradiated, through collimation or other beam restriction, is an extremely important method of controlling patient exposure dose.* *(Selman, p 530)*

8. **(A)** Radiation effects appearing days or weeks following exposure ("early effects") are in response to relatively high radiation doses. These never occur in diagnostic radiology today; they occur only in response to doses much greater than those used in diagnostic radiology. One of the effects that may be noted in such a circumstance is the hematological effect—reduced numbers of white blood cells, red blood cells, and platelets in the circulating blood. Immediate local tissue effects may include gonads (temporary sterility) and skin (epilation, erythema). Acute radiation lethality, or radiation death, occurs after an acute exposure and results in death in weeks or days. Radiation-induced malignancy, leukemia, and genetic effects are "late effects" (or stochastic effects) of radiation exposure. These can occur years after survival from an acute radiation dose, or after exposure to low levels of radiation over a long period of time. Late effects of radiation are the effects of which radiation workers need to be especially aware, because their exposure to radiation is usually low level over a long period of time. Occupational radiation protection guidelines are therefore based on late effects of radiation according to a linear, nonthreshold dose–response curve. *(Bushong, pp 521–522)*

9. **(A)** Mobile radiography is an area of *higher occupational exposure* (along with fluoroscopy and special procedures). With no lead barrier to retreat behind, distance becomes the best source of protection. The exposure switch of portable equipment must be manufactured to allow the technologist to stand *at least 6 feet* away from the patient and x-ray tube. Hospital personnel, visitors, and patients must also be protected from unnecessary radiation exposure. Therefore, the radiographer must request that these people leave the immediate area until after the exposure is made, and *announce* in a loud voice when the exposure is about to be made, allowing time for individuals to leave the area. The use of a short SID increases patient exposure and produces poor recorded detail. *(Carlton & Adler, p 526)*

10. **(A)** Because the established dose-limit formula guideline is used for occupationally exposed persons 18 years of age and older, guidelines had to be established in the event a student entered training prior to age 18. The guideline states that the occupational dose limit for students under age 18 is 0.1 rem (100 mrem or 1 mSv) in any given year. It is important to note that this 0.1 rem *is included* in the 0.5 rem dose-limit allowed the student as a member of the general public. *(Bushong, p 605)*

11. **(B)** Kilovoltage (kV) and half-value layer (HVL) affect a change in both the quantity and quality of the primary beam. *The principle qualitative factor of the primary beam is kV,* but an increase in kV will also affect an increase in the *number* of photons produced at the target. HVL, defined as *the amount of material necessary to decrease the intensity of the beam to one half,* therefore affects a change in both beam quality and quantity. MA is directly proportional to x-ray intensity (quantity) but is unrelated to the quality of the beam. *(Thompson, p 405)*

12. **(A)** It is our ethical responsibility to minimize radiation dose to our patients. X-rays produced at the target comprise a heterogeneous primary beam. There are many "soft" (low-energy) photons that, if not removed by filters, would only contribute to greater patient dose. They

are too weak to penetrate the patient and expose the film; they just penetrate a small thickness of tissue and are absorbed. *(Selman, p 164)*

13. **(B)** If mechanical restraint is impossible, a friend or relative accompanying the patient should be requested to hold the patient. If a parent is to perform this task, it is preferable to elect the father so as to avoid the possibility of subjecting a newly fertilized ovum to even scattered radiation. If a friend or relative is not available, a nurse or transporter may be asked for help. Protective apparel, such as lead apron and gloves, should be provided to the person(s) holding the patient. *Radiology personnel must NEVER assist in holding patients, and the individual assisting must NEVER be in the path of the primary beam. (Thompson, p 484)*

14. **(D)** A controlled area is one that is occupied by radiation workers trained in radiation safety and who wear radiation monitors. The exposure rate in a controlled area must not exceed 100 mR/wk; its occupancy factor is considered to be one (1)—indicating that the area may always be occupied, and therefore requiring maximum shielding. An uncontrolled area is one occupied by the general population; the exposure rate there must not exceed 10 mR/wk. Shielding requirements vary according to several factors, one being occupancy factor. *(Statkiewicz, p 221)*

15. **(A)** Genetically significant dose (*GSD*) illustrates that large exposures to a few people are cause for little concern when diluted by the total population. On the other hand, we all share the burden of that radiation received by the total population and, especially as the use of medical radiation increases, each individual's share of the total exposure increases. *(Wolbarst, p 289)*

16. **(D)** The fluoroscopist should release his or her foot from the exposure pedal at frequent intervals, as the image will fade only slowly from the screen, thus reducing total patient exposure. Field size plays an important role in fluoroscopy (as in radiography) in controlling patient dose. Focus-to-table distance is

extremely important in controlling patient exposure; the law states that this distance must be a minimum of 15 inches (preferable 18 inches) to decrease patient dose. *(Bushong, pp 630–631)*

17. **(C)** The dose-equivalent limit to the hands and skin of an occupationally exposed individual is 50 rem. The dose-equivalent limit to the lens of the eye is 5 rem. An occupationally exposed individual may receive up to 3 rem in a given calendar quarter, or 13-week period. But that individual may not exceed 5 rem in that particular year. If, for example, one received 3 rem during the first 3 months of a year, that individual must not receive more than 2 rem in the remaining 9 months. *(Thompson, p 459)*

18. **(B)** In the photoelectric effect, a relatively low-energy photon uses all its energy to eject an inner-shell electron, leaving a vacancy. An electron from the shell above drops down to fill the vacancy and in doing so gives up a characteristic ray. This type of interaction is most harmful to the patient, as all the photon energy is transferred to tissue. In *Compton scatter,* a high-energy incident photon ejects an outer-shell electron. In doing so, the incident photon is deflected with reduced energy, but it *usually retains most of its energy and exits the body as an energetic scattered ray.* This scattered ray will either contribute to film fog or will pose a radiation hazard to personnel, depending on its direction of exit. In classical scatter, a low-energy photon interacts with an atom but causes no ionization; the incident photon disappears into the atom, and is then immediately released as a photon of identical energy but changed direction. Thompson scatter is another name for classical scatter. *(Selman, pp 513–514)*

19. **(B)** Each time an x-ray photon scatters, there is a significant reduction in intensity. The first and most important scattering object is the patient. The second may be the protective curtain (during fluoro), a wall, or the control booth. If scattered radiation reaches the operator, it is practically insignificant. *(Selman, p 520)*

20. **(A)** As an x-ray source moves away from a detector, the x-ray intensity (quantity) decreases. Conversely, as the source of x-rays moves closer to the detector, intensity increases. This is a predictable relationship and may be calculated using the inverse square law, which states that the intensity (exposure rate) of radiation at a given distance from a point source is inversely proportional to the square of the distance. For example, if the distance from an x-ray source was doubled, the intensity of x-rays at the detector would be one fourth of its original value. This relationship is represented by the formula:

$$\frac{I_1}{I_2} = \frac{D_2^2}{D_1^2}$$

(Bushong, p 67)

21. **(B)** The entire population of the world is exposed to varying amounts of background (environmental) radiation. Sources of background radiation are either natural or man-made. Exposure to *natural* background radiation is a result of cosmic radiation from space, naturally radioactive elements within the earth's crust, and our own bodies. Naturally, the closer we are to the cosmic radiations from space, the greater will be our personal exposure; living at higher elevations, and air travel, expose us to greater amounts of radiation. Living or working in a building made of materials derived from the ground exposes us to some background radiation from the naturally radioactive elements found in the earth's crust. The food we eat, the water we drink, and the air we breath all contribute to a quantity of radiation we ingest and inhale. Man-made radiation, however, is the type of background radiation over which we have some control. Medical and dental x-rays contribute to our exposure to *man-made* background radiation. *(Selman, p 501)*

22. **(D)** One of the radiation protection guidelines for the occupationally exposed is that the x-ray beam should scatter twice before reaching the operator. Each time the x-ray beam scatters, its intensity at 1 meter from the scattering object is *one-thousandth* of its original intensity. Of course, the operator should be behind a shielded booth while making the exposure, but multiple scatterings further reduce danger of exposure from scatter radiation. *(Bushong, p 633)*

23. **(D)** Gonadal shielding should be used whenever appropriate and possible during radiographic and fluoroscopic exams. *Flat contact* shields are useful for simple recumbent studies, but when the exam necessitates that oblique, lateral, or erect projections be obtained, they become less efficient. *Shaped contact (contour)* shields are best because they enclose the male reproductive organs, remaining in position in oblique, lateral, and erect positions. *Shadow* shields that attach to the tube head are particularly useful for surgical sterile fields. *(Bushong, pp 667–668)*

24. **(B)** Several devices are used to detect the presence of radiation; each of the devices is particularly suited for its particular use. Film badges are used to detect the presence of radiation and are used as personal dosimeters. X-ray photons *ionize* the silver halide film emulsion. The degree of emulsion exposure is related to a particular amount of radiation exposure. In TLDs, ionization occurs within the crystalline lattice of lithium fluoride. Ionization chambers, such as those found in pocket dosimeters, contain a quantity of air which ionizes when exposed to radiation; the amount of ionization that occurs is equated to a quantity of radiation. *(Bushong, pp 193, 660–661)*

25. **(C)** *Secondary* radiation consists of *leakage and scattered* radiation. Leakage radiation can be emitted when a defect exists in the tube housing. A significant quantity of scattered radiation is generated within, and emitted from, the patient. Background radiation is naturally occurring radiation that is emitted from the earth and that also exists within our bodies. *(Statkiewicz, p 217)*

26. **(A)** Filters are used in radiography to remove soft (low-energy) radiation that con-

tributes only to patient dose. The filters are usually made of aluminum. Equipment operating above 70 kVp must have total filtration of 2.5 mm aluminum equivalent (inherent + added). *(Curry, p 78)*

27. **(A)** In the *photoelectric effect*, a relatively low-energy incident photon uses all of its energy to eject an inner-shell electron, leaving a vacancy. An electron from the shell above will drop to fill the vacancy, and *a characteristic ray is given up* in the transition. This type of interaction is more harmful to the patient, as all the photon energy is transferred to tissue. *(Bushong, pp 175–176)*

28. **(C)** The reduction in the intensity (quantity) of an x-ray beam as it passes through matter is called *attenuation. Absorption* occurs when an x-ray photon interacts with matter and disappears, as in the *photoelectric effect. Scattering* occurs when there is partial transfer of energy to matter, as in the *Compton effect.* *(Bushong, p 66)*

29. **(D)** Radiation quality determines degree of penetration and the amount of energy transferred to the irradiated tissue (LET). Certainly the larger the absorbed radiation dose, the greater the effect. Biologic effect is increased as the size of the irradiated area is increased. The nature of the effect is influenced by the location of irradiated tissue (bone marrow versus gonads, and so on). *(Selman, p 190)*

30. **(B)** X-ray photons produced in the x-ray tube can radiate in directions other than that desired. The *tube housing* is therefore constructed so that very little of this leakage radiation is permitted to escape. The regulation states that leakage radiation must not exceed 100 mR/hr at 1 meter while the tube is operated at maximum potential. *(Bushong, p 629)*

31. **(B)** Occupationally exposed individuals are required to use devices that will record and provide documentation of the radiation they receive over a given period of time, traditionally one month. The most commonly used personal dosimeters are the film badge and

the TLD. These devices must be worn *only* for documentation of occupational exposure. They must not be worn for any medical or dental x-rays one receives as a patient, and they are not used to measure naturallly occurring background radiation. *(Thompson, p 459)*

32. **(C)** There are a few different types of monitoring devices available for the occupationally exposed. Ionization is the fundamental principle of operation of both the film badge and pocket dosimeter. In the *film badge,* the film's silver halide emulsion is ionized by x-ray photons. The *pocket dosimeter* contains an ionization chamber (containing *air*), and the number of ions formed (of either sign) is equated to exposure dose. *Thermoluminescent dosimeters* (TLDs) are personnel radiation monitors that use lithium fluoride crystals. These crystals, once exposed to ionizing radiation and then heated, give off light proportional to the amount of radiation received. *(Selman, p 162)*

33. **(A)** $^{130}_{56}$Ba and $^{138}_{56}$Ba are isotopes of the same element, barium (Ba), because they have the *same atomic number* but different mass numbers (numbers of neutrons). *Isobars* are atoms with the same mass number but different atomic numbers. *Isotones* have the *same number of neutrons* but different atomic numbers. *Isomers* have the *same atomic number and mass number;* they are identical atoms existing at different energy states. *(Selman, p 44)*

34. **(C)** Electromagnetic radiations such as x and gamma are considered low LET radiations because they produce fewer ionizations than the highly ionizing particulate radiations such as alpha particles. *Alpha particles* are large and heavy (two protons and two neutrons), and although they possess a great deal of kinetic energy (approximately 5 MeV), their energy is rapidly lost through multiple ionizations (approximately 40,000 atoms/cm of air). As an *external source,* alpha particles are almost harmless (because they just ionize the air). But as *internal sources,* they ionize tissues, and are potentially very dangerous. It may be stated that the alpha particle has one of the highest LETs of all ionizing radiations. *(Bushong, pp 52–53; Travis, p 31)*

35. **(C)** *Somatic effects* are those induced in the irradiated body. *Genetic effects* of ionizing radiation are those that may not appear for many years (generations) following exposure. Formation of cataracts or cancer, and embryologic damage, are all possible long-term somatic effects of radiation exposure. A fourth is life-span shortening. Nausea and vomiting are early effects of exposure to large quantities of ionizing radiation. *(Bushong, pp 560, 584, 588)*

36. **(A)** The x-ray tube's glass envelope and oil coolant are considered inherent filtration. Thin sheets of aluminum are added to make *a total of 2.5 mm Al equivalent filtration in equipment operated above 70 kVp*. The function of aluminum filtration is to remove from the x-ray beam the soft (long-wavelength) x-ray photons that do not contribute to the radiographic image but do contribute to patient dose. These soft x-rays penetrate only a small thickness of tissue before being absorbed. *(Statkiewicz, pp 171–173)*

37. **(C)** The first noticeable skin response to excessive irradiation would be erythema: a reddening of the skin, very much like sunburn. Dry desquamation, a dry peeling of the skin, may follow. Moist desquamation is peeling with associated pus-like fluid. Epilation, or hair loss, may be temporary or permanent depending on sensitivity and dose. *(Bushong, pp 564–565)*

38. **(A)** The faster the intensifying screens used, the less the required mAs. Decreasing the intensity (mAs/quantity) of photons significantly contributes to reducing total patient dose. Decreasing the kV would increase patient dose because less penetrating photons would compose the primary beam. The importance of patient shielding is never diminished. *(Statkiewicz, p 185)*

39. **(B)** Cells are frequently identified by their stage of development. *Immature* cells may be referred to as *undifferentiated* or *stem* cells. Immature cells are much more radiosensitive than mature cells. *(Statkiewicz, pp 120–125)*

40. **(C)** *Roentgen* is the unit of exposure; it measures the quantity of ionizations in air. *Rad* is

an acronym for *r*adiation *a*bsorbed *d*ose; it measures the energy deposited in any material. *Rem* is an acronym for *r*adiation *e*quivalent *m*an; it includes the relative biologic effectiveness (RBE) specific to the tissue irradiated, thereby being a valid unit of measure of dose to biologic material. *(Bushong, p 16)*

41. **(C)** The development of male and female reproductive stem cells has important radiation protection implications. Male stem cells reproduce continuously. However, the female stem cells develop only during fetal life; females are born with all the reproductive cells they will ever have. It is exceedingly important to shield children whenever possible, as they have their reproductive futures ahead of them. *(Bushong, pp 566–567)*

42. **(D)** For radiation safety, the fluoroscopy exposure switch must be the *dead-man type*. When the foot is removed from the fluoro pedal, the dead-man switch will terminate the exposure immediately. There must also be a fluoroscopy timer that will either sound or interrupt exposure after 5 minutes of fluoroscopy. *(Bushong, p 554)*

43. **(C)** Radiation protection guidelines have established that primary radiation barriers must be 7 feet high. Primary radiation barriers are walls that the primary beam might be directed toward. They usually contain 1.5 mm lead, but this may vary depending on use factor, and so on. *(Statkiewicz, p 217)*

44. **(C)** The occupational MPD is valid for beta, x, and gamma radiations. Because alpha radiation is so rapidly ionizing, traditional personnel monitors will not record alpha radiation. Because alpha particles are capable of penetrating only a few cm of air, they are practically harmless as an external source. *(Statkiewicz, p 249)*

45. **(D)** Diagnostic x-ray photons interact with tissue in a number of ways, but most frequently they are involved in the *photoelectric effect* or in the production of *Compton scatter*. The photoelectric effect is pictured; it occurs when a relatively low-energy x-ray photon

uses all its energy to eject an inner-shell electron. That electron is ejected (photoelectron), leaving a "hole" in the K shell and producing a positive ion. An L shell electron then drops down to fill the K vacancy, and in doing so, emits a characteristic ray whose energy is equal to the difference between binding energies of the K and L shells. The photoelectric effect occurs with high atomic number absorbers such as bone and positive contrast media, and is responsible for the production of contrast. Therefore, its occurrence is helpful for the production of the radiographic image, but it contributes significantly to the dose received by the patient (because it involves complete absorption of the incident photon). Scattered radiation, which produces a radiation hazard to the radiographer (as in fluoroscopy), is a product of the Compton scatter interaction occurring with higher-energy x-ray photons. *(Selman, p 158)*

46. **(C)** Radiation effects to occupationally exposed individuals whose dose is kept below the recommended limits is very remote. Exposure to ionizing radiation always carries some risk, but studies have indicated that the risk is a very small one if established guidelines are followed. Potential hazards must be understood and proper precautions taken. *(Bushong, p 559)*

47. **(B)** The filters (usually aluminum and copper) serve to help measure radiation quality (energy). Only the most energetic radiation will penetrate the copper; radiation of lower levels will penetrate the aluminum, and the lowest energy radiation will pass readily through the unfiltered area. Thus radiation of different energy levels can be recorded, measured, and reported. *(Bushong, p 659)*

48. **(B)** *Rad* is an acronym for *radiation absorbed dose*; it measures the energy deposited in any material. *Roentgen* is the unit of exposure; it measures the quantity of ionizations in air. *Rem* is an acronym for *radiation equivalent man*; it includes the relative biologic effectiveness (RBE) specific to the tissue irradiated, thereby being a valid unit of measure of *dose* to biologic tissue. *(Bushong, p 15)*

49. **(B)** Bergonié and Tribondeau were French scientists who, in 1906, theorized what has now become verified law. Cells are more radiosensitive if they are *immature* (undifferentiated or stem) cells, if they are *highly mitotic* (having a high rate of proliferation), and if the irradiated tissue is young. Cells and tissues that are still undergoing development are more radiosensitive than fully developed tissues. *(Statkiewicz, p 120)*

50. **(D)** Artificial/manmade sources of radiation include radioactive fallout, industrial radiation, and medical and dental x-rays. Ninety percent of the general public's artificial radiation exposure is from medical and dental x-rays. It is our professional obligation, therefore, to keep our patient's radiation dose to a minimum *(Statkiewcz, p 6)*

51. **(D)** Skin responses to irradiation include all four choices. The first noticeable response would be *erythema*, a reddening of the skin very much like sunburn. *Dry desquamation* may follow; it is a dry peeling of the skin. *Moist desquamation* is peeling with associated pus-like fluid. *Epilation* is hair loss; it may be temporary or permanent depending on sensitivity and dose. *(Statkiewicz, p 45)*

52. **(C)** Tissue is most sensitive to radiation when *oxygenated*. *Anoxic* refers to tissue without oxygen; *hypoxic* refers to tissue with little oxygen. Anoxic and hypoxic tumors are typically avascular (with little or no blood supply) and therefore more radioresistant. *(Statkiewicz, pp 119–120)*

53. **(A)** The length of time from conception to birth, that is, pregnancy, is referred to as *gestation*. The term *congenital* refers to a condition existing at birth. *Neonatal* relates to the time immediately after birth and the first month of life. *In vitro* refers to something living outside a living body (as in a test tube) as opposed to *in vivo* (within a living system). *(Travis, pp 262, 266, 268, 273)*

54. **(D)** All the factors listed influence the effect of radiation on tissue. Larger quantities, of course, increase tissue effect. The energy

(quality, penetration) of the radiation determines whether the effects will be superficial (erythema) or deep (organ dose). Certain tissues such as blood-forming organs, the lens, and gonads are more radiosensitive than others (such as muscle and nerve). The length of time over which the exposure is spread (fractionation) is important; the longer that period of time, the less the tissue effects. *(Bushong, p 534)*

55. **(D)** The presence of radiation may be detected in several ways. It has an *ionizing* effect on air, the basic principle of the Roentgen as unit of measure. X-rays have a *photographic* effect on film emulsion, readily observable on radiographic images. The *fluorescent* effect on certain crystals, such as calcium tungstate and lanthanum, accounts for our use of these phosphors in intensifying screens. Radiation's *physiologic* effects have been demonstrated as genetic damage, erythema, and cataractogenesis, many of these noted by the early radiology pioneers. *(Selman, pp 53–54, 186)*

56. **(B)** The *rad* (acronym for *r*adiation *a*bsorbed *d*ose) is the unit of absorbed dose. It is equal to *100 ergs of energy per gram of any absorber.* The Roentgen measures quantity of ionization in air. The rem and RBE express radiation dose to biologic material. *(Selman, p 188)*

57. **(B)** In 1906, two scientists, Bergonié and Tribondeau, theorized that ionizing radiation more readily damages cells that are actively dividing, are undifferentiated, and have a long mitotic future. A mature, or *differentiated*, cell has a specialized function and/or structure. An *undifferentiated* cell is one that is immature. Its purpose is to divide, thereby producing more of its kind. Undifferentiated cells may be referred to as *immature, precursor*, or *stem* cells. *(Travis, p 68)*

58. **(B)** The cell cycle may be divided into two phases: mitosis and interphase. Interphase is the phase between mitotic events. Mitosis is divided into four stages: prophase, metaphase, anaphase, and telophase. During *metaphase*, chromosomes appear and line up along the nuclear equator; this is the most radiosensitive stage of mitosis. *(Bushong, p 531)*

59. **(B)** Genetic effects of radiation and some somatic effects, like leukemia, are plotted on a linear dose–response curve. The linear dose–response curve has *no threshold*, that is, *there is no dose below which radiation is absolutely safe.* The sigmoidal dose–response curve has a threshold and is thought to be generally correct for most somatic effects. *(Bushong, p 539)*

60. **(B)** Special arrangements are required for occupational monitoring of the pregnant radiographer. The pregnant radiographer will wear two dosimeters: one in its usual place at the collar, and the other, a "baby dosimeter," worn over the abdomen and under the lead apron during fluoroscopy. The baby dosimeter must be identified as such and must always be worn in the same place. Care must be taken not to mix the position of the two dosimeters. The dosimeters are read monthly, as usual. *(Bushong, p 609)*

61. **(A)** Radiographers are usually required to wear one dosimeter, positioned at their collar, and worn *outside* a lead apron. Special circumstances, however, warrant the use of a second monitor. During pregnancy, a second *"baby monitor"* is worn at the abdomen, and *under* any lead apron. During special vascular procedures, where the radiographer's upper extremities may receive a greater exposure (for example, when assisting during catheter introduction), a *ring badge* is often recommended. A second dosimeter is not required when performing mobile radiography. *(Bushong, p 663)*

62. **(B)** *Isotopes* are atoms of the same element (the same atomic number or number of protons) but a different mass number. They differ, therefore, in the number of neutrons. Atoms with the same mass number but different atomic number are *isobars*. Atoms with the same atomic number but different neutron number are *isotones*. Atoms with the same atomic number and mass number are *isomers*. *(Bushong, p 40)*

63. **(D)** Gamma rays are electromagnetic radiation. They are physically identical to x-rays, differing only in origin. Like x-rays, gamma rays have no mass and no charge. *(Bushong, p 53)*

64. **(D)** The x-ray photons produced at the tungsten target comprise a heterogeneous beam, a spectrum of photon energies. This is accounted for by the fact that the incident electrons have differing energies. Also, the incident electrons travel through several layers of tungsten target material, lose energy with each interaction, and therefore produce increasingly weaker photons. During characteristic x-ray production, vacancies may be filled in the K, L, or M shells, which differ from each other in binding energies, and therefore, varying-energy photons are emitted. *(Bushong, pp 154–155)*

65. **(C)** Bremsstrahlung (or Brems) radiation is one of the two kinds of x-rays produced at the tungsten target of the x-ray tube. The incident high-speed electron, passing through a tungsten atom, is attracted by the positively charged nucleus, and therefore is *deflected from its course with a resulting loss of energy*. This energy loss is given up in the form of an x-ray photon. *(Bushong, p 150)*

66. **(B)** *Scattering* occurs when there is partial transfer of energy to matter, as in the Compton effect. *Absorption* occurs when an x-ray photon interacts with matter and disappears, as in the photoelectric effect. The reduction in the intensity (quantity) of an x-ray beam as it passes through matter is termed *attenuation*. *(Bushong, pp 185–186)*

67. **(A)** In the *photoelectric effect*, a relatively low-energy photon uses all its energy to eject an inner-shell electron, leaving a vacancy. An electron from the shell above drops down to fill the vacancy and in doing so emits a characteristic ray. This type interaction is most harmful to the patient, as *all the photon energy is transferred to tissue*. In *Compton scatter*, a high-energy incident photon uses some of its energy to eject an outer-shell electron. In doing so the incident photon is deflected with reduced energy, but usually retains most of its energy and exits the body as an energetic scattered ray. The scattered radiation will either contribute to film fog or will pose a radiation hazard to personnel, depending on its direction of exit. In *classical scatter*, a low-energy photon interacts with an atom but causes no ionization; the incident photon disappears in the atom, and then immediately reappears and is released as a photon of identical energy but changed direction. *Thompson scatter* is another name for classical scatter. *(Selman, p 179)*

68. **(A)** The Inverse Square Law of radiation states that the intensity or exposure rate of radiation at a given distance from a point source is inversely proportional to the square of the distance. This illustrates that if distance from a radiation source is doubled, dose received will be one fourth the original intensity. Distance, therefore, is a very effective means of protection from radiation exposure. *(Selman, p 333)*

69. **(C)** The most radiosensitive cells in the fetus are those of the *central nervous system*. If abnormalities occur as a result of fetal irradiation during the first trimester of gestation, they are most likely to be CNS abnormalities such as microcephaly, stunted growth, and mental retardation. If disorders occur as a result of fetal irradiation during the second and third trimesters, they are more likely to be sterility, leukemia, and bone marrow syndrome. *(Travis, p 160)*

70. **(D)** Late effects of radiation can occur in cells that have survived a previous irradiation, months or years earlier. These late effects, such as carcinogenesis and genetic effects, are "all-or-nothing" effects—either the organism develops cancer or it does not. Most late effects do not have a threshold dose, that is, *any* dose, however small, theoretically can induce an effect. Increasing that dose will increase the likelihood of the occurrence, but will not affect its severity. Therefore, late, all-or-nothing type effects not exhibiting a threshold are termed stochastic. On the other hand, *non-stochastic* effects are those that will

not occur below a particular threshold dose, and that increase in severity as the dose increases. *(Statkiewicz, p 136)*

71. **(C)** *Roentgen* is the unit of exposure; it measures the quantity of ionization in air. *Rad* is an acronym for radiation absorbed dose; it measures the energy deposited in any material. *Rem* is an acronym for radiation equivalent man; it includes the relative biologic effectiveness *(RBE)* specific to the tissue irradiated, thereby being a valid unit of measure of dose to biologic material. *(Statkiewicz, p 52)*

72. **(C)** Thermoluminescent dosimeters (TLDs) are personnel radiation monitors that use lithium fluoride crystals. These crystals, once exposed to ionizing radiation and then heated, give off light proportional to the amount of radiation received. TLDs are very accurate personal monitors. *(Statkiewicz, p 255)*

73. **(B)** The patient is the most important radiation scatterer during both radiography and fluoroscopy. In general, at 1 meter from the patient, *the intensity is reduced by a factor of 1000,* to about 0.1 percent of the original intensity. Successive scatterings can render the intensity to unimportant levels. *(Bushong, p 633)*

74. **(D)** The *bucky slot cover* shields the opening at the side of the table, as the bucky tray is parked at the end of the table for the fluoro procedure; this is important because the opening created would otherwise allow scatter radiation to emerge at approximately gonad level of the operator. The *exposure switch* (usually a foot pedal) must be the dead-man type—that is, when the foot is released from the switch there is immediate termination of exposure. The *cumulative exposure timer* sounds or interrupts the exposure after 5 minutes of fluoro time, thus making the fluoroscopist aware of accumulated fluoro time. Additionally, *source-to-tabletop distance* is restricted to at least 15 inches for stationary equipment and at least 12 inches for mobile equipment. Increased source-to-tabletop distance increases source-to-patient distance, thereby decreasing patient dose. *(Bushong, p 631)*

75. **(D)** Photoelectric interaction in tissue involves complete absorption of the incident photon, whereas Compton interactions involve only partial transfer of energy. The larger the quantity of radiation and the greater the number of photoelectric interactions, the greater the patient dose. Radiation to more radiosensitive tissues such as gonadal tissue or blood-forming organs is more harmful than the same dose to muscle tissue. *(Selman, pp 513–514)*

76. **(A)** Protective apparel functions to protect the occupationally exposed *from scattered radiation only.* Lead aprons and lead gloves do not protect from the primary beam. No one in the radiographic room, except the patient, must ever be exposed to the primary beam. The occupationally exposed, and those (family and friends) who might assist a patient during an exam, must wear protective apparel and keep out of the way of the primary beam. *(Statkiewicz, p 215)*

77. **(B)** *Use factor* describes the percentage of time the primary beam is directed toward a particular wall. The use factor is one of the factors considered in determining protective barrier thickness. Another is *work load*, determined by the number of x-ray exposures made per week. *Occupancy factor* is a reflection of who occupies particular areas—radiation workers or non-radiation workers—and is another factor used in determining radiation barrier thickness. *(Statkiewicz, pp 222–223)*

78. **(B)** The film badge and thermoluminescent dosimeter (TLD) are the most commonly used personnel monitors. A pocket dosimeter is used primarily when working with large amounts of radiation and when a daily reading is desired. A cutie pie is a radiation survey instrument. *(Statkiewicz, pp 246–256)*

79. **(B)** As the size of the irradiated field decreases, scattered radiation production and patient hazard decreases. If the amount of scattered radiation decreases, then radiographic contrast would be higher (shorter scale). *(Statkiewicz, p 194)*

80. (C) The pregnant radiographer poses a special radiation protection consideration, for the safety of the unborn individual must be considered. It must be remembered that the developing fetus is particularly sensitive to radiation exposure. Therefore, established guidelines state that the occupational radiation exposure to the fetus must not exceed 0.5 rem (500 mrem, or 5 mSv) during the entire gestation period. *(Bushong, p 610)*

81. (B) Occupationally exposed individuals generally receive small amount of low-energy radiation over a long period of time. These individuals are therefore concerned with the potential *long-term* effects of radiation, such as *carcinogenesis* (including *leukemia*) and *cataractogenesis*. However, if a large amount of radiation is delivered to the whole body all at one time, the short-term early somatic effects must be considered. If the whole body receives 600 rad at one time, *acute radiation syndrome* is likely to occur. Early symptoms of acute radiation syndrome include *nausea* and *vomiting*, which occur in the first *(prodromal)* stage of acute radiation syndrome. *(Statkiewicz, p 130)*

82. (D) An approximate patient skin dose can be determined using the illustrated nomogram. First, mark 2.5 mm Al on the horizontal axis. Next, mark where a line drawn up from that point intersects the 60-kVp line. Draw a line straight across to the vertical axis; this should approximately reach the 3 mR/mAs point. Because 60 mAs was used for the exposure, the approximate skin dose is 180 mR. *(Bushong, p 652)*

83. (A) X-radiation used for diagnostic purposes is of relatively *low energy*. Kilovolts of up to 150 are used, as compared with radiations having energies up to several million volts. LET (linear energy transfer) refers to the rate at which energy is transferred to soft tissue from ionizing radiation. Particulate radiations, such as alpha particles, have mass and charge, and therefore lose energy rapidly as they penetrate only a few cm of air. X and gamma radiations, having no mass or charge, are *low-LET* radiations. *(Travis, p 31)*

84. (D) Radiation is most hazardous when received in a large dose, all at one time, to the whole body. When 100 R or more is received as a whole-body dose in a short time, biologic effects will appear within minutes to weeks (depending on the dose received). These immediate effects are known as *acute radiation syndrome. (Statkiewicz p 130)*

85. (B) In the photoelectric effect, a relatively low-energy photon uses all its energy to eject an inner-shell electron, leaving a vacancy. An electron from the shell above drops down to fill the vacancy and in doing so gives up a characteristic ray. This type of interaction is most harmful to the patient, as all the photon energy is transferred to tissue. In Compton scatter, a high-energy incident photon uses some of its energy to eject an outer-shell electron. In doing so, the incident photon is deflected with reduced energy, but it usually retains most of its energy and exits the body as an energetic scattered ray. This scattered ray will either contribute to film fog or will pose a radiation hazard to personnel, depending on its direction of exit. In classical scatter, a low-energy photon interacts with an atom but causes no ionization; the incident photon disappears into the atom, and then is immediately released as a photon of identical energy but with changed direction. Thompson scatter is another name for classical scatter. *(Selman, pp 513–514)*

86. (B) *Rem (dose-equivalent) is the only unit of measure that expresses dose–effect relationship.* The product of rad (absorbed dose) and the quality factor appropriate for the radiation type is expressed as rem or DE (dose equivalent), and may be used to predict the type and extent of response to radiation. *(Statkiewicz, p 53)*

87. (D) The *protective curtain*, usually made of leaded vinyl with at least 0.25 mm Pb equivalent, must be positioned between the patient and fluoroscopist to greatly reduce exposure of the fluoroscopist to energetic scatter from the patient. As with overhead equipment, fluoroscopic total *filtration* must be at least 2.5 mm Al equivalent to reduce excessive exposure to soft radiation. *Collimator*/beam alignment must be accurate to within 2%. *(Bushong, pp 630–631)*

88. **(B)** Gonadal shielding should be used whenever appropriate and possible during radiographic and fluoroscopic exams. *Flat contact* shields are useful for simple recumbent studies, but when the exam necessitates that oblique, lateral, or erect projections be obtained, they become less efficient. *Shaped-contact* (contour) shields are best because they enclose the male reproductive organs, remaining in position in oblique, lateral, and erect positions. *Shadow* shields that attach to the tube head are particularly useful for surgical sterile fields. *(Bushong, pp 667–668)*

89. **(A)** 3000 mrad is equal to 3 rad. If 3 rad were delivered in 15 minutes, then the dose rate must be 0.2 rad/min:

$$\frac{3\ \text{rad}}{15\ \text{min}} = \frac{x\ \text{rad}}{1\ \text{min}}$$
$$15\ x = 3$$
$$x = 0.2\ \text{rad/min}$$

(Selman, p 528)

90. **(C)** Sensitometry and densitometry are used in evaluation of the film processor and are just one portion of a complete *quality assurance* (QA) program. Modulation transfer function (MTF) is used to express spatial resolution—another component of the QA program. A complete QA program includes testing of all components of the imaging system: processors, focal spot, x-ray timers, filters, intensifying screens, beam alignment, and so on. *(Thompson pp 399–400)*

91. **(D)** *Beam restriction* is probably the single best method of protecting your patient from excessive radiation (it is also an important factor in obtaining quality radiographs because there will be less scatter radiation fog). *Shielding* areas not included in the radiograph, especially particularly radiosensitive areas, is another effective means of reducing patient dose. If the patient is subjected to less radiation exposure, then so is the operator. *Shielding, distance,* and

time are the three cardinal rules of radiation protection. *High kV, low mAs* exposure factors employ the use of fewer, but more penetrating x-rays. *(Gurley, p 230)*

92. **(D)** The amount of scattered radiation produced is dependent first on the kVp (beam quality) selected; *the higher the kVp,* the more scattered radiation produced. The size of the irradiated field has a great deal to do with the amount of scatter radiation produced; *the larger the field size,* the greater the amount of scattered radiation. Thickness and condition of tissue are important considerations; *the thicker the tissue,* the more scatter produced. If the condition of the tissue is such that *pathology* makes it more difficult to penetrate, more scattered radiation will be produced. *(Thompson, pp 280, 282)*

93. **(C)** Lead and distance are the two most important ways to protect from radiation exposure. Fluoroscopy can be particularly hazardous because the SID is so much shorter than in overhead radiography. Therefore it has been established that *fixed* (stationary) *and mobile* fluoroscopic equipment must provide at least 12 inches source to tabletop/skin distance for the protection of the patient. *(Carlton & Adler, p 548; NCRP report no. 102, p 25)*

94. **(B)** The effects of a quantity of radiation delivered to a body are dependent on the amount of radiation received, the size of the irradiated area, and how the radiation is delivered in time. If the radiation is delivered in portions over a period of time, it is said to be fractionated and has a less harmful effect than if the radiation was delivered all at once. Therefore, cells have an opportunity to repair and some recovery occurs between doses. *(Bushong, p 534)*

95. **(B)** When biologic material is irradiated, there are a number of modifying factors that determine what kind and how much response will occur in the material. One of these factors is *LET,* which expresses *the rate at which particulate or photon energy is transferred to the absorber.* Because different kinds of radiation have different degrees of penetration in different materials, it is also a use-

ful way of expressing the quality of the radiation. *(Thompson, p 419)*

96. **(B)** It is our ethical responsibility to minimize radiation exposure to our patients and ourselves, particularly during early pregnancy. One way is to inquire about the possibility of our female patients being pregnant, or for the date of their last menstrual period (to determine the possibility if irradiating a newly fertilized ovum). *The safest time for a woman of childbearing age to have elective radiographic exams is during the first 10 days following the onset of menstruation. (Thompson, p 487)*

97. **(B)** The Law of Bergonié and Tribondeau states that stem cells (which give rise to a specific type of cell, as in hematopoeisis) are particularly radiosensitive, as are *young cells* and tissues. It also states that *cells with a high rate of proliferation* (mitosis) are more sensitive to radiation. This law is historically important in that it was the first to recognize that some tissues have greater radiosensitivity than others (such as the fetus). *(Statkiewicz, p 120)*

98. **(A)** LET expresses the rate at which photon or particulate energy is transferred to (absorbed by) biologic material (through ionization-processes) and is dependent upon the type of radiation and absorber characteristics. RBE describes the degree of response or amount of biologic change one can expect of the irradiated material. *As the amount of transferred energy (LET) increases* (from interactions occurring between radiation and biologic material), *the amount of biologic effect/damage will also increase. (Thompson, pp 419–420)*

99. **(A)** *The shorter the SID, the greater the skin dose.* That is why there are specific source-to-skin distance restrictions in fluoroscopy. High kVp produces more penetrating photons, thereby decreasing skin dose. Filtration is used to remove the low-energy photons from the primary beam, which contribute to skin dose. *(Thompson, pp 293–294)*

100. **(B)** A 1984 review of radiation exposure data revealed that the average annual dose equiv-

alent for monitored radiation workers was approximately 2.3 mSv (0.23 rem). The fact that this is approximately one tenth of the recommended limit indicates that the limit is adequate for radiation protection purposes. Therefore the NCRP reiterates its 1971 recommended annual limit of 50 mSv (5 rem). *(NCRP report no. 105, pp 14–15)*

101. **(A)** mAs regulates the quantity of radiation delivered to the patient, and kVp regulates the quality (penetration) of the radiation delivered to the patient. Therefore, higher-energy (more penetrating) radiation—which is more likely to exit the patient—accompanied by lower mAs, is the safest combination for the patient. *(Thompson, p 275)*

102. **(C)** A quick survey of the distractors reveals that options A and B will increase exposure dose; thus, they are eliminated as possible correct answers. Both C and D will serve to reduce radiation exposure, as distance is increased and exposure time decreased in each case. It remains to be seen, then, which is the more effective. Using the Inverse Square Law of radiation, it is found that at 6 feet the individual will receive 40 mR in 2 minutes:

$$\frac{90}{x} = \frac{36}{16}$$

$$36\,x = 1440$$

$$x = 40 \text{ mR/2 min at 6 feet}$$

and that at 5 feet, the individual will receive 57.6 mR in 2 minutes:

$$\frac{90}{x} = \frac{25}{16}$$

$$25\,x = 1440$$

$$x = 57.6 \text{ mR/2 min at 5 feet}$$

Therefore, the individual will receive 28.8 mR in *1 minute* at 5 feet; the most effective option. *(Bushong, pp 67–68)*

103. **(C)** The most radiosensitive portion of the GI tract is the small bowel. Projecting from the lining of the small bowel are villi that are responsible for absorption of nutrients into the

bloodstream. Because cells from the villi are continually being cast off, new cells must continually arise from the crypts of Lieberkühn. Being highly mitotic, undifferentiated stem cells, they are radiosensitive. Thus, the small bowel is the most radiosensitive portion of the GI tract. In the adult, the CNS is the most radioresistant system. *(Travis, p 121)*

104. (C) *Erythema* is the reddening of skin as a result of exposure to large quantities of ionizing radiation. It was one of the first somatic responses to irradiation demonstrated to the early radiology pioneers. The effects of radiation exposure to the skin follow a *nonlinear, threshold dose–response relationship*. An individual's response to skin irradiation depends on the dose received, the period of time over which it was received, the size of the area irradiated, and the individual's sensitivity. The dose that it takes to bring about a noticeable erythema is referred to as the *skin erythema dose (SED)*. *(Bushong, p 566)*

105. (D) The photoelectric effect occurs when a relatively low-energy photon uses all its energy to eject an inner-shell electron. That electron flies off into space, leaving a hole in, for example, the K shell. An L shell electron then drops down to fill the K vacancy, and in doing so emits a characteristic ray whose energy equals the difference in binding energies for the K and L shells. The photoelectric effect occurs with high atomic number absorbers such as bone and positive contrast media. *(Selman, pp 513–514)*

106. (D) Inherent filtration is composed of materials that are a permanent part of the tube housing, that is, the glass envelope of the x-ray tube and the oil coolant. Added filtration, usually thin sheets of aluminum, is present in order to make a total of 2.5 mm Al equivalent for equipment operated above 70 kVp. Filtration is used to decrease patient dose by removing the weak x-ray having no value but contributing to skin dose. According to the Inverse Square Law of Radiation, exposure dose increases as distance from the source decreases, and vice versa. The effect of focal spot size is principally on radiographic sharpness, having no effect on patient dose. *(Statkiewicz, pp 171–173)*

107. (D) Any of the types of damage listed can occur as a result of DNA irradiation. In the *single-chain break*, one "side rail" of the helix is broken. If the opposite side is also broken, as in a *double-chain break*, the molecule is actually severed. *Cross-linking* can occur when a broken chain attempts to repair but attaches to another molecule. Another type of structural damage is when "rungs" of a DNA molecule are broken. Irradiation can also affect base damage. DNA damage will result in cell repair, cell death, or mutations. *(Bushong, p 546)*

108. (B) It is our professional responsibility to minimize exposure dose to patients and ourselves, and one of the most important ways is with a closely collimated field. Gonadal shielding should be used when the patient is of reproductive age or younger, when the gonads are in or near the collimated field, and when the clinical objectives will not be compromised. *(Bushong, p 669)*

109. (A) Restriction of field size is one important method of patient protection. However, the accuracy of the light field must be evaluated periodically as part of a QA program. Guidelines set forth for patient protection state that the collimator light and actual irradiated area must be accurate to within 2% of the focal-film distance. *(Thompson, p 403)*

110. (A) *The Roentgen measures ionization in air and is referred to as the unit of exposure*. Rad is an acronym for radiation absorbed dose and rem is an acronym for radiation equivalent man. RBE is relative biologic effectiveness and is used to determine biologic damage in living tissue. *(Thompson, p 456)*

111. **(D)** All of those listed are considered especially radiosensitive tissues. Excessive radiation to the reproductive organs can cause genetic mutations or sterility. Excessive radiation to blood-forming organs can cause leukemia or life span shortening. Lymphocyte cells are the most radiosensitive material in the body. *(Bushong, p 606)*

112. **(C)** Half-value layer (HVL) may be defined as the amount and thickness of absorber necessary to reduce the radiation intensity to half its original value. Thus, the first HVL would reduce the intensity to 50 percent of its original value, the second to 25 percent, the third to 12.5 percent, and the fourth to 6.25 percent of its original value. *(Bushong, p 165)*

113. **(B)** Radiation safety guidelines are valuable only if employed by radiation personnel. The radiation safety officer (RSO) is responsible for being certain that established guidelines are enforced and that personnel understand and employ radiation safety measures to protect themselves and their patients. The RSO is also responsible for performing routine equipment checks to ensure that all equipment meets radiation safety standards. *(NCRP report no. 405, p 34)*

114. **(B)** Radiographic and fluoroscopic equipment is designed to help decrease the exposure dose to patient and operator. One of the design features is the exposure cord. Exposure cords on fixed equipment must be short enough to prevent the exposure from being made outside the control booth. Exposure cords on mobile equipment must be long enough to permit the operator to stand at least 6 ft from the x-ray tube. *(Bushong, p 630)*

115. **(A)** The gradual decrease in exposure rate as radiation passes through matter is called *attenuation*. This is attributed to the two major types of interactions that occur in tissue between x-ray photons and matter. In the photoelectric effect, *absorption* and *secondary radi-*

ation take place. In the Compton effect, scattered radiation occurs. With each occurrence of absorption, scattered radiation, and secondary radiation, there is a decrease in the exposure rate that is termed *attenuation*. *(Thompson, p 209)*

116. **(B)** Whole-body dose is calculated to include all the especially radiosensitive organs. The gonads, lens of eye, and blood-forming organs are particularly radiosensitive. Some body parts, such as the skin and extremities, have a higher annual dose limit. *(Bushong, p 606)*

117. **(A)** The relationship between x-ray intensity and distance from the source is expressed in the Inverse Square Law of Radiation. The formula is:

$$\frac{I_1}{I_2} = \frac{D_2^2}{D_2^1}$$

Substituting known values:

$$\frac{12 \text{ R/min}}{x} = \frac{12 \text{ ft}^2}{5 \text{ ft}^2}$$

$$\frac{12}{x} = \frac{144}{25}$$

$$144x = 300$$

$$x = 2.08 \text{ R/min}$$

at 12 feet from the source

Note the inverse relationship between distance and dose. *As distance from the source of radiation increases, dose rate significantly decreases. (Bushong, pp 67–68)*

118. **(A)** The thermoluminescent dosimeter (TLD) can measure exposures as low as 5 mrem, whereas film badges will measure a minimum exposure only as low as 10 to 20 mrem. Pocket dosimeters are not generally used in diagnostic radiography but in areas of higher radiation exposure. A Geiger counter is not a personal monitor; it is a survey instrument used to detect areas of radioactive contamination on surfaces (such as countertops in a nuclear medicine lab). *(Thompson, p 461)*

119. **(C)** This is the definition of maximum permissible dose (MPD) equivalent. The annual occupational MPD for the whole body is 5000 mrem (50 mSv), not to exceed 3000 mrem (30 mSv) in a calendar quarter (13-week period). The MPD for the general population is 1/10th of that for the radiographer (0.5 rem or 500 mrem). Current literature refers to *dose-equivalent limit* rather than MPD. *(Statkiewicz, p 64)*

120. **(C)** The greatest effect/response to irradiation is brought about by a *large dose of radiation, to the whole body, delivered all at one time.* Whole-body radiation can depress many body functions. With a fractionated dose, the effects would be less severe because the body would have an opportunity to repair between doses. *(Bushong, p 534)*

121. **(A)** In the photoelectric effect, the incident (low-energy) photon is completely absorbed, and thus is responsible for producing contrast and contributing to patient dose. Photoelectric effect is the interaction between x-ray and tissue that predominates in the diagnostic range. In Compton scatter, only partial absorption occurs, and most energy emerges as scattered photons. In coherent scatter, no energy is absorbed by the part; it all emerges as scattered photons. Pair production occurs only at very high energy levels, at least 1.02 meV. *(Selman, pp 513–514)*

122. **(A)** Lead aprons require certain maintenance and care if they are expected to continue to provide protection from ionizing radiation. They can be kept clean with a damp cloth. It is very important that they be hung when not in use, rather than folded or left in a heap between examinations. A folded or crumpled position encourages the formation of cracks in the leaded vinyl. Lead aprons should be *fluoroscoped* (at about 120 kVp) at least *once a year* to check for development of any cracks. *(Bushong, p 665)*

123. **(B)** *Alpha and beta are particulate radiations;* alpha is composed of two protons and two neutrons, and beta is identical to an electron.

Gamma and x-radiation are electromagnetic, having wavelike fluctuations like other radiations of the electromagnetic spectrum (visible light, radio waves, and so on). *(Bushong, pp 52–53)*

124. **(B)** Compton scattering occurs when a relatively high-energy incident photon uses *part* of its energy to eject an outer-shell electron, and in doing so the incident photon changes direction (is scattered). The energy retained by the scattered photon depends on the angle formed by the ejected electron and the scattered photon: the greater the angle of deflection, the less the retained energy. Compton scatter is very energetic scatter. It emerges from the patient and is responsible for scatter radiation reaching the film in the form of fog. Characteristic radiation is associated with the photoelectric effect. *(Curry, pp 55–56)*

125. **(B)** Bergonié and Tribondeau theorized in 1906 that all precursor cells are particularly radiosensitive (for example, stem cells found in bone marrow). There are several types of stem cells in bone marrow and the different types differ in degree of radiosensitivity. Of these, red blood cell precursors, erythroblasts, are the most radiosensitive. White blood cell precursors, myelocytes, follow. Lastly, platelet precursor cells, megakaryocytes, are the least radiosensitive. Myocytes are mature muscle cells and are fairly radioresistant. *(Travis, p 115)*

126. **(D)** A thermoluminescent dosimeter (TLD) is used to measure monthly exposure to radiation, as does the film badge. Lithium fluoride chips are the thermoluminescent material used in TLDs. A pocket dosimeter (a small personal ionization chamber) measures quantity of ionizations occurring during the period worn, and reads out in mrem; it is used primarily when working with large quantities of radiation. *(Bushong, pp 661–662)*

127. **(B)** *Radiolysis* has to do with irradiation of water molecules and the formation of free radicals. Free radicals contain enough energy to damage other molecules some distance away. They can migrate to and damage a DNA molecule (indirect hit theory). *(Bushong, pp 547–548)*

128. (C) *Scattered and secondary radiations* are those that have been deviated in direction while passing through a part. Leakage radiation is that which emerges from the leaded tube housing in directions other than that of the useful beam. Tube head construction must keep leakage radiation to less than 0.1 R/hr at 1 meter from the tube. *Remnant radiation* is that which emerges from the patient to form the radiographic image. *(Curry, p 413)*

129. (B) Cell division in somatic cells is called mitosis. *Meiosis is a process of reduction division*, in which after the union of two germ cells (46 total chromosomes), the cell undergoes two (reduction) divisions to finish with cells of 23 chromosomes each. *(Bushong, p 531)*

130. (D) It is important to limit tabletop exposure during fluoroscopy, because the SSD (source-to-skin distance) is so much less than in overhead radiography—thus delivering a much higher skin dose to the patient. For this reason the tabletop exposure rate during fluoroscopy shall not exceed 10 R/min. *(Bushong, p 631)*

131. (A) Characteristic radiation is one of two kinds of x-rays produced at the tungsten target of the x-ray tube. The incident, or incoming, high-speed electron ejects a K shell electron. This leaves a hole in the K shell, and an L-shell electron drops down to fill the K vacancy. Because L electrons have a greater binding energy than do K-shell electrons, the L-shell electron gives up the difference in binding energy in the form of a photon, a "characteristic x-ray" (characteristic of the K shell). *(Bushong, pp 148–149)*

132. (B) *Genetic damage* and *cell death* are two major effects of DNA irradiation. *Malignant disease* is the third major effect, and is a result of abnormal metabolic activity. *Radiolysis* has to do with irradiation of water molecules and the formation of free radicals. The free radical then migrates to and damages a DNA molecule. *(Bushong, pp 546–548)*

133. (B) With greater beam restriction, less biologic material is irradiated, thereby reducing the possibility of harmful effects. If less tissue is irradiated, less scattered radiation is produced, resulting in improved film contrast. The total filtration is not a function of beam restriction, but rather is a radiation protection guideline aimed at reducing patient skin dose. *(Selman, pp 386–388)*

134. (D) Any of the doses listed may be used to express patient dose. *Skin dose* represents entrance exposure. *Gonadal dose* is sometimes used to express patient dose, and its determination is sometimes indicated in abdominal radiography. *Bone marrow dose* may be used to express the dose to a particular population in order to indicate late effects of radiation, such as leukemia. Another term, *midline dose*, is an expression of tissue dose used in mammography, where a fairly high skin dose or a very low exit dose could convey misleading information. *(Bushong, pp 550–551)*

135. (A) The relationship between x-ray intensity and distance from the source is expressed in the Inverse Square Law of Radiation. The formula is:

$$\frac{I_1}{I_2} = \frac{D_2^2}{D_1^2}$$

Substituting known values:

$$\frac{18 \text{ R/min}}{x} = \frac{5.0 \text{ m}^2}{2.0 \text{ m}^2}$$

$$\frac{18}{x} = \frac{25}{4}$$

$$25x = 72$$

$$x = 2.88 \text{ R/min at 5 m}$$

Distance has a profound effect on dose received and therefore is one of the cardinal rules of radiation protection. As distance from the source increases, dose received decreases. (Bushong, pp 67–68)

136. (B) The film badge and TLD are most frequently used to measure radiation exposure to radiographers. The pocket dosimeter may be employed by those radiation workers exposed to higher doses of radiation and needing a daily reading. A blood test is an unacceptable method of monitoring radiation dose effects, as a very large dose would have to be received before blood changes would occur. *(Statkiewicz, pp 246–256)*

137. (C) The relationship between x-ray intensity and distance from the source is expressed in the Inverse Square Law of Radiation. The formula is:

$$\frac{I_1}{I_2} = \frac{D_2^2}{D_1^2}$$

Substituting known values:

$$\frac{22 \text{ R/hr}}{8 \text{ R/hr}} = \frac{x \text{ ft}^2}{3 \text{ ft}^2}$$

$$\frac{22}{8} = \frac{x^2}{9}$$

$$8x^2 = 198$$

$$x^2 = 24.7$$

$$x = 5 \text{ ft at } 8 \text{ R/hr}$$

Note that in order for the exposure rate to decrease, the distance from the source of radiation must increase. (Bushong, pp 67–68)

138. (C) Biologic tissue is more sensitive to radiation when it is in an oxygenated state. A characteristic of many avascular (and therefore, hypoxic) tumors is their resistance to treatment with radiation. Hyperbaric (high-pressure oxygen) therapy is used in some therapy centers in an effort to increase the sensitivity of the tissues being treated. *(Bushong, p 535)*

139. (D) *Accurate positioning* helps decrease the number of retakes. *Patient shielding* should be used whenever appropriate and possible. The use of *high kV* and *low mAs* exposure factors limits the quantity of radiation delivered to the patient. Use of *rare earth screens* can enable the technologist to reduce mAs by as much as four to eight times. *(Statkiewicz, p 181, 184–185, 187)*

140. (C) The patient, as the first scatterer, is the most important scatterer. At 1 m from the patient, the intensity of the scattered beam is 0.1% of the intensity of the primary beam. Compton scatter emerging from the patient is almost as energetic as the primary beam entering the patient. *(Bushong, p 633)*

141. (D) A thermoluminescent dosimeter (TLD) is a sensitive and accurate device used in radiation dosimetry. It may be used as a personal dosimeter or to measure patient dose during radiographic examinations and therapeutic procedures. The TLD utilizes a thermoluminescent phosphor, usually lithium fluoride. When used as a personal monitor, the TLD is worn for 1 month. During this time it stores information regarding the radiation to which it has been exposed. It is then returned to the commercial supplier. In the laboratory, the phosphors are heated. They respond by emitting a particular quantity of light, which is in proportion to the quantity of radiation delivered to it. *(Statkiewicz, p 255)*

142. (B) We must be particularly careful to avoid radiating a newly fertilized ovum. This is precisely the time when a pregnancy may be unsuspected and fetal irradiation could be most damaging. The ICRP therefore recommends use of the "10-day rule": Elective abdominal radiography should be limited to the first 10 days following the onset of menses, a time when pregnancy is most improbable. *(Bushong, pp 612–613)*

143. (A) With large doses of radiation, especially delivered all at one time, the body's systems demonstrate characteristic changes. The *central nervous system* will be affected with doses of 5000 rad or more. The *gastrointestinal system* demonstrates characteristic effects after a dose of 600 rad. The *hematapoietic* system ("bone marrow syndrome") can demonstrate the effects of radiation after 100 rad, making it the most radiosensitive of the systems listed. *(Statkiewicz, pp 132–134)*

144. (D) According to Bergonié and Tribondeau, the most radiosensitive cells are undifferentiated, rapidly dividing cells such as lymphocytes, intestinal crypt (of Lieberkühn) cells, and spermatogonia. Liver cells are among the types of cells that are somewhat differentiated and capable of mitosis. These characteristics render them somewhat radiosensitive. Muscle cells, as well as nerve cells and red blood cells, are highly differentiated and do not divide. Therefore, in order of *increasing* sensitivity (from least to greatest sensitivity), the cells are: muscle, liver, and then intestinal crypt cells. *(Travis, pp 70–72)*

145. (A) The x-ray tube's glass envelope and oil coolant are considered inherent ("built-in") filtration. Thin sheets of aluminum are added to make *a total of at least 2.5 mm Al equivalent filtration in equipment operated above 70 kVp.* This is done in order to remove the low-energy photons that only serve to contribute to skin dose. *(Statkiewicz, pp 171–173)*

146. (A) X-rays produced at the target comprise a heterogeneous primary beam. Filtration serves to eliminate the softer, less-penetrating photons, leaving an x-ray beam of higher average energy. Filtration is important in patient protection, because unfiltered, low-energy photons not energetic enough to reach the film stay in the body and contribute to total patient dose. *(Statkiewicz, pp 171–173)*

147. (C) The radiograph illustrates testing done to evaluate the x-ray beam and light beam alignment. Light-localized collimators must be tested periodically and must be accurate to within 2% of the focal-film distance. Linearity means that a given mA, using different mA stations with appropriate exposure time adjustments, will provide consistent intensity. A star pattern would be used to evaluate focal spot resolution, and a parallel line type resolution pattern could also be used to evaluate spatial resolution. *(Bushong, p 431)*

148. (D) As the amount of time one spends in a controlled area decreases, radiation exposure should decrease. Radiation exposure is affected considerably by one's proximity to the radiation source, as defined by the Inverse Square Law. Barriers (shielding) are an effective means of reducing radiation exposure; primary barriers, such as walls, protect from the primary beam and secondary barriers, such as lead aprons, are used to protect from secondary radiation. *(Bushong, pp 13–15)*

149. (B) Even the smallest exposure to radiation can be harmful. It must, therefore, be every radiographer's objective to keep his or her occupational exposure as far below the MPD as possible. Radiology personnel should never hold patients during an x-ray examination. *(Bushong, p 607)*

150. (D) Gamma rays and x-ray photons are physically identical. They differ in their origin. *Gamma rays are nuclear radiation;* that is, they are emitted from the nucleus of a radioactive atom. *X-rays are extranuclear radiation;* they are produced outside the nucleus from interactions with electrons. *(Bushong, pp 56–57)*

Subspecialty List

49. Patient protection: biologic effects of radiation
50. Personnel protection: sources of radiation exposure
51. Patient protection: biologic effects of radiation
52. Patient protection: biologic effects of radiation
53. Patient protection: biologic effects of radiation
54. Patient protection: biologic effects of radiation
55. Radiation exposure and monitoring: dosimeters
56. Radiation exposure and monitoring: units of measurement
57. Patient protection: biologic effects of radiation
58. Patient protection: biologic effects of radiation
59. Patient protection: biologic effects of radiation
60. Personnel protection: NCRP recommendations for protective devices
61. Personnel protection: special considerations
62. Patient protection: biologic effects of radiation
63. Radiation exposure and monitoring: sources of radiation exposure
64. Radiation exposure and monitoring: sources of radiation exposure
65. Radiation exposure and monitoring: sources of radiation exposure
66. Radiation exposure and monitoring: sources of radiation exposure
67. Personnel protection: basic methods of protection
68. Radiation exposure and monitoring: properties of radiation
69. Personnel protection: sources of radiation exposure
70. Personnel protection: sources of radiation exposure
71. Personnel protection: sources of radiation exposure
72. Radiation exposure and monitoring: dosimeters
73. Radiation exposure and monitoring: basic properties of radiation
74. Personnel protection: special considerations
75. Patient protection: biologic effects of radiation
76. Personnel protection: basic methods of protection
77. Personnel protection: NCRP recommendations for protective devices
78. Radiation exposure and monitoring: dosimeters
79. Patient protection: minimizing patient exposure
80. Personnel protection: special considerations
81. Patient protection: biologic effects of radiation
82. Patient protection: biologic effects of radiation
83. Radiation exposure and monitoring: basic properties of radiation
84. Patient protection: biologic effects of radiation
85. Personnel protection: sources of radiation exposure
86. Radiation exposure and monitoring: units of measurement
87. Personnel protection: special considerations
88. Patient protection: minimizing patient exposure
89. Patient protection: minimizing patient exposure
90. Patient protection: minimizing patient exposure
91. Patient protection: minimizing patient exposure
92. Patient protection: minimizing patient exposure
93. Personnel protection: special considerations
94. Patient protection: biologic effects of radiation
95. Patient protection: biologic effects of radiation
96. Patient protection: minimizing patient exposure
97. Patient protection: biologic effects of radiation
98. Patient protection: biologic effects of radiation
99. Patient protection: minimizing patient exposure
100. Personnel protection: NCRP recommendations for protective devices
101. Patient protection: minimizing patient exposure
102. Personnel protection: basic methods of protection
103. Patient protection: biologic effects of radiation
104. Patient protection: biologic effects of radiation
105. Patient protection: biologic effects of radiation
106. Patient protection: minimizing patient exposure
107. Patient protection: biologic effects of radiation
108. Patient protection: minimizing patient exposure
109. Personnel protection: special considerations
110. Radiation exposure and monitoring: units of measurement
111. Patient protection: biologic effects of radiation
112. Patient protection: biologic effects of radiation
113. Radiation exposure and monitoring: NCRP recommendations for personnel monitoring
114. Personnel protection: special considerations
115. Patient protection: biologic effects of radiation

116. Patient protection: biologic effects of radiation
117. Radiation exposure and monitoring: basic properties of radiation
118. Radiation exposure and monitoring: dosimeters
119. Patient protection: biologic effects of radiation
120. Radiation exposure and monitoring: basic properties of radiation
121. Radiation exposure and monitoring: basic properties of radiation
122. Personnel protection: special considerations
123. Radiation exposure and monitoring: basic properties of radiation
124. Radiation exposure and monitoring: basic properties of radiation
125. Patient protection: biologic effects of radiation
126. Radiation exposure and monitoring: NCRP recommendations for personnel monitoring
127. Patient protection: biologic effects of radiation
128. Personnel protection: sources of radiation exposure
129. Patient protection: biologic effects of radiation
130. Personnel protection: NCRP recommendations for protective devices
131. Radiation exposure and monitoring: sources of radiation exposure
132. Patient protection: biologic effects of radiation
133. Patient protection: minimizing patient exposure
134. Radiation exposure and monitoring: NCRP recommendations for personnel monitoring
135. Personnel protection: basic methods of protection
136. Radiation exposure and monitoring: dosimeters
137. Personnel protection: basic methods of protection
138. Personnel protection: basic methods of protection
139. Patient protection: minimizing patient exposure
140. Radiation exposure and monitoring: basic properties of radiation
141. Radiation exposure and monitoring: dosimeters
142. Patient protection: minimizing patient exposure
143. Patient protection: biologic effects of radiation
144. Patient protection: biologic effects of radiation
145. Patient protection: minimizing patient exposure
146. Patient protection: minimizing patient exposure
147. Patient protection: minimizing patient exposure
148. Personnel protection: basic methods of protection
149. Radiation exposure and monitoring: NCRP recommendations for personnel monitoring
150. Radiation exposure and monitoring: sources of radiation exposure

CHAPTER 4

Image Production and Evaluation
Questions

DIRECTIONS (Questions 1 through 250): Each of the numbered items or incomplete statements in this section is followed by answers or by completions of the statement. Select the ONE lettered answer or completion that is BEST in each case.

1. Failure to add starter solution to the developer will result in

    (A) processor transportation problems
    (B) poor resolution
    (C) radiographic image too dark
    (D) radiographic image too light

2. If the radiographer is unable to achieve a short object-image distance due to structure of the body part or patient condition, which of the following adjustments may be made to minimize magnification distortion?

    (A) smaller focal spot size should be used
    (B) longer source-image distance should be used
    (C) faster intensifying screens should be used
    (D) slow intensifying screens should be used

3. Exposure-type artifacts include

    1. processor roller marks
    2. static electricity marks
    3. foreign object in cassette

    (A) 1 only
    (B) 1 and 2 only
    (C) 2 and 3 only
    (D) 1, 2, and 3

4. The quantity of x-ray photons delivered to the patient in a given exposure is PRIMARILY regulated by

    (A) mAs
    (B) kVp
    (C) SID
    (D) focal spot size

5. That portion of a characteristic curve generally representative of the diagnostic density range is the

    (A) toe portion
    (B) straight line portion
    (C) shoulder portion
    (D) solarization point

6. Which of the radiographs in Figure 4–1 should be rejected by the quality control technologist for excessively high contrast?

    (A) film A
    (B) film B
    (C) both film A and film B
    (D) neither film A nor film B

7. Of the following groups of exposure factors, which will produce the greatest radiographic density?

    (A) 400 mA, 0.075 sec, 72-inch SID
    (B) 200 mA, 0.075 sec, 36-inch SID
    (C) 200 mA, 0.150 sec, 36-inch SID
    (D) 400 mA, 0.150 sec, 72-inch SID

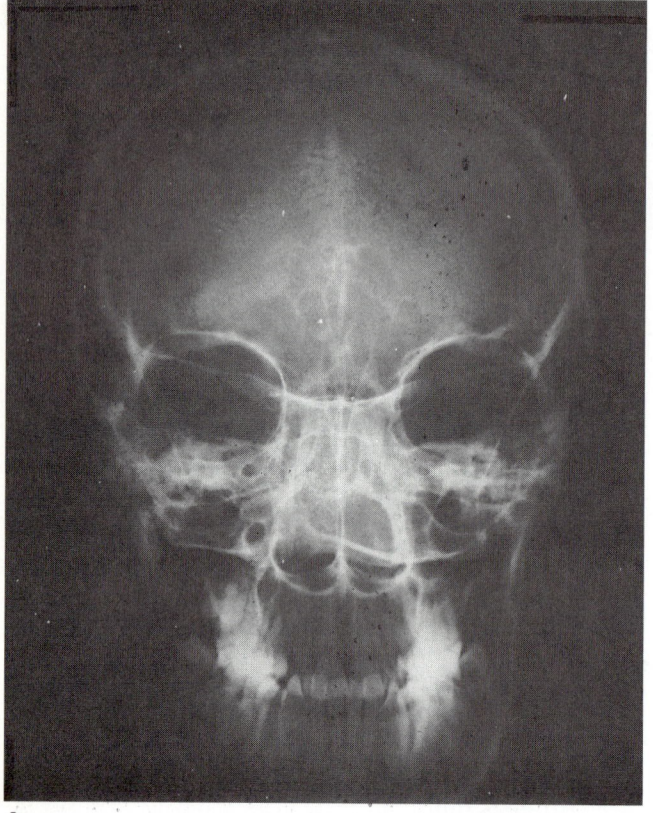
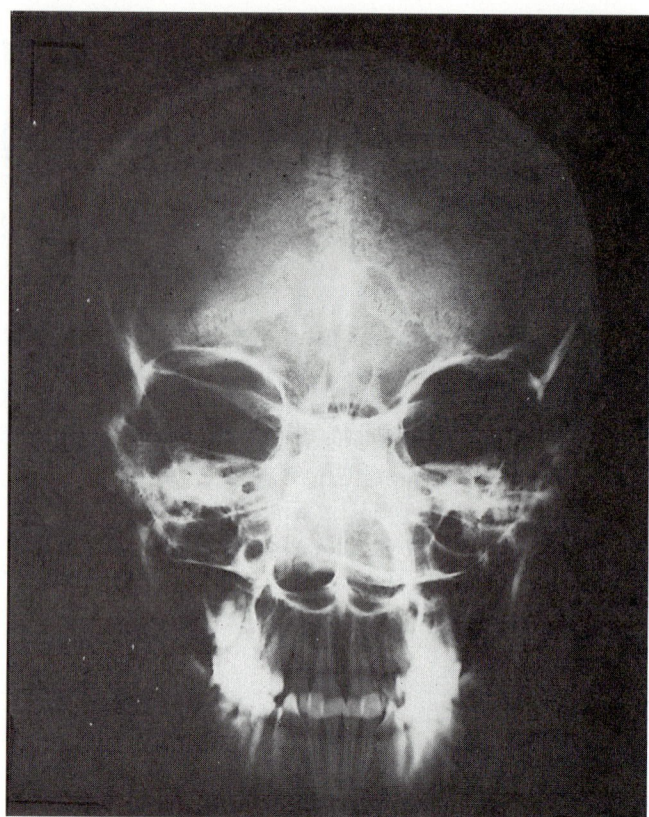

A   B

Figure 4–1. From the American College of Radiology Learning File. Courtesy of the ACR.

8. The exposure factors used for a particular non-grid chest radiograph were 300 mA, 0.01 sec, and 90 kVp. An additional radiograph using an 8:1 grid is requested. Which of the following groups of factors is most appropriate?

(A) 400 mA, 0.03 sec, 120 kVp
(B) 200 mA, 0.06 sec, 90 kVp
(C) 300 mA, 0.02 sec, 90 kVp
(D) 300 mA, 0.01 sec, 110 kVp

9. Which of the following is (are) tested, as part of a quality assurance program?

1. beam alignment
2. reproducibility
3. linearity

(A) 1 only
(B) 1 and 2 only
(C) 1 and 3 only
(D) 1, 2, and 3

10. Which of the following are methods of limiting the production of scattered radiation?

1. using the prone position for abdominal exams
2. using high kilovoltage levels
3. restricting the field size to the smallest practical size

(A) 1 and 2 only
(B) 1 and 3 only
(C) 2 and 3 only
(D) 1, 2, and 3

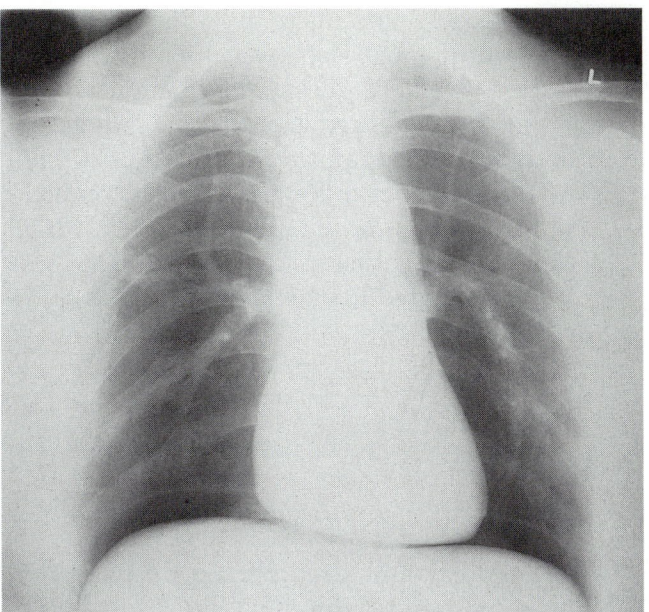

A

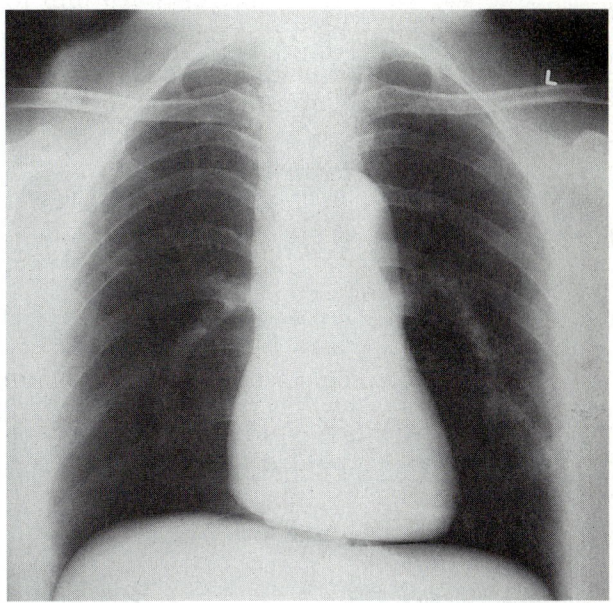

B

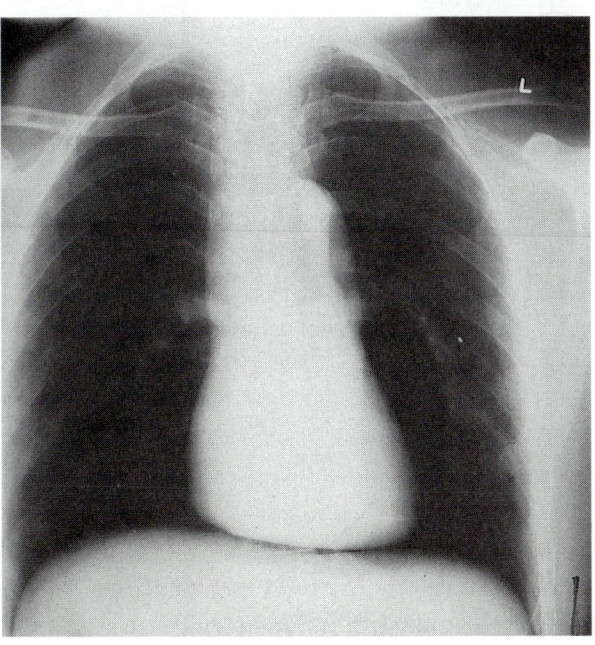

C

Figure 4–2. From the American College of Radiology Learning File. Courtesy of the ACR.

11. In order that a phosphor be suitable for use in intensifying screens, it should have which of the following characteristics?

1. high conversion efficiency
2. high x-ray absorption
3. high atomic number

(A) 1 only
(B) 3 only
(C) 1 and 2 only
(D) 1, 2, and 3

12. The three radiographs illustrated in Figure 4–2 were made with identical exposures, but one was developed at 90°F, one at the usual 95°F, and one at 100°F. Which is the radiograph made at 90°F?

(A) film A
(B) film B
(C) film C
(D) both films A and C

13. Which combination of exposure factors will MOST likely contribute to producing the longest scale contrast?

	mAs	kVp	Film/Screen System	Grid Ratio	Field Size
(A)	10	70	400	5:1	14 × 17″
(B)	12	90	200	8:1	14 × 17″
(C)	15	90	200	12:1	8 × 10″
(D)	20	80	400	10:1	11 × 14″

14. When green sensitive rare earth screens are properly matched with the correct film, what type of safelight should be used in the darkroom?

    (A) Wratten-6B
    (B) GBX or GS1
    (C) amber
    (D) none

15. The function of the developing solution is to

    (A) reduce the manifest image to a latent image
    (B) increase production of silver halide crystals
    (C) reduce the latent image to a manifest image
    (D) remove the unexposed crystals from the film

16. Which of the following statements is (are) true with respect to the magnification radiographs in Figure 4–3?

    1. film B was made using a larger focal spot than film A
    2. film B has a shorter scale contrast than film A
    3. film A has more focal spot blur than film B

    (A) 1 only
    (B) 1 and 2 only
    (C) 2 and 3 only
    (D) 1, 2, and 3

17. An exposure was made at 36 inches SID using 300 mA, 0.05-sec exposure, and 70 kVp with a 400 film/screen combination and an 8:1 grid. It is desired to repeat the radiograph and, in order to improve recorded detail, use 44 inches SID and 200 film/screen combination. Using a 0.22-sec exposure, and with all other factors remaining constant, what mA will be required to maintain the original radiographic density?

    (A) 100 mA
    (B) 200 mA
    (C) 300 mA
    (D) 400 mA

18. The reduction in x-ray photon intensity as it passes through material is termed

    (A) absorption
    (B) scattering
    (C) attenuation
    (D) divergence

19. Which of the following is (are) essential to high-quality mammographic examinations?

    1. small focal spot x-ray tube
    2. long scale contrast
    3. use of a compression device

    (A) 1 only
    (B) 1 and 2 only
    (C) 1 and 3 only
    (D) 1, 2, and 3

20. Which of the following affects both the quantity and quality of the primary beam?

    1. half-value layer
    2. kVp
    3. mA

    (A) 1 only
    (B) 2 only
    (C) 1 and 2 only
    (D) 1, 2, and 3

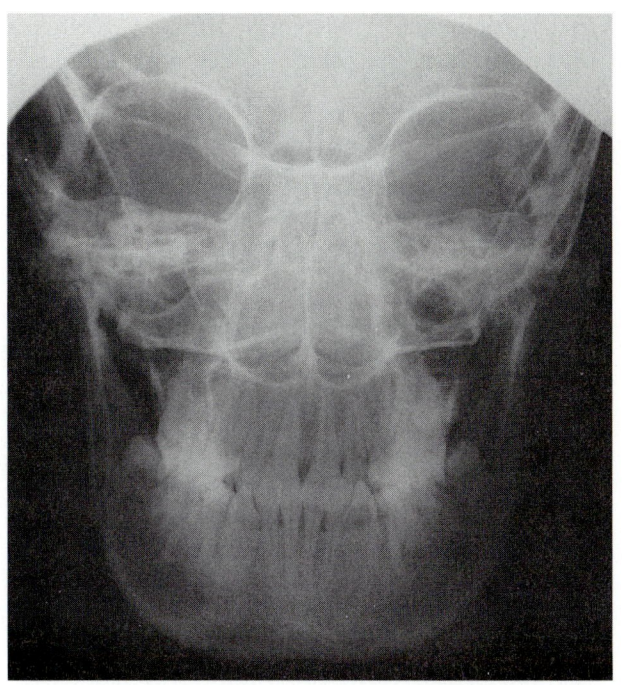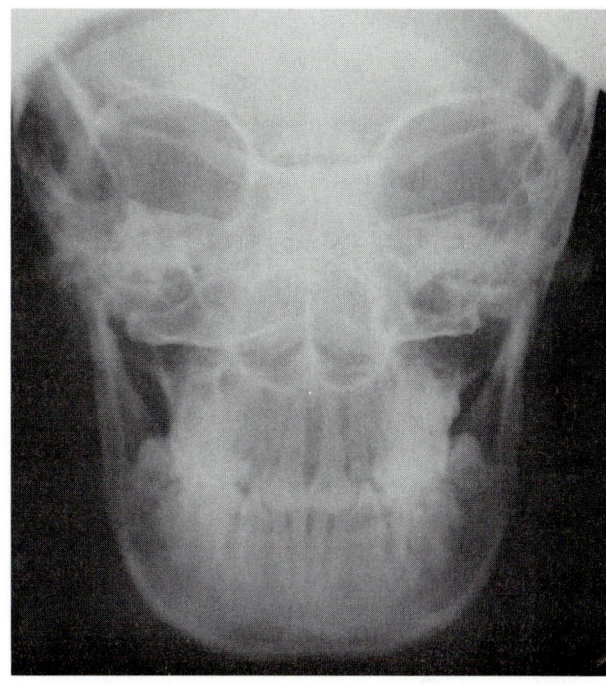

A   B

Figure 4–3. From the American College of Radiology Learning File. Courtesy of the ACR.

21. If the quantity of black metallic silver on a particular radiograph is such that it allows 10 percent of the illuminator light to pass through the film, that film has a density of

   (A) 0.01
   (B) 0.1
   (C) 1.0
   (D) 2.0

22. The term "latitude" describes

   1. an emulsion's ability to record a range of densities
   2. the degree of error tolerated with given exposure factors
   3. conversion efficiency of a given intensifying screen

   (A) 1 only
   (B) 1 and 2 only
   (C) 2 and 3 only
   (D) 1, 2, and 3

23. Exposure factors of 90 kVp and 6 mAs are used for a particular non-grid exposure. What should be the new mAs if an 8:1 grid is added?

   (A) 12 mAs
   (B) 18 mAs
   (C) 24 mAs
   (D) 30 mAs

24. What is added to the developer to prevent excessive softening of the emulsion and sticky processor rollers?

   (A) hydroquinone
   (B) glutaraldehyde
   (C) ammonium thiosulfate
   (D) potassium bromide

25. A radiograph of the upper extremity was made using 5 mAs, 60 kVp, at a 40 inch SID and having a 12-inch OID. What is the percent magnification if the exposure is repeated using the same exposure factors but with an 8-inch OID?

    (A) 20 percent
    (B) 25 percent
    (C) 50 percent
    (D) 75 percent

26. Unopened boxes of radiographic film should be stored away from radiation and

    (A) in the horizontal position
    (B) in the vertical position
    (C) stacked with the oldest on top
    (D) stacked with the newest on top

27. Which of the following may be used to reduce the effect of scattered radiation on the finished radiograph?

    1. grids
    2. collimators
    3. compression bands

    (A) 1 only
    (B) 1 and 3 only
    (C) 2 and 3 only
    (D) 1, 2, and 3

28. Factors that contribute to film fog include

    1. the age of the film
    2. excessive exposure to safelight
    3. processor chemistry

    (A) 1 only
    (B) 1 and 2 only
    (C) 1 and 3 only
    (D) 1, 2, and 3

29. Which of the following contribute to the radiographic contrast present on the finished radiograph?

    1. atomic number of tissues radiographed
    2. any pathologic processes
    3. degree of muscle development

    (A) 1 and 2 only
    (B) 1 and 3 only
    (C) 2 and 3 only
    (D) 1, 2, and 3

30. Which of the radiographs in Figure 4–4 most likely required a greater exposure?

    (A) film A
    (B) film B
    (C) they required identical exposures
    (D) insufficient information provided

31. Radiographic recorded detail is directly related to

    1. source-image distance
    2. object-image distance
    3. imaging system speed

    (A) 1 only
    (B) 1 and 2 only
    (C) 2 and 3 only
    (D) 1, 2, and 3

32. Compared to a low-ratio grid, a high-ratio grid will

    1. absorb more primary radiation
    2. absorb more scattered radiation
    3. allow more centering latitude

    (A) 1 only
    (B) 1 and 2 only
    (C) 2 and 3 only
    (D) 1, 2, and 3

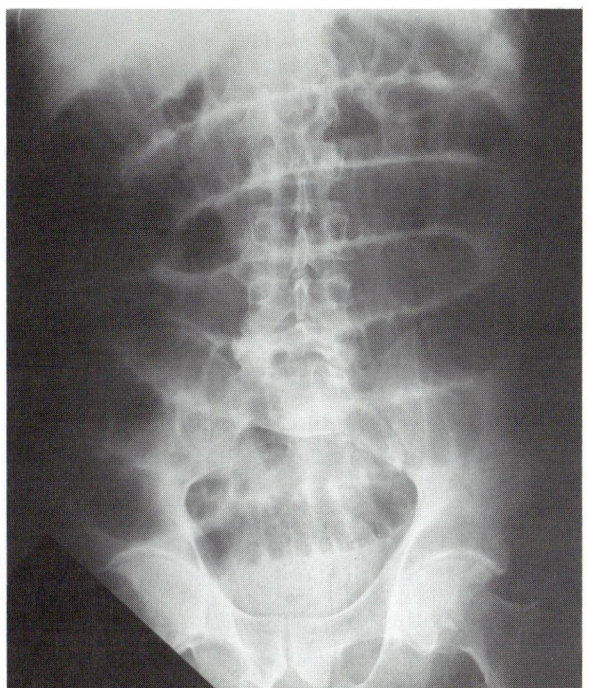 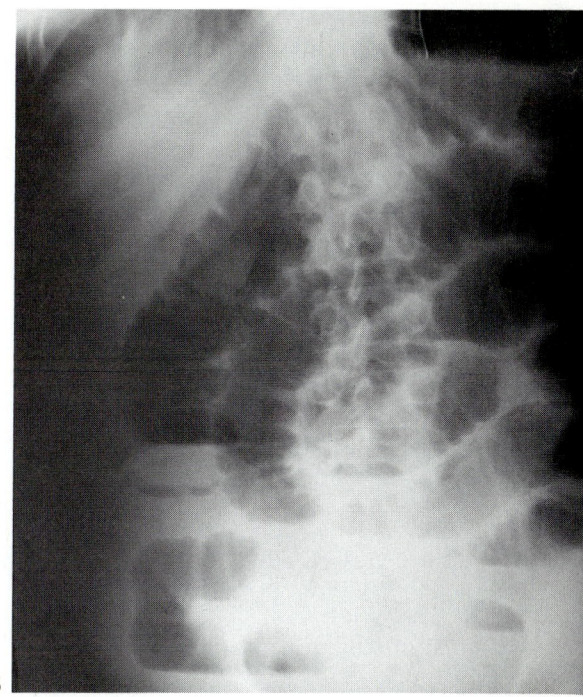

A   B

Figure 4–4. From the American College of Radiology Learning File. Courtesy of the ACR.

33. How is the mAs adjusted in an Automatic Exposure Control (AEC) system as the film/screen combination is decreased?

    (A) the mAs increases as film/screen speed decreases

    (B) both the mAs and kVp increase as film/screen speed decreases

    (C) the mAs decreases as film/screen speed decreases

    (D) the mAs remains unchanged as film/screen speed decreases

34. If a 6 inch object-image distance is introduced during a particular radiographic examination, what change in source-image distance will be necessary in order to overcome objectionable magnification?

    (A) the SID must be increased by 6 inches

    (B) the SID must be increased by 18 inches

    (C) the SID must be decreased by 6 inches

    (D) the SID must be increased by 42 inches

35. Screen-film imaging is one example of

    (A) an analog system

    (B) a digital system

    (C) an electromagnetic system

    (D) a direct action radiation system

36. Which portion of the characteristic curve represents overexposure of the film emulsion?

    (A) toe portion

    (B) straight line portion

    (C) shoulder portion

    (D) average gradient portion

37. Boxes of x-ray film should be stored in a

    (A) cool, humid area

    (B) cool, dry area

    (C) warm, humid area

    (D) warm, dry area

38. A lateral radiograph of the lumbar spine was made using 200 mA, 1 sec exposure, and 90 kVp. If the exposure factors were changed to 200 mA, ½ sec, and 104 kVp, there would be an obvious change in which of the following?

    1. radiographic density
    2. scale of radiographic contrast
    3. distortion

    (A) 1 only
    (B) 2 only
    (C) 2 and 3 only
    (D) 1, 2, and 3

39. The effect described as differential absorption is

    1. responsible for radiographic contrast
    2. a result of attenuating characteristics of tissue
    3. minimized by the use of high kVp

    (A) 1 only
    (B) 1 and 2 only
    (C) 1 and 3 only
    (D) 1, 2, and 3

40. Quantum mottle is MOST obvious when using

    (A) slow-speed screens
    (B) rare earth screens
    (C) fine-grain film
    (D) minimal filtration

41. Of the following groups of exposure factors, which will produce the shortest scale of radiographic contrast?

    (A) 500 mA, 0.040 sec, 70 kVp
    (B) 100 mA, 0.100 sec, 80 kVp
    (C) 200 mA, 0.025 sec, 92 kVp
    (D) 700 mA, 0.014 sec, 80 kVp

42. In general, as the intensification factor increases

    1. radiographic density increases
    2. screen resolution increases
    3. recorded detail increases

    (A) 1 only
    (B) 1 and 2 only
    (C) 2 and 3 only
    (D) 1, 2, and 3

43. An exposure was made using 300 mA, 0.04-sec exposure, and 85 kVp. Each of the following changes will serve to decrease the radiographic density by one half, EXCEPT

    (A) change to ⅙₀-sec exposure
    (B) change to 72 kVp
    (C) change to 10 mAs
    (D) change to 150 mA

44. The AP knee radiograph in Figure 4–5, made using automatic exposure control, exhibits which of the following?

    1. grid cutoff
    2. x-ray tube not centered
    3. Bucky tray not pushed in

    (A) 1 only
    (B) 1 and 2 only
    (C) 2 and 3 only
    (D) 3 only

45. Which of the following is recommended upon restarting an automatic film processor that has been shut off for a period of time?

    1. wipe down the crossover rollers
    2. let it run a few minutes to bring up the temperature
    3. run a clean-up film through it

    (A) 1 only
    (B) 1 and 2 only
    (C) 2 and 3 only
    (D) 1, 2, and 3

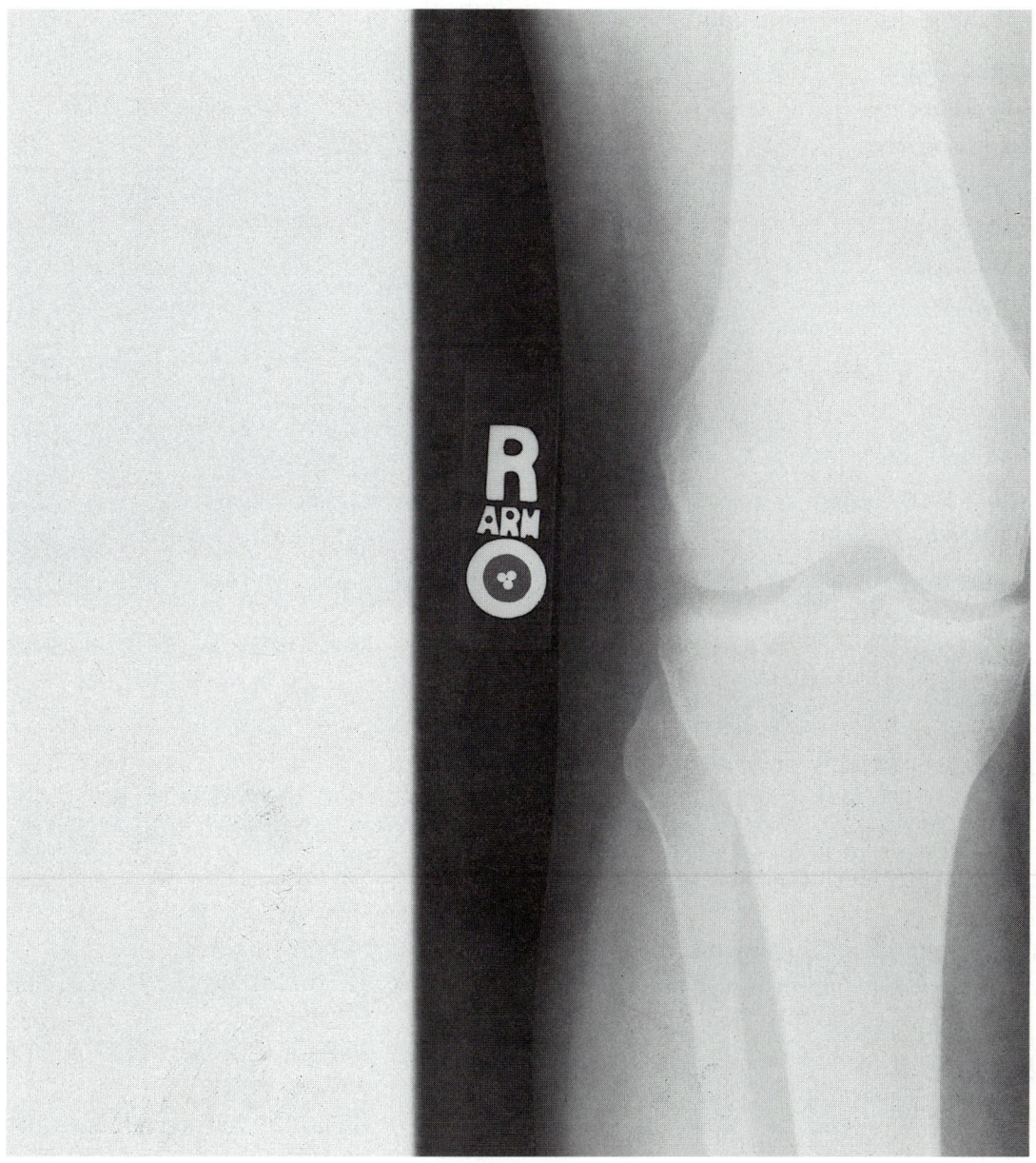

Figure 4–5. Courtesy of The Stamford Hospital, Department of Radiology.

46. In comparison to 60 kVp, 80 kVp will

   1. permit greater exposure latitude
   2. produce longer-scale contrast
   3. produce more scatter radiation

  (A) 1 only
  (B) 2 only
  (C) 1 and 2 only
  (D) 1, 2, and 3

47. A satisfactory radiograph was made using a 40 inch source-image distance, 10 mAs, and a 12:1 grid. If the exam must be repeated using the mobile x-ray unit at a distance of 48 inches and using an 8:1 grid, then in order to maintain density, what should be the new mAs?

  (A) 5.6 mAs
  (B) 8.8 mAs
  (C) 11.5 mAs
  (D) 14.4 mAs

48. Types of shape distortion include

   1. magnification

   2. elongation

   3. foreshortening

   (A) 1 only

   (B) 1 and 2 only

   (C) 2 and 3 only

   (D) 1, 2, and 3

49. Which of the following have an effect on recorded detail?

   1. focal spot size

   2. type of rectification

   3. source-image distance

   (A) 1 and 2 only

   (B) 1 and 3 only

   (C) 2 and 3 only

   (D) 1, 2, and 3

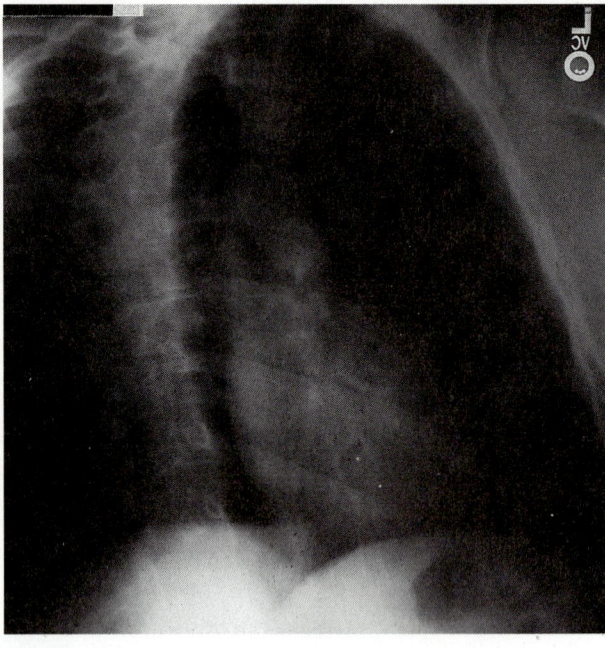

Figure 4–6. Courtesy of The Stamford Hospital, Department of Radiology.

50. Which of the following statements are true with respect to the radiograph shown in Figure 4–6?

   1. the radiograph exhibits poor recorded detail

   2. the radiograph exhibits high contrast

   3. the radiograph demonstrates motion

   (A) 1 and 2 only

   (B) 1 and 3 only

   (C) 2 and 3 only

   (D) 1, 2, and 3

51. If a particular grid has lead strips 0.40 mm thick, 4.0 mm high, and 0.25 mm apart, what is its grid ratio?

   (A) 8:1

   (B) 10:1

   (C) 12:1

   (D) 16:1

52. Which of the following groups of exposure factors will produce the greatest radiographic density?

   (A) 400 mA, 0.010 sec, 94 kVp, 100 speed screens

   (B) 500 mA, 0.008 sec, 94 kVp, 200 speed screens

   (C) 200 mA, 0.040 sec, 94 kVp, 50 speed screens

   (D) 100 mA, 0.020 sec, 80 kVp, 200 speed screens

53. Which of the following is NOT related to radiographic contrast?

   (A) photon energy

   (B) grid ratio

   (C) object-image distance

   (D) focal spot size

54. Greater latitude is available to the radiographer in which of the following circumstances?

    1. using high kVp techniques
    2. using slow film/screen combination
    3. using low-ratio grid

    (A) 1 only
    (B) 1 and 2 only
    (C) 2 and 3 only
    (D) 1, 2, and 3

55. The relationship between the intensity of light striking a film compared to the intensity of light transmitted by the film is an expression of which of the following?

    (A) radiographic contrast
    (B) radiographic density
    (C) recorded detail
    (D) radiographic filtration

56. Of the following groups of exposure factors, which will produce the greatest radiographic density?

    (A) 10 mAs, 74 kVp, 44 inches SID
    (B) 10 mAs, 74 kVp, 36 inches SID
    (C) 5 mAs, 85 kVp, 48 inches SID
    (D) 5 mAs, 85 kVp, 40 inches SID

57. Using a 48-inch source-image distance, how much object-image distance must be introduced to magnify an object two times?

    (A) 8 inch OID
    (B) 12 inch OID
    (C) 16 inch OID
    (D) 24 inch OID

58. High kilovoltage exposure factors are usually required for radiographic examinations using

    1. water-soluble, iodinated media
    2. a negative contrast medium
    3. barium sulfate

    (A) 1 only
    (B) 2 only
    (C) 3 only
    (D) 1 and 3 only

59. What determines the amount of fluorescent light emitted from a fluorescent screen?

    1. thickness of the active layer
    2. type of phosphor used
    3. kV range used for exposure

    (A) 1 only
    (B) 1 and 2 only
    (C) 2 and 3 only
    (D) 1, 2, and 3

60. In which of the following positions can the anode heel effect be an important consideration?

    1. lateral thoracic spine
    2. AP femur
    3. RAO sternum

    (A) 1 only
    (B) 1 and 2 only
    (C) 1 and 3 only
    (D) 1, 2, and 3

61. Which of the following technical changes would BEST serve to remedy the effect of widely different tissue densities?

    (A) use of high-speed screens
    (B) use of high-ratio grid
    (C) high kVp exposure factors
    (D) high mAs exposure factors

62. The amount of replenishment solution added to the automatic processor is determined by the

    1. size of the film
    2. position of film on tray feeding into processor
    3. length of time required for film to enter processor

    (A) 1 only
    (B) 1 and 2 only
    (C) 1 and 3 only
    (D) 1, 2, and 3

63. An exposure was made using 12 mAs and 60 kVp. If the kVp was changed to 70 in order to obtain longer scale contrast, what should be the new mAs?

(A) 3 mAs

(B) 6 mAs

(C) 18 mAs

(D) 24 mAs

64. The radiograph shown in Figure 4–7 is an example of

(A) linear tomography

(B) computed tomography

(C) grid cutoff

(D) poor screen/film contact

65. The device used to give a predetermined exposure to a film in order to test its response to processing is called the

(A) sensitometer

(B) densitometer

(C) step wedge

(D) spinning top

66. Which of the two film emulsions illustrated in Figure 4–8 possesses more latitude between densities 1.0 and 2.5?

(A) number 1 possesses more latitude

(B) number 2 possesses more latitude

(C) the films possess identical latitude

(D) latitude cannot be predicted from the illustration

67. The microswitch for controlling the amount of replenishment used in an automatic processor is located at the

(A) receiving bin

(B) crossover roller

(C) entrance roller

(D) replenishment pump

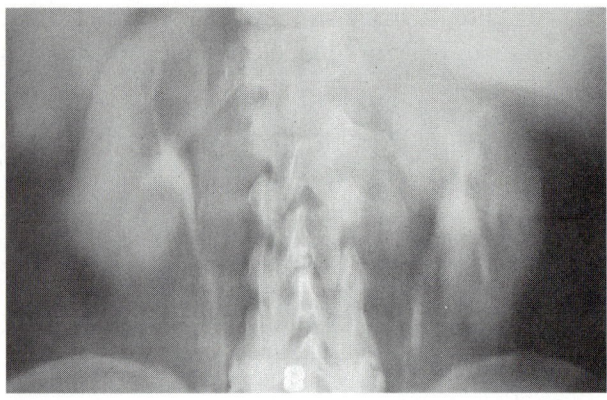

Figure 4–7. From the American College of Radiology Learning File. Courtesy of the ACR.

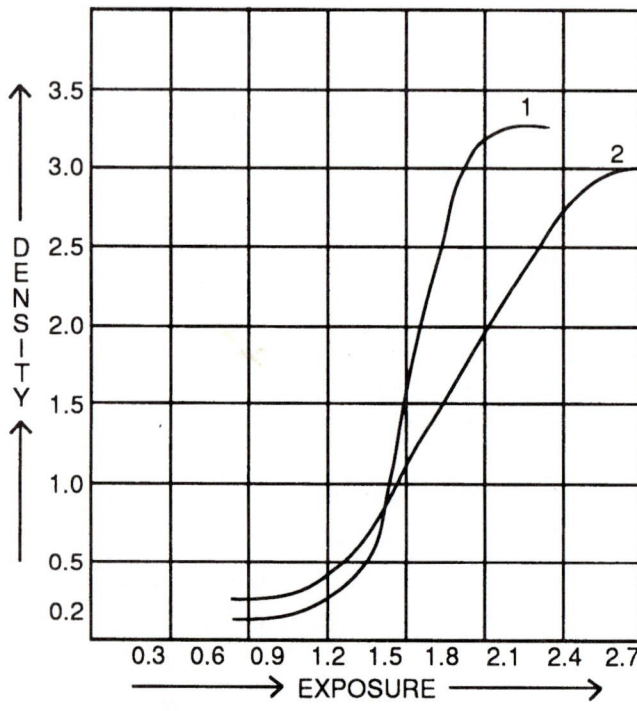

Figure 4–8. Courtesy of David Perri.

68. A film emerging from the automatic processor exhibits excessive density. This may be attributable to which of the following?

    1. developer temperature too low

    2. chemical fog

    3. over-replenishment

(A) 1 only

(B) 1 and 2 only

(C) 2 and 3 only

(D) 1, 2, and 3

69. Types of moving grid mechanisms include

    1. oscillating

    2. synchronous

    3. reciprocating

(A) 1 only

(B) 1 and 2 only

(C) 1 and 3 only

(D) 2 and 3 only

70. The cause of films coming from the automatic processor still damp can be

(A) air velocity too high

(B) unbalanced processing temperatures

(C) insufficient hardening action

(D) excessive hardening action

71. An exposure was made of a part using 200 mA and 0.03 sec using a 200 speed film/screen combination. An additional radiograph is requested using a 400 speed system in order to reduce motion unsharpness. Using 300 mA, all other factors remaining constant, what should be the new exposure time?

(A) 0.010 sec

(B) 0.015 sec

(C) 0.020 sec

(D) 0.025 sec

72. Most laser film must be handled

(A) under a Wratten 6B safelight

(B) in total darkness

(C) under a GBX safelight

(D) with high-temperature processors

73. Which of the following groups of exposure factors will produce the longest scale of contrast?

(A) 200 mA, 0.08 sec, 95 kVp, 12:1 grid

(B) 500 mA, 0.03 sec, 81 kVp, 8:1 grid

(C) 300 mA, 0.05 sec, 95 kVp, 8:1 grid

(D) 600 mA, $\frac{1}{40}$ sec, 70 kVp, 6:1 grid

74. The layer of phosphors coated on the intensifying screen's plastic or cardboard base is referred to as the

(A) active layer

(B) screen layer

(C) base layer

(D) protective layer

75. A particular radiograph was produced using 4 mAs and 96 kVp with an 8:1 ratio grid. The film is to be repeated using a 16:1 ratio grid. What should be the new mAs?

(A) 3 mAs

(B) 6 mAs

(C) 8 mAs

(D) 12 mAs

76. Which of the following are methods used for silver reclamation?

    1. electrolytic method

    2. metallic replacement

    3. photoelectric method

(A) 1 only

(B) 1 and 2 only

(C) 2 and 3 only

(D) 1, 2, and 3

77. What happens when a slow screen/film system is used with a fast/screen film automatic exposure control system?

(A) the resulting films too are light
(B) the resulting films too are dark
(C) the resulting films have improved detail
(D) the resulting films have poor detail

78. Which of the following is/are causes of grid cutoff when using reciprocating grids?

1. inadequate source-image-distance
2. x-ray tube off-center across the long axis of the lead strips
3. angling the beam in the direction of the lead strips

(A) 1 only
(B) 1 and 2 only
(C) 2 and 3 only
(D) 1, 2, and 3

79. What combination of exposure factors and image receptor speed would BEST function to reduce quantum mottle?

(A) decreased mAs, decreased kVp, fast-speed screens
(B) increased mAs, decreased kVp, slow-speed screens
(C) decreased mAs, increased kVp, fast-speed screens
(D) increased mAs, increased kVp, fast-speed screens

80. Penumbra, or edge gradient, is greatest

(A) directly along the course of the central ray
(B) toward the cathode end of the x-ray beam
(C) toward the anode end of the x-ray beam
(D) as the SID is increased

81. If a 4-inch collimated field is changed to a 14-inch collimated field, with no other changes, the radiographic image will possess

(A) more density
(B) less density
(C) more detail
(D) less detail

82. Factor(s) that may be used to regulate radiographic density is (are)

1. milliamperage
2. exposure time
3. kilovoltage

(A) 1 only
(B) 2 only
(C) 1 and 2 only
(D) 1, 2, and 3

83. Tree-like, branching black marks on a radiograph are usually due to

(A) bending the film acutely
(B) improper development
(C) improper film storage
(D) static electricity

84. Which of the following refers to a regular program of evaluation that ensures proper functioning of x-ray equipment, thereby ensuring radiographic reproducibility and protecting both radiation workers and patients?

(A) sensitometry
(B) densitometry
(C) quality assurance
(D) modulation transfer function

85. Which of the following is (are) methods that would enable the radiographer to reduce the exposure time required for a particular radiograph?

1. use higher mA
2. use higher kVp
3. use slower film/screen combination

(A) 1 only
(B) 1 and 2 only
(C) 2 and 3 only
(D) 1, 2, and 3

86. If 82 kVp, 500 mA, and 0.025 sec were used for a particular exposure using three-phase, twelve-pulse equipment, what mAs would be required, using single-phase equipment, to produce a similar radiograph?

    (A) 6 mAs
    (B) 10 mAs
    (C) 20 mAs
    (D) 25 mAs

87. A focal spot size of 0.3 mm or smaller is essential for which of the following procedures?

    (A) stereoradiography
    (B) magnification radiography
    (C) tomography
    (D) fluoroscopy

88. The radiograph in Figure 4–9 demonstrates evidence of

    (A) dust artifacts present on intensifying screens
    (B) chemical fog
    (C) excessive/incorrect collimation
    (D) precaution taken against undercutting

89. Materials that emit light when stimulated by x-ray photons are called

    (A) ions
    (B) electrodes
    (C) phosphors
    (D) crystals

90. Which of the following examinations might require the use of 120 kVp?

    1. AP abdomen
    2. air gap chest
    3. barium-filled stomach

    (A) 1 only
    (B) 2 only
    (C) 1 and 2 only
    (D) 2 and 3 only

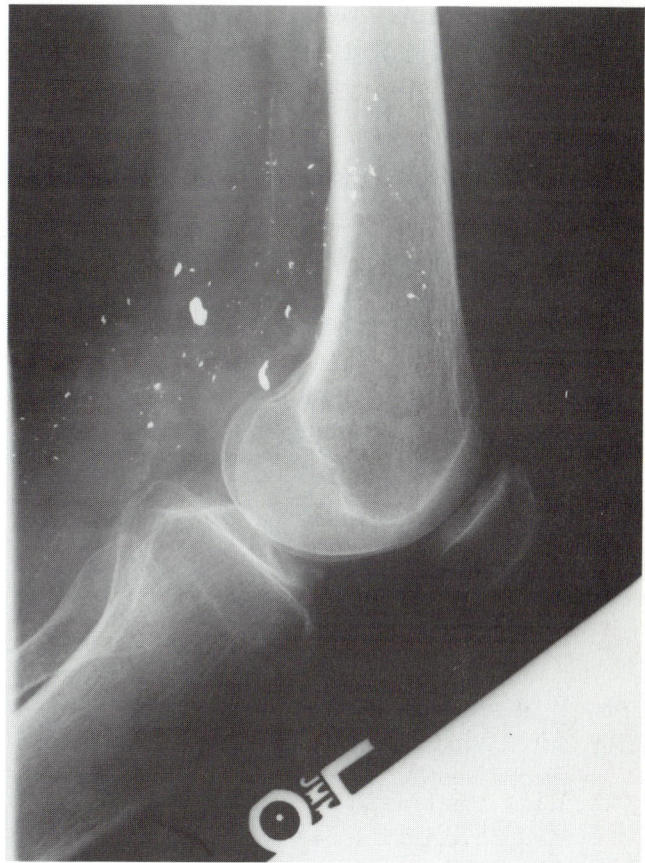

Figure 4–9. Courtesy of The Stamford Hospital, Department of Radiology.

91. The term "spectral matching" refers to the fact that film sensitivity must be matched with the

    (A) proper color screen fluorescence
    (B) correct kVp level
    (C) correct mA level
    (D) proper developer concentration

92. Which of the following is the correct order of radiographic film processing?

    (A) developer, wash, fixer, dry
    (B) fixer, wash, developer, dry
    (C) developer, fixer, wash, dry
    (D) fixer, developer, wash, dry

93. How are mAs and radiographic density related in the process of image formation?

(A) mAs and density are inversely proportional

(B) mAs and density are directly proportional

(C) mAs and density are related to image unsharpness

(D) mAs and density are unrelated

94. The variation in photon distribution between the anode and cathode ends of the x-ray tube is known as

(A) line focus principle

(B) anode heel effect

(C) Inverse Square Law

(D) Bohr's theory

95. If 1/40 sec was selected for a particular exposure, what mA would be necessary to produce 15 mAs?

(A) 900 mA

(B) 600 mA

(C) 500 mA

(D) 300 mA

96. The advantages of high kilovoltage chest radiography are that

1. the grid is no longer needed
2. it produces longer-scale contrast
3. it reduces patient dose

(A) 1 only

(B) 1 and 2 only

(C) 2 and 3 only

(D) 1, 2, and 3

97. The exposure factors of 400 mA, 0.04 sec, and 95 kVp were used to produce a particular radiographic density and contrast. A similar radiograph can be produced using 500 mA, 80 kVp, and

(A) 0.032 sec

(B) 0.064 sec

(C) 0.08 sec

(D) 0.12 sec

98. What is the best way to reduce magnification distortion?

(A) use a small focal spot

(B) increase the SID

(C) decrease the OID

(D) use a slow screen/film combination

99. Foreshortening may be caused by

1. the radiographic object being placed at an angle to the film
2. excessive distance between the object and the film
3. insufficient distance between the focus and the film

(A) 1 only

(B) 2 only

(C) 1 and 2 only

(D) 1, 2, and 3

100. Some of the causes of static electrical discharge in the x-ray darkroom include

1. low humidity
2. sliding films from cassette
3. use of rubber-soled shoes

(A) 1 only

(B) 1 and 2 only

(C) 2 and 3 only

(D) 1, 2, and 3

101. If 400 mA has been selected for a particular exposure, what exposure time would be required to produce 15 mAs?

(A) 1/60 sec

(B) 1/40 sec

(C) 1/30 sec

(D) 1/20 sec

102. Which of the following should be used for an examination of a patient suffering from Parkinson's disease?

    1. high-speed screens
    2. short exposure time
    3. compensating filtration

    (A) 1 only
    (B) 1 and 2 only
    (C) 1 and 3 only
    (D) 1, 2, and 3

103. Which of the following groups of exposure factors would be MOST appropriate to control involuntary motion?

    (A) 400 mA, 0.03 sec
    (B) 200 mA, 0.06 sec
    (C) 600 mA, 0.02 sec
    (D) 100 mA, 0.12 sec

104. A film emulsion having wide latitude is likely to exhibit

    (A) high density
    (B) low density
    (C) high contrast
    (D) low contrast

105. Misalignment of the tube-part-film relationship results in

    (A) shape distortion
    (B) size distortion
    (C) magnification
    (D) penumbra

106. Although the stated focal spot size is measured directly under the actual focal spot, focal spot size really varies along the length of the x-ray beam. At which portion of the x-ray beam is the effective focal spot the smallest?

    (A) at its outer edge
    (B) along the path of the central ray
    (C) at the cathode end
    (D) at the anode end

107. The radiograph pictured in Figure 4–10 demonstrates

    (A) underdevelopment
    (B) quantum noise
    (C) pi lines
    (D) grid cutoff

108. Which of the following will contribute to the production of shorter-scale radiographic contrast?

    1. an increase in kV
    2. an increase in grid ratio
    3. an increase in photon energy

    (A) 1 only
    (B) 2 only
    (C) 2 and 3 only
    (D) 1, 2, and 3

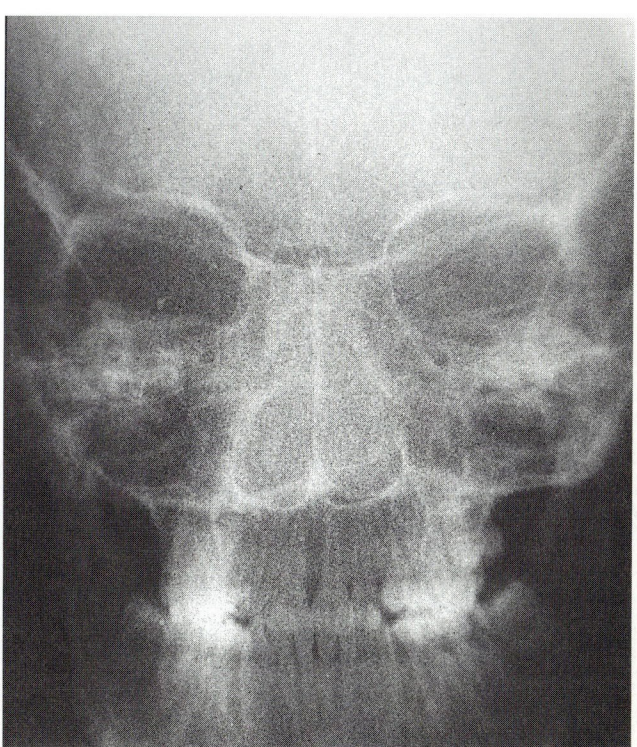

Figure 4–10. From the American College of Radiology Learning File. Courtesy of the ACR.

109. What apparatus is needed for the construction of a characteristic curve?

1. penetrometer
2. densitometer
3. electrolytic canister

(A) 1 and 2 only
(B) 1 and 3 only
(C) 2 and 3 only
(D) 1, 2, and 3

110. Due to the anode heel effect, the intensity of the x-ray beam is greatest along the

(A) path of the central ray
(B) anode end of the beam
(C) cathode end of the beam
(D) transverse axis of the film

111. The steeper the straight line portion of a characteristic curve for a particular film, the

1. slower the film speed
2. higher the film contrast
3. greater the exposure latitude

(A) 1 only
(B) 2 only
(C) 2 and 3 only
(D) 1, 2, and 3

112. Which of the following should be performed to check correctness of developing parameters?

(A) densitometry
(B) a thorough cleaning of rollers
(C) a warm-up procedure
(D) sensitometry

The following information should be used to answer questions 113 and 114: An AP radiograph of the hip was made using 300 mA, 0.05 sec, 76 kVp, 40 inch SID, 0.6 mm focal spot, 400 speed film/screen system.

113. Referring to the above information, and with all other factors remaining constant, which of the following exposure times would be required in order to maintain radiographic density at a 46 inch SID using the 500 mA station, and with an increase to 87 kVp?

(A) 0.015 sec
(B) 0.25 sec
(C) 0.02 sec
(D) 0.10 sec

114. Referring to the original factors, and with all other factors remaining constant, which of the following exposure times would be required in order to maintain radiographic density using the 400 mA station, 200 speed film/screen system, and with the addition of an 8:1 grid?

(A) 0.12 sec
(B) 0.18 sec
(C) 0.3 sec
(D) 0.6 sec

115. The Characteristic (H&D) Curve may be used to

1. identify automatic processing problems
2. determine film sensitivity
3. illustrate screen speed

(A) 1 only
(B) 1 and 2 only
(C) 2 and 3 only
(D) 1, 2, and 3

116. An AP radiograph of the lumbar spine was made at a 40 inch SID using 90 kVp. The 400 film/screen speed automatic exposure device (AED) terminated the exposure at 30 mAs. If the film is repeated using 200 speed film/screen combination to improve detail, the resulting radiograph will appear

(A) blurry
(B) magnified
(C) too light
(D) too dark

117. A decrease in recorded detail may be expected with a (an)

   1. decrease in screen speed
   2. increase in focal spot size
   3. decrease in source-image distance

   (A) 2 only
   (B) 1 and 2 only
   (C) 2 and 3 only
   (D) 1, 2, and 3

118. The cassette front may be made of which of the following materials?

   1. bakelite
   2. magnesium
   3. lead

   (A) 1 only
   (B) 1 and 2 only
   (C) 1 and 3 only
   (D) 1, 2, and 3

119. Which of the following errors is illustrated in Figure 4–11?

   (A) patient not centered to film
   (B) x-ray tube not centered to grid
   (C) inaccurate collimation
   (D) unilateral grid cutoff

120. Which of the following can result from improper film storage or darkroom conditions?

   1. safelight fog
   2. background radiation fog
   3. screen lag

   (A) 1 only
   (B) 1 and 2 only
   (C) 2 and 3 only
   (D) 1, 2, and 3

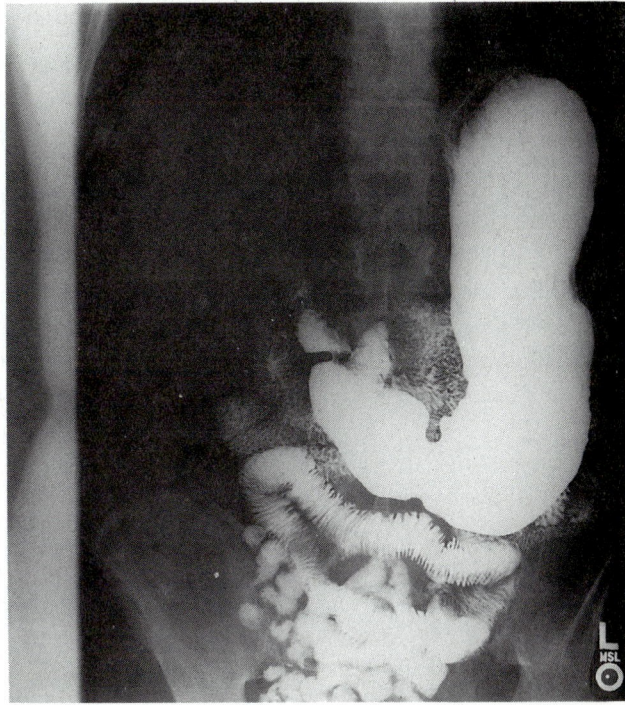

Figure 4–11. Courtesy of The Stamford Hospital, Department of Radiology.

121. Decreasing field size, from 14 × 17 inches to 8 × 10 inches will

   (A) decrease radiographic density and decrease the amount of scatter radiation produced within the part
   (B) decrease radiographic density and increase the amount of scatter radiation produced within the part
   (C) increase radiographic density and increase the amount of scatter radiation produced within the part
   (D) increase radiographic density and decrease the amount of scatter radiation produced within the part

122. Which of the following will result if developer replenisher is inadequate?

   (A) films with excessively high contrast
   (B) films with excessively low contrast
   (C) films with excessively high density
   (D) dry, brittle films

123. Foreshortening of an anatomic structure means that it is

(A) projected on the film shorter than its actual size

(B) imaged in an elongated fashion

(C) accompanied by significant geometric blur

(D) magnified

124. Recorded detail can be improved by

1. decreasing the source-image distance
2. decreasing the object-image distance
3. decreasing motion unsharpness

(A) 1 only

(B) 3 only

(C) 2 and 3 only

(D) 1, 2, and 3

125. The conversion of the invisible latent image into a visible manifest image takes place in the

(A) developer

(B) stop bath

(C) first half of the fixer process

(D) second half of the fixer process

126. Which of the following quantities of filtration is MOST likely to be used in mammography?

(A) 0.5 mm Mo

(B) 1.5 mm Al

(C) 1.5 mm Cu

(D) 2.0 mm Cu

127. The device shown in Figure 4–12 is used for

(A) tomographic quality assurance testing

(B) timer and rectifier testing

(C) mammography quality assurance testing

(D) kV calibration testing

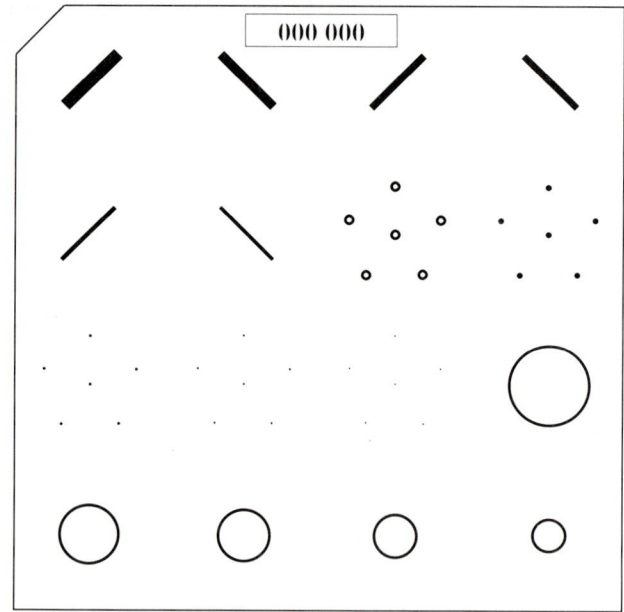

Figure 4–12. Reproduced with permission, compliments of Gammex/RMI, 2500 West Beltline Highway, Middletown, WI 53562.

128. Which of the following terms refers to light reflecting from one intensifying screen, through the film, to the opposite emulsion and screen?

(A) reflectance

(B) crossover

(C) scatter

(D) filtration

129. A wire mesh test is performed to diagnose

(A) screen lag

(B) screen contact

(C) screen resolution

(D) screen intensification

130. If an AP projection of the knee required 70 kVp, 300 mA, and ⅕ sec with single-phase equipment, what mAs would be required for the same knee using three-phase, six-pulse equipment?

(A) 18 mAs

(B) 13 mAs

(C) 10 mAs

(D) 8 mAs

131. The continued emission of light by a phosphor after the activating source has ceased is termed

(A) fluorescence
(B) phosphorescence
(C) image intensification
(D) quantum mottle

132. Which of the following radiographic accessories functions to produce uniform density on a radiograph?

(A) grid
(B) intensifying screens
(C) compensating filter
(D) penetrometer

133. A change from 100 speed screens to 200 speed screens would require what change in mAs?

(A) mAs should be increased by 15 percent
(B) mAs should be increased by 30 percent
(C) mAs should be doubled
(D) mAs should be halved

134. Which of the following is (are) classified as rare earth phosphors?

1. lanthanum oxybromide
2. gadolinium oxysulfide
3. cesium iodide

(A) 1 only
(B) 1 and 2 only
(C) 2 and 3 only
(D) 1, 2, and 3

135. Which of the following is MOST likely to produce a radiograph with a long scale of contrast?

(A) increased photon energy
(B) increased screen speed
(C) increased mAs
(D) increased SID

136. Which of the following pathologic conditions would require an increase in exposure factors?

(A) pneumoperitoneum
(B) obstructed bowel
(C) renal colic
(D) ascites

137. The degree to which an imaging system records an accurate and sharp radiographic image is expressed in line pairs per millimeter and is termed

(A) visibility
(B) resolution
(C) intensification factor
(D) quality assurance

138. Which of the two characteristic curves shown in Figure 4–13 will require less exposure to produce a density of 1.5 on the finished radiograph?

(A) film 1
(B) film 2
(C) they require identical exposures
(D) insufficient information is provided

139. The processor rollers that are out of solution and function to transfer the film from one solution to another are the

(A) turnaround assembly
(B) crossover rollers
(C) guide shoes
(D) deflector plates

140. Methods that help reduce the production of scattered radiation include using

1. compression
2. beam restriction
3. a grid

(A) 1 and 2 only
(B) 1 and 3 only
(C) 2 and 3 only
(D) 1, 2, and 3

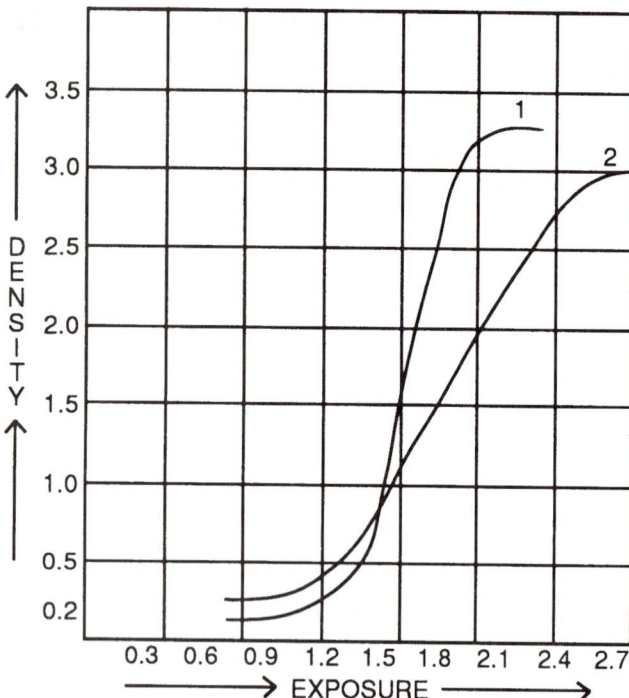

Figure 4–13. Courtesy of David Perri.

141. Which of the following devices is used to overcome severe variation in patient anatomy or tissue density, providing more uniform radiographic density?

(A) compensating filter
(B) grid
(C) collimator
(D) intensifying screen

142. The developer temperature in today's 90-sec automatic processors is usually about

(A) 75 to 80°F
(B) 80 to 85°F
(C) 85 to 90°F
(D) 90 to 95°F

143. An increase in kilovoltage will serve to

(A) produce a longer scale of contrast
(B) produce a shorter scale of contrast
(C) decrease the radiographic density
(D) decrease the production of scatter radiation

144. When the use of grids is indicated in mammography, which of the following would be MOST appropriate?

(A) 12:1 stationary grid
(B) 12:1 moving grid
(C) 5:1 stationary grid
(D) 5:1 moving grid

145. Crescent-shaped crinkle mark artifacts on a film are due to

(A) static electricity discharge
(B) processor transport problems
(C) bending film during handling
(D) increased darkroom moisture

146. A radiograph made using 300 mA, ⅕ sec, and 75 kVp exhibits motion unsharpness, but otherwise satisfactory technical quality. You therefore wish to decrease the exposure time. Using 86 kV and 500 mA, what should be the new exposure time?

(A) 0.12 sec
(B) 0.06 sec
(C) 0.03 sec
(D) 0.01 sec

147. Which of the following factors contribute(s) to the efficiency of a grid?

1. grid ratio
2. number of lead strips per inch
3. degree to which its use improves contrast

(A) 1 only
(B) 2 only
(C) 1 and 2 only
(D) 1, 2, and 3

148. Source-image distance affects recorded detail in which of the following ways?

(A) recorded detail is directly related to SID
(B) recorded detail is inversely related to SID
(C) as SID increases, recorded detail decreases
(D) SID is not a detail factor

149. The artifact pictured in Figure 4–14 usually results from

  (A) static electricity

  (B) chemical fog

  (C) misaligned guide shoe

  (D) acute bending of the film

150. Cassettes frequently have a lead foil layer behind the rear screen, which functions to

  (A) improve penetration

  (B) absorb backscatter

  (C) preserve resolution

  (D) increase the screen speed

151. A compensating filter is used to

  (A) absorb the harmful photons contributing only to patient dose

  (B) even out widely differing tissue densities

  (C) eliminate much of the scattered radiation

  (D) improve fluoroscopy

152. Focusing distance is associated with which of the following?

  (A) stereoradiography

  (B) chest radiography

  (C) magnification radiography

  (D) grids

153. Underexposure of a radiograph can be caused by all of the following, EXCEPT

  (A) insufficient mA

  (B) insufficient exposure time

  (C) insufficient kVp

  (D) insufficient SID

154. The relationship between the height of a grid's lead strips to the distance between them is referred to as grid

  (A) ratio

  (B) radius

  (C) frequency

  (D) focusing distance

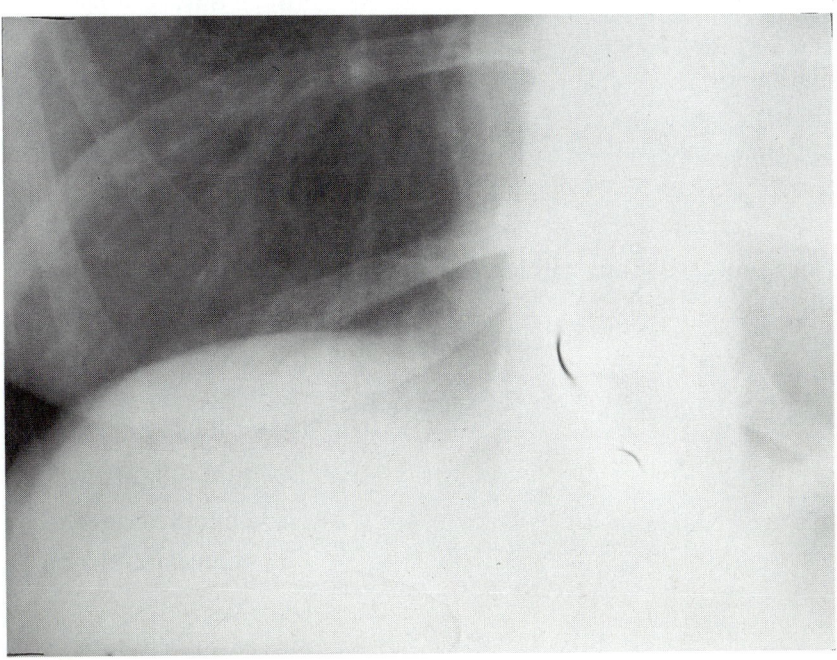

Figure 4–14. From the American College of Radiology Learning File. Courtesy of the ACR.

155. What are the effects of scatter radiation on the radiographic image?

1. it produces fog
2. it increases contrast
3. it increases grid cut-off

(A) 1 only
(B) 2 only
(C) 1 and 2 only
(D) 1, 2, and 3

156. Which of the following tests is performed to evaluate screen contact?

(A) spinning-top test
(B) wire mesh test
(C) penetrometer test
(D) star pattern test

157. Boxes of film stored in too warm an area may be subject to

(A) static marks
(B) film fog
(C) high contrast
(D) loss of density

158. The area of blurriness seen in the upper part of the radiograph shown in Figure 4–15 is MOST likely due to

(A) scatter radiation fog
(B) patient motion
(C) poor screen–film contact
(D) grid cutoff

159. An unexposed and processed film will have a density of about

(A) zero
(B) 0.1
(C) 1.0
(D) 2.5

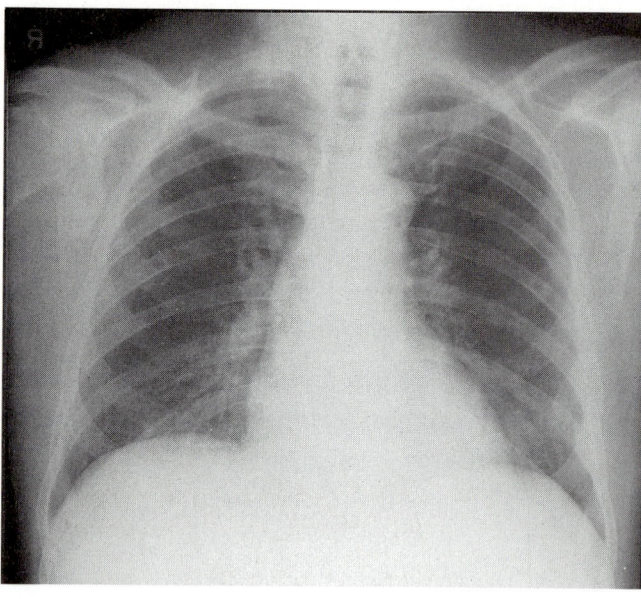

Figure 4–15. From the American College of Radiology Learning File. Courtesy of the ACR.

160. Poor screen–film contact can be caused by which of the following?

1. damaged cassette frame
2. foreign body in cassette
3. warped cassette front

(A) 1 only
(B) 2 only
(C) 1 and 3 only
(D) 1, 2, and 3

161. In which of the following examinations should 70 kVp NOT be exceeded?

(A) GI series
(B) BE
(C) IVP
(D) chest

162. Base-plus fog is a result of

1. blue tinted base
2. chemical development
3. manufacturing

(A) 1 only
(B) 1 and 2 only
(C) 1 and 3 only
(D) 1, 2, and 3

163. In radiography of a large abdomen, which of the following is (are) effective ways to minimize the amount of scatter radiation reaching the film?

1. use of close collimation
2. use of compression devices
3. use of a low-ratio grid

(A) 1 only
(B) 1 and 2 only
(C) 1 and 3 only
(D) 1, 2, and 3

164. Which of the following is (are) advantages of high kVp techniques?

1. greater exposure latitude
2. long-scale contrast
3. less scatter radiation fog

(A) 2 only
(B) 1 and 2 only
(C) 2 and 3 only
(D) 1, 2, and 3

165. All of the following are related to recorded detail, EXCEPT

(A) mA
(B) focal spot size
(C) screen speed
(D) object-image distance

166. With all other factors remaining the same, as grid ratio is increased

(A) recorded detail decreases
(B) radiographic density decreases
(C) penumbral distortion decreases
(D) the scale of contrast becomes longer

167. The function(s) of the fixer in film processing is (are) to

1. remove the unexposed silver bromide crystals
2. change the unexposed silver bromide crystals to black metallic silver
3. harden the emulsion

(A) 1 only
(B) 1 and 3 only
(C) 2 and 3 only
(D) 1, 2, and 3

168. The radiograph seen in Figure 4–16 illustrates

1. chemical fog
2. short scale contrast
3. low contrast

(A) 1 only
(B) 2 only
(C) 2 and 3 only
(D) 1, 2, and 3

169. X-ray film is packaged in a foil bag to protect it from

(A) excessive heat
(B) excessive humidity
(C) radiation fog
(D) dust

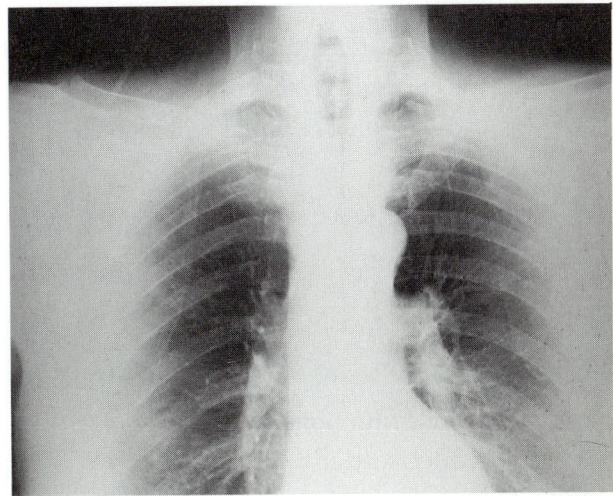

Figure 4–16. Reproduced with permission from the American College of Radiology Learning File. Courtesy of the ACR.

170. A satisfactory radiograph of the abdomen was made at a 36 inch SID using 300 mA, $\frac{1}{15}$ sec, and 80 kVp. If the distance is changed to 48 inches, what new mAs would be required?

(A) 11 mAs
(B) 15 mAs
(C) 27 mAs
(D) 35 mAs

171. With a given exposure, as intensifying screen speed increases, how is radiographic density affected?

(A) density decreases
(B) density increases
(C) density remains unchanged
(D) density is variable

172. The squeegee assembly in an automatic processor

1. functions to remove excess solution from films
2. is located near crossover rollers
3. helps establish the film's rate of travel

(A) 1 only
(B) 2 only
(C) 1 and 2 only
(D) 1, 2, and 3

173. An increase in kVp will have which of the following effects?

1. more scatter radiation will be produced
2. the exposure rate will increase
3. radiographic contrast will increase

(A) 1 only
(B) 1 and 2 only
(C) 2 and 3 only
(D) 1, 2, and 3

174. The major function of filtration is to

(A) reduce film noise
(B) reduce scatter radiation
(C) reduce operator dose
(D) reduce patient dose

175. An increase in the kilovoltage applied to the x-ray tube increases the

1. x-ray wavelength
2. exposure rate
3. patient absorption

(A) 1 only
(B) 2 only
(C) 2 and 3 only
(D) 1, 2, and 3

176. The interaction between x-ray photons and matter illustrated in Figure 4–17 is MOST likely to occur

1. when using a positive contrast medium
2. during radiographic examination of the abdomen
3. using high kV and low mAs exposure factors

(A) 1 only
(B) 1 and 2 only
(C) 2 and 3 only
(D) 1, 2, and 3

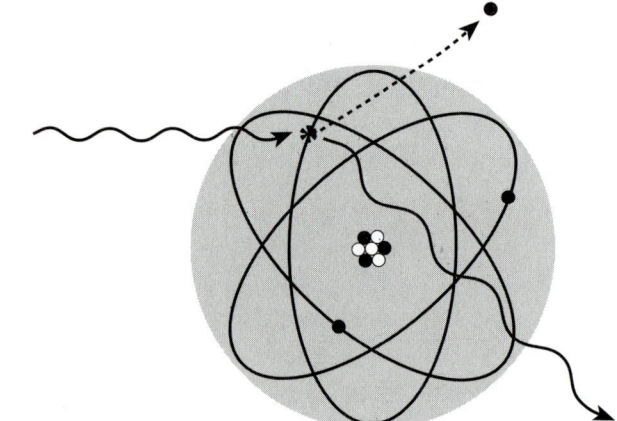

Figure 4–17. Reproduced with permission from Wolbarst AB. *Physics of Radiology.* East Norwalk, CT: Appleton & Lange, 1993.

177. The absorption of excessive primary radiation by a grid is called

(A) grid selectivity

(B) contrast improvement factor

(C) grid cut-off

(D) latitude

178. A lateral radiograph of the cervical spine was made at 40 inches using 3 mAs. If it is desired to increase the distance to 72 inches, what should be the new mAs, all other factors remaining the same?

(A) 1 mAs

(B) 1.6 mAs

(C) 5 mAs

(D) 10 mAs

179. Which of the following is (are) characteristics of a 16:1 grid?

1. absorbs a high percent of scattered radiation
2. has little positioning latitude
3. used with high kVp exposures

(A) 1 only

(B) 1 and 3 only

(C) 2 and 3 only

(D) 1, 2, and 3

180. Exposure rate increases with an increase in

1. mA
2. kVp
3. SID

(A) 1 only

(B) 1 and 2 only

(C) 2 and 3 only

(D) 1, 2, and 3

181. A particular mAs, regardless of the combination of mA and time, will reproduce the same radiographic density. This is a statement of the

(A) Line Focus Principle

(B) Inverse Square Law

(C) Reciprocity Law

(D) Law of Conservation of Energy

182. Which of the following is an abnormal intensifying screen action?

(A) fluorescence

(B) luminescence

(C) speed

(D) lag

183. X-ray photon energy is inversely proportional to

1. applied kVp
2. applied mA
3. photon wavelength

(A) 1 only

(B) 1 and 2 only

(C) 3 only

(D) 1, 2, and 3

184. Slow-speed screens are used

(A) to minimize patient dose

(B) to keep exposure time to a minimum

(C) to image fine anatomic details

(D) in pediatric radiography

185. Film base is currently made of which of the following materials?

(A) cellulose nitrate

(B) cellulose acetate

(C) polyester

(D) glass

186. Which of the following pathologic conditions would require a decrease in exposure factors?

(A) congestive heart failure

(B) pneumonia

(C) emphysema

(D) pleural effusion

187. The primary source of scatter radiation is the

(A) patient

(B) tabletop

(C) x-ray tube

(D) grid

188. Use of high ratio grids is associated with which of the following?

  1. increased patient dose
  2. higher contrast
  3. tissues of low density

(A) 1 only
(B) 1 and 2 only
(C) 1 and 3 only
(D) 1, 2, and 3

189. Radiographic contrast is a result of

  1. differential tissue absorption
  2. emulsion characteristics
  3. proper regulation of mAs

(A) 1 only
(B) 1 and 2 only
(C) 1 and 3 only
(D) 1, 2, and 3

190. In order to produce just a perceptible increase in radiographic density, the radiographer must increase the

(A) mAs by 30 percent
(B) mAs by 15 percent
(C) kVp by 15 percent
(D) kVp by 30 percent

191. All the following affect the exposure rate of the primary beam, EXCEPT

(A) mA
(B) kVp
(C) distance
(D) part thickness

192. The radiograph in Figure 4–18 exhibits an artifact caused by

(A) an inverted focused grid
(B) poor screen/film contact
(C) foreign body in cassette
(D) static electricity

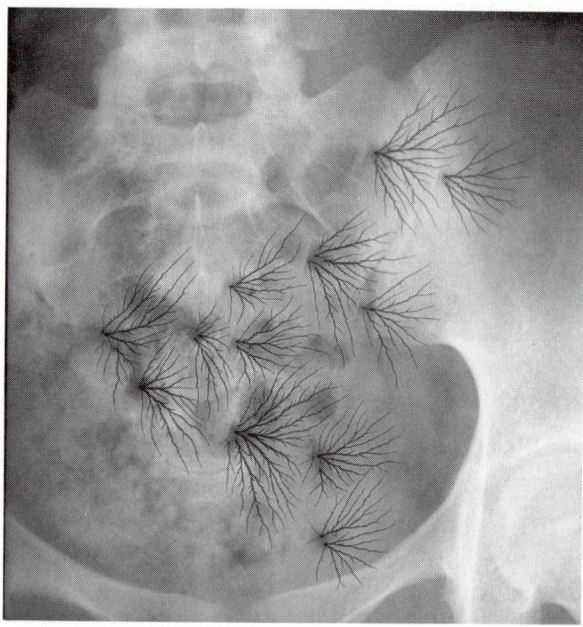

Figure 4–18. From the American College of Radiology Learning File. Courtesy of the ACR.

193. The function(s) of automatic beam limitation devices include

  1. reducing the production of scatter radiation
  2. absorption of scatter radiation
  3. changing the quality of the x-ray beam

(A) 1 only
(B) 2 only
(C) 1 and 2 only
(D) 1, 2, and 3

194. Which of the following statements is true with respect to the diagram in Figure 4–19?

(A) film number 2 has more sensitivity above the point of intersection
(B) film number 1 has more sensitivity below the point of intersection
(C) film number 1 has more sensitivity above the point of intersection
(D) film number 2 has less sensitivity below the point of intersection

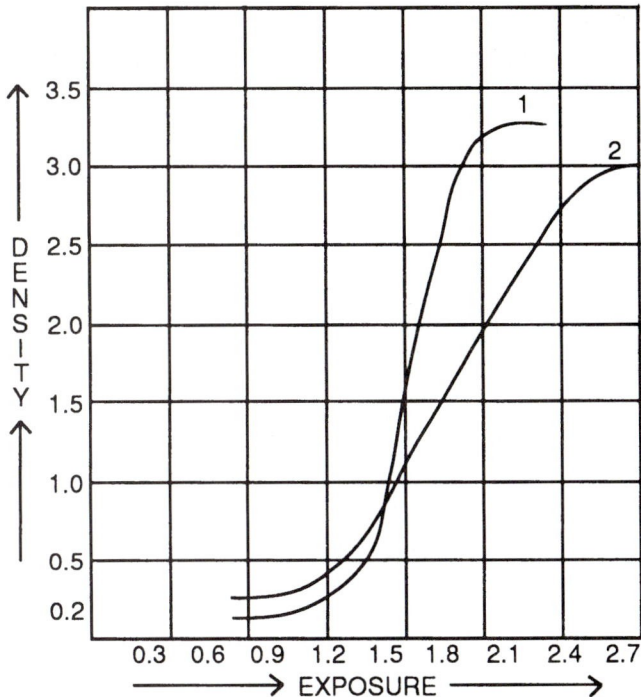

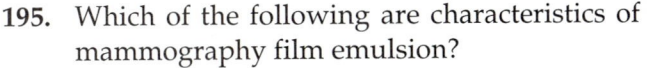

Figure 4–19. Courtesy of David Perri.

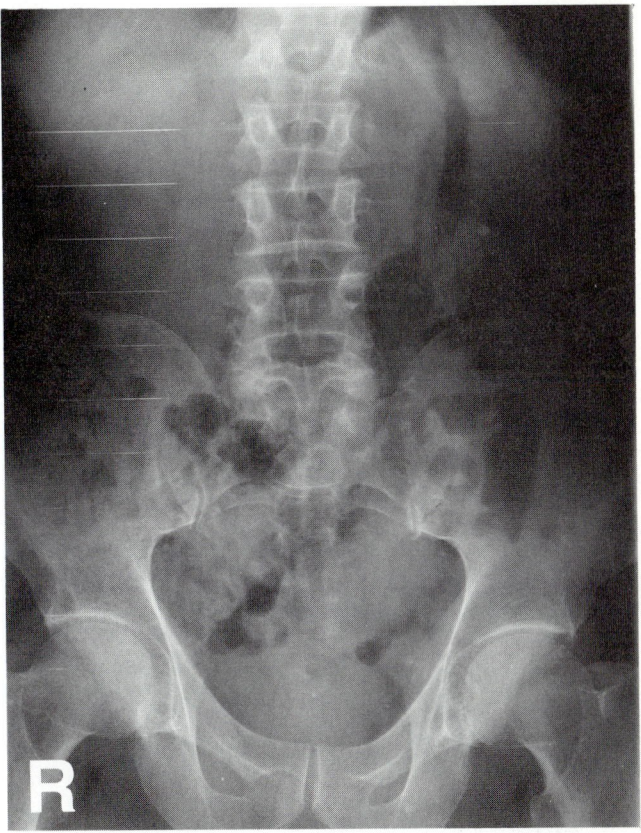

Figure 4–20. Courtesy of The Stamford Hospital, Department of Radiology.

195. Which of the following are characteristics of mammography film emulsion?

1. high contrast
2. fine grain
3. single emulsion

(A) 1 only
(B) 1 and 2 only
(C) 2 and 3 only
(D) 1, 2, and 3

196. When involuntary motion must be considered, the exposure time may be cut in half if the kVp is

(A) doubled
(B) increased by 15 percent
(C) increased by 25 percent
(D) increased by 35 percent

197. Which of the following groups of exposure factors would be MOST effective in eliminating prominent pulmonary vascular markings in an RAO position of the sternum?

(A) 500 mA, ⅟₃₀ sec, 70 kVp
(B) 200 mA, 0.04 sec, 80 kVp
(C) 300 mA, ⅟₁₀ sec, 80 kVp
(D) 25 mA, ⅞₀ sec, 70 kVp

198. If the developer temperature in the automatic processor is higher than normal, what will be the effect on the finished radiograph?

1. loss of contrast
2. increased density
3. wet, tacky films

(A) 1 only
(B) 1 and 2 only
(C) 2 and 3 only
(D) 1, 2, and 3

199. The artifacts on the radiograph in Figure 4–20 are called

   (A) pi lines
   (B) guide shoe marks
   (C) hesitation marks
   (D) reticulation

200. All of the following are related to recorded detail, EXCEPT

   (A) motion
   (B) screen speed
   (C) object-image distance
   (D) grid ratio

201. X-ray photon beam attenuation is influenced by

   1. tissue type
   2. subject thickness
   3. photon quality

   (A) 1 only
   (B) 3 only
   (C) 2 and 3 only
   (D) 1, 2, and 3

202. Using the fixed mAs, variable kVp exposure factor technique, each centimeter increase in patient thickness requires what adjustment in technique?

   (A) increase 2 kVp
   (B) decrease 2 kVp
   (C) increase 4 kVp
   (D) decrease 4 kVp

203. Why is a single intensifying screen and single emulsion film used for select radiographic examinations?

   (A) to decrease patient dose
   (B) to achieve longer-scale contrast
   (C) for better recorded detail
   (D) to decrease operating expenses

204. Which of the following may be used to determine the sensitivity of a particular film emulsion?

   (A) characteristic curve
   (B) dose–response curve
   (C) Reciprocity Law
   (D) Inverse Square Law

205. Recorded detail will be influenced by which of the following?

   1. screen speed
   2. screen/film contact
   3. penumbra

   (A) 1 and 2 only
   (B) 1 and 3 only
   (C) 2 and 3 only
   (D) 1, 2, and 3

206. Which of the following chemicals is used in the production of radiographic film emulsion?

   (A) sodium sulfite
   (B) potassium bromide
   (C) silver halide
   (D) chrome alum

207. Which of the following terms is used to describe unsharp edges of tiny radiographic details?

   (A) diffusion
   (B) mottle
   (C) penumbra
   (D) umbra

208. Which of the following materials may be used as grid interspace material?

   1. lead
   2. plastic
   3. aluminum

   (A) 1 only
   (B) 1 and 2 only
   (C) 2 and 3 only
   (D) 1, 2, and 3

209. The mottled appearance of the radiograph in Figure 4–21 is illustrative of

(A) Paget's disease
(B) osteoporosis
(C) safelight fog
(D) pillow artifacts

210. As grid ratio is increased

(A) the scale of contrast becomes longer
(B) the scale of contrast becomes shorter
(C) radiographic density increases
(D) radiographic distortion decreases

211. X-ray film emulsion is most sensitive to safe-light fog

(A) before exposure and development
(B) after exposure
(C) during development
(D) at low humidity

212. Glutaraldehyde is added to developer solution of automatic processors in order to

1. keep emulsion swelling to a minimum
2. decrease the possibility of a processor jam-up
3. remove unexposed silver halide crystals

(A) 1 only
(B) 1 and 2 only
(C) 2 and 3 only
(D) 1, 2, and 3

213. Very low humidity in the darkroom can lead to which of the following?

(A) crinkle marks
(B) static electrical discharge
(C) excessive emulsion swelling
(D) chemical fog

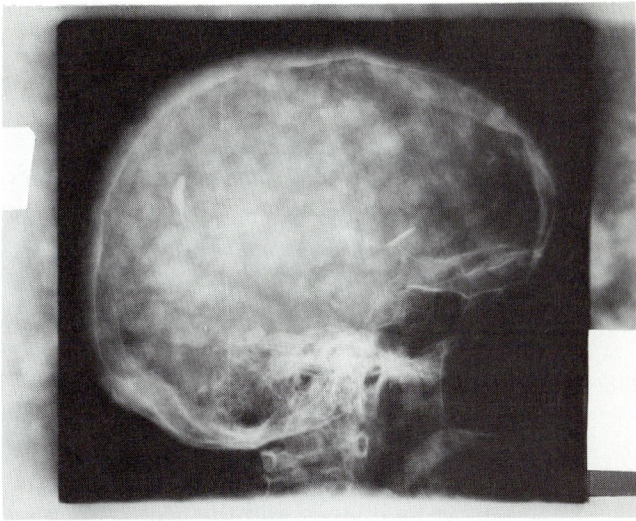

Figure 4–21. Courtesy of The Stamford Hospital, Department of Radiology.

214. The following can affect demonstration of the anode heel effect

1. SID
2. image recorder size
3. screen speed

(A) 1 only
(B) 1 and 2 only
(C) 2 and 3 only
(D) 1, 2, and 3

215. Chemical fog may be attributed to

1. excessive developer temperature
2. oxidized developer
3. insufficient replenishment

(A) 1 only
(B) 1 and 2 only
(C) 2 and 3 only
(D) 1, 2, and 3

216. Which of the following has the greatest effect on radiographic density?

(A) aluminum filtration
(B) kilovoltage
(C) source-image distance
(D) scatter radiation

217. Conditions contributing to poor radiographic archival quality include

   1. fixer retention
   2. insufficient developer replenishment
   3. poor storage conditions

   (A) 1 only
   (B) 3 only
   (C) 2 and 3 only
   (D) 1, 2, and 3

218. Shape distortion is influenced by the relationship between the

   1. x-ray tube and part to be imaged
   2. part to be imaged and the film
   3. film and the x-ray tube

   (A) 1 only
   (B) 1 and 2 only
   (C) 1 and 3 only
   (D) 1, 2, and 3

219. Which of the following is (are) associated with subject contrast?

   1. patient thickness
   2. tissue density
   3. kilovoltage

   (A) 1 only
   (B) 1 and 2 only
   (C) 1 and 3 only
   (D) 1, 2, and 3

220. If a radiograph, exposed using a 12:1 ratio grid, exhibits a loss of density at its lateral edges, it is probably because the

   (A) SID was too great
   (B) grid failed to move during the exposure
   (C) x-ray tube was angled in the direction of the lead strips
   (D) central ray was off center

221. Which of the following is MOST closely associated with short scale contrast?

   (A) large number of density differences
   (B) greater film latitude
   (C) more contrast
   (D) less contrast

222. How is source-image distance related to exposure rate and radiographic density?

   (A) as SID increases, exposure rate increases and radiographic density increases
   (B) as SID increases, exposure rate increases and radiographic density decreases
   (C) as SID increases, exposure rate decreases and radiographic density increases
   (D) as SID increases, exposure rate decreases and radiographic density decreases

223. Which of the following focal spot sizes should be employed for magnification radiography?

   (A) 0.2 mm
   (B) 0.6 mm
   (C) 1.2 mm
   (D) 2.0 mm

224. For which of the following examinations may the use of a grid NOT be necessary in the adult patient?

   (A) hip
   (B) knee
   (C) abdomen
   (D) lumbar spine

225. Grid cut-off due to off-centering would result in which of the following?

   (A) overall loss of density
   (B) one side of the film underexposed
   (C) overexposure under the anode end
   (D) underexposure under the anode end

226. The line focus principle expresses the relationship between

   (A) actual and effective focal spot
   (B) exposure given the film and resultant density
   (C) source-image distance used and resultant density
   (D) kilovoltage used and the resulting contrast

227. Which of the following pathologic conditions will probably require a decrease in exposure factors?

   (A) osteomyelitis
   (B) osteoporosis
   (C) osteosclerosis
   (D) osteochondritis

228. Disadvantage(s) of using low kV include

   1. insufficient penetration
   2. increased patient dose
   3. diminished latitude

   (A) 1 only
   (B) 1 and 2 only
   (C) 1 and 3 only
   (D) 1, 2, and 3

229. Which of the following statements is (are) true regarding Figure 4–22?

   1. film B was made with a higher ratio grid than film A
   2. film A was made with an inverted focused grid
   3. more exposure was used for film B than film A

   (A) 1 only
   (B) 1 and 2 only
   (C) 1 and 3 only
   (D) 1, 2, and 3

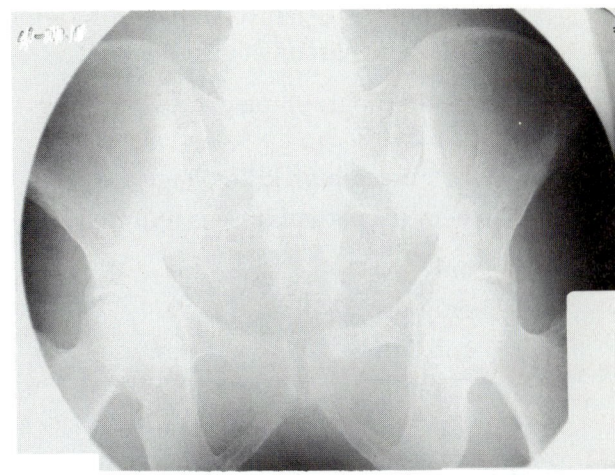

A

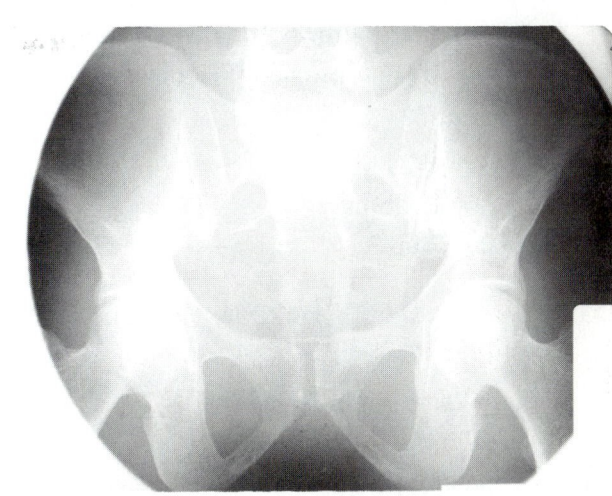

B

Figure 4–22. From the American College of Radiology Learning File. Courtesy of the ACR.

230. Geometric unsharpness is influenced by which of the following?

   1. object-image distance
   2. focal-object distance
   3. source-image distance

   (A) 1 only
   (B) 1 and 2 only
   (C) 1 and 3 only
   (D) 1, 2, and 3

231. A grid should is usually employed in which of the following circumstances?

1. when radiographing a large or dense body part
2. when using high kilovoltage
3. when less patient dose is required

(A) 1 only
(B) 3 only
(C) 1 and 2 only
(D) 1, 2, and 3

232. A quality assurance program serves to

1. keep patient dose to a minimum
2. keep radiographic quality consistent
3. ensure equipment efficiency

(A) 1 only
(B) 1 and 2 only
(C) 1 and 3 only
(D) 1, 2, and 3

233. The darkroom should be constructed and equipped to avoid

1. external light leaks
2. film bin light leaks
3. safelight fog

(A) 1 only
(B) 2 only
(C) 1 and 3 only
(D) 1, 2, and 3

234. What information, located on each box of film, is important to note and has a direct relationship to film quality?

(A) number of films in the box
(B) manufacturer's name
(C) expiration date
(D) emulsion lot

235. The radiograph illustrated in Figure 4–23 should be rejected by the quality control technologist because of

(A) inadequate penetration
(B) excessive density
(C) too little contrast
(D) rotation

236. Exposed silver halide crystals are changed to black metallic silver by the

(A) preservative
(B) reducers
(C) activators
(D) hardener

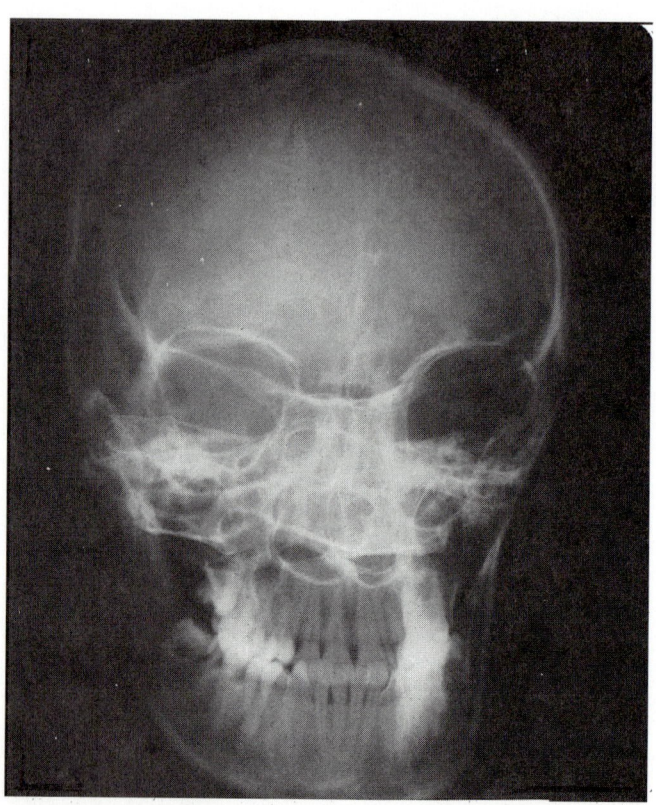

Figure 4–23. From the American College of Radiology Learning File. Courtesy of the ACR.

237. Acceptable method(s) of minimizing motion unsharpness is (are)

1. suspended respiration
2. short exposure time
3. patient instruction

(A) 1 only
(B) 1 and 2 only
(C) 1 and 3 only
(D) 1, 2, and 3

238. Compression of the breast during mammographic imaging improves the technical quality of the image because

1. geometric blurring is decreased
2. less scatter radiation is produced
3. patient motion is reduced

(A) 1 only
(B) 3 only
(C) 2 and 3 only
(D) 1, 2, and 3

239. Distortion can be caused by

1. tube angle
2. the position of the organ or structure within the body
3. radiographic positioning of the part

(A) 1 only
(B) 1 and 2 only
(C) 2 and 3 only
(D) 1, 2, and 3

240. Object-image distance (OID) is related to recorded detail in which of the following ways?

(A) radiographic detail is directly proportional to OID
(B) radiographic detail is inversely proportional to OID
(C) as OID increases, so does radiographic detail
(D) OID is unrelated to radiographic detail

241. The speed of an intensifying screen is influenced by which of the following factors?

1. active layer thickness
2. antihalation backing
3. phosphor type used

(A) 1 only
(B) 1 and 3 only
(C) 2 and 3 only
(D) 1, 2, and 3

242. Which of the following conditions would require an increase in exposure factors?

1. congestive heart failure
2. pleural effusion
3. emphysema

(A) 1 only
(B) 1 and 2 only
(C) 1 and 3 only
(D) 1, 2, and 3

243. Problems that can occur when using automatic exposure devices for very short exposure times and high mA include

1. grid lines
2. focal spot blooming
3. excessive density

(A) 1 only
(B) 2 only
(C) 1 and 2 only
(D) 1, 2, and 3

244. If a radiograph exhibits insufficient density, this may be attributed to

1. inadequate kVp
2. inadequate SID
3. grid cut-off

(A) 1 only
(B) 1 and 2 only
(C) 1 and 3 only
(D) 1, 2, and 3

245. Which of the following groups of exposure factors will produce the greatest radiographic density?

(A) 100 mA, 0.30 sec
(B) 200 mA, 0.10 sec
(C) 400 mA, 0.03 sec
(D) 600 mA, 0.03 sec

246. The use of which of the following is (are) essential in magnification radiography?

1. high-ratio grid
2. fractional focal spot
3. direct exposure film

(A) 1 only
(B) 2 only
(C) 1 and 3 only
(D) 1, 2, and 3

247. Which of the following factors influence the production of scatter radiation?

1. kilovoltage level
2. tissue density
3. size of field

(A) 1 only
(B) 1 and 2 only
(C) 1 and 3 only
(D) 1, 2, and 3

248. Which of the following groups of technical factors would be MOST appropriate for the radiographic examination shown in Figure 4–24?

(A) 400 mA, ⅟₃₀ sec, 72 kVp
(B) 300 mA, ⅟₆₀ sec, 82 kVp
(C) 300 mA, ⅟₁₂₀ sec, 94 kVp
(D) 50 mA, ¼ sec, 72 kVp

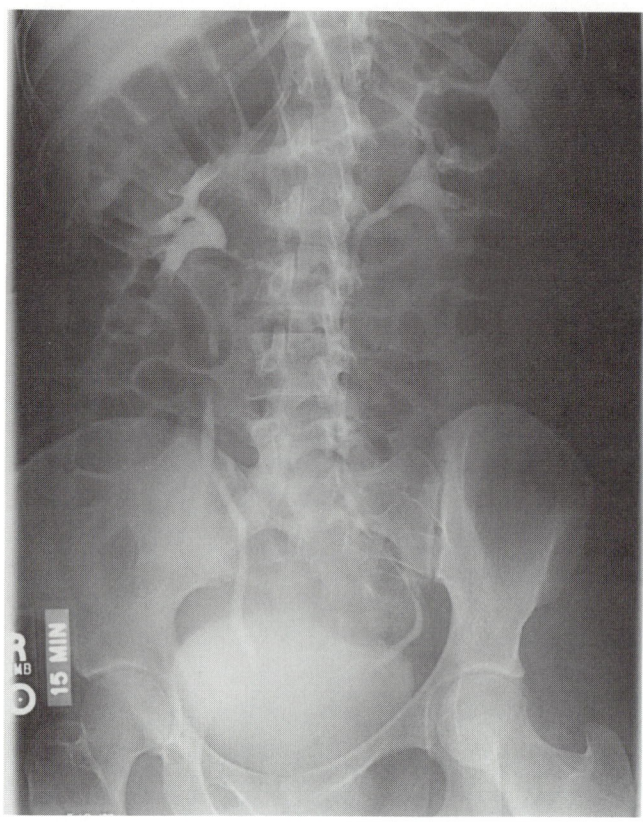

Figure 4–24. Courtesy of The Stamford Hospital, Department of Radiology.

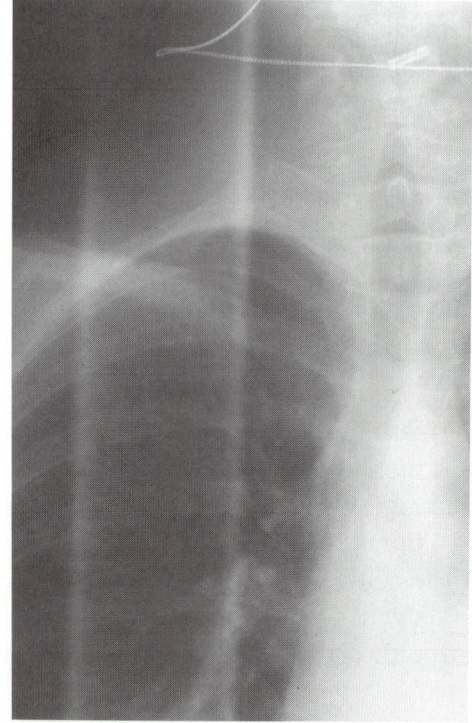

Figure 4–25. From the American College of Radiology Learning File. Courtesy of the ACR.

249. Why is a very short exposure time essential in chest radiography?

  (A) to avoid excessive penumbra
  (B) to maintain short scale contrast
  (C) to minimize involuntary motion
  (D) to minimize patient discomfort

250. Which of the following statements is (are) true regarding the large vertical foreign object seen in the erect PA projection of the chest shown in Figure 4–25?

  1. the object is located in the cassette
  2. the object is located between the patient and the cassette
  3. the object is located between the patient and x-ray tube

  (A) 1 only
  (B) 2 only
  (C) 1 and 2 only
  (D) 3 only

Answers and Explanations

1. **(C)** *Starter solution* is added to fresh developer solution to lower the pH (from 13 to 11). Starter solution is simply the restrainer, *potassium* (or sodium) *bromide*, which functions to prevent development of unexposed silver bromide crystals. Developmental *fog* is increased (resulting in a darker image) when the developer's potassium bromide level is low—that is, the unexposed crystals begin to develop. Potassium bromide is called starter solution because it is only needed in new, fresh developer. During the development process, bromine ions are released from the film emulsion into the developer, making potassium bromide unnecessary in *replenisher* solution. *(Carlton & Adler, p 304)*

2. **(B)** An increase in SID will help decrease the effect of excessive OID. For example, in the lateral projection of the cervical spine there is normally a significant OID. This effect is decreased by the use of a 72-inch SID. However, especially with larger body parts, increased SID usually requires a significant increase in exposure factors. Focal spot size and screen speed are unrelated to magnification. *(Selman, p 321)*

3. **(C)** Exposure-type artifacts are those that appear on the radiograph as a result of image formation processes. A *foreign body* in the cassette will cast its image on the radiographic film. As the exposed film is removed from the cassette, *static electrical discharge* will expose the film in a characteristic manner. These are *exposure-type artifacts*. Processor roller marks are also artifacts, but they are not placed on the film during image for-

mation; processor roller marks are classified as processor artifacts. *(Saia, PREP, pp 354, 356–357)*

4. **(A)** Although mAs is primarily used to regulate x-ray *quantity*, the number of x-ray photons produced in a given exposure, kVp is generally used to regulate x-ray *quality*, the energy or wavelength of x-ray photons produced at the target during a given exposure. Although source-image distance significantly *affects* radiographic density, it is not used to *regulate* the quantity of photons produced. Focal spot size is unrelated to radiation quantity or quality. *(Bushong, p 298)*

5. **(B)** A characteristic curve is the representation of a film emulsion's response to exposure by light or x-rays. Upon observation, it is seen that a characteristic curve does not begin at zero density (Fig. 4–26). That is because an "unexposed" and processed film has a small base-plus fog density (which should not exceed 0.2) due to base dye, environmental radiation fog during production, transportation and storage, and chemical fog from high processing temperatures. The initial ascent of the curve is called the "toe," and represents minimal density. The straight line portion of the curve represents the useful radiographic density range, lying approximately between densities 0.25 and 2.0. The characteristic curve levels off again at the "shoulder" portion of the curve, which represents maximum radiographic density (D-Max). Past the shoulder is the solarization point, where further exposure would actually reverse image densities. *(Cullinan, p 92)*

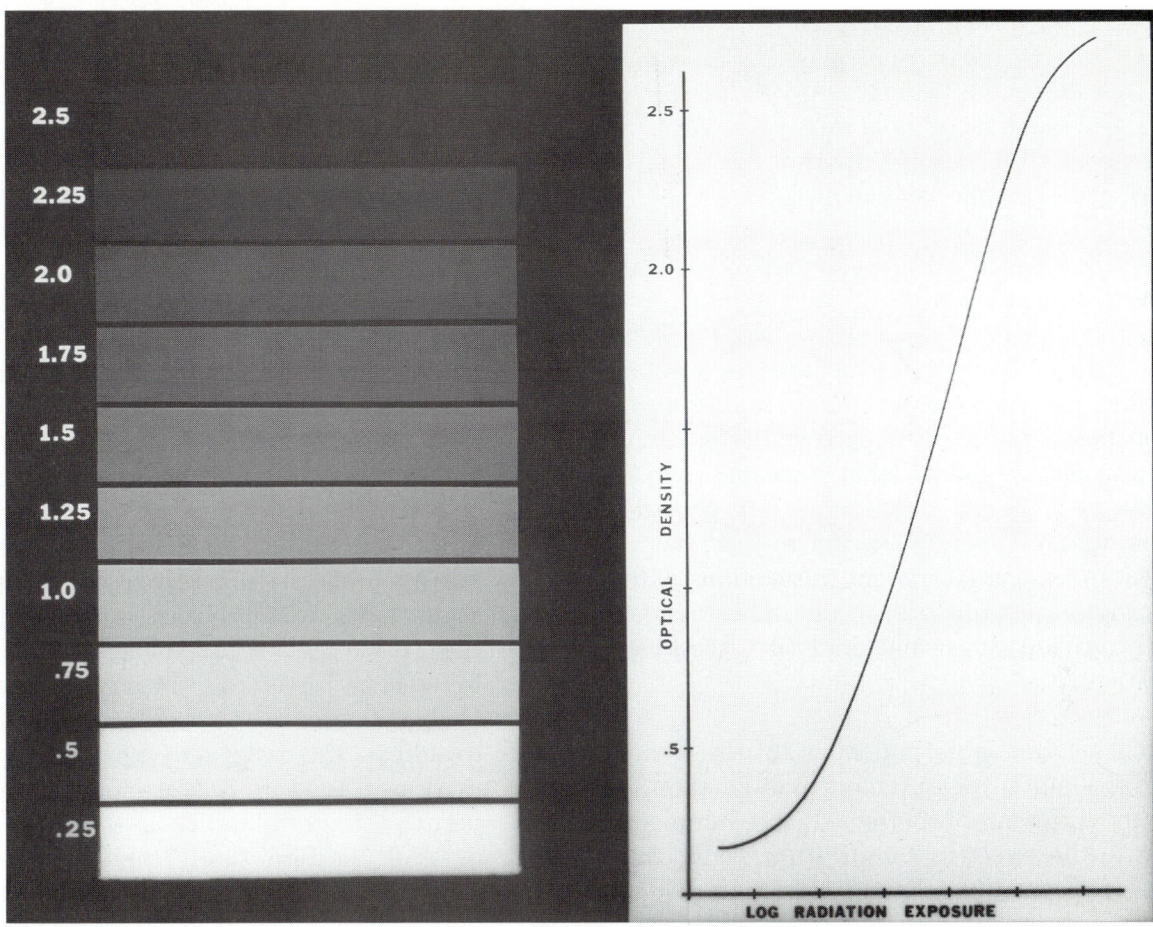

Figure 4–26. From the American College of Radiology Learning File. Courtesy of the ACR.

6. **(B)** The radiographs illustrated in Figure 4–1 would *both* be rejected by the quality control technologist, but for different reasons. Film A, while adequately penetrated, has too much background density; this film should be repeated at a lower mAs. Film B demonstrates areas of inadequate penetration resulting in excessively high contrast; this film should be repeated using higher kVp and appropriately lower mAs. *(Bushong, p 294)*

7. **(C)** Using the formula mA × sec = mAs, determine each mAs. The greatest radiographic density will be produced by the combination of greatest mAs and shortest SID. Groups B and D should produce identical radiographic density, according to the Inverse Square Law,

because group D is twice the distance and four times the mAs of group B. Group A has twice the distance of group B, but only twice the mAs; it has, therefore, less density than groups B and D. Group C has the same distance as group B, and twice the mAs—making C the group of technical factors that will produce the greatest radiographic density. *(Selman, pp 334–335)*

8. **(B)** The addition of a grid will help clean up the scatter radiation produced by high kVp, but it requires an mAs adjustment. According to the grid conversion factors listed below, the addition of an 8:1 grid requires that the original mAs be multiplied by a factor of four:

no grid = 1 × the original mAs
5:1 grid = 2 × the original mAs
6:1 grid = 3 × the original mAs
8:1 grid = 4 × the original mAs
12:1 grid = 5 × the original mAs
16:1 grid = 6 × the original mAs

The adjustment therefore requires 12 mAs at 90 kVp. *(Cullinan, p 292)*

9. **(D)** Each of the three is included in a good quality assurance program. *Beam alignment* must be accurate to 2 percent of the SID. *Reproducibility* means that repeated exposures at a given technique must provide consistent intensity. *Linearity* means that a given mAs, using different mA stations with appropriate exposure time adjustments, will provide consistent intensity. *(Bushong, pp 431–433)*

10. **(B)** If a fairly large patient is turned *prone*, the abdominal measurement will be significantly different from the AP measurement due to the effect of compression. Thus, the part is essentially "thinner" and less scatter radiation will be produced. If the patient remains supine and a compression band is applied, the same effect will take place. Because scattered radiation increases as *kilovoltage* increases, using a *lower* kV will serve to decrease the generation of scattered radiation. *Beam restriction* is probably the single most effective measure of reducing the production of scattered radiation. *(Cullinan, p 68)*

11. **(D)** Intensifying screen phosphors having *high atomic number* are more likely to *absorb* a high percentage of the incident x-ray photons and *convert* x-ray photon energy to fluorescent light energy. How efficiently the phosphors detect and interact with the x-ray photons is termed *quantum detection efficiency*. How effectively the phosphors make this energy conversion is termed *conversion efficiency*. *(Bushong, p 218)*

12. **(A)** *Radiograph B* was made of a chest phantom and processed using the recommended 95°F developer temperature. *Radiograph A* was exposed under the same conditions, but processed with developer temperature of 90°F.

The development process slows at the lower temperature, and a much lighter radiograph results. In order to produce the original density using a 90° developer, higher exposure factors would be required. *Radiograph C* was exposed under the same conditions as A and B but developed at 100°F; at this temperature the development process is accelerated and the radiograph is too dark. *(Carroll, p 457)*

13. **(B)** Review the groups of factors. First, because mAs has no effect on the scale of contrast produced, eliminate mAs from consideration by drawing a line through the column. Then check the two in each column most likely to produce long-scale contrast. For example, in the kVp column, because higher kVp will produce longer-scale contrast, place check marks next to each 90 kVp. In the Film/Screen column, the slower screens (200) will produce lower (longer-scale) contrast than the faster screens; place a check mark next to each. Because lower-ratio grids permit a larger quantity of scatter radiation to reach the film, the 5:1 and 8:1 grids will produce a longer scale of contrast than the higher-ratio grids; check them. As the volume of irradiated tissue increases, so does the amount of scatter radiation produced, and consequently, the longer the scale of radiographic contrast; therefore check the 14 × 17 inch field sizes. An overview shows that the factors in groups A and B have two check marks, whereas the factors in group B have four check marks—indicating that group B will produce the longest scale contrast. *(Cullinan, p 300)*

14. **(B)** The GBX or GS1 is a red filter that is safe with green-sensitive film emulsion. The amber colored Wratten 6B filter is safe for blue-sensitive film only. Although using no safelight is possible, it is not a practical way to function. *(Selman, p 294)*

15. **(C)** The latent image is the invisible image produced within the film emulsion as a result of exposure to radiation. The developing solution convert this to a visible, manifest image. The exposed silver halide grains in the emulsion undergo chemical change in the devel-

oper solution and the unexposed crystals are removed from the film during the fixing process. *(Wallace, p 144)*

16. **(A)** The magnification radiographs received almost identical exposures except that film A was made using a *fractional* (0.3-mm) focal spot. It is essential that *macroradiography* (magnification films) be performed using a 0.3-mm focal spot or smaller. Use of a larger focal spot (even of 0.6 mm) will produce objectionable focal spot blur and make the exam diagnostically useless. A fractional focal spot is essential for reproducing fine detail without penumbral blur. As the object image is magnified, so will be the associated penumbra unless the fractional focal spot is used. Magnification radiography may be used to delineate a suspected hairline fracture or enlarge tiny, contrast-filled blood vessels. It also has application in mammography. *(Cullinan, p 129)*

17. **(B)** A review of the problem reveals that three changes are being made: an increase in SID, a change from 400 speed system to 200 speed system, and an increase in exposure time (to be considered last). Because the original mAs was 15, cutting the speed of the system in half (from 400 to 200) will require a doubling of the mAs, to 30, in order to maintain density. Now we must deal with the distance change. Using the Inverse Square Law (and remembering that 30 is now the OLD mAs!), we find that the required new mAs at 44 inches is 44.8. Because the problem states that we are now using 0.22 sec exposure, it is left to determine what mA, used with 0.22 sec, will provide 44.8 mAs:

$$0.22x = 44.8$$
$$x = 203 \text{ mA}$$

(Selman, pp 332, 335–336)

18. **(C)** *Absorption* occurs when an x-ray photon interacts with matter and disappears, as in the photoelectric effect. *Scattering* occurs when there is partial transfer of energy to matter, as in the Compton effect. The reduction in the intensity of an x-ray beam as it passes through matter is called *attenuation*. *(Bushong, p 66)*

19. **(C)** Breast tissue has very low subject contrast, but microcalcifications and subtle density differences are imperative to visualize (Fig. 4–27). Fine detail is necessary to visualize the microcalcifications; therefore, a *small focal spot* tube is essential. *High* contrast (therefore, low kilovoltage) is needed to accentuate any differences in tissue density. A *compression device* serves to even out differences in tissue thickness (thicker at chest wall, thinner at nipple) and decrease object-image distance, and helps to decrease the production of scatter radiation. *(Selman, pp 344–346)*

20. **(C)** Kilovoltage (kVp) and half-value layer (HVL) affect a change in both quantity and quality of the primary beam. The principal qualitative factor of the primary beam is kVp, but an increase in kVp will also affect an increase in the *number* of photons produced at the target. HVL is defined as the amount of material necessary to decrease the intensity of the beam to one half its original value, thereby affecting a change in both beam quality and quantity. The mAs value are adjusted to regulate the number of x-ray photons produced at the target. X-ray beam quality is unaffected by changes in mAs. *(Bushong, p 167)*

21. **(C)** If a film was placed on an illuminator and 100 percent of the illuminator's light was transmitted through the film, that film must have a density of 0. According to the equation

$$\text{density} = \log \frac{\text{incident light intensity}}{\text{transmitted light intensity}}$$

if 10 percent of the illuminator's light passes through the film, that film has a density of 1. If 1 percent of the light passes through the film, that film has a density of 2. *(Bushong, p 270)*

22. **(B)** The term *latitude* may refer to either film latitude or exposure latitude. Exposure latitude refers to the margin of error inherent in a particular group of exposure factors. Selection of high kVp and low mAs factors will allow greater exposure latitude than low kVp and high mAs factors. Film latitude is chemi-

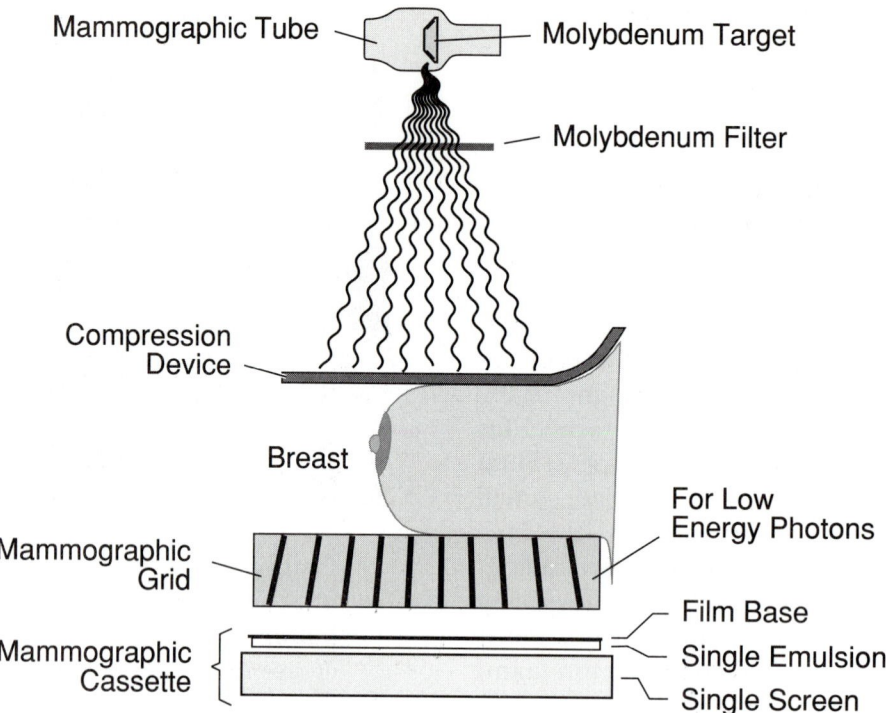

Mammographic Tube — Molybdenum Target

Molybdenum Filter

Compression Device

Breast

For Low Energy Photons

Mammographic Grid

Mammographic Cassette

Film Base
Single Emulsion
Single Screen

Figure 4–27. Reproduced with permission from Wolbarst AB. *Physics of Radiology*. East Norwalk, CT: Appleton & Lange, 1993.

cally built into the film emulsion and refers to the emulsion's ability to record a long range of densities from black to white (long-scale contrast). *(Cullinan, pp 166, 269)*

23. **(C)** In order to change non-grid to grid exposure, or to adjust exposure when changing from one grid ratio to another, recall the factor for each grid ratio:

no grid = 1 × the original mAs
5:1 grid = 2 × the original mAs
6:1 grid = 3 × the original mAs
8:1 grid = 4 × the original mAs
12:1 grid = 5 × the original mAs
16:1 grid = 6 × the original mAs

Therefore, to change from non-grid to 8:1 grid, multiply the original mAs by a factor of four. A new mAs of 24 is required. *(Saia, PREP, p 316)*

24. **(B)** Automatic processing chemically converts the invisible latent image to a visible, or manifest, image in a very short time—usually 90 sec. One of the ways it achieves this rapid process is by using higher solution temperatures (than that used in manual processing).

However, at higher temperatures the film emulsion tends to swell excessively, causing emulsion to stick to processor rollers. *Glutaraldehyde* is a hardener that is added to the developer solution to minimize emulsion swelling. *(Selman, p 313)*

25. **(B)** The formula for *magnification factor* is MF = SID/SOD. The formula for *percentage magnification* is OID/SOD × 100. In the stated problem the new OID is 8 inches and the SID is 40 inches; therefore the SOD is 32 inches. Substituting the known factors in the appropriate equation:

$$\frac{8}{32} \times 100$$

$$800 \div 32 = 25 \text{ percent magnification}$$

(Bushong, p 280)

26. **(B)** Boxes of x-ray film, especially the larger sizes, should be in the vertical (upright) position. If film boxes are stacked upon one another, the sensitive emulsion can be affected by pressure from the boxes above. *Pressure marks* are produced and result in loss of con-

trast in that area of the radiographic image. When retrieving x-ray film from storage, the oldest should always be used first. *(Saia, PREP, p 328)*

27. **(D)** Collimators restrict the size of the irradiated field, thereby limiting the volume of irradiated tissue, and hence less scatter radiation is produced. Once radiation has scattered and emerged from the body, it can be trapped by the grid's lead strips. Grids effectively remove much of the scatter radiation in the remnant beam before it reaches the x-ray film. A compression band may be applied to reduce the effect of excessive fatty tissue (eg, across the abdomen), in effect reducing the thickness of the part to be radiographed. *(Cullinan, pp 68–69)*

28. **(D)** Film *age* is an important consideration when determining causes of film fog. Outdated film will exhibit loss of contrast in the form of fog and loss of speed. A *safelight* is "safe" only for practical periods of time required for necessary handling of film. Films left out on the darkroom counter can be fogged by excessive exposure to the safelight. Remember, too, that film emulsion is much more sensitive to safelight fog after exposure. One cannot imagine being without *automatic processors*, but the high temperatures required for rapid processing are a source of film fog. Daily QA ensures that fog levels do not exceed the upper limit of 0.2 D. *(Selman, p 309)*

29. **(D)** The radiographic subject, the patient, is composed of many different tissue types, having varying densities, and resulting in varying degrees of photon attenuation and absorption. The *atomic number* of the tissues under investigation is directly related to their *attenuation coefficient*. This *differential absorption* contributes to the various shades of gray (scale of radiographic contrast) on the finished radiograph. Normal tissue density may be significantly altered in the presence of *pathologic processes*. For example, destructive bone disease can cause a dramatic decrease in tissue density (and subsequent increase in radiographic density). Abnormal accumulation of fluid (as in ascites) will cause a significant

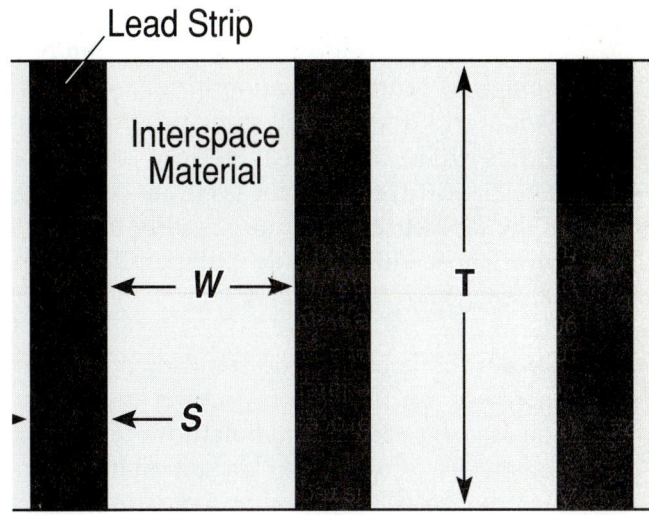

Lead Strip

Grid Ratio: *T/W*

Figure 4–28. Reproduced with permission from Wolbarst AB. *Physics of Radiology.* East Norwalk, CT: Appleton & Lange, 1993.

increase in tissue density. *Muscle* atrophy, or highly developed muscles, will similarly decrease or increase tissue density. *(Selman, p 340)*

30. **(B)** In the two radiographs illustrated, film A was made recumbent. Film B was made in the erect position; this may be discerned by the presence of clearly defined air/fluid levels in the lower abdomen. Abdominal viscera move to a lower position in the erect position, making the abdomen "thicker" and requiring an increase in exposure (usually the equivalent of 10 kVp). *(Saia, PREP, p 160)*

31. **(A)** As SID increases so does recorded detail, because magnification is decreased. OID is *inversely* related to recorded detail because as OID increases, recorded detail decreases. As screen speed increases, recorded detail decreases, due to greater diffusion of light. *Therefore, SID is directly related to recorded detail.* OID and screen speed are inversely related to recorded detail. *(Carroll, p 233)*

32. **(B)** Grid ratio is defined as the height of the lead strip to the width of the interspace material (Fig. 4–28). The higher the lead strips (or

smaller the distance between the strips), the greater the grid ratio and the greater the percentage of scatter radiation absorbed. However, a grid does absorb some primary radiation as well. The higher the lead strips, the more critical the need for accurate centering, as the lead strips will more readily trap photons whose direction do not parallel them. *(Selman, p 366)*

33. **(D)** As the speed of the film/screen system decreases, an increase in mAs is usually required in order to maintain radiographic density. *However,* when an automatic exposure device (phototimer or ionization chamber) is used, the system is programmed for the use of a particular film/screen speed. If a slower speed screen cassette is placed in the Bucky tray, the AEC has no way of recognizing it as different and will time the exposure for the system that it is programmed for. For example, if the system is programmed for a 400 film/screen combination, and if a 200 speed screen cassette was placed in the Bucky tray, the resulting radiograph would have half the required radiographic density.

34. **(D)** As object-image distance is increased, recorded detail is diminished due to magnification distortion. If the object-image distance cannot be minimized, an increase in SID is required in order to reduce the effect of magnification distortion. However, an equal relationship between OID and SID does not exist. In fact, to compensate for every 1 inch of OID an increase of 7 inches of SID is required. Therefore, an OID of 6 inches requires an SID increase of 42 inches. That is why a 6-inch air gap chest radiograph is performed at a 10-foot SID.

35. **(A)** Screen-film imaging consists of an exposure method of converting x-ray energy to light energy, and then converting light energy to electrochemical energy in the development process. Processing changes the invisible electrochemical image to a visible/manifest radiographic image. This process ends in *analog* data. *Digital* imaging is an electronic imaging system that allows data capture and manipulation in an electron pattern. It can be turned into

an analog image after going through several energy changes (electron to light to film or TV screen). The *direct action of x-rays* has very little influence on a radiographic image produced with intensifying screens (fluorescent light is responsible for the majority of film exposure). *(Pizzutiello & Cullinan, p 174)*

36. **(C)** A characteristic curve is used to demonstrate the relationship given the film and the resulting radiographic density. It has three main portions: the toe, the straight line portion or average gradient, and the shoulder. The *toe* is the lower portion, the beginning of the slope. Exposures made in the toe portion of the curve (Fig. 4–29A) will be underexposed. The *straight line portion* or *average gradient* is the region of correct exposure. The useful diagnostic density range is generally said to be between 0.25 and 2.5 density (Fig. 4–29B); this corresponds to the straight line portion of the characteristic curve. The *shoulder* is the rounded upper portion of the characteristic curve, often referred to as D max—the portion of maximum film exposure. Exposures made in the shoulder portion of the curve will be overexposed (Fig. 4–29C). *(Saia, PREP, p 312)*

37. **(B)** In order to adequately protect the sensitive silver emulsion, film must be stored in a cool (50 to 70°F), dry (50 percent humidity) area. Excessive temperatures can fog x-ray film. Excessive humidity will not affect unopened boxes of film, but can affect film in opened boxes. *(Bushong, p 223)*

38. **(B)** The original mAs (regulating radiographic density) was 200. The original kVp (regulating radiographic contrast) was 90. The mAs was cut in half, to 100, causing a decrease in density. The kVp was increased (by 15 percent) to compensate for the density loss, thereby increasing the scale of contrast. *(Cullinan, p 94)*

39. **(D)** *Differential absorption* refers to the x-ray absorption characteristics of neighboring anatomic structures. The radiographic representation of these structures is referred to as radiographic contrast, and may be enhanced

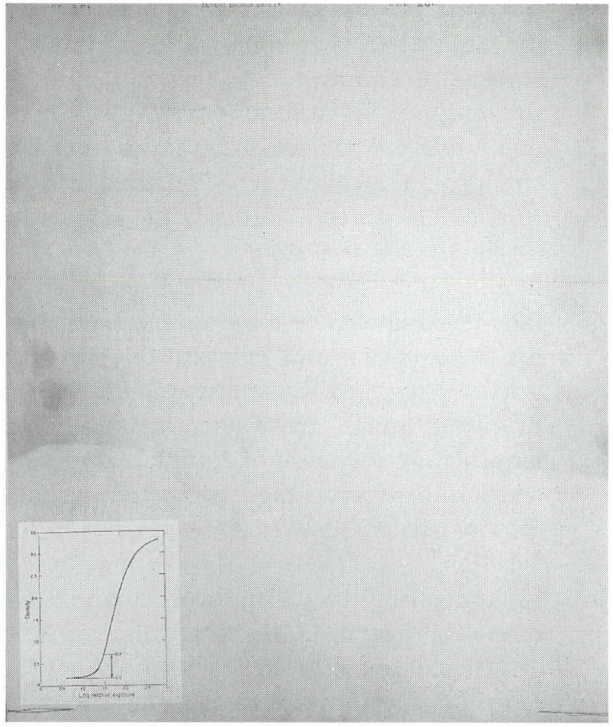

A

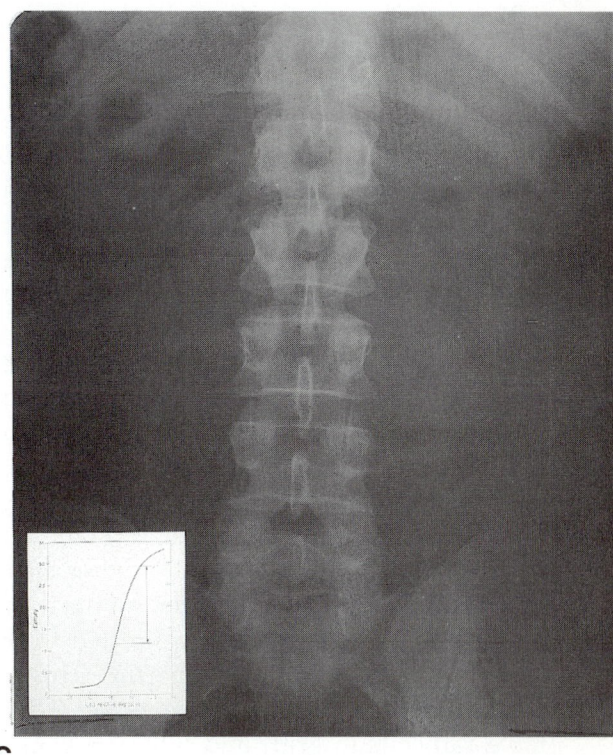

B

C

Figure 4–29. From the American College of Radiology Learning File. Courtesy of the ACR.

with high-contrast technical factors, especially using low kilovoltage levels. At low kilovoltage levels, the photoelectric effect predominates. *(Cullinan, p 9)*

40. **(B)** Quantum mottle is a grainy appearance on a finished radiograph, seen especially in very fast (rare earth) imaging systems. It is very similar in appearance to an enlarged photograph taken with fast film; it has a spotted or freckled appearance. Fast film and screens with low mAs and high kVp factors are most likely to be the cause of quantum mottle. *(Selman, p 328)*

41. **(A)** First, determine each individual mAs. Use the equation mA × sec = mAs: group A = 20 mAs; group B = 10 mAs; group C = 5 mAs; and group D = 9.8 mAs. Groups B and D are essentially the same. Group C has half the mAs and has compensated for that by increasing the kVp by 15 percent, resulting in a lower, or longer-scale, contrast. Group A is just the opposite: the mAs is doubled to 20 and, to compensate, the kVp is decreased by 15 percent, to 70 kVp. This results in fewer

grays (a higher, shorter-scale contrast). The single most important factor regulating radiographic contrast is kVp. The lower the kVp, the shorter the scale of contrast. All the mAs values in this problem have been adjusted for kVp changes in order to maintain density, but just a glance at each of the kilovoltages is often a good indicator of which will produce the longest scale or shortest scale contrast. *(Cullinan, p 300)*

42. **(A)** Factors that contribute to an increase in the intensification factor generally function to reduce resolution. Slow-speed (detail or "extremity") screens resolve more line-pairs per mm than much faster screens. The use of fast screens results in some loss of recorded detail. As intensification factor increases, radiographic density generally increases. *(Bushong, pp 223–224)*

43. **(C)** Radiographic density is directly proportional to mAs. If exposure time is *halved* from 0.04 (1/25) sec to 1/50 (0.02) sec, radiographic density will be cut in half. Changing to 150 mA will halve the mAs, effectively halving the radiographic density. If the kVp is decreased by 15 percent, from 85 to 72 kVp, radiographic density will be halved according to the 15 percent *rule*. In order to cut the density in half, the mAs must be reduced to 6 (rather than 10). *(Selman, p 332)*

44. **(B)** Although the AP knee radiograph was made using AEC, the resulting image appears underexposed. Its underexposure, and the fact that almost half the film was not exposed at all, indicates that (1) the knee was not centered to the photocell, and (2) the central ray was not directed to the center of the grid. The floating tabletop was not centered to the grid; then the CR was directed to the center of the *table* (rather than grid). This explains why half the film is unexposed (CR to table rather than grid) and why the knee is underexposed (grid cutoff as well as not being centered to the photocell). If the centering was correct and the Bucky tray was not pushed in, the knee would not have been underexposed. *(Wallace, pp 218–221)*

45. **(D)** Part of a good quality control program for automatic processing suggests that the startup procedure for a processor is to turn the water on, wait for the tanks to fill, wash the crossover rollers to remove any dried chemicals that may cause artifacts, and then run a cleanup film through to pick up any foreign material. *(Pizzutiello & Cullinan, p 197)*

46. **(D)** The higher the kVp range, the greater the exposure latitude (margin of error in exposure). Higher kVp is more penetrating and produces more grays on the radiograph, lengthening the scale of contrast. As kVp increases, the percentage of scatter radiation also increases. *(Bushong, p 277)*

47. **(C)** According to the Inverse Square Law of Radiation (the mAs/distance formula), if the SID is changed to 48 inches, 14.4 mAs is required in order to maintain the original radiographic density.

$$\frac{(old\ mAs)\ 10}{(new\ mAs)\ x} = \frac{(old\ D^2)\ 40^2}{(new\ D^2)48^2}$$

$$\frac{10}{x} = \frac{1600}{2304}$$

$$1600x = 23040$$
$$x = 14.4\ mAs\ at\ 48\ inch\ SID$$

Then, to compensate for changing from a 12:1 grid to an 8:1 grid, the mAs becomes 11.5.

$$\frac{(old\ mAs)\ 14.4}{(new\ mAs)\ x} = \frac{(old\ grid\ factor)\ 5}{(new\ grid\ factor)\ 4}$$

$$5x = 57.6$$
$$x = 11.5\ mAs\ with\ 8{:}1\ grid\ at\ 48''\ SID$$

Thus, 11.5 mAs is required to produce a film density similar to the original radiograph.

The following are the factors used for mAs conversion from non-grid to grid:

no grid = 1 × the original mAs
5:1 grid = 2 × the original mAs
6:1 grid = 3 × the original mAs
8:1 grid = 4 × the original mAs
12:1 grid = 5 × the original mAs
16:1 grid = 6 × the original mAs

(Cullinan, p 292)

48. **(C)** *Size* distortion (magnification) is inversely proportional to SID and directly proportional to OID. Increasing the SID and decreasing the OID decreases size distortion. Aligning the tube, part, and film so as to be parallel reduces *shape* distortion. There are two types of shape distortion, *elongation* and *foreshortening*. Angulation of the part with relation to the film results in foreshortening of the object. Tube angulation causes elongation of the object. *(Cullinan, p 129)*

49. **(B)** *Focal spot size* affects recorded detail by its effect on penumbra: the larger the focal spot size the greater the penumbra produced. Recorded detail is significantly affected by distance changes because of their effect on magnification. As *SID* increases, magnification decreases and recorded detail increases. The method of rectification has no impact on recorded detail. Single-phase, rectified units produce "pulsed" radiation, whereas three-phase units produce almost constant potential. *(Selman, p 319)*

50. **(B)** The radiograph seen in Figure 4–6 demonstrates *motion* blur. Motion obliterates *detail*. Patients who are in pain are often unable to cooperate as fully as patients who are not in pain. Careful positioning and patient instruction are often helpful, but it remains exceedingly useful to employ the shortest exposure time possible to help overcome the effects of involuntary motion. *(Selman, p 327)*

51. **(D)** Grid ratio is defined as the ratio between height of the lead strips to the width of the distance between them (that is, their height divided by the distance between them). If the height of the lead strip is 4.0 mm and the lead strips are 0.25 mm apart (4.0 divided by 0.25), the grid ratio must be 16:1. The thickness of the lead strip is unrelated to grid ratio. *(Selman, p 367)*

52. **(B)** Each mAs is determined (A = 4; B = 4; C = 8; D = 2) and numbered in order of greatest to least density (C = 1; A and B = 2; D = 3). Then the kilovoltages are reviewed and also numbered in order of greatest to least density (A, B, and C = 1; D = 2;). Next, screen speeds are numbered from greatest density-producing to least density-producing (D and B = 1; A = 2; C = 3). Lastly, the numbers assigned to mAs, kVp, and screen speed are added up for each of the four groups (B = 4; A and C = 5; D = 6); the *lowest* total (B) indicates the group of factors that will produce the greatest radiographic density. This process is illustrated as follows:

A. 4 mAs (2) + 94 kVp (1) + 100 screens (2) = (5)
B. 4 mAs (2) + 94 kVp (1) + 200 screens (1) = (4)
C. 8 mAs (1) + 94 kVp (1) + 50 screens (3) = (5)
D. 2 mAs (3) + 80 kVp (2) + 200 screens (1) = (6)

(Cullinan, p 300)

53. **(D)** As photon energy increases, more penetration and greater production of scatter radiation occurs, producing a longer scale of contrast. As grid ratio increases, more scatter radiation is absorbed, producing a higher contrast. As OID increases, the distance between the part and film acts as a grid and consequently less scatter radiation reaches the film, producing a higher contrast. Focal spot size is related only to recorded detail. *(Cullinan, pp 121–122)*

54. **(D)** In the low kilovoltage ranges, a difference of just a few kVp makes a very noticeable radiographic difference. High kVp techniques offer much greater margin for error, as do slow film/screen combinations. Lower-

ratio grids offer more tube-centering latitude (leeway, margin for error) than high-ratio grids. *(Carroll, pp 112–113)*

55. **(B)** The greater the quantity of black metallic silver deposited on a film, the greater the radiographic density. The greater the degree of radiographic density (degree of blackening), the *less* the quantity of illuminator light transmitted through the film. Therefore, the relationship between the amount of illuminator light striking the film and the amount of light transmitted through the film is an expression of radiographic density. It is expressed by the formula:

$$density = log \frac{incident\ light\ intensity}{transmitted\ light\ intensity}$$

(Selman, p 331)

56. **(B)** If A and B are reduced to 5 mAs for mAs consistency, the kVp would increase in both cases to 85 kVp, thereby balancing radiographic densities. Thus, the greatest density is determined by the shortest SID (greatest exposure rate). *(Cullinan, p 295)*

57. **(D)** Magnification radiography may be used to delineate a suspected hairline fracture or enlarge tiny, contrast-filled blood vessels. It also has application in mammography. In order to magnify an object twice its actual size, the part must be placed midway between the focal spot and film. *(Selman, p 352)*

58. **(C)** Positive contrast medium is radiopaque; negative contrast material is radioparent. Barium sulfate (radio-opaque, positive contrast material) is most frequently used for exams of the intestinal tract, and high kVp exposure factors are used in order to penetrate (to see through and behind) the barium. Water-base iodinated contrast media (Conray, Amipaque) are also positive contrast agents. However, the K-edge binding energy of iodine prohibits the use of much greater than 70 kVp with these materials. Higher kVp values will obviate the effect of the contrast agent. Air is an example of a negative contrast agent, and high kVp factors are clearly not indicated. *(Cullinan, pp 142–144)*

59. **(D)** The thicker the active layer of phosphors, the more fluorescent light is emitted from the screen. Different types of phosphors have different *conversion efficiencies*; rare earth phosphors emit more light during a given exposure than do calcium tungstate phosphors. As the kVp level is increased, so is the amount of fluoroscopic light emitted by intensifying screen phosphors. *(Selman, pp 282–284)*

60. **(B)** The heel effect is characterized by a variation in beam intensity, *gradually increasing from anode to cathode*. This can be effectively put to use when performing radiographic exams on large body parts of uneven tissue density. For example, the AP thoracic spine is thicker caudally than cranially, and so the thicker portion is best placed under the cathode. However, in the lateral projection of the thoracic spine, the upper portion is thicker because of superimposed shoulders, and therefore is best placed under the cathode end of the beam. The femur is also uneven in density, particularly in the AP position, and can benefit from use of the heel effect. However, the sternum and its surrounding anatomy are fairly uniform in thickness and would not benefit from use of the anode heel effect. *The heel effect is most pronounced when using large-size films, at short SIDs, and with an anode having a steep (small) target angle.* *(Bushong, pp 124–125)*

61. **(C)** When tissue densities within a part vary greatly, the radiographic result can be unacceptably high contrast. In order to "even out" these densities and produce a more appropriate scale of grays (as with a chest x-ray), exposure factors using high kVp should be employed. Radiographic contrast generally increases with an increase in screen speed. The higher the grid ratio, the higher the contrast. Exposure factors using high mAs generally result in excessive film density, frequently obliterating much of the gray scale. *(Selman, pp 339–340)*

62. **(D)** When a film first enters the processor from the feed tray, a microswitch signals the replenishment pump to begin sending replenisher solution into the processor. Replenishment continues until the microswitch senses the end (edge) of the film and terminates pump action. So as long as film is being fed into the processor, replenishment solution will be added. There is therefore more replenisher added with larger-size films. There is also more replenisher added when rectangular films are fed into the processor "the long way" because the processor is sensing, for example, 17 inches of film rather than 14 inches of film. Film should be put through the processor consistently according to the particular department's preference/routine. A change in film direction can lead to over/under replenishment and thereby a change in film density. *(Bushong, p 213; McKinney, p 106)*

63. **(B)** According to the *15 percent rule*, if the kVp is increased by 15 percent, radiographic density will be doubled. Therefore, to compensate for this change, and to maintain radiographic density, the mAs should be reduced to 6 mAs. *(Bushong, p 307)*

64. **(A)** Body section tomography functions to provide an image of a particular plane (objective plane) of tissues within the body, blurring out everything above and below the plane of interest. The radiograph shown in Figure 4–7 is an example of linear tomography; the x-ray tube moves in one direction while the x-ray film moves in the opposite direction. The two pivot at a fixed fulcrum which corresponds to the objective plane and is therefore the level of no motion. A variety of x-ray tube motions (circular, hypocycloidal, and so on) is available with more complex tomographic equipment. *(Wolbarst, p 226)*

65. **(A)** In order to test a film's response to processing, the film must first be given a predetermined exposure with a *sensitometer*. The film is then processed and the densities are read using a *densitometer*. Any significant variation

from the expected densities is further investigated. A step wedge is used to evaluate the effect of kVp on contrast, and a spinning top test is used to check timer accuracy. *(Cullinan, p 91)*

66. **(B)** A characteristic curve is representative of a film's response to light or x-rays. A slow film emulsion (one with greater latitude and lower contrast) responds more gradually than does a fast film. In general, the more gentle or gradual the slope of a particular film's characteristic curve, the slower it is, the longer the scale of contrast it will produce, and the more latitude it possesses. A line is first drawn horizontally from density 0.5 to intersect films 1 and 2, and is seen to intersect at the same point, exposure 1.5. Then, a line is drawn horizontally from density 2.5 to intersect films 1 and 2. Film 1 is intersected at approximately exposure 1.9. Film 2 is intersected at approximately exposure 2.3. The distance between the two points of intersection of film 1 illustrates its degree of exposure latitude. Similarly, the distance between the two points of intersection of film 2 illustrates its degree of exposure latitude. Because the distance associated with film 2 is greater, it possesses a greater exposure latitude. A quick glance at the two curves, noting the "gentler" slope of film 2, indicated it to be the emulsion with greater exposure latitude. *(Cullinan, p 92)*

67. **(C)** The *wider* dimension of the x-ray film is usually placed on the feed tray and fed into the processor. The *entrance roller* is the first roller of the transport system located at the end of the feed tray; this is where the microswitch that determines amount of replenishment is located. The *length* of the film (the shorter dimension) activates the microswitch and replenisher is added according to the length of the film; a 10 × 12 will receive less replenisher than will a 14 × 17 film. Crossover rollers are located between the different tanks. The receiving bin is where the films exit the processor. The replenishment pump is activated by the microswitch. *(McKinney, p 106)*

68. (C) Excessive radiographic density may be a result of overdevelopment. Overdevelopment may be due to excessive developer temperature, resulting in chemical fog. Excessive density can also be a result of over-replenishment due to faulty microswitches or from feeding film into the processor "the long way" rather than "the wide way." *(McKinney, p 190)*

69. (C) Grids are devices constructed of alternating strips of lead foil and radiolucent interspacing material. They are placed between the patient and film, and function to remove scatter radiation from the remnant beam before it forms the latent image. *Stationary* grids will efficiently remove scatter radiation from the remnant beam; however, their lead strips will be imaged on the radiograph. If the grid is made to *move* (usually in a direction perpendicular to the lead strips) during the exposure, the lead strips will be effectively blurred. The motion of a moving grid, or Potter-Bucky diaphragm, may be *reciprocating* (equal strokes back and forth), *oscillating* (almost circular direction), or *catapult* (rapid forward motion and slow return). Synchronous refers to a type of x-ray timer. *(Cullinan, p 78)*

70. (C) If the fixer fails to sufficiently harden the gelatin emulsion, water will remain within the still swollen emulsion. The dryer mechanism will be unable to completely rid the emulsion of wash water, and the film will emerge from the processor damp and tacky. On the other hand, excessive hardening action may produce brittle radiographs. High air velocity usually encourages more complete drying. Unbalanced processing temperatures can result in blistering of the emulsion. *(McKinney, p 189)*

71. (A) High-speed imaging systems are valuable for reducing patient exposure and patient motion. However, some detail must be sacrificed, and quantum mottle can cause further image impairment. In general, doubling the film/screen speed doubles the radiographic density, thereby requiring that the mAs be halved in order to maintain the origi-

nal radiographic density. Changing from 200 to 400 screens requires halving the mAs value to 3 mAs. Then the new exposure time, using 300 mA, is determined:

$$300x = 3$$
$$x = 0.01 \text{ sec}$$

(Wallace, p 165)

72. (B) Most laser film is sensitive to both the Wratten 6B and the GBX safelight filters. Laser film will fog if handled under these safelight conditions. Most laser film is loaded into a film magazine in total darkness. Processing temperatures are the same for laser film as for regular x-ray film. *(Pizzutiello & Cullinan, p 172)*

73. (C) Of the given factors, kilovoltage and grid ratio will have a significant effect on radiographic contrast. The mAs values are almost identical. Because an increased kilovoltage and low-ratio grid combination would allow the greatest amount of scattered radiation to reach the film, thereby producing more gray tones, C is the best answer. D also uses a low-ratio grid, but the kV is too low to produce as many gray tones as C. *(Cullinan, p 300)*

74. (A) An intensifying screen consists of microscopic crystals called *phosphors*, a high grade of cardboard or plastic called the base (which is coated with a layer of phosphors), and a smooth abrasion resistant material called the *protective coat*. The layer of phosphors coating the base is referred to as the *active layer*. This is the layer that fluoresces when activated by x-rays. The *thickness* of the active layer is the principle factor in determining screen speed. *(Selman, pp 275–276)*

75. (B) In order to change non-grid exposures to grid exposures, or to adjust exposure when changing from one grid ratio to another, you must remember the factor for each grid ratio:

no grid = 1 mAs
5:1 grid = 2 mAs
6:1 grid = 3 mAs
8:1 grid = 4 mAs

12:1 grid = 5 mAs
16:1 grid = 6 mAs

To adjust exposure factors, you simply compare the old with the new:

$$\frac{4 \text{ (old mAs)}}{x \text{ (new mAs)}} = \frac{4 \text{ (old grid factor)}}{6 \text{ (new grid factor)}}$$

$$\frac{4}{x} = \frac{4}{6}$$

$$4x = 24$$

$$x = 6 \text{ mAs using a 16:1 grid}$$

(Cullinan, p 300)

76. **(B)** About half the silver in film emulsion remains to form the image. The other half is removed from the film during the fixing process. Fixer solution, therefore, has a high silver content. Silver is a *toxic metal* and cannot simply be disposed of into the public sewer system. As silver is expensive to mine, it also becomes financially wise to recycle the silver removed from x-ray film. The three most commonly used silver recovery systems are the *electrolytic, metallic replacement*, and *chemical precipitation methods*. In *electrolytic* units, an electric current is passed through the fixer solution. Silver ions are attracted to, and become plated onto, the negative electrode of the unit. The plated silver is periodically scraped from the cathode, and accurately measured so that the hospital can be appropriately reimbursed. The electrolytic method is a practical recovery system for moderate and high-use processors. *Metallic replacement* (or displacement) method of silver recovery uses a steel mesh/steel wool type cartridge that traps silver as fixer is run through it. This system is useful for low-volume processors. *Chemical precipitation* adds chemicals that release electrons into the fixer solution. This causes the metallic silver to precipitate out, falling to the bottom of the tank, forming a recoverable sludge. This method is used principally by commercial silver dealers. *(Carlton & Adler, pp 309–311)*

77. **(A)** When an automatic exposure device (phototimer or ionization chamber) is used, the system is programmed for the use of a particular film/screen speed (for example, 400 speed). If a slower speed screen cassette is placed in the Bucky tray, the AEC has no way of recognizing it as different, and will time the exposure for the system that it is programmed for. For example, if the system is programmed for a 400 film/screen combination, and if a 200 speed screen cassette was placed in the Bucky tray, the resulting radiograph will have half the required radiographic density.

78. **(B)** If the source-image distance is above or below the recommended focusing distance, the primary beam at the lateral edges will not coincide with the angled lead strips. Consequently, there will be absorption of the primary beam, termed grid cut-off. If the central ray is off center, the lead strips are no longer parallel with the divergent x-ray beam and there will be loss of density due to grid cutoff. Central ray angulation in the direction of the lead strips is appropriate and will not cause grid cut-off. *(Pizzutiello & Cullinan, p 62)*

79. **(B)** Quantum mottle is a grainy appearance on a finished radiograph, seen especially in very fast imaging systems. It is very similar in appearance to an enlarged photograph taken with fast film; it has a spotted or freckled appearance. Fast film and fast screens with low mAs and high kVp factors are most likely to be the cause of quantum mottle. *(Wolbarst, p 154)*

80. **(B)** Penumbra, or geometric blur, is caused by photons emerging from a large focal spot. Because the projected focal spot is greatest at the cathode end of the x-ray tube, geometric blur is also greatest at the corresponding part (cathode end) of the radiograph. The projected focal spot size becomes progressively smaller toward the anode end of the x-ray tube. *(Bushong, p 289)*

81. **(A)** More scattered radiation is generated within a part as the kV is increased, the size of the field is increased, and the thickness and density of tissue increases. As the quantity of scattered radiation increases from any

of these sources, more densities are added to the radiographic image. *(Wolbarst, p 170)*

82. **(D)** Factors that regulate the number of x-ray photons produced at the target may be used to control radiographic density, namely mA and exposure time (mAs). Radiographic density is directly proportional to mAs; if the mAs is cut in half, the radiographic density will decrease by one half. Although kilovoltage is used primarily to regulate radiographic contrast, it may also be used to regulate radiographic density in variable kVp techniques, according to the 15 percent rule. *(Selman, p 332)*

83. **(D)** X-ray film is sensitive and requires proper handling and storage. Several kinds of artifacts can be produced by careless handling during production of the radiographic image. Tree-like, branching black marks on a radiograph are usually due to static electrical discharge. Problems with static electricity are especially prevalent during cold, dry weather and can be produced by simply removing a sweater in the darkroom. *(Selman, p 309)*

84. **(C)** Sensitometry and densitometry are used in the evaluation of the film processor, which is just one part of a complete quality assurance (QA) program. Modulation transfer function (MTF) is used to express spatial resolution, another component of the QA program. A complete QA program includes testing of all components of the imaging system: processors, focal spot, x-ray timers, filters, intensifying screens, beam alignment, and so on. *(Noz, pp 120–125)*

85. **(B)** If it is desired to reduce the exposure time for a particular radiograph, as it might be when radiographing children and others who are unable to cooperate fully, the mA must be increased sufficiently in order to maintain the original mAs, and thus radiographic density. A higher kV could be useful because it would allow further reduction of the mAs (exposure time) according to the 15 percent rule. Use of *higher*-speed film/screen combination also helps reduce mAs (exposure time) through more efficient con-

version of photon energy to fluorescent light energy. *(Cullinan, pp 292, 295–296)*

86. **(D)** With three-phase equipment, the voltage never drops to zero and x-ray intensity is significantly greater. When changing *from single-phase to three-phase, six-pulse equipment, two-thirds the original mAs is required* to produce a radiograph with similar density. When changing *from single-phase to three-phase, twelve-pulse equipment, only one-half the original mAs is required*. In this problem we are changing from three-phase, twelve-pulse to single-phase equipment; therefore the mAs should be doubled (from 12.5 to 25 mAs). *(Cullinan, p 147)*

87. **(B)** A fractional focal spot of 0.3 mm or smaller is essential for reproducing fine detail without penumbral blurring in magnification radiography. As the object image is magnified, so will be the associated penumbra unless the fractional focal spot is used. Tomographic and fluoroscopic procedures would probably cause great wear on a fractional focal spot. Use of the fractional focal spot is not essential in stereoradiography. *(Selman, p 352)*

88. **(D)** The illustrated radiograph is that of a knee with an old gunshot wound. Some anatomic parts not conforming well with the collimator configuration (for example, lateral knee, lateral lumbar spine, AP shoulder) frequently suffer loss of contrast from "undercutting" of the image. The primary beam, interacting only with the tabletop, gives rise to scatter radiation, which can only degrade contrast and image quality. If a lead rubber shield is placed in the area where there is no anatomic absorber (anterior to the flexed knee/tibia, behind the lumbar spine), image quality can be vastly improved. Dust on the intensifying screens would cause tiny "pinhole" type minus density artifacts. Chemical fog would appear as a dull, gray image lacking contrast. *(Pizzutiello & Cullinan, p 68)*

89. **(C)** *Materials that emit light when stimulated by x-ray photons are called phosphors.* Phosphors

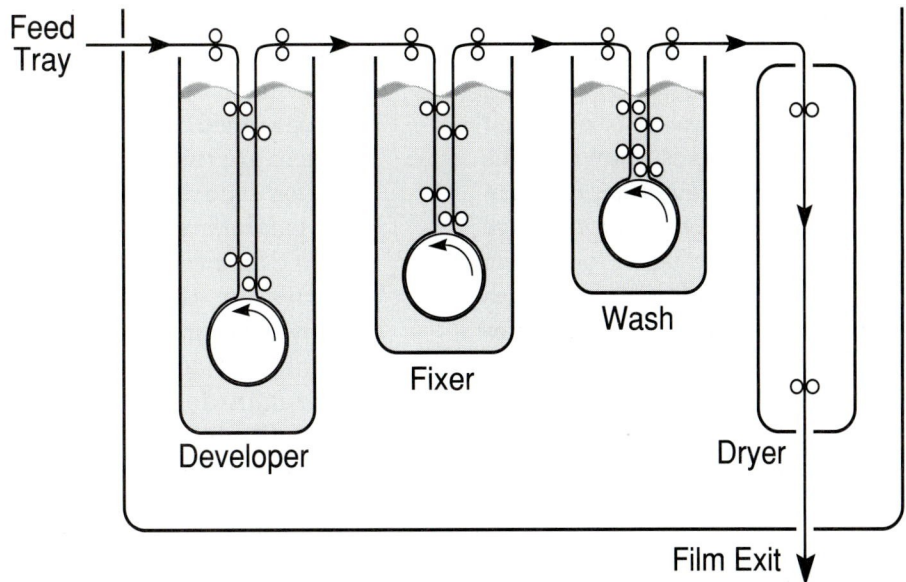

Feed Tray

Developer

Fixer

Wash

Dryer

Film Exit

Figure 4–30. Reproduced with permission from Wolbarst AB. *Physics of Radiology.* East Norwalk, CT: Appleton & Lange, 1993.

are used in intensifying screens, and function to absorb x-ray photon energy and convert it to visible light energy. Typically, for each x-ray photon absorbed, many light photons are emitted; intensifying screens serve to amplify the action of x-rays. *(Bushong, pp 217–218)*

90. **(D)** High kV factors are frequently used to even out densities in anatomic parts of high tissue contrast. But, as high kV produces added scatter radiation, it generally must be used with a grid. It would be inappropriate to perform an AP abdomen with high kV because it has such low subject contrast. Barium-filled structures are frequently radiographed using 120 kV or more in order to penetrate the barium—to see through to structures behind. A 10-foot air gap (6 to 10 inches OID) chest permits the use of high kilovoltage, as most of the scatter radiation produced is eliminated by the grid-like effect of the air gap. *(Cullinan, pp 142–143)*

91. **(A)** Different types of intensifying screens are available for radiographic use and they can differ greatly. Some intensifying screens emit a blue and others a green fluorescent light. Films are manufactured to be sensitive to one of these. This is termed spectral matching; if the film and screens are in-

correctly matched, speed will be reduced. *(Bushong, p 226)*

92. **(C)** During automatic processing (Fig. 4–30), radiographic film is first immersed in the *developer* solution, which functions to reduce the exposed silver bromide crystals in the film emulsion to black metallic silver (which constitutes the image). Film next goes directly into the *fixer*, which functions to remove the *unexposed* silver bromide crystals from the emulsion. The film is then transported to the *wash* tank, where chemicals are removed from the film; and then into the *dryer* section, where it is dried before leaving the processor. *(Selman, p 312)*

93. **(B)** Radiographic density is described as the *overall degree of blackening of a radiograph or a part of it.* The mAs regulates the number of x-ray photons produced at the target, and thus, regulates radiographic density. If it is desired to double radiographic density, one simply doubles the mAs; therefore mAs and radiographic density are directly proportional. *(Selman, p 332)*

94. **(B)** Because the focal spot (track) of an x-ray tube is along the anode's beveled edge, photons produced at the target are able to di-

verge toward the cathode end of the tube, but are absorbed by the "heel" of the anode at the opposite end of the tube. This results in a greater number of x-ray photons distributed toward the cathode end and is known as the *anode heel effect*. The *line focus principle* is a geometric principle illustrating that the effective focal spot is smaller than the actual focal spot. The *Inverse Square Law of Radiation* deals with the relationship between distance and radiation intensity. *Bohr's theory* refers to an atom's resemblance to the solar system. *(Selman, p 394)*

95. **(B)** The formula for mAs is mA × sec = mAs. Substituting known values:

$$\frac{1}{40} x = 15 \text{ mAs}$$

$$x = 600 \text{ mA}$$

(Selman, p 332)

96. **(C)** The chest is composed of widely differing tissue densities (bone and air). In an effort to "even out" these tissue densities and better visualize pulmonary vascular markings, high kV is generally used. This produces more uniform penetration and results in a *longer scale of contrast* with visualization of the pulmonary vascular markings as well as bone (which is better penetrated) and air densities. The increased kV also affords the advantage of greater exposure latitude (an error of a few kV will make little if any difference). The fact that the kV is increased means that the mAs is accordingly reduced, and thus patient *dose is reduced* as well. A grid is usually used whenever high kilovoltage is required. *(Cullinan, p 118)*

97. **(B)** First, evaluate the change(s): the kVp was decreased 15 percent (95 − 15 percent = 80.7). A 15 percent decrease in kVp will cut the radiographic density in half; therefore, it is necessary to use twice the original mAs to maintain the original density. The original mAs was 16; we now need 32 mAs, using the 500-mA station. Because mA × sec = mAs, then:

$$500x = 32$$
$$x = 0.064 \text{ sec}$$

(Cullinan, p 292)

98. **(C)** There are two types of distortion, size and shape. *Shape distortion* relates to the alignment of the x-ray tube, the part to be radiographed, and the film/image recorder. There are two kinds of shape distortion, *elongation* and *foreshortening*. *Size distortion* is *magnification*, and is related to the OID and SID. Magnification can be reduced by either increasing the SID or decreasing the OID. However, an increase in SID must be accompanied by an increase in mAs, in order to maintain density. It is therefore preferable, in the interest of exposure, to reduce OID whenever possible. *(Selman, pp 348, 356)*

99. **(A)** Size distortion (magnification) is inversely proportional to SID and directly proportional to OID. Decreasing the SID and increasing the OID serve to increase size distortion. Aligning the tube, anatomic part, and film so as to be parallel reduces shape distortion. Angulation of the long axis of the part with respect to the film results in foreshortening of the object. Tube angulation causes elongation of the part. *(Bushong, pp 282–283)*

100. **(D)** Static electricity builds up in nonconductive materials, such as plastic and synthetic fibers, until it reaches a point where it discharges the electricity in the form of a spark. Because gelatin is semiconductive it will accept the electrical discharge, and static electricity artifacts will appear on the radiographic image. Darkroom personnel should be instructed to wear leather-soled shoes and nonsynthetic clothing (such as cotton). Polyesters and nylons will encourage the production of static electricity. Low humidity also encourages static electrical build-up. Sliding films in and out of cassettes causes friction and is another means of creating a static electrical build-up. *(McKinney, pp 6–9, 190)*

101. **(C)** The mAs is the exposure factor that regulates radiographic density. The equation used to determine mAs is mA × sec = mAs. Substituting known factors:

$$400x = 15$$
$$x = 0.037 \; (\tfrac{1}{30}) \text{ sec}$$

(Cullinan, p 392)

102. **(B)** *The shortest possible exposure should be used as a matter of routine.* Parkinson's disease is characterized by uncontrollable tremors. The resulting unsharpness can destroy image detail. It is therefore necessary to use as fast an imaging system as possible. High-speed (preferably rare earth) intensifying screens will permit a considerable reduction in mAs (specifically, exposure time). Compensating filtration is unrelated to the problem and not indicated here. *(Bushong, pp 292–293)*

103. **(C)** Control of motion, both voluntary and involuntary, is an important part of radiography. Patients are unable to control types of motion such as heart action, peristalsis, and muscle spasm. In these circumstances it is essential to use the shortest possible exposure time in order to have a "stop action" effect. *(Selman, p 332)*

104. **(D)** Every film emulsion has a characteristic curve representative of that film's speed, contrast, and latitude. A gentle curve (as opposed to a steep curve) usually indicates a

film having slow speed, low contrast, and more latitude. *(Cullinan, p 92)*

105. **(A)** Shape distortion (foreshortening, elongation) is caused by improper alignment of the tube, part, and film. Size distortion, or magnification, is caused by too great an object-image distance or too short a source-image distance. Penumbra is caused principally by use of a large focal spot. *(Bushong, p 283)*

106. **(D)** X-ray tube targets are constructed according to the *line focus principle*—the focal spot is angled (usually 12 to 17°) to the vertical (Fig. 4–31). As the actual focal spot is projected downward it is foreshortened; thus, the effective focal spot is always smaller than the actual focal spot. As it is projected toward the cathode end of the x-ray beam it becomes larger and approaches its actual size. As it is projected toward the anode end, it gets smaller because of the anode "heel" effect. *(Bushong, p 289)*

107. **(B)** Quantum *noise*, or mottle, is a grainy appearance on a finished radiograph, seen especially in very fast imaging systems. It is very similar in appearance to a photograph taken with fast film and enlarged; it has a spotted or freckled appearance. Fast film and screens with low mAs and high kVp factors

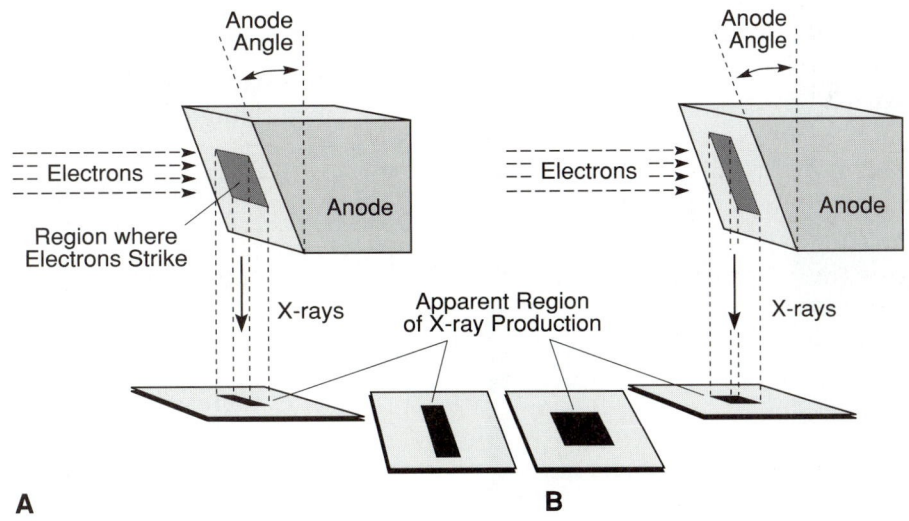

A   **B**

Figure 4–31. Reproduced with permission from Wolbarst AB. *Physics of Radiology*. East Norwalk, CT: Appleton & Lange, 1993.

are most likely to be the cause of quantum noise/mottle. *Pi lines* appear as plus density lines running perpendicular to the direction of film travel, and are sometimes seen in new processors or after a complete maintenance/overhaul. *Grid cutoff* is absorption of the primary beam by the grid and usually results in loss of density and visibility of grid lines. *(Wolbarst, pp 153–154)*

108. (B) Increasing grid ratio will result in a larger percentage of scattered radiation being absorbed and hence a shorter scale of contrast. Increased photon energy is caused by an increase in kVp, resulting in more penetration of the part and a longer-scale contrast. *(Bushong, p 295)*

109. (A) Only two pieces of apparatus are needed to construct a characteristic curve (Fig. 4–32). First, a *penetrometer* (aluminum step wedge) is used to expose a film. Once the film is processed, a *densitometer* is needed to read the resulting densities. Log relative exposure is charted along the x (horizontal) axis; an increase in log relative exposure of 0.3 results from doubling the exposure. Optical density is plotted on the y (vertical) axis and represents the amount of light transmitted through a film compared to the amount of light striking the film (expressed as a logarithm). *(Bushong, p 269)*

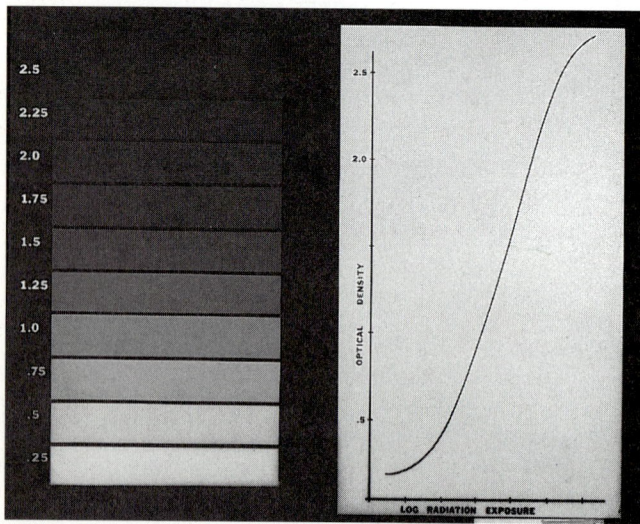

Figure 4–32. From the American College of Radiology Learning File. Courtesy of the ACR.

110. (C) Because the anode's focal track is beveled (angled, facing the cathode), x-ray photons can freely diverge toward the cathode end of the x-ray tube. However, the "heel" of the focal track prevents x-ray photons from diverging toward the anode end of the tube. This results in varying intensity from anode to cathode, fewer photons at the anode end, and more photons at the cathode end. The anode heel effect is most noticeable using large film sizes, short SIDs, and steep target angles. *(Bushong, p 289)*

111. (B) The steepness of the characteristic curve is representative of film contrast. The steeper the curve, the greater the density *differences* and the *higher* the contrast. The speed of the film is determined by the curve's position on the log relative scale: when comparing two or more characteristic curves, the *faster* film lies farthest to the left. The faster the film speed, the less the exposure latitude. *(Bushong, pp 273–275)*

112. (D) *Sensitometry* is a method of quality control for daily monitoring of the automatic film processor. A *densitometer* is a device used to read optical density. Crossover rollers should be *cleaned daily* to prevent crystallized solution buildup on the rollers. A *warm-up* procedure is performed on an x-ray tube for safe operation after prolonged disuse. *(Selman, p 314)*

113. (C) The original mAs was 15 (300 mA × 0.05 sec). Using the mAs/distance formula the new mAs must be determined for the distance change from 40 to 46 inches of SID:

$$\frac{15}{x} = \frac{1600 \ (40^2)}{2116 \ (46^2)}$$

$$1600x = 31{,}740$$
$$x = 20 \text{ mAs at a 46 inch SID}$$

A 15 percent increase in kilovoltage was made, increasing the kV to 87. Because the kV change effectively doubles the radiographic density, the mAs must be cut in half (from 20 to 10) to compensate. Then, if 500 is the new mA, we must determine what exposure time is required to achieve 10 mAs:

$$500x = 10$$
$$x = 0.02 \text{ sec}$$

(Selman, p 335)

114. **(C)** If the imaging system speed is cut in half, half of the original density will occur on the radiograph. Therefore, to maintain the original density, the mAs must be doubled from the original 15 to 30 mAs. Grids are used to absorb scatter radiation from the remnant beam before it can contribute to the latent image. Because scatter (and some primary) radiation is removed from the beam, an increase in exposure factors is required. The amount of increase is dependent upon the grid ratio: the higher the grid ratio, the higher the correction factor. The correction factor for an 8:1 grid is 4; therefore, the mAs (30) is multiplied by 4 to arrive at the new required mAs (120). Using the mAs equation mA × time = mAs, it is determined that 0.3 sec will be required at the 400 mA station:

$$400x = 120$$
$$x = 0.3 \text{ sec}$$

(Cullinan, p 296)

115. **(B)** The *Characteristic (H&D) Curve* is used to illustrate the relationship between the exposure given the film and the resulting film density. It can be used to predict a particular film emulsion's response (speed, sensitivity) by determining how long it takes to record a particular density. The Characteristic Curve is used in sensitometry to monitor automatic processing efficiency and consistency. A film is given a series of predetermined exposures and processed. The resulting densities are plotted, and the resulting curve is compared with a known correct curve. Any deviation between the two may indicate processing difficulties. The Characteristic Curve illustrates the effects of exposure and processing on radiographic film emulsion and is unrelated to film speed. *(Burns, p 169)*

116. **(C)** When an automatic exposure device (phototimer or ionization chamber) is used, the system is programmed for the use of a particular film/screen speed (such as 400 speed). If a slower-speed screen cassette is placed in the Bucky tray, the AEC has no way of recognizing it as different, and will time the exposure for the system that it is programmed for. For example, if the system is programmed for a 400 film/screen combination, and if a 200 speed screen cassette is placed in the Bucky tray, the resulting radiograph will have half the required radiographic density.

117. **(C)** A decrease in screen speed is associated with less diffusion of fluorescent light and therefore an increase in recorded detail. Increasing the focal spot size will decrease the recorded detail because larger focal spot sizes are associated with more penumbral (edge gradient) blur. Decreasing the source-image distance will reduce recorded detail by increasing magnification. *(Selman, pp 319–320, 327)*

118. **(B)** Cassette front material must not attenuate the remnant beam, yet must be sturdy enough to withstand daily use. Bakelite has long been used as material for tabletops and cassette fronts. Magnesium is also frequently used. Lead would not be a suitable material, as it would absorb the remnant beam, and no image would be formed. *(Bushong, pp 224–225)*

119. **(B)** The illustrated radiograph demonstrates a 1½-inch unexposed strip along the length of the film. This occurred because, although the patient was centered correctly to the collimator light, the x-ray tube was not centered to the grid. If the patient was off center, the entire film would be exposed and the patient's spine would be off center. Grid cutoff would not appear as such a sharply delineated line, but rather a gradually decreasing density. *(Wallace, 218–221)*

120. **(B)** If the safelight bulb is greater wattage than it should be, the safelight filter incorrect

for the film type, or if the filter is cracked, film fog can occur. If film is not stored in a radiation safe area, it can be fogged by background radiation. Screen lag is not caused by improper film storage conditions, but rather by aged or defective intensifying screens. *(Selman, p 294)*

121. **(A)** Limiting the size of the radiographic field serves to limit the amount of scatter radiation developed within the anatomic part. As the percentage of scatter radiation produced within the part decreases, so does the resultant density within the radiographic image. Hence, beam restriction is a very effective means of reducing the quantity of non-information-carrying scatter radiation (fog) produced, resulting in a shorter scale of contrast with fewer radiographic densities. *(Bushong, p 295)*

122. **(B)** As films are developed, the developer solution becomes weaker and oxidation products are produced in solution. If sufficient replenishment of new developer solution does not take place, and the older solution activity decreases, chemical fog is produced. Films lack contrast, and have a flat, gray appearance. *(Bushong, p 208)*

123. **(A)** If a structure of a given length is not positioned parallel to the recording medium (film), it will be projected smaller than its actual size (foreshortened). An example of this can be a lateral projection of the third digit. If the finger is positioned so as to be parallel to the film, no distortion will occur. If, however, the finger is positioned so that its distal portion rests on the film while its proximal portion remains a distance from the film, foreshortening will occur. *(Bushong, p 283)*

124. **(C)** Motion, voluntary or involuntary, is most detrimental to good recorded detail. Though all other detail factors may be adjusted to maximize detail, if motion occurs during exposure, detail is lost. The most important ways to reduce the possibility of motion are by using the shortest possible exposure time, by careful patient instruction (for suspended respiration), and by adequate im-

mobilization when necessary. Minimizing magnification through the use of *increased* SID and *decreased* OID functions to improve recorded detail. *(Selman, pp 319–320, 327)*

125. **(A)** The invisible silver halide image is composed of exposed silver grains. These are "reduced" to a visible black metallic silver image in the developer solution. The fixer solution functions to remove unexposed silver halide crystals from the film. *(Bushong, p 206)*

126. **(A)** Soft tissue radiography requires the use of long-wavelength, low-energy x-ray photons. Very little filtration is used in mammography. Certainly anything more than 1.0 mm aluminum would remove the useful soft photons and the desired high contrast could not be achieved. Dedicated mammographic units usually have molybdenum targets (for the production of soft radiation) and a small amount of molybdenum filtration. *(Bushong, p 553)*

127. **(C)** Quality control in mammography includes scrupulous testing of virtually all component parts of the mammographic imaging system. It includes processor checks, screen maintenance, accurate and consistent viewing conditions, and evaluation of phantom images, to name a few. The illustration pictured is that of the structures to be imaged within a mammography phantom. A mammographic phantom contains mylar fibers, simulated masses, and specks of simulated calcifications. The ACR accreditation criteria state that a minimum of ten objects (four fibers, three specks, three masses) must be visualized on test films. Changes in any part(s) of the imaging system (film, screens, cassettes, x-ray equipment, filtration, viewbox) can result in unsuccessful results. *(Cullinan, p 262)*

128. **(B)** If fluorescent light from one intensifying screen passes through the film to the opposite emulsion and intensifying screen, the associated diffusion creates a type of distortion called crossover. Intensifying screens do need a degree of reflectance to enhance their speed. Scatter and filtration are unrelated to intensifying screens. *(Cullinan, pp 96–97)*

129. **(B)** A wire mesh supported between two rigid pieces of clear plastic is used to evaluate screen/film contact. The mesh is placed on a cassette and radiographed. Upon viewing, any areas that appear unsharp or blurry are indicative of poor screen/film contact. A screen lag test is performed by radiographing a phantom using an empty cassette, loading it with film and leaving it a few minutes. If, after processing, there is any indication of an image, there is most probably screen lag. *(Bushong, p 232)*

130. **(B)** Single-phase equipment is much less efficient than three-phase equipment, because voltage amplitude is always changing from zero to peak value and back to zero again. *With three-phase equipment, voltage never drops to zero; x-ray intensity is significantly greater.* When changing from single-phase to three-phase, six-pulse equipment, only two thirds of the original mAs is used to produce a similar radiograph ($\frac{2}{3} \times 20 = 13$ mAs). When changing from single-phase to three-phase, twelve-pulse equipment, only one half the original mAs is required. *(Cullinan, p 147)*

131. **(B)** Fluorescence is when an intensifying screen absorbs x-ray photon energy, emits light, and then ceases to emit light as soon as the energizing source ceases. Phosphorescence is when an intensifying screen absorbs x-ray photon energy, emits light, and continues to emit light for a short time after the energizing source ceases. Quantum mottle is the freckle-like appearance on some radiographs made using a very fast imaging system. *(Selman, p 277)*

132. **(C)** When the anatomic part is of greatly differing densities, a compensating filter is frequently helpful. Compensating filters can be accommodated for by tracks in the tube head. They can be wedge-shaped, the thinner part of the wedge paralleling the thinner body part, thus *compensating* for greater or lesser tissue densities (as in a large decubitus abdomen). A grid is used to absorb scatter radi-

ation before it reaches the film; intensifying screens amplify the action of x-rays; and a penetrometer (Al step wedge) is used to illustrate the effect of kVp on contrast. *(Selman, pp 396–397)*

133. **(D)** As screen speed is increased, exposure factors must be decreased in order to maintain the original film density. A change from 100 to 200 speed usually requires that the mAs be reduced by one half. If screen speed was changed from 400 to 200 speed, the mAs would need to be doubled. *(Cullinan, p 296)*

134. **(B)** Rare earth phosphors have a greater conversion efficiency than do other phosphors. Lanthanum oxybromide is a blue-emitting phosphor and gadolinium oxysulfide is a green-emitting phosphor. Cesium iodide is the phosphor used on the input screen of image intensifiers; it is not rare earth. *(Cullinan, p 99)*

135. **(A)** An increase in photon energy accompanies an increase in kilovoltage. Kilovoltage regulates the penetrability of x-ray photons; it regulates their wavelength—the amount of energy with which they are associated. The higher the related energy of an x-ray beam, the greater its penetrability (kV and photon energy are directly related; kV and wavelength are inversely related). Adjustments in kV have a big impact upon radiographic contrast: as kV (photon energy) is increased, the number of grays increases, thereby producing a longer scale of contrast. In general, as screen speed increases, so does contrast (resulting in a shorter scale of contrast). An increase in mAs is frequently accompanied by an appropriate decrease in kV, which would also shorten the contrast scale. SID and radiographic contrast are unrelated. *(Cullinan, p 121)*

136. **(D)** Because pneumoperitoneum is an abnormal accumulation of air or gas in the peritoneal cavity, it would require a decrease in exposure factors. Obstructed bowel usually involves distended, air- or gas-filled bowel loops, again requiring a decrease in exposure factors. With ascites there is an abnormal accumulation of fluid in the abdominal cavity, necessitating an increase in exposure factors. Renal colic is the pain associated with the passage of renal calculi; no change from the normal exposure factors is usually required. *(Carroll, pp 158–161)*

137. **(B)** Recorded detail is evaluated by how sharply tiny anatomic details are imaged (or "resolved") on the radiograph. This is termed *resolution*, and may be measured with a line-pair test pattern. The greater the number of line-pairs imaged the better the resolution. Intensification factor indicates exposure difference between non-screen exposure and screen exposure. Quality assurance is a system of testing, evaluation, and preventive maintenance employed to ensure consistent radiographic results and minimum patient exposure. *(Selman, p 326)*

138. **(A)** Locate density 1.5 on the vertical axis. Follow it across to where it intersects with film 1, then to where it intersects with film 2. At each intersection, follow the vertical line down and note the corresponding log relative exposure. Film 1 requires an exposure of about 1.6 to record a density of 1.5, while film 2 requires an exposure of about 1.8 to record the same density. Film 1 is clearly the faster film. The faster film always occupies the position furthest to the left in a comparison of two or more films. *(Cullinan, p 92)*

139. **(B)** Turnarounds are located at the bottom of each solution tank and function to direct the film from a downward to an upward motion. Guide shoes, also called deflector plates, serve to keep the film on its proper course by directing or guiding it around corners. The crossovers are located at the top of the processor, out of solution, and direct the film from one solution tank to the next. These are the racks that need daily cleaning to avoid chemical or emulsion build-up on their surface. *(Carroll, pp 472–474)*

140. **(A)** Limiting the *size of the irradiated field* is a most effective method of decreasing the production of scatter radiation. The smaller the *volume of tissue* irradiated, the smaller the

amount of scatter radiation produced; this can be accomplished using *compression* (prone position instead of supine). Use of a grid does not affect the production of scatter radiation, but rather removes it once it has been produced. *(Bushong, pp 235, 238)*

141. **(A)** A compensating filter is used when the part to be radiographed is of uneven thickness or density (in the chest, mediastinum versus lungs). The filter (made of aluminum or lead acrylic) is constructed so as to absorb much of the primary radiation that would expose the low-tissue-density area, while allowing the primary radiation to pass unaffected to the high-tissue-density area. A collimator is used to decrease the production of scatter radiation by limiting the volume of tissue irradiated. The grid functions to trap scatter radiation before it reaches the film, thus reducing scatter radiation fog. *(Bushong, pp 168–169)*

142. **(D)** The advantages of automatic processors are quicker, more efficient operation and consistent results. Quicker operation is attained with increased solution temperatures. The usual temperature of a 90-sec developer is 90 to 95°. Excessively high developer temperature can cause chemical fog. *(Bushong, p 212)*

143. **(A)** An increase in kilovoltage increases the overall average energy of the x-ray photons produced at the target, thus giving them greater penetrability. (This can increase the incidence of Compton interaction and therefore production of scatter radiation.) Greater penetrability of all tissues serves to lengthen the scale of contrast. However, excessive scatter radiation reaching the film will cause a fogged, gray appearance and carry no useful information. *(Selman, p 339)*

144. **(D)** Grids are frequently used today to improved visibility of detail in mammographic examinations. They are especially useful when examining breast tissue that is likely to produce more scatter radiation than usual. Large breasts unable to be compressed to a thickness of 6 cm or less, and particularly

dense breasts, are examples where a grid is especially useful. Because low kilovoltage is employed in mammography, only a low-ratio grid such as a 4:1 or 5:1 is generally used. The grid must be a moving grid, because a stationary grid would image very objectionable grid lines. *(Cullinan, p 207)*

145. **(B)** An *artifact* is an unnatural feature on a radiograph. Several kinds of artifacts can be produced from careless handling during production of a radiographic image. A crinkle mark appears as a *crescent-shaped artifact* and is usually a result of *bending* the film sharply, as over one's finger when removing from the cassette. Static electrical discharge causes a characteristic (usually black, branching, tree-like) artifact on radiographic film. Processor transport problems can cause excessive density and/or artifacts on the film. Excessive humidity can also be damaging to film, although no characteristic artifact is produced. *(Selman, p 309)*

146. **(B)** The mAs formula is mA × time = mAs. With two known factors, the third can be determined. To find the mAs that was originally used, substitute known values:

$$mA \times time = mAs$$
$$300 \times \tfrac{1}{5} = x$$
$$x = 300 \times \tfrac{1}{5}$$
$$x = 60 \text{ mAs}$$

We have increased the kilovoltage to 86, an increase of 15 percent, which has an effect similar to doubling the mAs. Therefore, only 30 mAs is now required due to the kV increase:

$$mA \times time = mAs$$
$$500x = 30$$
$$x = 0.06 \text{ sec}$$

You may check for correctness:

$$mA \times time = mAs$$
$$500 \times 0.06 \text{ sec} = 30 \text{ mAs}$$

(Selman, p 333)

147. **(D)** *Grid ratio is defined as the ratio of the height of the lead strip to the width of the interspace material*; the higher the lead strip, the more scatter radiation it will trap and the greater its efficiency. The greater the number of lead strips per inch, the thinner and less visible they will be. The function of a grid is to absorb scatter radiation in order to improve contrast. *(Selman, pp 367–370)*

148. **(A)** As the distance from focal spot to film (SID) increases, so does recorded detail. Because the part is being exposed by more perpendicular (less divergent) rays, less magnification and penumbral blur are produced. Though the best recorded detail is obtained using a long SID, the necessary increase in exposure factors and resulting increased patient exposure becomes a problem. An optimum 40 inches SID is used for most radiography, with the major exception being chest examinations. *(Selman, p 319)*

149. **(D)** The crescent-shaped *kink marks* seen on the radiographic image are caused by acutely bending the x-ray film, usually *after* exposure. It must be remembered that x-ray film emulsion is very sensitive to handling artifacts. Static electrical discharge often occurs from the friction of sliding film in or out of the cassette. Guide shoe marks are minus density scratches. Chemical fog is an overall gray appearance frequently caused by incorrect development. *(Saia, PREP, p 356)*

150. **(B)** Many cassettes have a thin lead foil layer behind the rear screen to absorb backscatter radiation that is energetic enough to exit the rear screen, strike the metal back, and bounce back to fog the image. In this way, the cassette's metal hinges or straps may be imaged in high kVp radiography. *(Selman, p 275)*

151. **(B)** A compensating filter is used to make up for widely differing tissue densities. For example, in the thoracic cavity it is difficult to obtain a satisfactory image of the mediastinum and lungs simultaneously without the use of a compensating filter to "even out densities." With this device, the chest is radiographed using mediastinal factors and a

trough-shaped filter (thicker laterally) is used to absorb excess photons that would overexpose the lungs. The middle portion of the filter lets the photons pass to the mediastinum almost unimpeded. Filters that absorb photons contributing to skin dose are the inherent and added filters. Compensating filtration is unrelated to elimination of scatter radiation or fluoroscopy. *(Selman, pp 396–398)*

152. (D) Focusing distance is the term used to specify the optimum source-image distance used with a particular focused grid. It is usually expressed as *focal range*, indicating the minimum and maximum SID workable with that grid. Distances lesser or greater can result in grid cutoff. Although proper distance is important in stereoradiography, chest, and magnification radiography, focusing distance is unrelated to them. *(Selman, pp 373–374)*

153. (D) Insufficient mA and/or exposure time will result in lack of radiographic density. Insufficient kVp results in underpenetration and excessive contrast. Insufficient SID, however, will result in increased exposure rate and radiographic *overexposure*. *(Selman, pp 331–333)*

154. (A) Grids are used in radiography to trap scatter radiation that would otherwise cause fog on the radiograph. *Grid ratio* is defined as the height of the lead strips to the distance between them. *Grid frequency* refers to the number of lead strips per inch. *Focusing distance* and *grid radius* are terms denoting the distance range with which a focused grid may be used. *(Selman, p 369)*

155. (A) Scatter radiation is produced as x-ray photons travel through matter, interact with atoms, and are scattered (change direction). If these scattered rays are energetic enough to exit the body, they will strike the film from all different angles. They therefore do not carry useful information and merely produce a flat, gray (low-contrast) fog over the image. Grid cut-off increases contrast and is caused by improper relationship between the x-ray tube and grid. *(Selman, p 183)*

156. (B) Perfect film/screen contact is essential to sharply recorded detail. Screen contact can be evaluated with a wire mesh test (Fig. 4–33). A spinning top test is used to evaluate timer accuracy and valve tube operation. A penetrometer (aluminum step wedge) is used to illustrate the effect of kV on contrast. A star pattern is used to measure resolving power of the imaging system. *(Bushong, p 232)*

157. (B) X-ray film emulsion is sensitive and requires proper handling and storage. It should be stored in a cool (40 to 60°) dry (40 to 60 percent humidity) place. Exposure to excessive temperatures or humidity can lead to film fog and loss of contrast. Static marks are a result of low humidity. *(Carroll, p 486)*

158. (C) The radiograph is an illustration of poor film–screen contact. Motion and scatter radiation fog can be ruled out because the blurriness is seen only in the apical region. Screen film contact is evaluated using a wire mesh which is placed on the questionable cassette and radiographed. Any areas of unsharpness

Figure 4–33. From the American College of Radiology Learning File. Courtesy of the ACR.

represent poor contact which can result from warped screens, foreign body in the cassette, or damaged cassette frame. *(Bushong, p 232)*

159. **(B)** Film that is unexposed and processed will not be completely clear. The blue-tinted base contributes a small measure of density. A small but measurable amount of exposure from background radiation may also be present; and processing itself produces a small amount of density from chemical fog. Together, this is expressed as base-plus fog and should never exceed 0.2 density. *(Carroll, p 504)*

160. **(D)** Perfect contact between the intensifying screens and film is essential in order to maintain image sharpness. Any separation between them allows diffusion of fluorescent light and subsequent blurriness and loss of detail. Screen/film contact can be diminished if the cassette frame is damaged and misshapen, if the front is warped, or if there is a foreign body between the screens elevating them. *(Bushong, p 232)*

161. **(C)** Iodine-base contrast material used in intravenous urography gives optimum opacification at 60 to 70 kVp. Use of higher kVp will negate the effect of the contrast medium. A lower contrast will be produced and poor visualization of the renal collecting system will result. GI and BE exams employ high kVp exposure factors (above 120 kVp) to see through the barium. In chest radiography, high kVp factors are preferred for maximum visualization of pulmonary vascular markings, made visible through long-scale contrast. *(Cullinan, p 147)*

162. **(D)** Every film emulsion has a particular base-plus fog, which should not exceed 0.2. Base density is a result of the manufacturing process (environmental radiation) and the blue tint added to the base to reduce glare. The remaining fog density is a result of the chemical development process, when exposed silver bromide grains are converted to black metallic silver. *(Carroll, pp 504–506)*

163. **(B)** One way to minimize scatter radiation reaching the film is to use optimum kilovolt-age; excessive kVp increases the production of scatter radiation. Close collimation is also important because the smaller the volume of irradiated material, the less scatter radiation is produced. Using compression bands or the prone position in a large abdomen has the effect of making the abdomen "thinner" and thus producing less scatter radiation. Low-ratio grids allow a greater percentage of scatter radiation to reach the film. Use of a *high*-ratio grid will clean up a greater amount of scatter radiation before it reaches the film. *(Cullinan, pp 68–69, 75)*

164. **(B)** The use of high kilovoltage increases penetration of the part being radiographed, and therefore a longer scale of contrast results. The higher the kVp used, the greater the margin for error, or exposure latitude. However, as kVp increases, there is a corresponding increase in the percentage of scatter radiation produced. *(Carroll, p 115)*

165. **(A)** The focal spot size selected will determine the amount of penumbra, or geometric blur, produced in the image. Different screen speeds will create differing degrees of light diffusion, affecting recorded detail. Object-image distance is responsible for image magnification, and hence, recorded detail. The mA is unrelated to recorded detail, affecting only the quantity of x-ray photons produced and thus the radiographic density. *(Selman, pp 319–327)*

166. **(B)** Because lead content increases as grid ratio increases, more scatter radiation (and primary radiation) is trapped before reaching the film. There are, therefore, fewer photons striking the film, with a resultant decrease in radiographic density as grid ratio increases. Scale of contrast would decrease with an increase in grid ratio. *(Selman, pp 367–371)*

167. **(B)** Developing agents change the exposed silver bromide crystals to black metallic silver, thus producing a manifest image. The fixer solution removes the unexposed silver bromide crystals from the emulsion and hardens the gelatin emulsion, thus ensuring permanence of the radiograph. *(Selman, p 307)*

168. **(B)** The illustrated chest radiograph has a very "black and white" appearance, possessing *short scale*, or *high*, contrast. Chest radiographs frequently will have this appearance if too low a kVp value is used. Chest radiographs should be adequately penetrated and demonstrate a relatively long range of gray tones. The opposite appearance is low contrast—a very flat, gray appearance sometimes caused by chemical or scattered radiation fog. *(Carroll, p 109)*

169. **(B)** Unopened film is sealed in a moisture-proof package to protect it from excessive humidity. The unopened bag will not protect it from radiation or excessive heat. Once the foil bag is open, it is much more susceptible to excessive humidity. *(Carroll, p 486)*

170. **(D)** According to the Inverse Square Law of Radiation, as the distance between the radiation source and film increases, the exposure rate decreases. Therefore, an increase in technical factors is indicated. The following formula is used to determine new mAs values, when changing distance:

$$\frac{mAs_1}{mAs_2} = \frac{D_1^2}{D_2^2}$$

Substituting known values:

$$\frac{20}{x} = \frac{36^2}{48^2}$$

$$\frac{20}{x} = \frac{1296}{2304}$$

$$1296x = 46,080$$

$$x = 35.5 \text{ mAs at a 48 inch SID}$$

(Selman, p 335)

171. **(B)** As intensifying screen speed increases, more fluorescent light is emitted from the phosphors. If more fluorescent light strikes the film emulsion, a greater number of silver halide crystals are changed to black metallic silver in the developer, and hence an increase in radiographic density. As intensifying screen speed increases, so does radiographic density. *(Cullinan, p 100)*

172. **(C)** An exposed radiographic film contains an invisible (latent) image. Only through processing can this image be converted to a permanent, visible (manifest) image. As the film exits the developer section, it passes through the crossover assembly, and before entering the fixer section it passes through the squeegee assembly. The squeegee assembly rollers function to remove excess developer solution from the emulsion before the film enters the fixer. This process helps maintain fixer strength/activity. The rate of travel through the processor is determined by the transport mechanism, ie, the speed of rollers as established at time of manufacture. *(Burns, p 49)*

173. **(B)** An increase in kilovoltage (photon energy) will result in a greater number of scattered photons (Compton interaction). These scattered photons carry no useful information and contribute to radiation fog, thus decreasing radiographic contrast. *(Selman, p 364)*

174. **(D)** *X-rays produced at the target comprise a heterogeneous primary beam.* There are many "soft," low-energy photons that, if not removed, would only contribute to greater patient dose. They do not have enough energy to penetrate the patient and expose the film; they just penetrate a small thickness of the patient's tissue and are absorbed. These photons are removed by aluminum filters. *(Carlton & Adler, p 57)*

175. **(B)** As the kilovoltage is increased, a *greater number* of electrons are driven across to the anode with *greater force*. Therefore as energy conversion takes place at the anode, *more high-energy* (short-wavelength) photons are produced. However, because they are higher-energy photons, there will be *less* patient absorption. *(Selman, p 364)*

176. **(C)** Diagnostic x-ray photons interact with tissue in a number of ways, but mostly they are involved in the production of *Compton scatter* or in the *photoelectric effect. Compton scatter* is pictured; it occurs when a relatively *high-energy* photon uses *some* of its energy to eject an *outer* shell electron. In doing so, the photon is deviated in direction and becomes a scattered photon. Compton scatter causes objectionable scatter radiation fog and poses a radiation hazard to personnel during procedures such as fluoroscopy. In the *photoelectric effect* a relatively *low*-energy x-ray photon uses *all* its energy to eject an *inner* shell electron, leaving a "hole" in the K shell. An L shell electron then drops down to fill the K vacancy, and in so doing emits a characteristic ray whose energy is equal to the difference between binding energies of the K and L shells. The photoelectric effect occurs with high-energy atomic absorbers such as bone and positive contrast media, and is responsible for the production of radiographic contrast. It is helpful for the production of the radiographic image, but contributes to the dose received by the patient (because it involves complete absorption of the incident photon). *(Saia, PREP, pp 195–197)*

177. **(C)** Grids are used in radiography to absorb scatter radiation before it reaches the film, thus improving radiographic contrast. Contrast obtained with a grid compared to contrast without a grid is termed contrast improvement factor. The greater the percentage of scatter radiation absorbed compared to absorbed primary radiation, the greater the "selectivity" of the grid. If a grid absorbs an abnormally large amount of primary radiation due to improper centering, tube angle, or tube distance, grid cutoff occurs. *(Selman, p 370)*

178. **(D)** When exposure rate decreases (as a result of increased SID), an appropriate increase in mAs is required in order to maintain the original radiographic density. Unless exposure is increased, the resulting radiograph will be underexposed. The formula used to determine the new mAs is:

$$\frac{mAs_1}{mAs_2} = \frac{D_1^2}{D_2^2}$$

Substituting known values:

$$\frac{3}{x} = \frac{40^2}{72^2}$$

$$\frac{3}{x} = \frac{1600}{5184}$$

$$1600x = 15{,}552$$

$$x = 9.72 \text{ mAs at a 72 inch SID}$$

(Selman, p 335)

179. **(D)** High-kilovoltage exposures produce large amounts of scatter radiation and therefore high-ratio grids are used in an effort to trap more of this scatter radiation. However, accurate centering and positioning become more critical in order to avoid grid cutoff. *(Selman, pp 379–380)*

180. **(B)** The *quantity* of x-ray photons produced at the target is the function of mAs. The *quality* (wavelength, penetration, energy) of x-ray photons produced at the target is the function of kVp. The kVp also has an effect on exposure rate, because an increase in kVp will increase the number of high-energy photons produced at the target. Exposure rate decreases with an increase in SID. *(Selman, pp 332–333)*

181. **(C)** The *Reciprocity Law* states that a particular mAs, regardless of the mA and exposure time used, will provide identical radiographic density. This holds true with direct exposure techniques, but does fail somewhat with the use of intensifying screens. However, the fault is so slight as to be unimportant in most radiographic procedures. *(Selman, p 333)*

182. **(D)** Luminescence is the production of energy in the form of light. Two types of luminescence are fluorescence and phosphorescence. Fluorescence occurs when an intensifying (radiographic) screen absorbs x-ray photon energy, emits light and ceases to emit light as soon as the energizing source

ceases. Fluoroscopic screens continue to emit light for a short time after the exposure has terminated. This characteristic (phosphorescence) is a desirable quality in fluoroscopic screens. Lag occurs when an intensifying (radiographic) screen continues to fluoresce after the x-ray stimulation has terminated. This characteristic is undesirable and causes excessive density. *(Selman, pp 276–277)*

183. (C) As kVp is increased, more *high-energy* photons are produced and the overall energy of the primary beam is increased. *Photon energy is inversely proportional to wavelength*: that is, as photon energy increases, wavelength decreases. An increase in mA serves to increase the number of photons produced at the target, but is unrelated to their energy. *(Selman, p 177)*

184. (C) The slower the screen speed the smaller the quantity of fluorescent light emitted during x-ray exposure. Therefore, slow-speed screens *require more x-ray exposure* to provide adequate radiographic density and cannot be used when exposure reduction or fast exposure time is essential. However, because they are associated with less diffusion of fluorescent light, they *produce better recorded detail* and are used to image structures requiring excellent recorded detail. *(Selman, pp 286–287)*

185. (C) Film base functions to support the silver halide emulsion. Today's film base is made of tough, nonflammable polyester. Cellulose nitrate was used in the past, but was highly flammable. Cellulose acetate was not flammable, but was not as durable as polyester. The earliest supports for emulsion were plates of glass (hence the term "flat plate"). *(Wallace, p 140)*

186. (C) Emphysema is abnormal distention of the pulmonary alveoli (or tissue spaces) with air. The presence of abnormal amounts of air makes it necessary to decrease from normal exposure factors to avoid excessive density. Congestive heart failure, pneumonia, and pleural effusion all involve abnormal amounts of fluid in the chest and would

therefore require an increase in exposure factors. *(Cullinan, pp 166–167)*

187. (A) The scatterer between the target and film is the patient. After the radiation has scattered once, it has been significantly attenuated. The intensity of scatter radiation 1 m from the patient is approximately 0.1 percent of the primary beam. *(Bushong, p 633)*

188. (B) Grids are used to absorb scatter radiation produced with high kilovoltage and tissues of greater thickness and density. They require an adjustment in exposure factors: two to six times the mAs used without a grid, and therefore a considerable increase in patient dose. Because they also absorb scatter radiation, they contribute to a higher radiographic contrast. *(Selman, pp 379–380)*

189. (B) Radiographic contrast is defined as the degree of difference between adjacent densities. These density differences represent sometimes very subtle differences in absorbing properties of adjacent body tissues. The type of film emulsion used also brings with it its own contrast characteristics. Different types of film emulsions have different degrees of contrast "built into" them chemically. The technical factor used to regulate contrast is kilovoltage. Radiographic contrast is unrelated to mAs. *(Cullinan, p 121)*

190. (A) If a radiograph lacks sufficient blackening, an increase in mAs is required. The mAs regulates the number of x-ray photons produced at the target. An increase or decrease of at least 30 percent in mAs is necessary to produce a perceptible effect. Increasing the kVp 15 percent will have about the same effect as *doubling* the mAs. *(Carlton & Adler, p 370)*

191. (D) Exposure rate is regulated by mA. Distance significantly affects the exposure rate, according to the Inverse Square Law of Radiation. Kilovoltage also has an effect on exposure rate, because an increase in kVp will increase the number of high-energy photons produced at the target. The thickness of the part determines the degree of attenuation the primary beam will undergo, but is unrelated to the original exposure rate. *(Selman, p 164)*

192. **(D)** *Static electricity* is a problem especially in cool, dry weather. *Sliding* the film in and out of the cassette can be the cause of a static electrical discharge. Removing one's sweater in the darkroom on a dry winter day can cause static electrical sparking. The film exposed by a large static discharge ("tree static") frequently exhibits black, branching artifacts such as those illustrated. Poor screen/film contact results in very blurry areas of the finished radiograph. A foreign body in the cassette will be sharply imaged on the finished radiograph. An inverted focused grid will result in an area of exposure down the middle of the film and grid cutoff everywhere else. *(Saia, PREP, pp 327, 357)*

193. **(A)** Beam restrictors function to limit size of the irradiated field. In doing so, they limit the volume of tissue irradiated (thereby decreasing the percentage of scatter radiation built up in the part) and help reduce patient dose. Beam restrictors do not affect the quality (energy) of the x-ray beam—that is the function of kVp and filtration. Beam restrictors do not absorb scatter radiation—that is a function of grids. *(Cullinan, pp 22–23)*

194. **(C)** The answer to this question conforms to the general rule that when two or more characteristic curves are being compared, the fastest film emulsion is the one furthest to the left. The one difference is that there are intersecting characteristic curves here. Simply see which curve is further to the left *above the intersection* (number 1) and which is further to the left below the intersection (number 2). As you can see, film 1 has more sensitivity (speed) above the point of intersection. *(Carroll, p 506)*

195. **(D)** Because breast tissue is of such uniform density, every effort must be made to enhance what little subject contrast does exist. High-contrast film helps to accomplish this. It is essential that any existing microcalcifications be clearly imaged; therefore, fine grain

is also an important emulsion characteristic. Mammography film is single emulsion, used with a single intensifying screen cassette, for best detail. *(Selman, p 273)*

196. **(B)** If the exposure time is cut in half, one would normally double the mA in order to maintain the same mAs and, consequently, the same radiographic density. However, the kVp may be increased by 15 percent to give a similar effect. For example, if the original kVp was 85, 15 percent of that is 13, and therefore the new kVp would be 98. The same percentage value would be used to cut the radiographic density in half (reduce kVp by 15 percent). *(Cullinan, p 137)*

197. **(D)** In the RAO position, the sternum must be visualized through the thorax and heart. Prominent pulmonary vascular markings can hinder good visualization. A method frequently used to overcome this problem is to use an mAs with a long exposure time. The patient is permitted to breathe normally during the (extended) exposure and by doing so, blurs out the prominent vascularities. *(Cullinan, p 157)*

198. **(B)** Higher than normal developing temperatures cause overdevelopment of the less-exposed silver halide crystals, producing chemical fog. The resulting radiograph will appear very gray, exhibiting loss of contrast and increased density. Wet, tacky films are usually due to lower than normal dryer temperature or developer underreplenishment. *(Selman, p 309)*

199. **(B)** Guide shoes are found at crossover and turnaround assemblies and function to direct the film around corners as it changes direction. If a guide shoe becomes misaligned, it will scratch the emulsion and leave the characteristic "guide shoe marks" running in the direction of film travel, as seen in the pictured radiograph. Pi lines appear as plus density lines running perpendicular to the direction of film travel and are sometimes seen in new processors or after a complete maintenance/overhaul. Hesitation marks are plus density lines occurring as a result of pauses,

or hesitations, in a faulty roller transport system. *(McKinney, p 91; Sweeney, pp 287, 291)*

200. **(D)** Motion is said to be the greatest enemy of recorded detail because it completely obliterates image sharpness. Screen speed can reduce recorded detail according to the degree of light diffusion from the phosphors. Object-image distance causes magnification and blurriness of recorded detail. Grid ratio is related to scatter radiation cleanup; it is unrelated to detail. *(Wallace, p 6)*

201. **(D)** Attenuation (weakening) of the x-ray beam is a result of its original energy and its interactions with different types and thicknesses of tissue. The greater the original energy (the higher the kV) of the incident beam, the less the attenuation. The greater the effective atomic number of the tissues (tissue type determines absorbing properties), the greater the beam attenuation. The greater the volume of tissue (subject density and thickness), the greater the beam attenuation. *(Cahoon, p 135)*

202. **(A)** Using the variable kVp method, a particular mAs is assigned to each body part. As part thickness increases, the kVp (penetration) is also increased. The body part being radiographed must be carefully measured, and for each centimeter of increase in thickness, 2 kVp is added to the exposure. *(Bushong, p 314)*

203. **(C)** The diffusion of fluorescent light from intensifying screens is responsible for loss of recorded detail on double-emulsion film. Therefore, by changing the system to include only one intensifying screen and single-emulsion film, as in mammographic systems, light diffusion is reduced and better recorded detail results. Patient dose is somewhat greater than with a two-screen cassette system, but the advantage of significantly improved recorded detail greatly offsets this. *(Selman, pp 327–328)*

204. **(A)** The Characteristic Curve is used to show the relationship between the exposure given the film and the resulting film density. It can therefore be used to evaluate a particular film emulsion's response (speed, sensitivity) by determining how long it takes to record a particular density. A dose–response curve is used in radiation protection and illustrates the quantity of dose required to produce a particular effect. The reciprocity law states that a particular mAs, regardless of the combination of mA and time, should produce the same degree of blackening. The Inverse Square Law illustrates the relationship between distance and radiation intensity. *(Burns, p 169)*

205. **(D)** The faster the imaging system, the greater the sacrifice of image clarity (recorded detail). As intensifying screen speed increases, recorded detail decreases. Perfect screen/film contact is essential for good detail. Any areas of poor contact result in considerable blurriness in the radiographic image. Penumbral blur is related to focal spot size; smaller focal spots produce less penumbra and thus, better recorded detail. *(Selman, p 283)*

206. **(C)** Film emulsion consists of silver halide crystals suspended in gelatin. Sodium sulfite is a film processing preservative and potassium bromide is a developer restrainer. Potassium and chrome alum are emulsion hardeners used in fixer solution. *(Selman, p 271)*

207. **(C)** Recorded detail is evaluated by how sharply tiny anatomic details are imaged on the radiograph. The area of blurriness that may be associated with small image details is termed *penumbra* or geometric blur. The blurriness can be produced using a large focal spot, or from diffused fluorescent light from intensifying screens. The image proper (ie, without penumbra) is termed *umbra*. *Mottle* is a grainy appearance caused by fast imaging systems. *(Selman, p 320)*

208. **(C)** A grid is composed of alternate strips of lead and interspace material. The lead strips serve to trap scatter radiation before it fogs the film. The interspace material must be radiolucent and is usually plastic or sturdier aluminum. Cardboard was formerly used as interspace material but had the disadvantage of being affected by humidity (moisture). *(Selman, p 365)*

209. **(D)** The lateral skull radiograph pictured illustrates the result of great care taken by the radiographer to secure patient comfort. A skull examination was requested for this patient of advanced years. The radiographer positioned the patient, taking care to position and collimate accurately, all the while trying to ensure the patient's comfort by letting her keep her pillow! The foam stuffing of the pillow is nicely imaged. Although the artifacts somewhat resemble Paget's disease, note that they extend outside the collimated field. Safelight fog would be shown as a more uniform "blanket" of fog. *(Sweeney, p 362)*

210. **(B)** Because lead content increases and grid ratio increases, a greater amount of scatter radiation is trapped before reaching the film. Fewer grays are therefore recorded, and a shorter scale of contrast results. Radiographic density would decrease with an increase in grid ratio. *(Selman, pp 370–371)*

211. **(B)** X-ray film emulsion becomes more sensitive to safelight fog following exposure to fluorescent light from intensifying screens. Care must be taken not to leave exposed film on the darkroom workbench for any length of time, as its sensitivity to safelight fog is now greatly heightened. *(Saia, PREP, p 328)*

212. **(B)** The distance between transport rollers in an automatic processor is extremely critical and allows for the exact film thickness with minimum emulsion swelling. If the emulsion is allowed to swell excessively (from excessive temperature or inadequate replenishment), the emulsion will stick to the rollers and cause a processor jam-up. Glutaraldehyde is a hardener added to the developer to keep the emulsion swelling to a minimum. Unexposed silver halide crystals are removed in the fixer solution. *(Bushong, p 208)*

213. **(B)** X-ray film emulsion is sensitive and requires proper handling and storage. Several kinds of artifacts can be produced by careless handling. Tree-like, black branching marks are usually due to static electrical discharge especially prevalent during cold, dry weather. Crinkle marks are produced by acute bending of the film. Excessive emulsion swelling and chemical fog can be caused by excessive solution temperature or insufficient replenishment. *(Bushong, p 200)*

214. **(B)** Because the focal spot (track) of an x-ray tube is along the anode's beveled edge, photons produced at the target are able to diverge considerably toward the cathode end of the tube but are absorbed by the heel of the anode at the opposite end of the tube. This results in a greater number of x-ray photons distributed toward the cathode end, and is known as the *anode heel effect*. The effect of this restricting heel is most pronounced when the x-ray photons are required to *diverge* more, as would be the case with *short source-image distance, large size films,* and *steeper (smaller) target angles. (Saia, PREP, p 297)*

215. **(D)** If developer temperature is too high, some of the lesser-exposed or unexposed silver halide crystals may be reduced, thus creating chemical fog. If the developer solution has become oxidized from exposure to air, chemical fog also results. If developer replenisher is insufficient, inadequate new solution is replacing the deteriorated developer and chemical fog is again the result. *(McKinney, p 190)*

216. **(C)** Radiographic density is greatly affected by changes in the source-image distance, as expressed by the Inverse Square Law of Radiation. As distance from the radiation source increases, exposure rate decreases and radiographic density decreases. Exposure rate is inversely proportional to the square of the SID. Aluminum filtration, kilovoltage, and scatter radiation all have a significant effect on density, but are not the primary controlling factors. *(Selman, pp 333–334)*

217. **(D)** The archival quality of a film refers to its ability to retain its image for a long period of time. Many states have laws governing how long a patient's medical records, including films, must be retained. Very importantly, they must be retained in their original condition. Archival quality is poor if radiographic films begin to show evidence of stain after being stored for only a short time. Probably the most common cause of stain, and hence poor archival quality, is retained fixer within the emulsion. Fixer may be retained due to poor washing, or because there was insufficient hardener (underreplenishment) in the developer, thus permitting fixer to be retained by the swollen emulsion. A test for quantity of retained fixer in film emulsion is often included as part of a quality control program. Stain may also be caused by poor storage conditions. Storage in a hot, humid place will cause even the smallest amount of retained fixer to react with silver, causing stain. *(McKinney, p 79)*

218. **(D)** Shape distortion is caused by misalignment of the x-ray tube, part to be radiographed, and film. An object can be falsely imaged (foreshortened, elongated) by incorrect placement of tube, body part, or film. Only one of the three need be misaligned for distortion to occur. *(Bushong, p 283)*

219. **(D)** *Radiographic contrast is the sum of film contrast and subject contrast.* Subject contrast has by far the greatest influence on radiographic contrast. Several factors influence subject contrast, each as a result of beam attenuation differences in the irradiated tissues. As patient thickness and tissue density increase, attenuation increases and subject contrast is increased. As kilovoltage increases, higher-energy photons are produced, beam attenuation is decreased, and subject contrast decreases. *(Selman, p 337)*

220. **(A)** If the source-image distance is above or below the recommended focusing distance, the primary beam at the lateral edges will not coincide with the angled lead strips. Consequently, there will be absorption of the primary beam, termed grid cutoff. If the grid failed to move during the exposure, there would be grid lines throughout. Central ray angulation in the direction of the lead strips is appropriate and will not cause grid cutoff.

If the central ray was off center, there would be uniform loss of density. *(Pizzutiello & Cullinan, p 61)*

221. **(C)** A radiograph described as possessing short-scale contrast possesses relatively few different radiographic densities. It most likely presents black areas, a couple shades of gray, and many white areas. The number of different shades is few. The differences between the shades of densities is rather striking, and thus the radiograph possesses high contrast—more contrast than a radiograph having many shades of grays (described as low or less contrast). Film latitude refers to the radiographic contrast obtained as a result of the chemical properties of the film emulsion. *(Selman, p 338)*

222. **(D)** According to the Inverse Square Law of Radiation, the intensity or exposure rate of radiation from its source is inversely proportional to the square of the distance. Thus, as distance from the source of radiation increases, exposure rate decreases. Because exposure rate and radiographic density are directly proportional, if the exposure rate of a beam directed to a film is decreased, the resultant radiographic density would be decreased proportionally. *(Selman, p 333)*

223. **(A)** Proper use of focal spot size is of paramount importance in magnification radiography. A magnified image that is diagnostic can be obtained only by using a fractional focal spot of 0.3 mm or smaller. The amount of penumbra or geometric unsharpness produced by focal spots larger in size render the radiograph undiagnostic. *(Selman, p 352)*

224. **(B)** The abdomen is a thick structure containing many structures of similar density, and thus requires increased exposure and a grid to absorb scatter radiation. The lumbar spine and hip are also dense structures requiring increased exposure and use of a grid. The knee, however, is frequently small enough to be radiographed without a grid. The general rule is that structures measuring more than 10 cm should be radiographed with a grid. *(Saia, PREP, p 282)*

225. (A) Grids are composed of alternate strips of lead and interspace material and are used to trap scatter radiation after it emerges from the patient and before it reaches the film. Accurate centering of the x-ray tube is required. If the x-ray tube is off center, but within the recommended focusing distance, there will be an overall loss of density. Over- or underexposure under the anode is usually the result of exceeding the focusing distance limits in addition to being off center. *(Pizzutiello & Cullinan, p 61)*

226. (A) The line focus principle is a geometric principle illustrating that the *actual focal spot is larger than the effective (projected) focal spot.* The actual focal spot (target) is larger, to accommodate heat over a larger area, and angled so as to *project* a smaller focal spot, thus maintaining recorded detail by reducing penumbra. The relationship between the exposure given the film and the resulting density is expressed in the Inverse Square Law. The relationship between the kVp used and the resulting contrast may be illustrated using an aluminum step wedge. *(Bushong, p 124)*

227. (B) Osteoporosis is a condition often seen in the elderly, marked by increased porosity and softening of bone. Bones are much less dense, and thus a decrease in exposure is required. Osteomyelitis and osteochondritis are inflammatory conditions usually having no effect on bone density. Osteosclerosis is abnormal hardening of the bone, and an increase in exposure factors would be required. *(Laudicina, pp 138, 145, 159)*

228. (D) As the kilovoltage is decreased, x-ray beam energy (that is, penetration) is also decreased. Consequently, a shorter scale of contrast is obtained and, at lower kV levels, there is less exposure latitude (less margin for error in exposure). As kV is reduced, the mAs must be increased accordingly in order to maintain adequate density. This increase in mAs results in greater patient dose. *(Cahoon, p 135)*

229. (C) Film A was made using a 5:1 ratio grid at 180 mAs. Film B was made using a 12:1 ratio

grid using 400 mAs. A grid is the most effective way to remove scattered photons from those exiting the patient. Grids are designed to selectively absorb scattered radiation while absorbing as little of the primary radiation as possible. A radiographic grid is composed of alternate strips of lead and a radiolucent material such as plastic, fiber, or aluminum. It is positioned between the patient and image receptor; the radiolucent strips are aligned parallel with the x-ray beam (Fig. 4–34). Most of the primary photons will pass through the radiolucent strips. Because the direction of the scattered photons is different from that of the primary photons, the scattered photons will strike the sides of the lead strips and be absorbed. However, because the grid does absorb some primary radiation, and because there is a big decrease in scattered radiation, a significant increase in exposure to the patient is required in order to produce adequate radiographic density. *(Wallace, pp 112–113)*

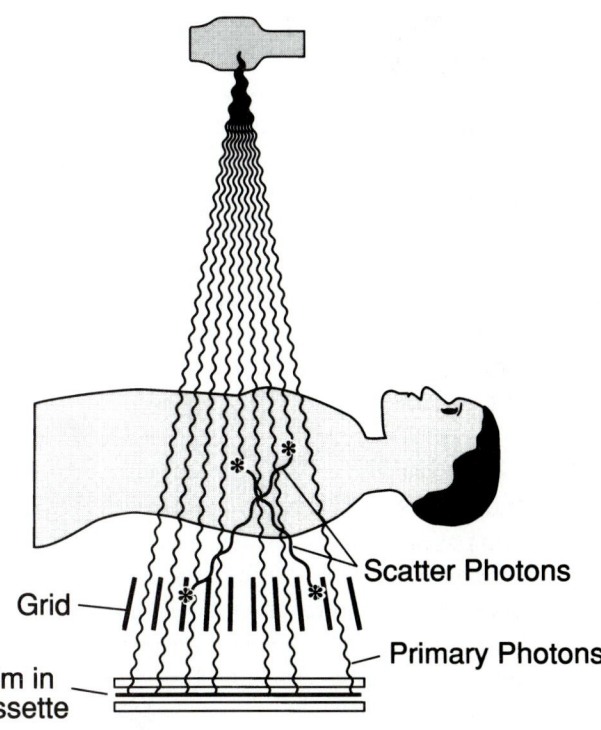

Figure 4–34. Reproduced with permission from Wolbarst AB. *Physics of Radiology.* East Norwalk, CT: Appleton & Lange, 1993.

230. **(D)** Geometric unsharpness is affected by all three factors listed. As object-image distance increases, so does magnification. As focal object and source-image distance decrease, so does magnification. Object-image distance may be said to be directly proportional to magnification. Focal object and source-image distance are inversely proportional to magnification. (*Selman, pp 319–321*)

231. **(C)** Significant scatter radiation is produced when radiographing large or dense body parts and when using high kilovoltage. A radiographic grid is made of alternate lead strips and interspace material, and is placed between the patient and film to absorb energetic scatter emerging from the patient. Although a grid prevents much scatter radiation fog from reaching the radiograph, its use does necessitate a significant increase in exposure to the patient. (*Bushong, pp 234–235*)

232. **(D)** A quality assurance program includes regular overseeing of all components of the imaging system: film and cassettes, processor, x-ray equipment, and so on. With regular maintenance, testing, and repairs, equipment should operate efficiently and consistently. In turn, radiographic quality will be consistent and repeat exposures will be minimized, thereby reducing patient exposure. (*Cullinan, p 272*)

233. **(D)** The darkroom must be constructed so as to be free from any white light leaks. The film bin should be secure and have a sign warning against opening in white light. Safelight bulbs must be of the correct wattage, and the filter should be appropriate for the type film used. (*Selman, pp 290–291*)

234. **(C)** Every box of film comes with a noted expiration date. Film used after the expiration date will usually suffer loss of speed and contrast and will exhibit fog. Film should be ordered in quantities that will ensure use before becoming outdated, and should be rotated in storage so that the oldest is used first. (*Selman, p 274*)

235. **(D)** The illustrated radiograph exhibits *rotation*. The distances between the lateral border of the orbits and outer margin of the skull should be symmetrical. All the cranial and facial structures are adequately penetrated, demonstrating a diagnostic scale of contrast and adequate background density. (*Ballinger, vol 2, p 245*)

236. **(B)** As the film emulsion is exposed to light or x-rays, latent image formation takes place. The exposed silver halide crystals are reduced to black metallic silver in the developer solution. Automatic processor developer agents are hydroquinone and phenidone. The preservative—sodium sulfite or cycon—helps prevent oxidation. The activator provides the necessary alkalinity for the developer solution, and hardener is added to developer in automatic processing to keep emulsion swelling to a minimum. (*Bushong, p 206*)

237. **(D)** *The shortest possible exposure time should be used to minimize motion unsharpness.* Motion causes unsharpness that destroys detail. Careful and accurate patient instruction is essential for minimizing voluntary motion. Suspended respiration eliminates respiratory motion. Using the shortest possible exposure time is essential to decreasing involuntary motion. Immobilization is also very useful in eliminating motion unsharpness. (*Selman, p 327*)

238. **(D)** Compression of the breast tissue during mammographic imaging improves the technical quality of the image for several reasons. Compression brings breast structures in closer contact with the film, thus reducing geometric blur and improving detail. As the breast tissue is compressed, and essentially becomes thinner, less scatter radiation is produced. Compression serves as excellent immobilization as well. (*Cullinan, pp 206–207*)

239. **(D)** Distortion is caused by improper alignment of the tube, body part, and film. Anatomic structures within the body are rarely parallel to the film in a simple recumbent position. In an attempt to overcome this distortion, we position the part to be parallel with the film, or angle the central ray to "open up" the part. Examples of this technique are obliquing the pelvis to place the ilium parallel to the film, or angling the central ray cephalad in order to "open up" the sigmoid colon. *(Bushong, pp 282–286)*

240. **(B)** As the distance from the object to the film (OID) increases, so does magnification distortion. Some magnification is inevitable in radiography, as it is not possible to place anatomic structures directly on the film. However, our understanding of how to minimize magnification distortion is an important part of our everyday work. *(Saia, PREP, p 252)*

241. **(B)** Rare earth phosphors have a much higher conversion efficiency (and therefore speed) and have all but replaced the older calcium tungstate screens. The larger the phosphor and thicker the layer of phosphors (active layer), the greater the light emission and therefore speed. Antihalation backing is a component of single-emulsion film that prevents "crossover" of fluorescent light within a cassette. *(Selman, pp 299–301)*

242. **(B)** Emphysema is abnormal distention of alveoli (or tissue spaces) with air. The presence of abnormal amounts of air makes it necessary to decrease from normal exposure factors. Congestive heart failure and pleural effusion involve abnormal amounts of fluid in the chest and thus require an increase in exposure factors. *(Saia, PREP, p 267)*

243. **(D)** Occasionally, very short exposure times are required to "stop motion" when patients are unable to cooperate. One problem encountered is that sometimes the Bucky diaphragm is not as fast as the exposure time; it is unable to oscillate rapidly enough to blur out the grid lines and is "captured in motion." Focal spot "blooming" is an actual increase in the size of the focal spot as a result of the excessive heat produced using high mA stations. Additionally, automatic exposure devices are sometimes unable to "time" very short exposures. The AED's minimum response time may be shorter than the required exposure time (that is, the shortest time it is able to give may be longer than that required), resulting in an excessively long exposure and excessive radiographic density. *(Cullinan, pp 22–25)*

244. **(C)** As kVp is reduced, the number of high-energy photons produced at the target is reduced; therefore, a decrease in radiographic density occurs. If a grid has been used improperly (off-centering or out of focal range), the lead strips will absorb excessive amounts of primary radiation, resulting in grid cut-off and loss of radiographic density. If the source-image distance is inadequate, excessive radiographic density will occur. *(Carroll, p 296)*

245. **(A)** The mAs is the exposure factor governing radiographic density. Using the equation mA time = mAs, determine each mAs: A = 30 mAs; B = 20 mAs; C = 12 mAs; D = 18 mAs. Group A will produce the greatest radiographic density. *(Selman, p 333)*

246. **(B)** Magnification radiography is used to enlarge details to a more perceptible degree. Hairline fractures and minute blood vessels are candidates for magnification radiography. The problem of magnification unsharpness is overcome by using a fractional focal spot; larger focal spot sizes will produce excessive penumbral unsharpness. Grids are usually unnecessary in magnification radiography because of the air gap effect produced by the object-image distance. Direct exposure technique would not likely be used because of the excessive exposure required. *(Selman, p 352)*

247. **(D)** As photon energy (kVp) increases, so does the production of scatter radiation. The greater the density of the irradiated tissues, the greater the production of scatter radiation. As the size of the irradiated field increases, radiation has a greater opportunity for collision and the percentage of scatter again increases. Beam restriction is the single most important way to limit the amount of scatter radiation produced. *(Selman, pp 386–387)*

248. **(A)** A 15-minute oblique film of an intravenous urogram is pictured. Intravenous urography requires the use of iodinated contrast media. Low kilovoltage (about 70) is usually employed to enhance the photoelectric effect and, in turn, better visualize the renal collecting system. High kilovoltage will produce excessive scatter radiation and obviate the effect of the contrast agent. A higher mA with a shorter exposure time is preferred in order to decrease the possibility of motion. *(Cullinan, p 147)*

249. **(C)** Radiographers are usually able to stop voluntary motion using suspended respiration, careful instruction, and immobilization. However, *involuntary* motion must also be considered. In order to have a "stop action" effect on the heart when radiographing the chest, it is essential to use a short exposure time. *(Cullinan, pp 165–166)*

250. **(D)** The large vertical foreign body seen in the figure cannot be located either within the cassette or between the cassette and patient because its edges indicate significant *magnification* blur. The structure is located between the patient and x-ray tube—it is an IV pole that was not moved from within the x-ray field! Note, however, the neck chain that the radiographer conscientiously requested the patient to hold in the mouth out of the way of anatomic structures! *(Saia, PREP, p 358)*

Subspecialty List

61. Selection of technical factors: contrast
62. Film processing and quality assurance
63. Selection of technical factors: contrast
64. Evaluation of radiographs
65. Film processing and quality assurance
66. Film processing and quality assurance
67. Film processing and quality assurance
68. Film processing and quality assurance
69. Selection of technical factors: film, screen, and grid selection
70. Film processing and quality assurance
71. Selection of technical factors: film, screen, and grid selection
72. Film processing and quality assurance
73. Selection of technical factors: contrast
74. Selection of technical factors: film, screen, and grid selection
75. Selection of technical factors: film, screen, and grid selection
76. Film processing and quality assurance
77. Selection of technical factors: manual versus AEC
78. Selection of technical factors: film, screen, and grid selection
79. Evaluation of radiographs
80. Selection of technical factors: recorded detail
81. Evaluation of radiographs
82. Selection of technical factors: density
83. Film processing and quality assurance
84. Film processing and quality assurance
85. Evaluation of radiographs
86. Evaluation of radiographs
87. Selection of technical factors: recorded detail
88. Evaluation of radiographs
89. Selection of technical factors: film, screen, and grid selection
90. Evaluation of radiographs
91. Film processing and quality assurance
92. Film processing and quality assurance
93. Selection of technical factors: density
94. Selection of technical factors: density
95. Selection of technical factors: density
96. Evaluation of radiographs
97. Selection of technical factors: density
98. Selection of technical factors: recorded detail
99. Selection of technical factors: recorded detail
100. Film processing and quality assurance
101. Selection of technical factors: density
102. Selection of technical factors: technique charts
103. Selection of technical factors: technique charts
104. Film processing and quality assurance
105. Selection of technical factors: recorded detail
106. Selection of technical factors: recorded detail
107. Evaluation of radiographs
108. Selection of technical factors: contrast
109. Selection of technical factors: film, screen, and grid selection
110. Selection of technical factors: density
111. Selection of technical factors: film, screen, and grid selection
112. Film processing and quality assurance
113. Selection of technical factors: density
114. Selection of technical factors: film, screen, and grid selection
115. Selection of technical factors: film, screen, and grid selection
116. Selection of technical factors: film, screen, and grid selection
117. Selection of technical factors: recorded detail
118. Film processing and quality assurance
119. Evaluation of radiographs
120. Film processing and quality assurance
121. Selection of technical factors: contrast
122. Film processing and quality assurance
123. Selection of technical factors: recorded detail
124. Selection of technical factors: recorded detail
125. Film processing and quality assurance
126. Selection of technical factors: contrast
127. Film processing and quality assurance
128. Selection of technical factors: film, screen, and grid selection
129. Film processing and quality assurance
130. Evaluation of radiographs
131. Selection of technical factors: film, screen, and grid selection
132. Selection of technical factors: density
133. Selection of technical factors: film, screen, and grid selection
134. Selection of technical factors: film, screen, and grid selection
135. Selection of technical factors: contrast
136. Selection of technical factors: technique charts
137. Selection of technical factors: recorded detail
138. Selection of technical factors: film, screen, and grid selection
139. Film processing and quality assurance
140. Selection of technical factors: contrast
141. Selection of technical factors: contrast
142. Film processing and quality assurance
143. Selection of technical factors: contrast
144. Selection of technical factors: film, screen, and grid selection

145. Film processing and quality assurance
146. Evaluation of radiographs
147. Selection of technical factors: film, screen, and grid selection
148. Selection of technical factors: recorded detail
149. Film processing and quality assurance
150. Film processing and quality assurance
151. Selection of technical factors: density
152. Selection of technical factors: film, screen, and grid selection
153. Selection of technical factors: density
154. Selection of technical factors: film, screen, and grid selection
155. Selection of technical factors: contrast
156. Selection of technical factors: film, screen, and grid selection
157. Film processing and quality assurance
158. Evaluation of radiographs
159. Selection of technical factors: film, screen, and grid selection
160. Film processing and quality assurance
161. Selection of technical factors: contrast
162. Film processing and quality assurance
163. Evaluation of radiographs
164. Selection of technical factors: contrast
165. Selection of technical factors: recorded detail
166. Selection of technical factors: film, screen, and grid selection
167. Film processing and quality assurance
168. Evaluation of radiographs
169. Film processing and quality assurance
170. Selection of technical factors: density
171. Selection of technical factors: density
172. Film processing and quality assurance
173. Selection of technical factors: contrast
174. Selection of technical factors: contrast
175. Selection of technical factors: contrast
176. Evaluation of radiographs
177. Selection of technical factors: film, screen, and grid selection
178. Evaluation of radiographs
179. Selection of technical factors: film, screen, and grid selection
180. Selection of technical factors: density
181. Selection of technical factors: density
182. Selection of technical factors: film, screen, and grid selection
183. Selection of technical factors: contrast
184. Selection of technical factors: contrast
185. Selection of technical factors: film, screen, and grid selection
186. Film processing and quality assurance
187. Selection of technical factors: technique charts
188. Selection of technical factors: contrast
189. Selection of technical factors: film, screen, and grid selection
190. Selection of technical factors: contrast
191. Selection of technical factors: density
192. Selection of technical factors: density
193. Evaluation of radiographs
194. Selection of technical factors: technique charts
195. Selection of technical factors: film, screen, and grid selection
196. Selection of technical factors: film, screen, and grid selection
197. Selection of technical factors: contrast
198. Film processing and quality assurance
199. Film processing and quality assurance
200. Selection of technical factors: recorded detail
201. Selection of technical factors: contrast
202. Selection of technical factors: technique charts
203. Selection of technical factors: film, screen, and grid selection
204. Film processing and quality assurance
205. Selection of technical factors: recorded detail
206. Film processing and quality assurance
207. Selection of technical factors: recorded detail
208. Selection of technical factors: contrast
209. Evaluation of radiographs
210. Selection of technical factors: film, screen, and grid selection
211. Film processing and quality assurance
212. Film processing and quality assurance
213. Film processing and quality assurance
214. Selection of technical factors: density
215. Film processing and quality assurance
216. Selection of technical factors: density
217. Film processing and quality assurance
218. Selection of technical factors: recorded detail
219. Selection of technical factors: contrast
220. Selection of technical factors: film, screen, and grid selection
221. Selection of technical factors: contrast
222. Selection of technical factors: density
223. Selection of technical factors: recorded detail
224. Selection of technical factors: film, screen, and grid selection
225. Selection of technical factors: film, screen, and grid selection
226. Selection of technical factors: recorded detail
227. Selection of technical factors: technique charts
228. Selection of technical factors: contrast

229. Evaluation of radiographs
230. Selection of technical factors: recorded detail
231. Selection of technical factors: film, screen, and grid selection
232. Film processing and quality assurance
233. Film processing and quality assurance
234. Film processing and quality assurance
235. Evaluation of radiographs
236. Selection of technical factors: film, screen, and grid selection
237. Selection of technical factors: recorded detail
238. Selection of technical factors: contrast
239. Selection of technical factors: recorded detail
240. Selection of technical factors: recorded detail
241. Selection of technical factors: film, screen, and grid selection
242. Selection of technical factors: technique charts
243. Selection of technical factors: manual versus AEC
244. Selection of technical factors: density
245. Selection of technical factors: density
246. Selection of technical factors: recorded detail
247. Selection of technical factors: contrast
248. Selection of technical factors: technique charts
249. Selection of technical factors: recorded detail
250. Evaluation of radiographs

CHAPTER 5

Equipment Operation and Maintenance
Questions

DIRECTIONS (Questions 1 through 150): Each of the numbered items or incomplete statements in this section is followed by answers or by completions of the statement. Select the ONE lettered answer or completion that is BEST in each case.

1. The advantages of large-format spot film cameras, such as 100 mm and 105 mm, over smaller-format cameras such as 70 mm and 90 mm, include

    1. improved image quality
    2. decreased patient dose
    3. decreased x-ray tube heat load

    (A) 1 only
    (B) 1 and 2 only
    (C) 2 and 3 only
    (D) 1, 2, and 3

2. How is the thickness of the tomographic section related to the tomographic angle?

    (A) the greater the tomographic angle, the thicker the section
    (B) the greater the tomographic angle, the thinner the section
    (C) the lesser the tomographic angle, the thinner the section
    (D) tomographic angle is unrelated to section thickness

3. Which of the following must be selected by the radiographer when using the automatic exposure control?

    1. photocell(s)
    2. backup time
    3. minimum response time

    (A) 1 only
    (B) 1 and 2 only
    (C) 1 and 3 only
    (D) 1, 2, and 3

4. Which of the following fluoroscopic recording methods is most likely to result in the greatest patient dose?

    (A) 35-mm cine, 60 frames/sec
    (B) 16-mm cine, 30 frames/sec
    (C) 105-mm spot film
    (D) cassette-loaded spot film

5. Which of the following circuit devices operate(s) on the principle of self-induction?

    1. autotransformer
    2. choke coil
    3. high-voltage transformer

    (A) 1 only
    (B) 1 and 2 only
    (C) 2 and 3 only
    (D) 1, 2, and 3

6. Together, the filtering effect of the x-ray tube's glass envelope and its oil coolant are referred to as

(A) inherent filtration
(B) added filtration
(C) compensating filtration
(D) port filtration

7. Which of the following will occur as a result of a decrease in the anode target angle?

1. anode heel effect will be less pronounced
2. effective focal spot size will decrease
3. greater photon intensity toward the cathode side of the x-ray tube

(A) 1 only
(B) 1 and 2 only
(C) 2 and 3 only
(D) 1, 2, and 3

8. Which of the wave forms illustrated in Figure 5–1 represents single-phase full-wave equipment?

(A) figure 1
(B) figure 2
(C) figure 3
(D) figure 4

9. The energy of electrons comprising the tube current is measured in

(A) keV
(B) kVp
(C) mA
(D) mAs

10. Proper care of leaded apparel includes

1. periodic check for cracks
2. careful folding following each use
3. routine laundering with soap and water

(A) 1 only
(B) 1 and 2 only
(C) 1 and 3 only
(D) 1, 2, and 3

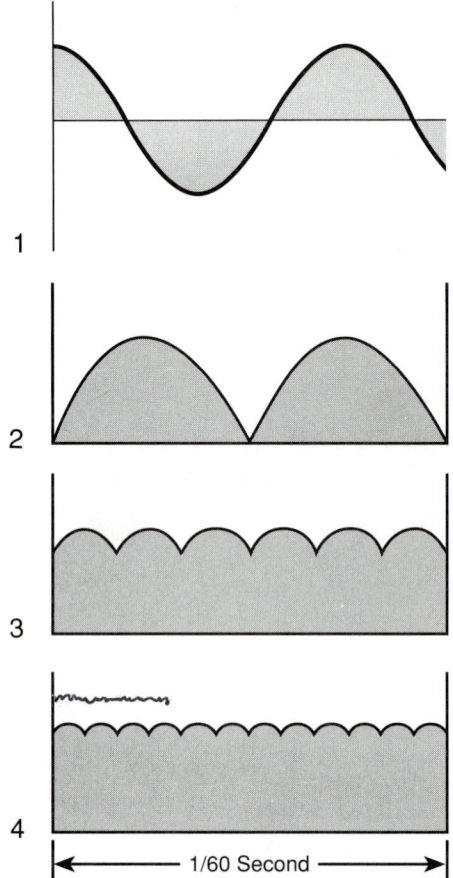

Figure 5–1. Reproduced with permission from Wolbarst AB. *Physics of Radiology.* East Norwalk, CT: Appleton & Lange, 1993.

11. Inadequate collimation, when using automatic exposure control, often results in

1. an underexposed radiographic image
2. an overexposed radiographic image
3. the timer not terminating soon enough

(A) 1 only
(B) 1 and 3 only
(C) 3 only
(D) 2 and 3 only

12. The source of electrons within the x-ray tube is a result of which of the following processes?

(A) electrolysis
(B) thermionic emission
(C) rectification
(D) photosynthesis

13. A three-phase timer can be tested for accuracy using a synchronous spinning top. The resulting image looks like a

(A) series of dots or dashes, each representative of a radiation pulse

(B) solid arc, the angle (in degrees) representative of the exposure time

(C) series of gray tones, from white to black

(D) multitude of small mesh-like squares of uniform sharpness

14. Which of the following statements regarding transformer laws is (are) correct?

1. the voltage and current values are increased with a step-up transformer

2. the voltage is directly related to the number of turns in the two coils

3. the product of voltage and current in the two circuits must be equal

(A) 1 only

(B) 1 and 2 only

(C) 2 and 3 only

(D) 1, 2, and 3

15. X-ray tube life may be extended by

1. using low mAs/high kVp exposure factors

2. avoiding lengthy anode rotation

3. avoiding exposures to a cold anode

(A) 1 only

(B) 1 and 2 only

(C) 1 and 3 only

(D) 1, 2, and 3

16. The device used to give a predetermined exposure to a film in order to test its response to processing is called a

(A) sensitometer

(B) densitometer

(C) step wedge

(D) spinning top

17. The total number of x-ray photons produced at the target is contingent upon

1. tube current

2. target material

3. square of the kilovoltage

(A) 1 only

(B) 1 and 2 only

(C) 2 and 3 only

(D) 1, 2, and 3

18. If an AP hip technique required 92 kVp with single-phase x-ray equipment, what kVp would produce a comparable image with three-phase equipment?

(A) 69 kVp

(B) 75 kVp

(C) 81 kVp

(D) 103 kVp

19. Fractional focus tubes, with 0.3 mm focal spot or smaller, have special application in

(A) magnification radiography

(B) fluoroscopy

(C) tomography

(D) image intensification

20. Which of the following combinations will present the greatest heat-loading capability?

(A) 17° target angle, 1.2 mm actual focal spot

(B) 10° target angle, 1.2 mm actual focal spot

(C) 17° target angle, 0.6 mm actual focal spot

(D) 10° target angle, 0.6 mm actual focal spot

21. Which of the following electrical devices will prevent equipment damage in the event of circuit overload?

(A) resistor

(B) rheostat

(C) rectifier

(D) circuit breaker

22. Which of the following devices is (are) components of a typical fluoroscopic video display system?

 1. videotape recorder
 2. TV camera
 3. TV monitor

(A) 1 only
(B) 1 and 3 only
(C) 2 and 3 only
(D) 1, 2, and 3

23. If the primary coil of the high-voltage transformer is supplied by 220 volts and has 200 turns, and the secondary coil has 100,000 turns, what is the voltage induced in the secondary coil?

(A) 40 kilovolts
(B) 110 kilovolts
(C) 40 volts
(D) 110 volts

24. In the production of Bremsstrahlung radiation, the incident electron

(A) ejects an inner-shell tungsten electron
(B) ejects an outer-shell tungsten electron
(C) is deflected with resulting energy loss
(D) is deflected with resulting energy increase

25. What is the relationship between kilovoltage (kV) and half value layer (HVL)?

(A) as kV increases, HVL increases
(B) as kV decreases, HVL decreases
(C) if the kV is doubled, the HVL doubles
(D) if the kV is doubled, the HVL is squared

26. Which of the following techniques is used to evaluate the dynamics of a part?

(A) fluoroscopy
(B) stereoscopy
(C) tomography
(D) phototiming

27. In which of the following portions of the x-ray circuit is a step-down transformer located?

(A) high-voltage side
(B) filament circuit
(C) rectification system
(D) secondary side

28. Which of the following is (are) essential to high-quality mammographic examinations?

 1. small focal spot x-ray tube
 2. high radiographic contrast
 3. use of a compression device

(A) 1 only
(B) 1 and 2 only
(C) 1 and 3 only
(D) 1, 2, and 3

29. With three-phase equipment, the voltage across the x-ray tube

 1. drops to zero every 180°
 2. is 87 to 96 percent of the maximum value
 3. is nearly constant potential

(A) 1 only
(B) 2 only
(C) 1 and 2 only
(D) 2 and 3 only

30. The photoelectric process is an interaction between an x-ray photon and

(A) an inner-shell electron
(B) an outer-shell electron
(C) a nucleus
(D) another photon

31. Which of the following devices converts electrical energy into mechanical energy?

(A) motor
(B) generator
(C) stator
(D) rotor

32. The basic function of the phototimer is to

(A) provide a brighter fluoroscopic image
(B) automatically restrict the field size
(C) terminate the x-ray exposure once the film is correctly exposed
(D) automatically increase or decrease incoming line voltages

33. A device used to regulate the amount of resistance in a circuit is the

(A) autotransformer
(B) rheostat
(C) circuit breaker
(D) transformer

34. If one rectifier is malfunctioning, the resultant mAs would be

(A) one fourth of that expected
(B) one half of that expected
(C) twice that expected
(D) four times that expected

35. A spinning top device can be used to evaluate

1. timer accuracy
2. rectifier failure
3. effect of kVp on contrast

(A) 1 only
(B) 2 only
(C) 1 and 2 only
(D) 1, 2, and 3

36. Which of the following will improve the spatial resolution of image-intensified images?

1. a very thin coating of cesium iodide on the input phosphor
2. a small input phosphor
3. increased total brightness gain

(A) 1 only
(B) 1 and 2 only
(C) 1 and 3 only
(D) 1, 2, and 3

37. Which of the following voltage ripples is (are) produced by three-phase equipment?

1. 100 percent voltage ripple
2. 13 percent voltage ripple
3. 3.5 percent voltage ripple

(A) 1 only
(B) 2 only
(C) 2 and 3 only
(D) 1, 2, and 3

38. Decreased x-ray tube output can be attributed to

1. roughened focal track
2. tungsten deposits on the glass envelope
3. high-speed anode rotation

(A) 1 only
(B) 2 only
(C) 1 and 2 only
(D) 1, 2, and 3

39. Which of the following occurs during Bremsstrahlung radiation production?

(A) an electron makes a transition from an outer to an inner electron shell
(B) an electron approaching a positive nuclear charge changes direction and loses energy
(C) a high-energy photon ejects an outer-shell electron
(D) a low-energy photon ejects an inner-shell electron

40. When a pair of intensifying screens are mounted inside a cassette, a thicker screen may be mounted

(A) inside the front of the cassette
(B) inside the lid (rear) of the cassette
(C) with a water-soluble paste
(D) using a more pliable adhesive

41. Methods of recording the fluoroscopic image include

    1. spot films
    2. cinefluorography
    3. videotape

    (A) 1 only
    (B) 1 and 2 only
    (C) 2 and 3 only
    (D) 1, 2, and 3

42. The x-ray tube's negative electrode is the

    (A) anode
    (B) cathode
    (C) diode
    (D) triode

43. All of the following devices may be located on the typical x-ray unit control panel, EXCEPT

    (A) mA meter
    (B) kVp selector
    (C) timer
    (D) filament ammeter

44. As the x-ray tube filament ages, it becomes progressively thinner due to evaporation. The vaporized tungsten is frequently deposited on the window of the glass envelope. This may

    1. act as an additional filter
    2. reduce tube output
    3. result in arcing and tube puncture

    (A) 1 only
    (B) 1 and 2 only
    (C) 2 and 3 only
    (D) 1, 2, and 3

45. The rear intensifying screen within a cassette may intentionally be made somewhat thicker to

    (A) compensate for photon absorption within the front screen
    (B) improve resolution capability of the screens
    (C) reduce the amount of light reflectance
    (D) minimize "crossover"

46. A film emerging from the automatic processor exhibits excessive density. This may be attributed to

    1. overexposure
    2. overdevelopment
    3. overreplenishment

    (A) 1 only
    (B) 1 and 2 only
    (C) 2 and 3 only
    (D) 1, 2, and 3

47. Poor screen/film contact can be caused by

    1. damaged cassette frame
    2. foreign body in cassette
    3. warped cassette front

    (A) 1 only
    (B) 2 only
    (C) 1 and 3 only
    (D) 1, 2, and 3

48. The method used to record successive physiologic events occurring during fluoroscopy is

    (A) cassette spot films
    (B) cinefluorography
    (C) photofluorography
    (D) image-intensified fluoroscopy

49. If exposure factors of 85 kVp, 400 mA, and 0.12 sec yield an output exposure of 150 mR, what is the mR/mAs?

    (A) 0.32 mR/mAS
    (B) 3.1 mR/mAs
    (C) 17.6 mR/mAs
    (D) 31 mR/mAs

50. In the production of characteristic radiation at the tungsten target, the incident electron

(A) ejects an inner-shell tungsten electron
(B) ejects an outer-shell tungsten electron
(C) is deflected with resulting energy loss
(D) is deflected with resulting energy increase

51. An automatic exposure device can operate on which of the following principles?

1. a photomultiplier tube charged by a fluorescent screen
2. a parallel plate ionization chamber charged by x-ray photons
3. motion of magnetic fields inducing current in a conductor

(A) 1 only
(B) 2 only
(C) 1 and 2 only
(D) 1, 2, and 3

52. Which of the following circuit devices must be connected in parallel?

(A) filament ammeter
(B) milliammeter
(C) voltmeter
(D) rectifiers

53. Which of the following is (are) characteristics of the x-ray tube?

1. target material should have a high atomic number and melting point
2. the useful beam emerges from the port window
3. the filament receives both low and high voltages

(A) 1 only
(B) 2 only
(C) 1 and 2 only
(D) 1, 2, and 3

54. Which of the following quantities of filtration is MOST likely to be used in mammography?

(A) 0.5 mm Mo
(B) 1.5 mm Al
(C) 1.5 mm Cu
(D) 2.0 mm Cu

55. The advantage(s) of collimators over aperture diaphragms and cones and cylinders include(s)

1. a variety of field sizes available
2. illuminated field with exact center indicated
3. clean up of scatter radiation

(A) 1 only
(B) 1 and 2 only
(C) 1 and 3 only
(D) 2 and 3 only

56. Dedicated radiographic units are available for

1. chest radiography
2. head radiography
3. mammography

(A) 1 only
(B) 2 only
(C) 1 and 2 only
(D) 1, 2, and 3

57. What is the device that directs the light emitted from the image intensifier to various viewing and imaging apparatus?

(A) output phosphor
(B) beam splitter
(C) spot film changer
(D) automatic brightness control

58. All of the following is (are) associated with the anode, EXCEPT

(A) line focus principle
(B) heel effect
(C) focal track
(D) thermionic emission

59. Which of the following contribute(s) to inherent filtration?

1. x-ray tube glass envelope
2. x-ray tube port window
3. aluminum between tube housing and collimator

(A) 1 only
(B) 1 and 2 only
(C) 1 and 3 only
(D) 1, 2, and 3

60. A quality control program includes checks on which of the following radiographic equipment conditions?

1. reproducibility
2. linearity
3. positive beam limitation

(A) 1 only
(B) 1 and 2 only
(C) 1 and 3 only
(D) 1, 2, and 3

61. The functions of the autotransformer include(s) providing

1. a selection of kilovoltage values at the control panel
2. voltage to the filament circuit
3. voltage to the primary coil of the step-up transformer

(A) 1 only
(B) 1 and 2 only
(C) 1 and 3 only
(D) 1, 2, and 3

62. The advantages of spot film cameras over cassette-loaded spot films include

1. decreased patient dose
2. decreased fluoro tube heat load
3. shortened exam time

(A) 1 only
(B) 1 and 2 only
(C) 1 and 3 only
(D) 1, 2, and 3

63. If the distance from the focal spot to the center of the collimator's mirror is 6 inches, what distance should the illuminator's light bulb be from the center of the mirror?

(A) 3 inches
(B) 6 inches
(C) 9 inches
(D) 12 inches

64. Which of the following systems function(s) to compensate for changing patient/part thicknesses during fluoroscopic procedures?

1. automatic brightness control
2. automatic gain control
3. automatic resolution control

(A) 1 only
(B) 1 and 2 only
(C) 2 and 3 only
(D) 1, 2, and 3

65. Disadvantages of moving grids over stationary grids include which of the following?

1. they can prohibit the use of very short exposure times
2. they increase patient radiation dose
3. they can cause phantom images when anatomic parts parallel their motion

(A) 1 only
(B) 1 and 2 only
(C) 2 and 3 only
(D) 1, 2, and 3

66. Which of the following devices is used to control voltage by varying resistance?

(A) autotransformer
(B) high-voltage transformer
(C) rheostat
(D) fuse

67. Which of the following imaging techniques is associated with the term "object plane?"

   (A) image intensification
   (B) cinefluorography
   (C) stereoradiography
   (D) tomography

68. How many half-value layers will it take to reduce an x-ray beam, whose intensity is 78 R/min, to an intensity of less than 10 R/min?

   (A) 2
   (B) 3
   (C) 4
   (D) 8

69. Double-focus x-ray tubes have two

   (A) port windows
   (B) anodes
   (C) filaments
   (D) rectifiers

70. The image intensifiers input phosphor differs from the output phosphor in that the input phosphor

   (A) is much larger than the output phosphor
   (B) emits electrons, and the output phosphor emits light photons
   (C) absorbs electrons, and the output phosphor absorbs light protons
   (D) is a fixed size, and the output phosphor size can vary

71. A mobile C-arm unit can provide

   1. fluoroscopic images
   2. dynamic images
   3. static images

   (A) 1 only
   (B) 1 and 2 only
   (C) 2 and 3 only
   (D) 1, 2, and 3

72. Periodic equipment calibration includes testing of the

   1. focal spot
   2. mA
   3. kVp

   (A) 1 only
   (B) 1 and 3 only
   (C) 2 and 3 only
   (D) 1, 2, and 3

73. Several types of exposure timers may be found on x-ray equipment. Which of the following types of timers functions to accurately duplicate radiographic densities?

   (A) synchronous
   (B) impulse
   (C) electronic
   (D) phototimer

74. Advantages of 105-mm strip, or 100-mm cut, spot filming over cassette spot filming include

   1. less patient exposure
   2. lower film cost
   3. practical storage

   (A) 1 only
   (B) 1 and 2 only
   (C) 2 and 3 only
   (D) 1, 2, and 3

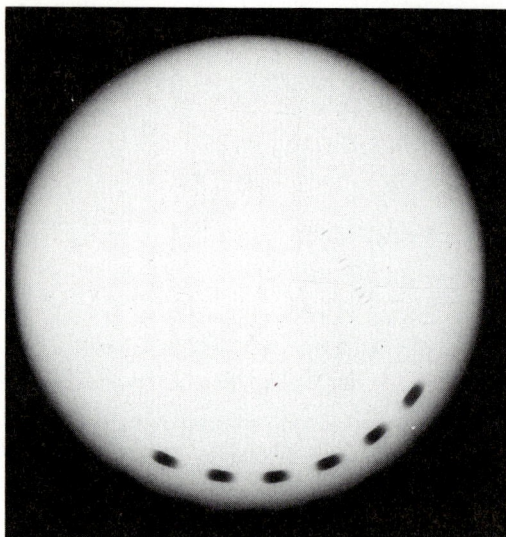

Figure 5–2. Courtesy of The Stamford Hospital, Department of Radiology.

75. The radiograph illustrated in Figure 5–2 was made using a single-phase, full-wave rectified unit with a timer and rectifiers known to be accurate and functioning correctly. What exposure time was used to produce this image?

(A) ⅒ sec
(B) 0.05 sec
(C) ½ sec
(D) 0.025 sec

76. Radiation-sensitive automatic exposure devices are known as

(A) automatic beam restrictors
(B) ionization chambers
(C) sensors
(D) backup timers

77. Advantages of battery-powered mobile x-ray units include their

1. ability to store a large quantity of energy
2. ability to store energy for extended periods of time
3. lightness and ease of maneuverability

(A) 1 only
(B) 1 and 2 only
(C) 2 and 3 only
(D) 1, 2, and 3

78. The total brightness gain of an image intensifier is a result of

1. flux gain
2. minification gain
3. focusing gain

(A) 1 only
(B) 2 only
(C) 1 and 2 only
(D) 1 and 3 only

79. The type(s) of radiation produced at the target is (are)

1. photoelectric
2. characteristic
3. Bremsstrahlung

(A) 1 only
(B) 1 and 2 only
(C) 2 and 3 only
(D) 1, 2, and 3

80. Which of the following tests is performed to evaluate screen contact?

(A) spinning top test
(B) wire mesh test
(C) penetrometer test
(D) star pattern test

81. The minimum response time of an automatic exposure device

(A) is the time required to energize the intensifying phosphors
(B) is its shortest possible exposure time
(C) functions to protect the patient from overexposure
(D) functions to protect the tube from excessive heat

82. A star pattern is used to measure

    1. focal spot resolution
    2. intensifying screen resolution
    3. SID resolution

    (A) 1 only
    (B) 1 and 2 only
    (C) 1 and 3 only
    (D) 1, 2, and 3

83. The procedure whose basic operation involves reciprocal motion of the x-ray tube and film is

    (A) cinefluorography
    (B) spot filming
    (C) tomography
    (D) image intensification

84. In order to be used more efficiently by the x-ray tube, alternating current is changed to unidirectional current by the

    (A) filament transformer
    (B) autotransformer
    (C) high-voltage transformer
    (D) rectifiers

85. All of the following x-ray circuit devices are located between the incoming power supply and the primary coil of the high-voltage transformer, EXCEPT

    (A) timer
    (B) kV meter
    (C) mA meter
    (D) autotransformer

86. In order to maintain image clarity, the path of electron flow from photocathode to output phosphor is controlled by

    (A) the accelerating anode
    (B) electrostatic lenses
    (C) the vacuum glass envelope
    (D) the input phosphor

87. The largest application of cinefluorography is with

    (A) GI examinations
    (B) ERCP procedures
    (C) cardiac catheterization
    (D) surgical radiography

88. Characteristic x-rays are produced when

    (A) high-speed electrons are attracted by a tungsten nucleus and decelerated
    (B) orbital electrons move from an outer shell to fill an inner-shell vacancy
    (C) orbital electrons move from an inner shell to fill an outer-shell vacancy
    (D) an outer-shell electron is ejected from orbit

89. The device used to test the accuracy of the x-ray timer is the

    (A) densitometer
    (B) sensitometer
    (C) penetrometer
    (D) spinning top

90. Which of the following meters will NOT register until the exposure is made?

    (A) kVp major meter
    (B) kVp minor meter
    (C) mA meter
    (D) line voltage meter

91. Automatic exposure devices (AEDs) function to

    1. reduce patient dose
    2. reduce repeat rates
    3. achieve consistent long scale contrast

    (A) 1 only
    (B) 1 and 2 only
    (C) 2 and 3 only
    (D) 1, 2, and 3

92. Features of x-ray tube targets that function to determine heat capacity include

1. rotation of the anode
2. diameter of the anode
3. size of the focal spot

(A) 1 only
(B) 1 and 2 only
(C) 1 and 3 only
(D) 1, 2, and 3

93. The voltage ripple associated with a three-phase, twelve-pulse rectified generator is about

(A) 100 percent
(B) 32 percent
(C) 13 percent
(D) 3 percent

94. Which of the following information is NOT necessary to determine the maximum safe exposure using a radiographic tube rating chart?

(A) the type rectification
(B) focal spot size
(C) anode rotation speed
(D) source–image distance

95. The contrast improvement factor of a grid will be greatest with

(A) high ratio grids with many lines per inch
(B) low ratio grids with many lines per inch
(C) high ratio grids with fewer lines per inch
(D) low ratio grids with fewer lines per inch

96. Circuit devices that will conduct electrons in only one direction are

1. resistors
2. valve tubes
3. solid-state diodes

(A) 1 only
(B) 1 and 3 only
(C) 2 and 3 only
(D) 1, 2, and 3

97. The brightness level of the fluoroscopic image can vary with

1. milliamperage
2. kilovoltage
3. patient thickness

(A) 1 only
(B) 1 and 2 only
(C) 1 and 3 only
(D) 1, 2, and 3

98. Which of the following information is necessary to determine the maximum safe kVp, using the appropriate x-ray tube rating chart?

1. mA and exposure time
2. focal spot size
3. imaging system speed

(A) 1 only
(B) 1 and 2 only
(C) 2 and 3 only
(D) 1, 2, and 3

99. Exposures less than the minimum response time of an automatic exposure device may be required when

1. using high mA
2. using fast film/screen combinations
3. examining large patients or body parts

(A) 1 only
(B) 1 and 2 only
(C) 2 and 3 only
(D) 1, 2, and 3

100. Cassettes frequently have a lead foil layer behind the rear screen, which functions to

(A) improve penetration
(B) absorb backscatter
(C) preserve resolution
(D) increase the screen speed

101. During cinefluorography, the camera shutter is open

(A) between frames
(B) when the film moves
(C) during exposures
(D) all the time

102. If a high-voltage transformer has 100 primary turns, 35,000 secondary turns, and is supplied by 220 volts and 75 amps, what is the secondary voltage and current?

(A) 200 amps and 77 volts
(B) 200 mA and 77 kVp
(C) 20 amps and 77 volts
(D) 20 mA and 77 kVp

103. Rare earth phosphors that may be used in intensifying screens include

1. cesium iodide
2. gadolinium oxysulfide
3. lanthanum oxybromide

(A) 1 only
(B) 1 and 2 only
(C) 2 and 3 only
(D) 1, 2, and 3

104. Electrical devices that allow current to flow easily in only one direction are called

(A) resistors
(B) rheostats
(C) rectifiers
(D) transformers

105. Radiation that passes through the tube housing in directions other than that of the useful beam is termed

(A) scattered radiation
(B) secondary radiation
(C) leakage radiation
(D) remnant radiation

106. Which of the following refers to a regular program of evaluation that ensures proper functioning of x-ray equipment, thereby protecting both patients and radiation workers?

(A) sensitometry
(B) densitometry
(C) quality assurance
(D) modulation transfer function

107. The most commonly used types of photocells are the

1. ion chamber
2. photomultiplier tube
3. cathode ray tube

(A) 1 and 2 only
(B) 1 and 3 only
(C) 2 and 3 only
(D) 1, 2, and 3

108. A light-absorbing dye is frequently incorporated during the manufacture of screens to

(A) reduce the diffusion of fluorescent light
(B) increase film contrast
(C) increase screen speed
(D) increase the useful life of the screen

109. As electrons impinge on the target surface, more than 99 percent of their kinetic energy is changed to

(A) x-rays
(B) heat
(C) gamma rays
(D) recoil electrons

110. The number 2 in Figure 5–3 indicates the

(A) anode
(B) anode stem
(C) focal spot/track
(D) cathode

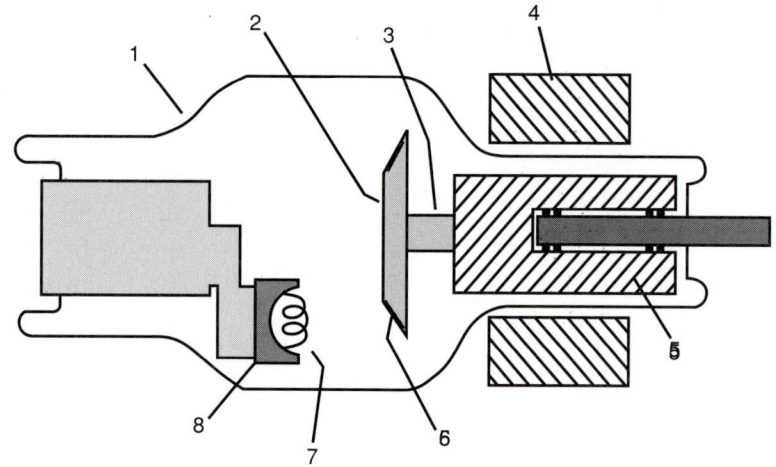

Figure 5–3. Reproduced with permission from Wolbarst AB. *Physics of Radiology.* East Norwalk, CT: Appleton & Lange, 1993.

111. The number *8* in Figure 5–3 indicates the

(A) filament

(B) focusing cup

(C) focal track

(D) cathode stem

112. Which of the following are features of fluoroscopic equipment, designed especially to eliminate unnecessary radiation to patient and personnel?

1. protective curtain
2. filtration
3. focal spot

(A) 1 only

(B) 1 and 2 only

(C) 1 and 3 only

(D) 1, 2, and 3

113. Which of the following terms describes the amount of electric charge flowing per second?

(A) voltage

(B) current

(C) resistance

(D) capacitance

114. Capacitor-discharge mobile x-ray units

1. use a grid-controlled x-ray tube
2. are typically charged before the day's work
3. provide a constant potential output

(A) 1 only

(B) 2 only

(C) 1 and 3 only

(D) 1, 2, and 3

115. Which of the following functions to protect the x-ray tube and patient from overexposure in the event the phototimer fails to terminate an exposure?

(A) circuit breaker

(B) fuse

(C) backup timer

(D) rheostat

116. The transition of orbital electrons from outer to inner shells gives rise to

(A) Compton scatter

(B) pair production

(C) Bremsstrahlung radiation

(D) characteristic radiation

117. Which of the following imaging modalities will, in general, require the MOST radiation exposure to a given part?

(A) cassette spot films

(B) 105-mm strip spot films

(C) 35-mm cine

(D) 16-mm cine

118. Off-focus radiation may be minimized by

(A) avoiding the use of very high kilovoltages

(B) restricting the x-ray beam as close to its source as possible

(C) using compression devices to reduce tissue thickness

(D) avoiding extreme collimation

119. When the radiographer selects kilovoltage on the control panel, which device is adjusted?

(A) transformer

(B) autotransformer

(C) filament circuit

(D) rectifier circuit

120. A steep, or small, target angle will have an effect on the

1. severity of the heel effect
2. focal spot size
3. heat load capacity

(A) 1 only

(B) 2 only

(C) 1 and 2 only

(D) 1, 2, and 3

121. Incorrect relationship between the primary beam and the center of a focused grid results in

1. an increase in scatter radiation production
2. grid cutoff
3. insufficient radiographic density

(A) 1 only

(B) 1 and 2 only

(C) 2 and 3 only

(D) 1, 2, and 3

122. When using the smaller field in a dual-field image intensifier

1. a smaller patient area is viewed
2. the image is magnified
3. the image is less bright

(A) 1 only

(B) 1 and 3 only

(C) 2 and 3 only

(D) 1, 2, and 3

123. A backup timer for the automatic exposure device serves to

1. protect the patient from overexposure
2. protect the x-ray tube from excessive heat
3. increase or decrease master density

(A) 1 only

(B) 1 and 2 only

(C) 2 and 3 only

(D) 1, 2, and 3

124. Which of the following combinations would pose the LEAST hazard to a particular anode?

(A) 1.2 mm focal spot, 92 kVp, 1.5 mAs

(B) 0.6 mm focal spot, 80 kVp, 3 mAs

(C) 1.2 mm focal spot, 70 kVp, 6 mAs

(D) 0.6 mm focal spot, 60 kVp, 12 mAs

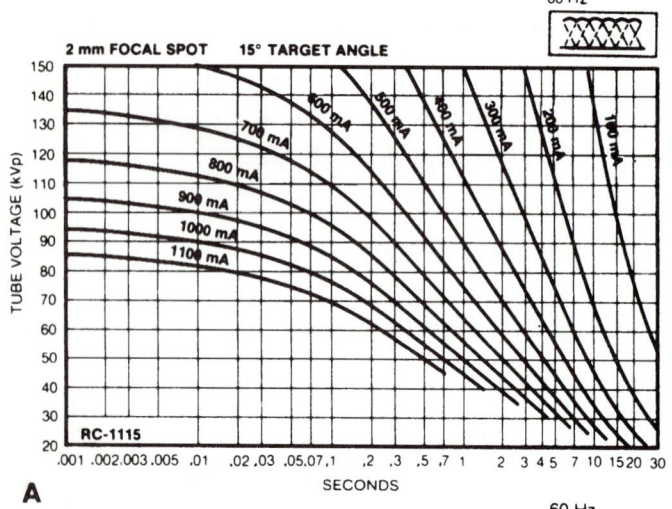

A

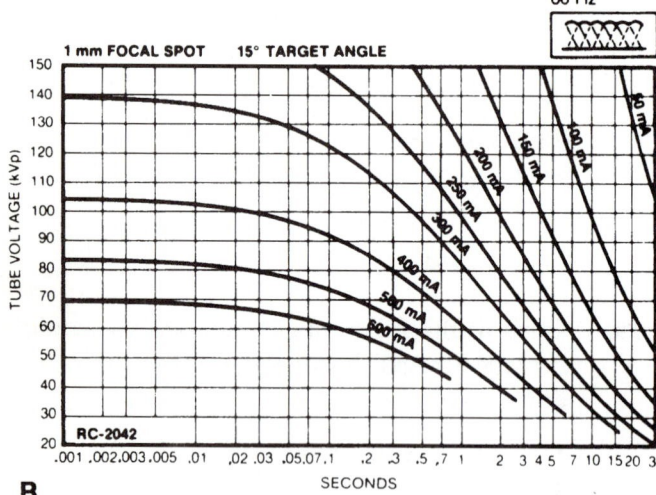

B

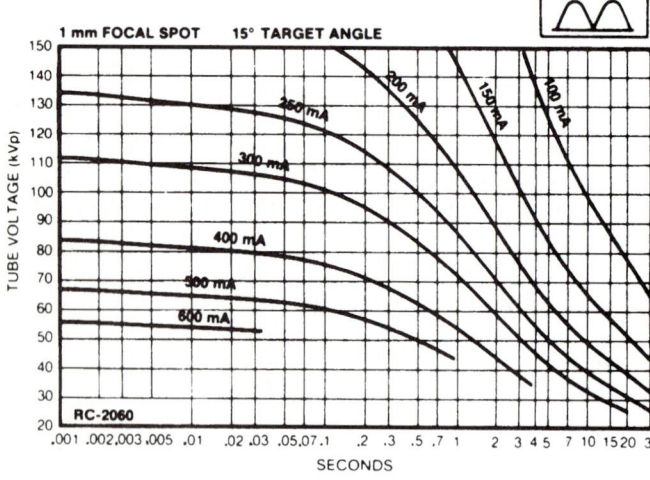

C

Figure 5–4. Reproduced with permission from Dunlee Tech Data Publication 50014 5/88.

125. In the radiographic rating charts shown in Figure 5–4, what is the maximum safe kVp that may be used with the 1-mm focal spot, single-phase x-ray tube, using 300 mA and 1/50 sec exposure?

(A) 80 kVp

(B) 95 kVp

(C) 105 kVp

(D) 112 kVp

126. In Figure 5–4, which of the illustrated x-ray tubes permit(s) an exposure of 400 mA, 0.1 sec, and 80 kVp?

1. tube A

2. tube B

3. tube C

(A) 1 only

(B) 1 and 2 only

(C) 2 and 3 only

(D) 1, 2, and 3

127. In Figure 5–4, what is the maximum safe mA that may be used with 0.1 sec exposure and 120 kVp, using the three-phase, 2-mm focal spot x-ray tube?

(A) 400 mA

(B) 500 mA

(C) 600 mA

(D) 700 mA

128. Referring to the anode cooling chart in Figure 5–5, if the anode has accumulated 250,000 heat units, how long will it take for it to cool to 100,000 heat units?

(A) 5 min

(B) 6.5 min

(C) 8 min

(D) 12.5 min

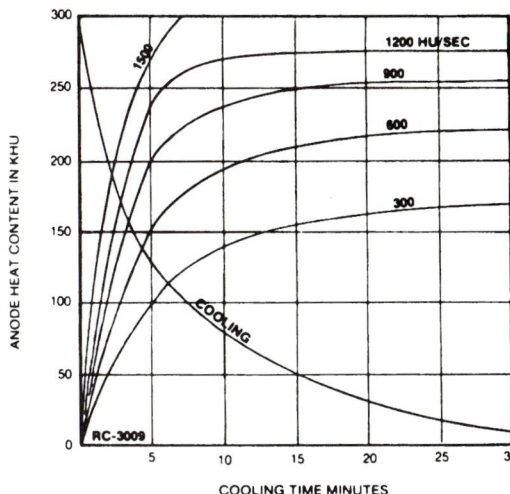

Figure 5–5. Reproduced with permission from Dunlee Tech Data Publication 50014 5/88.

129. The device used to ensure reproducible radiographs, regardless of tissue density variations, is the

(A) phototimer
(B) penetrometer
(C) grid
(D) rare earth screen

130. If 92 kilovolts and 12 mAs were used for a particular abdominal exposure with single-phase equipment, what mAs would be required with three-phase, six-pulse equipment to produce a similar radiograph?

(A) 36 mAs
(B) 24 mAs
(C) 8 mAs
(D) 6 mAs

131. Light-sensitive automatic exposure devices are known as

(A) phototimers
(B) ionization chambers
(C) sensors
(D) back-up timers

132. Delivery of large exposures to a cold anode, or use of exposures exceeding tube limitation, can result in

1. increased tube output
2. cracking of the anode
3. rotor bearing damage

(A) 1 only
(B) 1 and 2 only
(C) 2 and 3 only
(D) 1, 2, and 3

133. Tungsten alloy is the usual choice of target material for radiographic equipment because it

1. has a high atomic number
2. has a high melting point
3. can readily dissipate heat

(A) 1 only
(B) 1 and 2 only
(C) 2 and 3 only
(D) 1, 2, and 3

134. To determine how quickly an x-ray tube will disperse its accumulated heat, the radiographer uses a(n)

(A) technique chart
(B) radiographic rating chart
(C) anode cooling curve
(D) spinning top test

135. Which of the following formulas would the radiographer use to determine the total number of heat units produced with a given exposure using three-phase, six-pulse equipment?

(A) mA × time × kVp
(B) mA × time × kVp × 3.0
(C) mA × time × kVp × 1.35
(D) mA × time × kVp × 1.41

136. A high-speed electron entering the tungsten target is attracted to the positive nucleus of a tungsten atom and, in the process, is decelerated. This results in

(A) characteristic radiation
(B) Bremsstrahlung radiation
(C) Compton scatter
(D) photoelectric effect

137. Radiographs from a particular single-phase, full-wave rectified x-ray unit were overexposed, using known correct exposures. A spinning top test was performed at 200 mA, 0.05 second, and 70 kVp, and eight dots (dashes) were visualized on the finished film. Which of the following is indicated?

(A) the 0.05 sec time station is inaccurate
(B) the 200 mA station is inaccurate
(C) a rectifier is not functioning
(D) the processor needs servicing

138. The device that receives the remnant beam, converts it into light, and then increases the brightness of that light is the

(A) cine camera
(B) spot film camera
(C) image intensifier
(D) television monitor

139. If a radiograph, exposed using an automatic exposure device, is overexposed because an exposure shorter than the minimum response time was required, the radiographer generally should

(A) decrease the mA
(B) use the minus density
(C) use the plus density
(D) decrease the kVp

140. Which of the following would be appropriate cassette front material?

1. tungsten
2. magnesium
3. bakelite

(A) 1 only
(B) 1 and 2 only
(C) 2 and 3 only
(D) 1, 2, and 3

141. A technique chart should be prepared for each AEC x-ray unit and should contain the following information for each type of examination

1. photocell(s) used
2. optimum kVp
3. backup time

(A) 1 only
(B) 1 and 2 only
(C) 2 and 3 only
(D) 1, 2, and 3

142. The technique used to improve diagnostic quality by removing superimposed anatomic details from an x-ray image is called

(A) xeroradiography
(B) subtraction
(C) computed tomography
(D) magnetic resonance imaging

143. Which of the following causes pitting, or many small surface melts, of the anode's focal track?

(A) vaporized tungsten on the glass envelope
(B) loss of anode rotation
(C) large amount of heat to a cold anode
(D) repeated, frequent overloading

144. Correct operation of the automatic exposure control device is dependent upon

1. thickness and density of the object
2. positioning of the object with respect to the photocell
3. beam restriction

(A) 1 only
(B) 1 and 2 only
(C) 2 and 3 only
(D) 1, 2, and 3

145. The image intensifier's input phosphor is generally composed of

(A) cesium iodide
(B) zinc cadmium sulfide
(C) gadolinium oxysulfide
(D) calcium tungstate

146. Which of the following will serve to increase the effective energy of the x-ray beam?

1. increase in added filtration
2. increase in kilovoltage
3. increase in milliamperage

(A) 1 only
(B) 2 only
(C) 1 and 2 only
(D) 1, 2, and 3

147. A parallel plate ionization chamber receives a particular charge as x-ray photons travel through it. This is the operating principle of which of the following devices?

(A) automatic exposure device
(B) image intensifier
(C) cine film camera
(D) spot film camera

148. Excessive anode heating can cause vaporized tungsten to be deposited on the port window. This can result in

1. decreased tube output
2. tube failure
3. electrical sparking

(A) 1 only
(B) 2 only
(C) 1 and 2 only
(D) 1, 2, and 3

149. The input phosphor of the image intensifier tube functions to convert

(A) kinetic energy to light
(B) x-ray to light
(C) electrons to light
(D) fluorescent light to electrons

150. Radiographs from a particular three-phase, full-wave rectified x-ray unit were underexposed, using known correct exposures. A synchronous spinning top test was performed using 200 mA, $\frac{1}{2}$ sec, and 70 kVp, and a 20° arc is observed on the test film. Which of the following is MOST likely the problem?

(A) the $\frac{1}{2}$ sec time station is inaccurate
(B) the 200 mA station is inaccurate
(C) a rectifier is not functioning
(D) the processor needs servicing

Answers and Explanations

1. **(A)** Spot film cameras are rapidly replacing conventional spot film cassettes. A significant advantage of spot film cameras is the big reduction in patient dose that their use permits. However, as the film format increases, so does image quality, patient dose, and heat production. Patient dose, however, is still so much smaller than the dose with conventional spot film cassettes that it is almost insignificant when considering the small improvement in image quality afforded by cassette spot films. *(Bushong, p 364)*

2. **(B)** Tomography is a procedure that uses reciprocal motion between x-ray tube and film to image structures at a particular level in the body, while blurring everything above and below that level. The thickness of the level visualized can be varied by changing the tube angle (amplitude). In general, the greater the tube angle, the thinner the section imaged. Thinner sections may be used for imaging delicate or intricate structures. *(Bushong, p 322)*

3. **(B)** In order to achieve predictable and diagnostic results, the radiographer must use the AED in the correct manner. The position of the *photocell*(s) must be selected to correspond to the anatomy in question; that anatomy must cover the entire photocell, or under- or overexposure will occur. The radiographer must also select a *backup time* on the manual timer that is somewhat longer than the anticipated automatic exposure time. This is done so that, in case of AED failure, the manual timer will terminate the exposure and prevent an excessively long exposure to the patient and x-ray tube. *(Carroll, pp 333–335)*

4. **(A)** Of all the fluoroscopic imaging techniques, cinefluorography results in the greatest patient dose. In general, the larger the film format (35 versus 16 mm) and the greater the number of frames per second, the greater the patient exposure. The larger film format, however, produces image quality superior to that achieved with 16-mm format. Cassette loaded spot films result in a larger patient dose than that from spot film cameras (such as 105 and 70 mm). Again, the larger the film format, the larger the patient dose, but the better the image quality. *(Thompson, pp 372–373)*

5. **(B)** The principle of self induction is an example of the second law of electromagnetics (Lenz's Law), which states that an induced current within a conductive coil will oppose the direction of the current that induced it. It is important to note that self induction is a characteristic of alternating current *only*. The fact that alternating current is constantly changing direction accounts for opposing current set-up in the coil. Two x-ray circuit devices operate on the principle of self induction. The *autotransformer* operates on the principle of self induction and enables the radiographer to vary the kilovoltage. The *choke coil* also operates on the principle of self induction; it is a type of variable resistor that may be used to regulate filament current. The high-voltage transformer operates on the principle of mutual induction. *(Selman, pp 130, 133)*

6. **(A)** The x-ray beam emitted from the target has a heterogeneous nature. The low-energy photons within it must be removed because

they are not penetrating enough to contribute to the image and because they do contribute to the patient's skin dose. The glass envelope and oil coolant provide approximately 0.5 to 1.0 mm Al equivalent filtration, referred to as *inherent* because it is a built-in, permanent part of the tube head. *(Curry, p 77)*

7. **(C)** Target angle has a pronounced geometric effect on the effective, or projected, focal spot size. As target angle *decreases*, the effective (projected) focal spot becomes smaller. This is advantageous because it will improve radiographic detail without creating a heat-loading crisis at the anode (as would be the case if the actual focal spot size were reduced to produce a similar detail improvement). There are disadvantages, however. With a smaller target angle the anode heel effect increases; photons are more noticeably absorbed by the "heel" of the anode, resulting in a smaller percentage of x-ray photons at the anode end of the x-ray beam and a concentration of x-ray photons at the cathode end of the radiograph. *(Sprawls, pp 134–135)*

8. **(B)** Four wave forms are illustrated. Number *1* represents unrectified alternating current having constantly changing amplitude and periodically changing polarity; only the positive half cycle is useful. Number *2* represents single-phase, full-wave rectified current; the negative half cycle is rectified to a useful positive half cycle. Numbers *3* and *4* represent three-phase rectified current; number *3* is three phase, six pulse, and number *4* is three phase, twelve pulse. *(Wolbarst, p 81)*

9. **(A)** The energy of each electron within the tube current is measured in kiloelectron volts (keV). The energy used to *accelerate* those electrons is measured in *kVp*. It is unlikely that the electrons comprising the tube current have an energy equal to that of the applied voltage used to accelerate them, because the applied voltage is pulsating (varying) between a maximum value and a value somewhat lower (depending on the type of rectification). The mA and mAs are units of measure describing *quantity* of tube current. *(Curry, p 29)*

10. **(A)** Protective lead aprons and gloves are made of lead-impregnated vinyl or leather. They should be checked for cracks radiographically from time to time. Otherwise, minimal care is required. Lead aprons and gloves should always be hung on appropriate hangers. Glove supports permit air to circulate within the glove. Apron hangers provide convenient storage without folding. If lead aprons are folded (or just left in a heap!) cracks are more likely to form. If lead aprons or gloves become soiled, cleaning with a damp cloth and appropriate solution is all that is required. Excessive moisture should be avoided. *(Bushong, p 665)*

11. **(A)** The automatic exposure control (AEC) automatically terminates the exposure when the proper density has been recorded on the film. The important advantage of the phototimer, then, is that it can accurately duplicate radiographic densities. It is very useful in providing accurate comparison in follow-up examinations, and in decreasing patient exposure dose by decreasing the number of "retakes" due to improper exposure. Remember that proper functioning of the phototimer depends on accurate positioning by the radiographer. The correct photocell(s) must be selected and the anatomic part of interest must completely cover the photocell in order to achieve the desired density. If collimation is inadequate, and a field size larger than the part is used, excessive scatter radiation from the body or tabletop can cause the AEC to terminate the exposure prematurely, resulting in an underexposed radiograph. *(Carroll, p 335)*

12. **(B)** The thoriated tungsten filament of the cathode is heated by its own filament circuit. The x-ray tube filament is made of thoriated tungsten and is part of the cathode assembly. Its circuit provides current and voltage to heat it to incandescence, at which time it undergoes *thermionic emission*—the liberation of valence electrons from the filament atoms. *Electrolysis* describes the chemical ionization effects of an electric current. *Rectification* is the process of changing AC to unidirectional current. Photosynthesis is the process by which

plants take in carbon dioxide and release oxygen as a byproduct. *(Carlton, p 79)*

13. **(B)** Using a spinning top to test the efficiency of a single-phase timer, the result is a *series of dots* or dashes, with each representing a pulse of radiation. With full-wave rectified current, and a possible $\frac{1}{20}$ dots (pulses) available per second, one should visualize 12 dots at $\frac{1}{10}$ sec, 24 dots at $\frac{1}{5}$ sec, 6 dots at $\frac{1}{20}$ sec, and so on. But because three-phase equipment is almost constant potential, a synchronous spinning top must be used, and the result is a *solid arc* (rather than dots). The number of degrees formed by the arc is measured and equated to a particular exposure time. A multitude of small mesh-like squares describes a screen contact test. An aluminum step wedge (penetrometer) may be used to demonstrate the effect of kVp on contrast (demonstrating a series of gray tones from white to black), with a greater number of grays demonstrated at higher kVp levels. *(Saia, Program Review and Exam Preparation or PREP, pp 384–385)*

14. **(C)** Transformers are used in the x-ray circuit to change the value of the supplied voltage to a value appropriate for the production of x-rays. Although the incoming voltage supply may be 220 volts, thousands of volts are required for the production of x-rays. A step-up transformer consists of two coils, a primary and a secondary, in close proximity to each other (for example, side by side). As the primary coil is supplied with AC, a magnetic field rises up around the coil and proceeds to "cut" the secondary coil. Because the current supply is AC, the magnetic field is constantly expanding and contracting (rising and falling) and continuously "cutting" the secondary coil. This interaction between magnetic field and conductive coils induces a current in the secondary coil *proportional* to the number of turns in the secondary coil. As the number of turns in the secondary coil increases, so does the induced voltage. However, because we cannot just create energy, only change its form, *the amperage in the second coil is proportionally less*. For example, if the voltage in the secondary coil increased 10 times, the current would decrease by a factor

of 10. Therefore, in keeping with the law of conservation of energy, the product of the voltage and current in one coil must equal the product of voltage and current in the second coil. *(Curry, p 38)*

15. **(D)** X-ray tube life may be extended by using exposure factors that produce a minimum of heat, that is, a lower mAs and higher kVp combination whenever possible. When the rotor is activated, the filament current is increased to produce the required electron source (thermionic emission). Prolonged rotor time, then, can lead to shortened filament life due to early vaporization. Large exposures to a cold anode will heat the anode surface, and the big temperature difference can cause cracking of the anode. This can be avoided by proper warming of the anode prior to use, thereby allowing sufficient dispersion of heat through the anode. *(Selman, pp 377–378)*

16. **(A)** In order to test a film's response to processing, the film must first be given a predetermined exposure with a *sensitometer*. The film is then processed and the densities are read using a *densitometer*. Any significant variation from the expected densities is further investigated. An *aluminum step wedge* is used to evaluate the effect of kVp on contrast, and a *spinning top test* is used to check timer accuracy. *(Wolbarst, p 256)*

17. **(D)** The greater the number of electrons comprising the electron stream and bombarding the target, the greater the number of x-ray photons produced. Although kV is usually associated with the energy of the x-ray photons, because *a greater number of more energetic electrons* will produce more x-ray photons, an increase in kV will also increase the *number* of photons produced. Specifically, the quantity of radiation produced increases as the square of the kilovoltage. The material composition of the tube target also plays an important role in the number of x-ray photons produced. The higher the atomic number, the denser and more closely packed the atoms comprising the material; therefore, the greater the chance of an interaction between

a high-speed electron and target material. *(Curry, pp 33–35)*

18. **(C)** Single-phase x-ray equipment is much less efficient than three-phase equipment because voltage amplitude is always changing from zero to peak value and back to zero again. *With three-phase equipment, voltage never drops to zero*, and x-ray emission is 12 percent greater. Therefore, if 92 kVp was used with single-phase equipment, 81 kVp will be required (12 percent less) with three-phase equipment. *(Bushong, p 154)*

19. **(A)** Magnification radiography may be used to demonstrate small, delicate structures that are difficult to image with conventional radiography. Because OFD is an integral part of magnification radiography, the problem of magnification unsharpness arises. The use of a fractional focal spot *(0.3 mm or smaller)* is essential to the maintenance of image sharpness in magnification films. *Radiographic rating charts* should be consulted, as the heat load to the anode may be critical in magnification radiography. The long exposures typical of image-intensified fluoroscopy and tomography make the use of a fractional focal spot generally impractical and hazardous to the anode. *(Selman, pp 252–253)*

20. **(B)** The smaller the focal spot, the more limited the anode is with respect to the quantity of heat it can safely accept. As target angle decreases, the actual focal spot can be increased while still maintaining a small effective focal spot. Therefore, group B offers the greatest heat-loading potential, with a steep target angle and large actual focal spot. It must be remembered, however, that *a steep target angle increases the heel effect*, and film coverage may be compromised. *(Curry, p 13)*

21. **(D)** Excessive current passing through a circuit can damage the electrical devices in its path. A *circuit breaker* simply opens the circuit in the event of a current surge, thereby preventing the flow of damagingly high currents. A *fuse* provides the same function, except that a fuse needs to be replaced if it opens the circuit, while a circuit breaker is

simply reset. A *rheostat* is a type of variable resistor, and a *rectifier* changes AC to unidirectional current. *(Selman, p 244)*

22. **(C)** The image on the image intensifier's output phosphor may be displayed for viewing through the use of either a *series of lenses* or by a *fiber optic* link. The two devices needed in order to view the image are a *TV camera tube* and a *TV monitor*. The TV camera tube (usually a Plumbicon or Vidicon) converts the output phosphor image into an electrical signal. The TV monitor (a cathode ray tube) then converts the electrical signal into a visible light image. *(Thompson, p 370)*

23. **(B)** The high-voltage, or step-up, transformer functions to *increase voltage* to the necessary kilovoltage. It *decreases the amperage* to milliamperage. The amount of increase or decrease *depends on the transformer ratio*, that is, the number of turns in the primary coil to the number of turns in the secondary coil. The transformer law is as follows.

To determine secondary V:

$$\frac{V_s}{V_p} = \frac{N_s}{N_p}$$

To determine secondary I:

$$\frac{I_s}{I_p} = \frac{V_p}{V_s}$$

Substituting known values:

$$\frac{x}{220} = \frac{100,000}{200}$$

$$200x = 22,000,000$$
$$x = 110,000 \text{ volts } (110 \text{ kV})$$

(Selman, pp 125–126)

24. **(C)** Bremsstrahlung (or Brems) radiation is one of the two kinds of x-rays produced at the tungsten target of the x-ray tube. The incident high-speed electron, passing through a tungsten atom, is attracted by the positively charged nucleus; therefore it is *deflected from its course with a resulting loss of energy*. This

energy loss is given up in the form of an x-ray photon. *(Selman, pp 156–157)*

25. **(A)** The HVL of a particular beam is defined as that thickness of an material that will reduce the exposure rate to one-half its original value. The more energetic the beam (the higher the kV), the greater the HVL thickness required to cut its intensity in half. *Therefore, it may be stated the kV and HVL have a direct relationship: as kV increases, HVL increases. (Selman, pp 171–175)*

26. **(A)** The dynamics, or motion, of a part must be studied during a "real-time" examination such as *fluoroscopy* affords. *Stereoscopy* is a technique used to produce a radiographic third dimension. *Tomography* produces sectional images of body parts by blurring superimposed structures above and below the section, or level, of interest. A *phototimer* is one type of automatic exposure device. *(Bushong, p 352)*

27. **(B)** Transformers are used to change the value of alternating current, and operate on the principle of mutual induction. The secondary coil of the step-up transformer is located in the high-voltage (secondary) side of the x-ray circuit. The step-down transformer, or filament transformer, is located in the filament circuit and serves to regulate the voltage and current provided to heat the x-ray tube filament. The rectification system is also located on the high voltage, or secondary, side of the x-ray circuit. *(Selman, p 245)*

28. **(D)** Breast tissue has very low subject contrast, but microcalcifications and subtle density differences are imperative to visualize. Fine detail is necessary to visualize microcalcifications; therefore, a *small focal spot* tube is essential. *High contrast* (therefore, low kilovoltage) is needed to accentuate any differences in tissue density. A *compression device* serves to even out differences in tissue thickness (thicker at chest wall, thinner at nipple), decrease object-film distance, and decrease the production of scatter radiation. *(Selman, pp 344–347)*

29. **(D)** With single-phase, full-wave equipment, the voltage is constantly changing from 0 to 100 percent of its maximum value. It drops to 0 every 180° (of the AC waveform); that is, there is 100% voltage ripple. With three-phase equipment the voltage ripple is significantly smaller. Three-phase, six-pulse equipment has a 13 percent voltage ripple, and three-phase, twelve-pulse equipment has a 3.5 percent ripple. Therefore *the voltage never falls below 87 to 96.5 percent of its maximum value with three-phase equipment*, and closely approaches constant potential (DC). *(Selman, p 254)*

30. **(A)** In the photoelectric effect, a relatively low-energy incident photon uses all its energy to eject an inner-shell electron, leaving a vacancy. An electron from the shell above will drop to fill the vacancy, and a characteristic ray is given up in the transition. This type of interaction is more harmful to the patient, as *all the photon energy is transferred to tissue. (Bushong, p 176)*

31. **(A)** A motor is the device used to convert electrical energy into mechanical energy. The stator and rotor are the two principle parts of an induction motor. A generator converts mechanical energy into electrical energy. *(Selman, p 115)*

32. **(C)** A *phototimer* is a type of automatic exposure device (AED) used to automatically terminate the x-ray exposure once the film is correctly exposed. Another type of AED is the *ionization chamber*. An image intensifier functions to provide a brighter fluoroscopic image, and positive beam limitation (PBL), or automatic collimation, serves to restrict field size to the size of the cassette used in the bucky tray. The line voltage compensator automatically adjusts the incoming line voltage to the x-ray machine to correct for any voltage drops or surges. *(Selman, pp 241–242)*

33. **(B)** A *rheostat* is a type of variable resistor and functions to regulate the amount of voltage within the circuit. It can operate on AC or DC. An *autotransformer* is used to select the

amount of voltage sent to the high-voltage *transformer* to be "stepped up." A *circuit breaker* is a device used to protect the circuit from current surges. *(Selman, p 135)*

34. **(B)** When radiographs demonstrate half the expected density and the mA meter registers half the selected mA, rectifier failure is generally suspected. The equipment is said to be "half-waving" because the inverse voltage is not being rectified to useful voltage. The equipment may be tested for rectifier failure with the use of a spinning top (synchronous type for three-phase equipment) test. *(Carroll, p 357)*

35. **(C)** The spinning top test is used to evaluate *timer accuracy* or *rectifier failure*. With single-phase, full-wave rectified equipment (120 pulses/sec), for example, 12 dots should be visualized when using the ⅒-sec station. A few more or less indicates timer inaccuracy. If the test demonstrated five dots, one might suspect rectifier failure. With three-phase equipment a special synchronous spinning top (or oscilloscope) is used and a solid black arc is obtained rather than dots. The length of this arc is measured and compared with the known correct arc. *(Selman, p 233)*

36. **(B)** An image's spatial resolution refers to its recorded detail. The effect of the input phosphor's phosphor layer is similar to the effect of phosphor layer thickness in intensifying screens; that is, as the phosphor layer can be made thinner, recorded detail increases. Also, the smaller the input phosphor diameter, the greater the spatial resolution. A brighter image is easier to see, but does not affect resolution. *(Cullinan, p 215)*

37. **(C)** With single-phase, full-wave rectified equipment the voltage drops to zero every 180° (of the AC waveform); that is, there is *100 percent voltage ripple*. With three-phase equipment the voltage ripple is significantly smaller. Three-phase, six-pulse equipment has a *13 percent voltage ripple*, and three-phase, twelve-pulse has only a *3.5 percent ripple*. Three-phase, twelve-pulse equipment

comes closest to constant potential, as the voltage never falls below 96.5 percent of maximum value. *(Bushong, pp 202–203)*

38. **(C)** As the filament ages, vaporized tungsten is deposited on the port window of the glass envelope. It acts as a filter and decreases tube output. A roughened focal track decreases tube output because x-ray photons have to penetrate that uneven surface as they leave the focus. High-speed anode rotation increases tube-heating capacity. *(Selman, pp 208, 221)*

39. **(B)** In the production of Brems ("braking") radiation, a high-speed electron is attracted to the positive nuclear charge of a tungsten atom. In doing so, it is "braked" and gives up energy in the form of an x-ray photon. Most of the primary beam is made of Brems radiation. If the incident electron were to eject a K shell electron, an L shell electron moves in to fill the vacancy. It releases a photon (K characteristic ray) whose energy equals the difference between the K and L shell energy levels. This is characteristic radiation and is responsible for only a small portion of the primary beam. *(Bushong, pp 149–150)*

40. **(B)** There is often significant attenuation of the x-ray beam as it traverses the front screen and film. Rare earth phosphors absorb significantly more x-ray photons than did calcium tungstate screens. Consequently, rear screen fluorescence may be diminished. To compensate for the photon loss the rear screen may be thicker (faster). Water-soluble paste is never used to mount screens; rubber cement or a special adhesive tape are recommended. *(Wolbarst, p 184)*

41. **(D)** A fluoroscopic examination enables the user to visualize "live" or "real-time" body activity. An operator may record a particular event or image in a number of ways. Conventional *spot film cassettes* have been used for many years, but are being replaced by *100-mm cut film* or *105-mm strip film spot film cameras*. These spot film cameras significantly reduce patient dose while permitting single or

multiframe exposures. *Cinefluorography* is used to record rapid events that can be reviewed many times over at normal or slow speed. The fluoroscopic image can also be transferred to *videotape*; its advantage is instant playback without processing. *(Cullinan, p 217)*

42. **(B)** The x-ray tube is a glass vacuum tube with two electrodes: the anode and the cathode. The anode is the positive electrode and the cathode is the negative electrode. The cathode filament is heated to white hot and electrons are liberated. When a high voltage is applied to the cathode, these electrons are driven across to the tungsten target of the anode, where x-rays are produced on rapid deceleration of the electrons. A tube with two electrodes is termed a diode. A triode tube has three electrodes and has special application in radiography. *(Selman, p 154)*

43. **(D)** Because the radiographer must be able to select a different mA, kVp, and exposure time for each patient, the typical control panel will have selector switches for each function. An mA meter on the control panel functions to give a read-out of the mA for each exposure. It is good practice to get a glimpse of this meter during each exposure to ascertain that there indeed was an exposure and that the meter registers the selected mA. An mA meter will not have time to register, say, 300 mA during a very fast exposure; rather, just a slight movement of the needle will be observed. In such a case, an mAs meter would be needed to check the accuracy of the exposure. The filament ammeter regulates the amount of current to the x-ray tube filament circuit, and any required adjustment must be made by the equipment serviceperson. *(Thompson, p 167)*

44. **(D)** Through the action of thermionic emission, as the tungsten filament continually gives up electrons, it gradually becomes thinner with age. This evaporated tungsten is frequently deposited on the inner surface of the glass envelope at the tube window. As such, *it acts as an additional filter* of the x-ray beam, thereby *reducing tube output*. Also, the tung-

sten deposit may actually attract electrons from the filament, constituting a tube current, and causing *puncture of the glass envelope*. *(Selman, p 208)*

45. **(A)** There can be significant attenuation of the beam as it traverses the front screen and film. Rare earth phosphors absorb significantly more x-ray photons than did calcium tungstate screens. Consequently, rear screen fluorescence may be diminished. To compensate for the loss of photons, the rear screen may be thicker (faster). *(Wolbarst, p 184)*

46. **(D)** Excessive radiographic density may be a result of incorrect exposure factors. It can also be caused by overdevelopment. Overdevelopment may be due to excessive developer temperature, overreplenishment, or prolonged length of development. *(McKinney, p 190)*

47. **(D)** Perfect contact between the intensifying screens and film is essential in order to maintain image sharpness. Any separation between them allows diffusion of fluorescent light and subsequent blurriness and loss of detail. Screen/film contact can be diminished if the cassette frame is damaged and misshapen, if the front is warped, or if there is a foreign body between the screens elevating them. *(Selman, pp 285–286)*

48. **(B)** When rapidly occurring events are visualized in fluoroscopy, there may be insufficient time to evaluate them carefully (as in cardiac catheterizations). Recording the fluoroscopic events on 16 or 35-mm film permits review without further patient exposure. Cassette spot films, 100-mm cut film, and 105-mm strip spot films can record a sequence of events but do not permit the number of frames per second available in cine, nor do they allow motion to be appreciated. Photofluorography is a method of recording fluoroscopic images (usually chests) on single-emulsion film. *(Bushong, pp 352–353)*

49. **(B)** Determining mR/mAs output is often done to determine linearity among x-ray ma-

chines. But the equipment being compared must be of the same type (for example, all single phase; all three phase, six pulse). If there is linearity among these machines, then identical technique charts can be used. In the example given, 400 mA and 0.12 sec were used, equaling *48 mAs*. If the output for 48 mAs was 150 mR, then 1 mAs is equal to 3.1 mR (150 mR ÷ 48 mAs = 3.1 mR/mAs). *(Carroll, p 360)*

50. **(A)** Characteristic radiation is one of two kinds of x-rays produced at the tungsten target of the x-ray tube. The incident, or incoming, high-speed electron ejects a K shell electron. This leaves a hole in the K shell and an L shell electron drops down to fill the K vacancy. Because L electrons have a greater binding energy than do K shell electrons, the L shell electron gives up the difference in binding energy in the form of a photon, a "characteristic x-ray" (characteristic of the K shell). *(Selman, pp 157–158)*

51. **(C)** A *phototimer* is one type of *automatic exposure device (AED)* that actually measures light. As x-ray photons penetrate and emerge from a part, a fluorescent screen beneath the cassette glows and the fluorescent light charges a photomultiplier tube. Once a predetermined charge has been reached, the exposure automatically terminates. A parallel plate *ionization chamber* is another type of AED. A radiolucent chamber is beneath the patient (between the patient and film). As photons emerge from the patient, they enter the chamber and ionize the air within. Once a predetermined charge has been reached, the exposure is automatically terminated. Motion of magnetic fields inducing a current in a conductor refers to the principle of mutual induction. *(Wolbarst, p 205)*

52. **(C)** Voltmeters must be connected in parallel within a circuit, so as to be able to measure the potential difference between two points. Ammeters are connected in series. Rectifiers are located between the secondary coil of the high-voltage transformer and the x-ray tube and function to change AC to unidirectional current. *(Selman, p 77)*

53. **(D)** Anode target material of high atomic number produces higher energy x-rays more efficiently. Because a great deal of heat can be produced at the target, the material should have a high melting point so as to avoid damage to the target surface. The cathode filament receives low-voltage current to heat it to the point of thermionic emission. Then high voltage is applied to drive the electrons across to the focal track. *(Selman, pp 204–211)*

54. **(A)** Soft tissue radiography requires the use of long-wavelength, low-energy x-ray photons. Therefore, very little filtration is used in mammography. Certainly anything more than 1.0 mm Al would remove the useful soft photons and the desired high contrast could not be achieved. Dedicated mammographic units have molybdenum targets (for the production of soft, low-energy radiation) and a small amount of added molybdenum filtration. *(Selman, p 345)*

55. **(B)** There are three types of beam restrictors: aperture diaphragms, cones and cylinders, and collimators. The most practical and efficient is the collimator. Its design makes available an infinite combination of field size variations not available with the other types of beam restrictors. Aperture diaphragms, cones, and cylinders each have a fixed aperture size and shape. The collimator offers an illuminated field so that the radiographer knows with certainty the location and size of the field to be irradiated, and the exact center of that field is indicated by cross-hairs. Although the collimator assembly contributes approximately 1.0 mm Al equivalent to the added filtration of the x-ray tube (because of the plastic exit portal and silver-coated reflective mirror), its functions are unrelated to the cleanup of scatter radiation. This is because the *patient* is the principle scatterer and grids function to clean up scatter radiation produced by the patient. *(Selman, pp 386–391)*

56. **(D)** Special units have been designed to accommodate examinations with high patient volume. Dedicated chest units are available that will transport a piece of unexposed film from its magazine into position between a

pair of intensifying screens, make a photo-timed exposure, and transport the exposed film to the automatic processor. Dedicated head units are available for easy positioning of the skull, sinuses, mastoids, and so on. A great deal of attention has been called to the importance of high-quality mammographic examinations. Dedicated mammographic units are available with molybdenum targets and other beneficial features. *(Saia, PREP, p 367)*

57. **(B)** The light image emitted from the output phosphor of the image intensifier is directed to the TV monitor for viewing and sometimes to recording devices such as a spot film camera or cine film. The light is directed to these places by a *beam splitter* located between the output phosphor and TV camera tube. The majority of the light will go to the recording device, while a small portion goes to the TV so that the procedure may continue to be monitored during filming. *(Carroll, p 394)*

58. **(D)** The rotating anode has a target (or focal spot) on its beveled edge, which forms the target angle. As the anode rotates, it constantly turns a new face to the incoming electrons; this is the focal track. That portion of the focal track bombarded by electrons is the actual focal spot, and because of the target's angle, the effective or projected focal spot is always smaller. The electrons impinging on the target have "boiled off" the cathode filament as a result of thermionic emission. *(Selman, pp 208–213)*

59. **(B)** Inherent filtration is that which is "built into" the construction of the x-ray tube. Before exiting the x-ray tube, x-ray photons must pass through the tube's glass envelope and port window; the photons are filtered somewhat as they do so. This inherent filtration is usually the equivalent of 0.5 mm Al. Aluminum filtration *placed* between the x-ray tube housing and collimator is added in order to contribute to the total necessary requirement of 2.5 mm Al equivalent. The collimator itself is considered part of the added filtration (1.0 mm Al equiv.) because of the silver surface of the mirror within. It is important to remember that as aluminum filtra-

tion is added to the x-ray tube, the half-value layer (HVL) increases. *(Bushong, p 168)*

60. **(D)** The accuracy of all three are important to ensure adequate patient protection. *Reproducibility* means that repeated exposures at a given technique must provide consistent intensity. *Linearity* means that a given mAs, using different mA stations with appropriate exposure time adjustments, will provide consistent intensity. *Positive beam limitation (PBL)* is automatic collimation and must be accurate to 2 percent of the SID. Light-localized collimators must be available and must be accurate to within 2 percent. *(Bushong, pp 431, 433)*

61. **(D)** If the voltage supplied to the x-ray room was not variable, we would be limited to one kV value. For example, if 110 volts was incoming to a step-up transformer with 1000:1 ratio, we would be restricted to the use of 110 kV for every examination. However, the incoming voltage is actually supplied to an autotransformer, and we may select a portion of that voltage to send to the primary coil of the high-voltage transformer. In that way, the autotransformer permits us to vary our kVp at the control panel. The autotransformer is also connected to the filament circuit and supplies voltage to the primary coil of the filament transformer. *(Curry, p 39)*

62. **(D)** *Spot film cameras* are rapidly replacing conventional spot film cassettes. One of their most significant advantages is the big *reduction in patient dose*. Consequently, there is a significant decrease in fluoroscopic tube heat buildup as well. Because spot film cameras can expose a single frame of film, or multiple frames, there is no interruption of the fluoroscopic procedure to change spot film cassettes and the total exam time is reduced. *(Bushong, p 364)*

63. **(B)** The collimator assembly includes a series of lead shutters, a mirror, and a light bulb. The mirror and light bulb function to project the size, location, and center of the irradiated field. The bulb's emitted beam of light is deflected by a mirror placed at an angle of 45°

in the path of the light beam. *In order for the projected light beam to be the same size as the x-ray beam, the focal spot and the light bulb must be exactly the same distance from the center of the mirror. (Curry, pp 94–95)*

64. **(B)** Parts being examined during fluoroscopic procedures change in thickness and density as the patient is required to change positions, and as the fluoroscope is moved to examine different regions of the body that have varying thickness and tissue densities. The *automatic brightness control* (ABC) functions to vary the required mAs and/or kVp as necessary. With this method patient dose varies, and image quality is maintained. The *automatic gain control* (AGC) adjusts the degree of amplification of the electronic signal (gain). With this method patient dose does not vary but, if increased amplification (gain) is required, image noise can increase. *(Carroll, p 395)*

65. **(B)** One generally thinks in terms of moving grids being totally superior to stationary grids because moving grids function to blur the images of lead strips from the radiographic image. Moving grids do, however, have several disadvantages. First, their complex mechanism is expensive and subject to malfunction. Second, today's sophisticated x-ray equipment makes possible the use of extremely short exposures, a valuable feature whenever motion may be a problem (as in pediatric radiography). However, grid mechanisms are frequently not able to oscillate rapidly enough for the short exposure times, and as a result the grid motion is "stopped" and lead strips are imaged. Third, patient dose is increased with moving grids. Because the CR is not always centered to the grid because it is in motion, lateral decentering occurs (resulting in diminished density), and consequently an increase in exposure to compensate (either manually or via automatic exposure control). *(Curry, p 111)*

66. **(C)** The *autotransformer* operates on the principle of self-induction and functions to select, or "tap off," the correct voltage to be sent to the *high-voltage transformer* to be "stepped up." The high-voltage transformer increases the voltage and decreases the current. The *rheostat* is a type of variable resistor used to change voltage or current values. It is frequently found in the filament circuit. A *fuse* is a circuit device used to protect the circuit elements from overload by opening the circuit in the event of a power surge. *(Selman, p 135)*

67. **(D)** Because conventional radiographs lack the depth dimension, structures are superimposed on one another. For better visualization, the patient is turned in different positions and/or the x-ray tube is angled. Another way to separate superimposed images is through tomography. Tomography uses the principle of reciprocal motion between x-ray tube and film to image structures at a particular level, or "object plane," within the body while blurring everything above and below the object plane. Image-intensified fluoroscopy is used to observe the dynamics, or motion, of a part, and cinefluorography records these events. *(Selman, pp 418–419)*

68. **(B)** Half-value layer (HVL) may be used to express the quality of an x-ray beam. *The HVL of a particular beam is that thickness of an absorber that will decrease the intensity of the beam to one half its original value.* If the original intensity of the beam was 78 R/min, the first HVL will reduce it to 39 R/min, the second HVL will reduce it to 19.5 R/min, and the third HVL will reduce the intensity to 9.75 R/min. *(Selman, pp 172–174)*

69. **(C)** A double-focus tube has two focal spot sizes available. These focal spots are actually two *paths* available on the focal track. There are, however, two filaments. As the small focal spot is selected, the small filament is heated and electrons are driven across to the smaller portion of the focal track. As the large focal spot is selected, the large filament is heated and electrons are driven across to the larger portion of the focal track. *(Selman, p 206)*

70. **(A)** The image intensifier's input phosphor is six to nine times larger than the output phosphor. It receives the remnant radiation emerging from the patient and converts it into a fluorescent light image. Very close to the input phosphor, separated only by a thin transparent layer, is the photocathode. The photocathode is made of a photoemissive alloy, usually a cesium and antimony compound. The fluorescent light image strikes the photocathode and is converted to an electron image that is focused by the electrostatic lenses to the small output phosphor. *(Bushong, pp 256–257)*

71. **(D)** A mobile C-arm is used to obtain fluoroscopic images in areas such as surgery, emergency department, and critical care units. The images can be dynamic (moving), and static (still) images can be obtained as well. Many new C-arms have additional capabilities such as subtraction and recall and storage of previous images. *(Thompson, p 375)*

72. **(D)** Radiographic results should be consistent and predictable not only in positioning accuracy but with respect to exposure factors and image sharpness as well. X-ray equipment should be calibrated periodically as part of an ongoing quality assurance program. The quantity (mAs) and quality (kVp) of the primary beam have a big impact on the quality of the finished radiograph. The focal spot should be tested periodically to evaluate its impact on image sharpness. *(Bushong, p 430)*

73. **(D)** The synchronous timer is an older type timer that doesn't permit very precise, short exposures. The impulse timer permits a shorter more precise exposure, and the electronic timer may be used for exposures as short as 0.001 sec. The phototimer, however, automatically terminates the exposure when the proper density has been recorded on the film. The important advantage of the phototimer, then, is that it can accurately duplicate radiographic densities. It therefore is very useful in providing accurate comparison in follow-up examinations, and in decreasing patient exposure dose by decreasing the number of "retakes" due to improper exposure. Remember that proper functioning of

the phototimer depends on accurate positioning by the radiographer. *(Selman, pp 239–243)*

74. **(D)** Strip or cut spot films are obtained photographically, from the light emitted from the image intensifier's output phosphor. Consequently, much less patient exposure is involved. Because the film is smaller than cassette film, the exposure is also less. Their size (about 4×4 inches) decreases the necessary filing space. Because multiple frames may be exposed per second, their use also decreases exam time because the interruption of changing spot cassettes is eliminated. *(Bushong, p 364)*

75. **(B)** Using a spinning top to test the timer efficiency of full-wave rectified single-phase equipment, the result is a *series of dots* or dashes, with each dot representing a pulse of radiation. With full-wave rectified current, and a possible ¹⁄₂₀ dots (pulses) available per second, one should visualize 12 dots at ¹⁄₁₀ sec, 6 dots at 0.05 sec, 10 dots at ¹⁄₁₂ sec, and 3 dots at 0.025 sec. Because three-phase equipment is almost constant potential, a synchronous spinning top must be used for timer testing, and the result is a *solid arc* (rather than dots). The number of degrees formed by the arc is measured and equated to a particular exposure time. *(Saia, PREP, p 385)*

76. **(B)** Automatic exposure devices are used in today's equipment and serve to produce consistent and comparable radiographic results. In one type of AED there is an *ionization chamber* just beneath the tabletop above the cassette. The part to be examined is centered to it (the sensor) and radiographed. When a predetermined quantity of ionization has occurred (equal to the correct density), the exposure automatically terminates. In the other type of AED, the *phototimer*, a small fluorescent screen is positioned beneath the cassette. When remnant radiation emerging from the patient exposes the film and exits the cassette, the fluorescent screen emits light. Once a predetermined amount of fluorescent light is "seen" by the photocell sensor, the expo-

sure is terminated. *In either case, the manual timer should be used as a backup timer.* In case of AED malfunction, the exposure would terminate, thus avoiding patient overexposure and tube overload. *(Saia, PREP, p 215)*

77. **(B)** There are two main types of mobile x-ray equipment: capacitor discharge and battery powered. Although capacitor discharge units are light and therefore fairly easy to maneuver, the battery-powered mobile unit is very heavy (largely because it carries its heavy-duty power source). It is, however, capable of storing a large mAs capacity for extended periods of time. These units frequently have a capacity of 10,000 mAs, with 12 hours required for a full charge. *(Curry, p 52)*

78. **(C)** The brightness gain of image intensifiers is 5000 to 20,000. This increase is accomplished in two ways. First, as the electron image is focused to the output phosphor, it is accelerated by high voltage (this is *flux gain*). Second, the output phosphor is only a fraction of the size of the input phosphor and this image size decrease represents another brightness gain, termed *minification gain. Total brightness gain is equal to the product of minification gain and flux gain.* *(Bushong, p 357)*

79. **(C)** X-ray photons are produced in two ways as high-speed electrons interact with target atoms. First, if the high-speed electron is attracted by the nucleus of a tungsten atom and changes its course, the energy given up as the electron is "braked" in the form of an x-ray photon. This is called Bremsstrahlung ("braking") radiation and is responsible for the majority of x-ray photons produced at the conventional tungsten target. Second, a high-speed electron may eject a tungsten K shell electron, leaving a vacancy in the shell. An electron from the next energy level, the L shell, drops down to fill the vacancy, emitting the difference in energy as a K characteristic ray. Characteristic radiation comprises only about 15 percent of the primary beam. *(Bushong, pp 148–150)*

80. **(B)** Perfect film/screen contact is essential to sharply recorded detail. Screen contact can be

evaluated with a wire mesh test. A spinning top test is used to evaluate timer accuracy and rectifier operation. A penetrometer (aluminum step wedge) is used to illustrate the effect of kVp on contrast. A star pattern is used to measure resolving power of the imaging system. *(Selman, pp 234–235)*

81. **(B)** *The minimum response time, or minimum reaction time,* is the length of the shortest exposure possible with a particular automatic exposure device (AED). If less than the minimum response time is required for a particular exposure, the radiograph will exhibit excessive density. The problem may become apparent when using fast film/screen combinations, high milliamperage, or with small or easily penetrated body parts. The *backup timer* functions to protect the patient from overexposure and the x-ray tube from overload. *(Saia, PREP, p 217)*

82. **(A)** A quality control program requires the use of a number of devices to test the efficiency of various parts of the imaging system (Fig. 5–6). A star pattern is a resolution testing device used to test focal spot size. A parallel line type resolution test pattern is used to test the resolution capability of intensifying screens. *(Selman, p 326)*

83. **(C)** Structures that we wish to visualize are frequently superimposed on other structures of lesser interest. Tomography uses reciprocal motion between x-ray tube and film to image structures at a particular level in the body, while blurring everything above and below that level. The thickness of the level visualized can be changed by changing the tube angle (amplitude). The greater the tube angle, the thinner the section imaged. *(Selman, pp 412–413)*

84. **(D)** Rectifiers (solid state, or the older valve tubes) permit the flow of current in only one direction. They serve to change the AC, needed in the low-voltage side of the x-ray circuit, to unidirectional current. Unidirectional current is necessary for the efficient operation of the x-ray tube. *The rectification system is located between the secondary coil of the*

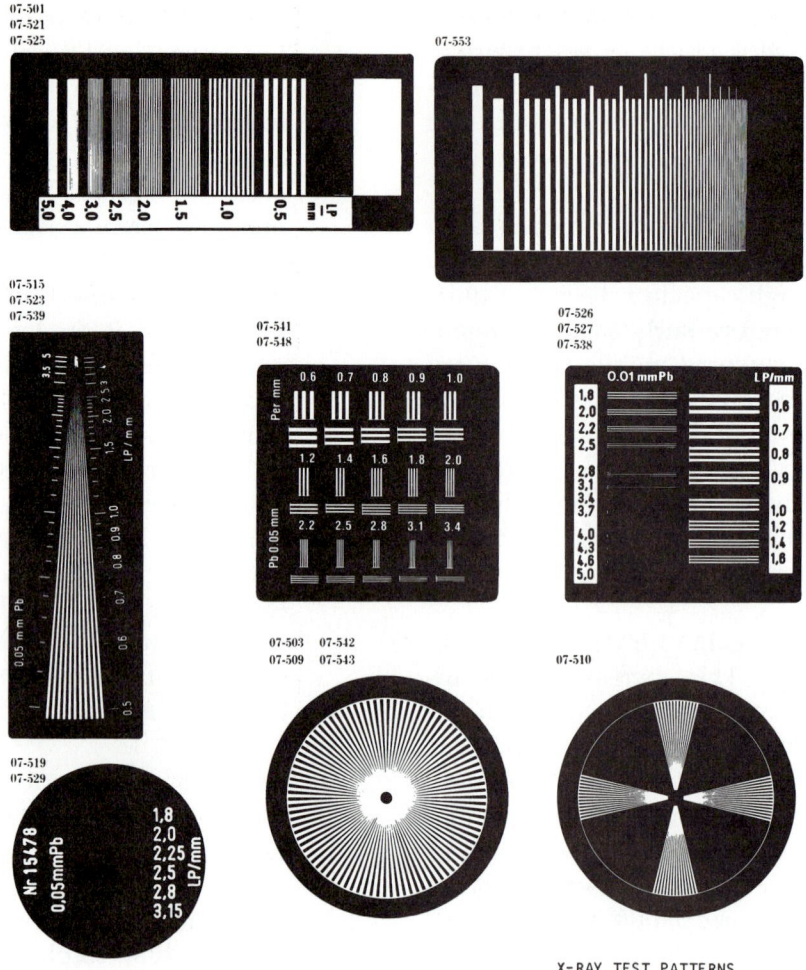

Figure 5–6. Courtesy of Nuclear Associates.

high-voltage transformer and the x-ray tube. The filament transformer functions to adjust the voltage and current going to heat the x-ray tube filament. The autotransformer varies the amount of voltage being sent to the primary coil of the high-voltage transformer in order that the appropriate kVp be obtained. The high-voltage transformer "steps up" the voltage to the required kilovoltage and steps down the amperage to milliamperage. *(Selman, pp 139–140)*

85. **(C)** All circuit devices located before the primary coil of the high-voltage transformer are said to be on the primary or low-voltage side of the x-ray circuit. The timer, autotransformer, and (prereading) kV meter are all located in the low-voltage circuit. The mA me-

ter, however, is connected at the midpoint of the secondary coil of the high-voltage transformer. When studying a diagram of the x-ray circuit, it will be noted that the mA meter is grounded at the midpoint of the secondary coil (where it is at zero potential). In this way, it may be safely placed in the control panel. *(Selman, p 251)*

86. **(B)** The *input phosphor* of an image intensifier receives remnant radiation emerging from the patient and converts it to a fluorescent light image. Directly adjacent to the input phosphor is the *photocathode*, which is made of a photoemissive alloy (usually a cesium and antimony compound). The fluorescent light image strikes the photocathode and is converted to an electron image. The electrons

are carefully focused, to maintain image resolution, by the *electrostatic focusing lenses*, through the *accelerating anode* and to the *output phosphor* for conversion back to light. *(Bushong, pp 356–357)*

87. **(C)** Cinefluorography is valuable when recording rapidly occurring events. Cardiac catheterizations lend themselves very nicely to cinefluorographic recording. Patient dose in cinefluorography is greater than that in video recording, but the image quality is better. *(Bushong, p 364)*

88. **(B)** Characteristic radiation is produced at the target as a high-energy incident electron ejects a K shell tungsten electron. Because the L electron is at a higher energy level than the K electron, it liberates the excess energy in the form of a K characteristic x-ray as it makes its transit the K shell. Characteristic radiation makes up only about 15 to 20 percent of the primary beam. Bremsstrahlung radiation is produced when high-speed electrons are attracted by a tungsten nucleus and decelerated. Brems radiation makes up the majority of the primary beam (about 80 to 85 percent). *(Selman pp 156, 158)*

89. **(D)** The spinning top test may be used to test timer accuracy in single-phase equipment. A spinning top is a metal disc with a small hole placed in its outer edge, and placed on a pedestal about 6 inches high. An exposure is made (for example, 0.1 sec) while the top spins. Because a full-wave rectified unit produces 120 x-ray photon impulses per second, in 0.1 sec the film should record 12 dots (if the timer is accurate). Because three-phase equipment produces almost constant potential—rather than pulsed radiation—the standard spinning top cannot be used. An oscilloscope or synchronous spinning top must be employed to test timers of three-phase equipment. *(Selman, p 233)*

90. **(C)** The kV meters/selectors usually appear as a kV major selector switch (which changes kV in increments of 10) and a kV minor selector (which changes kV in increments of 2). These meters indicate the selected kV. The

mA meter reads zero until the exposure is made, at which time it should read out the mA selected for that exposure. When the exposure time is so short that the needle has no time to register the mA, just slight movement of the needle will be observed. In such a case an mAs meter would be needed to indicate the selected mAs. The line voltage meter/compensator indicates the level of the incoming voltage and reflects any drop or surge in voltage. The radiographer can adjust the incoming voltage by simply turning the knob to bring the needle (and therefore, voltage) to the correct level. Newer equipment frequently has a built-in line voltage compensator that automatically adjusts itself. *(Selman, p 253)*

91. **(B)** Automatic exposure devices (AEDs) are also called automatic exposure controls (AECs). AEDs were originally developed in order to achieve more consistent and reproducible film densities. This consistency reduces the number of retakes, thereby reducing patient exposure dose. *(Carroll, p 332)*

92. **(D)** Each time an x-ray exposure is made, less than 1 percent of the total energy is converted to x-rays and the remainder (more than 99 percent!) of the energy is converted to heat. Thus the importance of using target material of high atomic number and high melting point. The larger the actual focal spot size, the larger the area over which the generated heat is spread and the more tolerant the x-ray tube. Heat is particularly damaging to the target if it is concentrated or limited to a small area. A target that rotates during the exposure is spreading the heat over a large area, the entire surface of the focal track. If the diameter of the anode is greater, the focal track will be longer and heat will be spread over an even larger area. *(Curry, pp 13–15)*

93. **(D)** Voltage ripple refers to the percentage drop from maximum voltage each pulse of current experiences. In single-phase rectified equipment the entire pulse (half cycle) is used; therefore, there is first an increase to maximum (peak) voltage value and a subsequent decrease to zero potential (90° past

peak potential). The entire waveform is used; if 100 kV were selected, the actual average kV output would be approximately 70. Three-phase rectification produces almost constant potential with just small ripples (drops) in maximum potential between pulses. The operation of three-phase, six-pulse generators is characterized by approximately a 13.5 percent voltage ripple (drop from maximum value). Three-phase, twelve-pulse generators have about a 3.5 percent voltage ripple. (*Curry, pp 49–50, 60; Selman, p 254*)

94. **(D)** A radiographic rating chart enables the operator to determine the maximum safe mA, time, and kVp for a given exposure. Because the heat load that a particular anode will safely accept varies with the size focal spot, type rectification, and anode rotation, these variables must also be identified. *Each x-ray tube has its own radiographic rating chart.* FFD has no impact on anode heat load, but rather on patient dose and film density. (*Selman, pp 217–219*)

95. **(C)** Contrast improvement factor is the ratio of radiographic contrast obtained using a grid to that obtained without a grid. In general, the greater the lead content of the grid, the better the "cleanup" of scatter radiation and the higher the contrast improvement factor. The greater the number of lead strips per inch, the thinner they must be and the less visible they will be. However, the thinner the lead strips, the greater the likelihood of energetic scatter radiation passing through them and reaching the film. Therefore, in order to maintain grid efficiency and an adequate contrast improvement factor, a higher-ratio grid with fewer lines per inch is preferred. (*Selman, pp 370–371*)

96. **(C)** *Rectifiers change AC into unidirectional current by allowing current to flow through them in only one direction. Valve tubes* are vacuum rectifier tubes found in older equipment. *Solid-state diodes* are the types of rectifiers used in today's x-ray equipment. Rectification systems are found between the secondary coil of the high-voltage transformer and the x-ray tube. *Resistors*, such as rheostats or choke coils, are circuit devices used to vary voltage or current. (*Selman, pp 139–141*)

97. **(D)** The thicker and more dense the anatomic part being studied, the less bright will be the fluoroscopic image. Both mA and kVp affect the fluoroscopic image similar to the way they affect the radiographic image. For optimum contrast, and especially in consideration of patient dose, higher kVp and lower mA are generally preferred. (*Selman, pp 403–405*)

98. **(B)** Given the mA and exposure time, a radiographic rating chart enables the radiographer to determine the maximum safe kVp for a particular exposure. Because the heat load an anode will safely accept varies with the size of the focal spot and type of rectification, these variables must be identified. Each x-ray tube has its own radiographic rating chart. The speed of the imaging system has no impact on the use of a radiographic rating chart. (*Selman, pp 217–219*)

99. **(B)** The minimum response time, or minimum reaction time, is the length of the shortest exposure possible with a particular automatic exposure device (AED). If less than the minimum response time is required for a particular exposure, the radiograph will exhibit excessive density. This problem becomes apparent when making exposures that require very short exposures, such as when using high mA and fast film/screen combinations. To resolve this problem the radiographer should decrease the mA rather than the kVp, in order to leave contrast unaffected. (*Saia, PREP, pp 216, 319*)

100. **(B)** Many cassettes have a thin lead foil layer behind the rear screen to absorb backscatter radiation that is energetic enough to exit the rear screen, strike the metal backing, and bounce back to fog the image. In this way, the cassette's metal hinges or straps may be imaged in high kVp radiography. (*Carroll, p 212*)

101. **(C)** The most frequent use of cinefluorography is for cardiac catheterizations. Cine film

may be 16 or 35 mm in size. Cine cameras operate very much like movie cameras; the x-ray tube, camera shutter device, and camera motor are all *synchronized* so that there is *x-ray exposure only during the shutter-open phase.* Patient dose from cinefluorography is high, so it is very important that there be no unnecessary exposure, such as during times when the film shutter is closed. *(Carroll, pp 413–414)*

102. **(B)** The high-voltage, or step-up, transformer functions to increase voltage to the necessary kilovoltage. It decreases the amperage to milliamperage. The amount of increase or decrease is dependent on the transformer ratio—the number of turns in the primary coil to the number of turns in the secondary coil. The transformer law is as follows.

To determine secondary V:

$$\frac{V_s}{V_p} = \frac{N_s}{N_p}$$

To determine secondary I:

$$\frac{I_s}{I_p} = \frac{V_p}{V_s}$$

Substituting known factors:

$$\frac{x}{220} = \frac{35,000}{100}$$
$$100x = 7,700,000$$
$$x = 77,000 \text{ volts} = 77 \text{ kV}$$

$$\frac{x}{75} = \frac{220}{77,000}$$
$$77000x = 16,500$$
$$x = 0.214 \text{ amp} = 214 \text{ mA}$$

(Selman, pp 125–126)

103. **(C)** Rare earth phosphors are not scarce; they are difficult to separate from other materials with which they are combined in the earth. Rare earth phosphors are much more efficient than calcium tungstate in absorbing x-ray photons and converting their energy into fluorescent light. Examples of rare earth phos-

phors are gadolinium oxysulfide and lanthanum oxybromide. Cesium iodide is the phosphor of preference for the input phosphor of an image intensifier. *(Selman, p 283)*

104. **(C)** The primary, or low-voltage, side of the x-ray circuit requires AC for operation, but the x-ray tube operates most efficiently on current that flows in only one direction (unidirectional). Therefore, the high-voltage side of the circuit contains the rectification system; here the current is changed to unidirectional just before reaching the x-ray tube. *Solid-state diodes* are the *rectifiers* used in x-ray equipment because they allow current flow in only one direction. *Resistors*, such as rheostats, may be used to vary circuit current or voltage. *Transformers* operate on the principle of mutual induction and function to change the current and voltage values. *(Bushong, p 104)*

105. **(C)** Scattered and secondary radiations have been deviated in direction while passing through a part. Leakage radiation is that which emerges from the leaded tube housing in directions other than that of the useful beam. Tube head construction must keep leakage radiation to less than 0.1 R/hr at 1 meter from the tube. Remnant radiation is that which emerges from the patient to form the radiographic image. *(Curry, p 413)*

106. **(C)** Sensitometry and densitometry are used in evaluation of the film processor, just one part of a complete quality assurance (QA) program. Modulation transfer function (MTF) is used to express spatial resolution, another component of the QA program. A complete QA program includes testing of all components of the imaging system: processors, focal spot, x-ray timers, filters, intensifying screens, beam alignment, and so on. *(Noz, pp 120–125)*

107. **(A)** Automatic exposure devices (AEDs) are also called automatic exposure controls (AECs). AEDs were originally developed in order to achieve more consistent and reproducible film densities. This consistency reduces the number of retakes, thereby reducing patient exposure dose. That part of an AED that

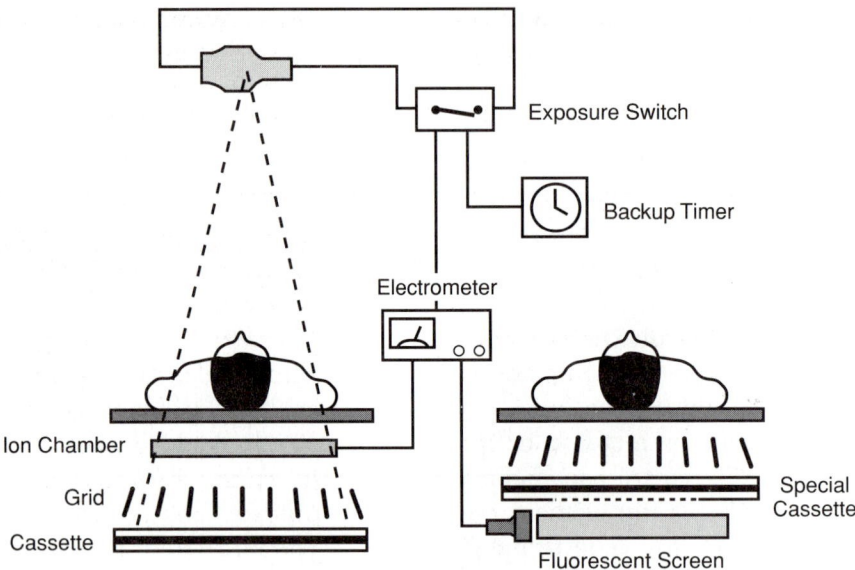

Figure 5–7. Reproduced with permission from Wolbarst AB. *Physics of Radiology.* East Norwalk, CT: Appleton & Lange, 1993.

detects the radiation is called the *photocell.* The two most commonly used photocells are the *photomultiplier* tube and the *ion* (or ionization) *chamber.* The ion chamber is positioned between the table and cassette, whereas the photomultiplier is located below the cassette (Fig. 5–7). *(Carroll, p 332)*

108. **(A)** Remnant radiation emerging from the patient causes fluorescence of the cassette's front intensifying screen. This fluorescent light not only exposes the adjacent film emulsion, but can pass through the film base to the opposite emulsion. Because the fluorescent light diffuses as it travels to the opposite emulsion, there is a decrease in image sharpness. This occurrence, termed crossover, is minimized by the incorporation of a light-absorbing dye in the active layer. *(Carroll, p 200)*

109. **(B)** The vast majority of target interactions involve the incident electrons and outer-shell tungsten electrons. No ionization occurs and the energy loss is reflected in heat generation. The production of x-rays is an amazingly inefficient process; more than 99 percent of the electrons' kinetic energy is changed to heat energy and less than 1 percent into x-ray photon energy. This presents a serious heat

buildup problem in the anode, as heat production is directly proportional to tube current. *(Selman, pp 155–156)*

110. **(C) and 111. (B)** The figure shows the component parts of a rotating anode x-ray tube enclosed within a glass envelope (no. *1*) to preserve the *vacuum* necessary for x-ray production. Number *2* depicts the anode, *3* is its stem, and *4* is the stator and *5* the rotor of the induction motor that rotates the anode. Number *6* is the target for the electrons, called the focal spot; it forms a band within the beveled edge of the anode, forming the focal track. Number *7* is the cathode's filament that functions to liberate electrons when heated to white hot, and *8* is the molybdenum focusing cup which focuses the electrons toward the focal spot. *(Saia, PREP, pp 376–377)*

112. **(B)** The protective curtain, usually made of leaded vinyl with at least 0.25 mm lead equivalent, must be positioned between the patient and fluoroscopist to greatly reduce exposure to energetic scatter from the patient. As with overhead equipment, fluoroscopic total filtration must be at least 2.5 mm Al equivalent to reduce excessive exposure to soft radiation. Focal spot size is unrelated to patient or personnel exposure. *(Selman, pp 630–631)*

113. (B) *Current* is defined as the amount of electric charge flowing per second. *Voltage* is the potential difference existing between two points. *Resistance* is the property of a circuit that opposes current flow. *Capacitance* describes a quantity of stored electricity. *(Selman, pp 66–67)*

114. (C) There are two main types of mobile x-ray units: capacitor discharge and battery powered. The capacitor discharge units consist of a capacitor, or condenser, which is given a charge and then stores energy until the x-ray tube uses it to produce x-rays. The charge may not be stored for extended periods, however, because the charge tends to "leak" away; the capacitor must be charged just before the exposure is made. Its x-ray tube is grid controlled, permitting very fast (short) exposure times. Capacitors discharge a direct current (as opposed to single- or three-phase pulsating current) in which the kilovoltage decreases by a value of approximately 1 kV per mAs. So, although the value at the onset of the exposure may be 20 mAs and 80 kVp, at the end of the exposure the kV value will be approximately 60. Additionally, capacitor discharge units permit only limited mAs values, usually 30 to 50 mAs per charge. *(Curry, p 51)*

115. (C) A phototimer is one type of automatic exposure device (AED). When it is installed in the x-ray unit, it is calibrated to produce radiographic densities as required by the radiologist. Once the part being radiographed has been exposed to produce the proper film density, the phototimer automatically terminates the exposure. The manual timer should be used as a backup timer should the phototimer fail to terminate the exposure, thus protecting the patient from overexposure and the x-ray tube from excessive heat load. Circuit breakers and fuses are circuit devices used to protect circuit elements from overload. In case of current surge, the circuit will be broken, thus preventing equipment dam-

age. A rheostat is a type of variable resistor. *(Saia, PREP, p 215)*

116. (D) When a low-energy x-ray photon interacts with tissue, a K shell orbital electron may be ejected from orbit. Because the K shell must be filled, an L shell electron drops down to fill the K vacancy. Because the L electron is at a higher energy level, it liberates the excess energy in the form of a K characteristic x-ray as it makes its transit to the K shell. K characteristic radiation is soft radiation and readily absorbed by tissue. This is a description of the photoelectric effect in which all of the incident photon energy is absorbed by tissue and results in high radiographic contrast. Characteristic radiation is also produced at the target when an incident electron ejects a K shell tungsten electron; this is responsible for only about 15 percent of the photons produced at the target. *(Selman, pp 156–158)*

117. (A) Film exposed by the output phosphor of the image intensifier requires less patient exposure than film exposed in cassettes by x-ray stimulation of phosphors. This partially explains the increased use of 100-mm cut and 105-mm strip film spot film cameras and the decrease in the use of cassette spot films. Cinefluorography can record many more frames per second than the strip or cut spot films, but the patient exposure is greater. Additionally, the size of the film influences patient dose; 35 mm requires more exposure than 16 mm. *(Curry, pp 224–225)*

118. (B) Off-focus radiation is produced as electrons strike metal surfaces other than the focal track and produce x-rays that emerge with the primary beam at a variety of angles. This radiation is responsible for indistinct images outside the collimated field. Off-focus radiation is minimized by mounting a pair of shutters as close to the source as possible. *(Selman, p 168)*

119. (B) Because the high-voltage transformer is a fixed ratio, there must be a means of changing the voltage sent to its primary coil; otherwise there would be a fixed kVp. The autotransformer makes these changes possible. As kVp is selected on the control panel, the radiographer is actually adjusting the auto-transformer and selecting the amount of voltage to send to the high-voltage transformer to be stepped up. The filament circuit supplies proper current and voltage to the x-ray tube filament for proper thermionic emission. The rectifier circuit is responsible for changing AC to unidirectional current. *(Selman, pp 130–133)*

120. **(D)** As the target angle is decreased (made steeper), a larger actual focal spot may be used, while still maintaining the same small effective focal spot. Because the actual focal spot is larger, it can accommodate a greater heat load. However, with steeper (smaller) target angles, the anode heel effect is accentuated and can compromise film coverage. *(Bushong, pp 124–125)*

121. **(C)** The lead strips of a focused grid are angled to correspond to the configuration of the divergent x-ray beam. Thus, any radiation changing direction, as is typical of scatter radiation, will be trapped by the lead foil strips. Therefore, if the CR and grid center do not correspond, lead strips will absorb primary radiation. The absorption of primary radiation is termed "cutoff" and results in diminished radiographic density. *(Curry, p 106)*

122. **(D)** When a dual-field image intensifier is switched to the smaller field, the electrostatic focusing lenses are given a greater charge to focus the electron image more tightly. The focal point, then, moves further from the output phosphor (the diameter of the electron image is therefore smaller as it reaches the output phosphor) and the brightness gain is somewhat diminished. Hence, the patient area viewed is somewhat smaller and magnified. However, the minification gain has been reduced and the image is somewhat less bright. *(Wolbarst, p 235)*

123. **(B)** A phototimer is one type of automatic exposure device (AED). When it is installed in the x-ray circuit, it is calibrated to produce radiographic densities as required by the radiologist. Once the part being radiographed has been exposed to produce the correct film density, the phototimer automatically terminates the exposure. The manual timer should be used as a backup timer, in case the AED fails to terminate the exposure, thus protecting the patient from overexposure and the x-ray tube from excessive heat load. The master density is generally set on normal to produce the required densities. In special cases, when this produces excessive or insufficient density, the master density may be adjusted to plus or minus density. *(Saia, PREP, p 215)*

124. **(A)** Radiographic rating charts enable the operator to determine the maximum safe mA, exposure time, and kVp for a particular exposure using a particular x-ray tube. An exposure made safely with the large focal spot may not be safe for use with the small focal spot of the same x-ray tube. The total number of heat units (HU) an exposure generates also influences the amount of stress (in the form of heat) imparted to the anode. Heat units are determined by the product of mAs and kVp. Group A produces 138 HU, group B produces 240 HU, group C produces 420 HU, and group D produces 720 HU. Group D is also delivering its heat load to the small focal spot making this the *most hazardous* group of technical factors. The *least hazardous* group of technical factors is, therefore, group A. *(Selman, pp 217–220)*

125. **(C)** A radiographic rating chart enables the radiographer to determine the maximum safe mA, exposure time, and kVp for a given exposure using a particular x-ray tube. Because the heat load an anode will safely accept varies with the size of the focal spot, type rectification, and anode rotation, these variables must also be identified. Each x-ray tube has its own characteristics and its own rating chart. First, find the chart with the identifying single-phase sine wave in the upper-right corner of the chart, and the correct focal spot size in the upper-left corner of the chart (chart C). Once the correct chart has been identified, locate $\frac{1}{50}$ (0.02) sec on the horizontal axis and follow its line up to where it intersects with the 300 mA curve. Then draw a

line to where this point meets the vertical (kVp) axis; it meets about between 100 and 110 kVp, at approximately 107 kVp. This is the maximum permissible kVp exposure at the given mAs for this x-ray tube. The radiographer should always use somewhat less than the maximum exposure. This same procedure is followed to answer the next two questions. *(Saia, PREP, pp 379–381)*

126. **(B)** Only x-ray tubes A and B, the three-phase rectified x-ray tubes, will safely permit this exposure. Locate 0.1 sec on the horizontal axis and follow it up to where it intersects with the 400 mA curve. X-ray tube A will permit over 150 kVp safely, while x-ray tube B will safely permit only about 92 kVp. Notice the significant difference between the two, solely due to the difference in focal spot size! X-ray tube C will only permit about 75 kVp at the given mAs. *(Saia, PREP, pp 379–381)*

127. **(C)** Find the correct chart for the three-phase, 2-mm focal spot x-ray tube. Locate 0.1 sec on the horizontal (seconds) axis and follow it up to where it intersects with the 120-kVp line on the vertical (kVp) axis. They intersect midway between the 600 and 700 mA curves, at approximately 650 mA. Thus, 600 mA is the maximum safe mA for this particular group of exposure factors and x-ray tube. *(Saia, PREP, pp 379–381)*

128. **(B)** Each x-ray exposure made by the radiographer produces hundreds or thousands of heat units at the target. If the examination requires several consecutive exposures, or if two radiographers are working together and many exams are being completed quickly, the potential for extreme heat load is increased. Just as each x-ray tube has its own radiographic rating chart, so does it have its own anode cooling curve to describe its unique heating and cooling characteristics. An x-ray tube generally cools most rapidly the first 2 min of nonuse. First, locate the 250,000 HU point at the top of the vertical axis. Next, find 100,000 HU on the same axis and follow it across horizontally until it intersects with the cooling curve. Now subtract the cooling time in minutes found opposite

250,000 HU (1) from that found opposite 100,000 HU (7.5). Thus, 6.5 min are required for the anode to cool from 250,000 HU to 100,000 HU. *(Bushong, p 143)*

129. **(A)** Radiographic reproducibility is an important concept in producing high-quality diagnostic films. Radiographic results should be consistent and predictable, not only in positioning accuracy but with respect to exposure factors as well. Automatic exposure devices (phototimers and ionization chambers) automatically terminate the x-ray exposure once a predetermined quantity of x-ray has penetrated the patient, thus ensuring consistent results. *(Cullinan, pp 21–22)*

130. **(C)** Single-phase radiographic equipment is much less efficient than three-phase equipment because it has a 100 percent voltage ripple. *With three-phase equipment, voltage never drops to zero*, and x-ray intensity is significantly greater. In order to produce similar density, only two thirds of the original mAs would be used for three-phase, six-pulse equipment (⅔ × 12 = 8 mAs). With three-phase, twelve-pulse equipment, the original mAs would be cut in half. *(Saia, PREP, pp 373–374)*

131. **(A)** Automatic exposure devices (AEDs) are used in today's equipment and serve to produce consistent and comparable results. In one type of AED there is an ionization chamber beneath the tabletop above the cassette. The part to be examined is centered to it (the sensor) and radiographed. When a predetermined quantity of ionization has occurred (equal to the correct density), the exposure automatically terminates. With the second type of AED, the phototimer, a small fluorescent screen is positioned *beneath* the cassette. When remnant radiation emerging from the patient exposes the film and exits the cassette, the fluorescent screen emits light. Once the predetermined amount of fluorescent light is "seen" by the photocell sensor, the exposure is automatically terminated. In either case the manual timer should be used as a *backup timer*; in case of AED malfunction, the exposure would terminate, thus avoiding pa-

tient overexposure and tube overload. (*Saia, PREP, p 215*)

132. (C) A large quantity of heat applied to a cold anode can cause enough surface heat to crack the anode. Excessive heat to the target can cause pitting or localized melting of the focal track. Localized melts can result in vaporized tungsten deposits on the glass envelope, which can cause a filtering effect, decreasing tube output. Excessive heat can be conducted to the rotor bearings, causing increased friction and tube failure. (*Selman, p 216–221*)

133. (D) The x-ray anode may be a molybdenum disc coated with a tungsten-rhenium alloy. Tungsten, with a *high atomic number* (74), produces high-energy x-rays more efficiently. Since a great deal of heat is produced at the target, its *high melting point* (3410°C) helps avoid damage to the target surface. Heat produced at the target should be dissipated readily, and tungsten's *conductivity is similar to that of copper*. Therefore, as heat is applied to the focus, it can be conducted throughout the disc to equalize the temperature and thus avoid pitting, or localized melting, of the focal track. (*Selman, p 159*)

134. (C) A *radiographic rating chart* is used to determine if the selected mA, exposure time, and kVp are within safe tube limits. An *anode cooling curve* identifies how many heat units the anode can accommodate and the length of time required for adequate cooling between exposures. A *technique chart* is used to determine the correct exposure factors for a particular part of the body of a given thickness. A *spinning top test* is used to test for timer inaccuracy or rectifier failure. (*Bushong, pp 142–143*)

135. (C) The number of *heat units* produced during a given exposure with single-phase equipment is determined by multiplying mA × sec × kVp. Correction factors are required with three-phase equipment. Unless the equipment manufacturer specifies otherwise, *three-phase, six-pulse* heat units are determined by multiplying mA × sec × kVp × 1.35. Three-phase, twelve-pulse heat units are de-

termined by multiplying mA × sec × kVp × 1.41. (*Selman, p 220*)

136. (B) The incident electron has a certain amount of energy as it approaches the tungsten target. If the positive nucleus of a tungsten atom attracts the electron, changing its course, a certain amount of energy is released during the "braking" action. This energy is given up in the form of an x-ray photon called Bremsstrahlung ("braking") radiation. Characteristic radiation is also produced at the target (less frequently) when an incident electron ejects a K shell electron and an L electron drops into its place. Energy is liberated in the form of a characteristic ray, and its energy is representative of the difference in energy levels. Compton scatter and photoelectric effect are interactions between x-ray photons and tissue atoms. (*Selman, pp 156–159*)

137. (A) The spinning top test is used to test timer accuracy or valve tube operation. Because single-phase, full-wave rectified current has 120 useful impulses per second, a 1-sec exposure of the spinning top should demonstrate 120 dots. Therefore, a *0.05*-sec exposure should demonstrate *six* dots. Anything more or less than that indicates that the time station needs calibration. If exactly *one half* of the expected number of dots appeared, one should suspect *rectifier* failure. (*Saia, PREP, p 385*)

138. (C) The visual apparatus responsible for visual acuity and contrast perception are the *cones* within the retina. Cones are also used for daylight vision. Therefore, the most desirable condition for fluoroscopic viewing is to have a bright enough image that will permit cone (daylight) vision, for better detail perception. The image intensifier accomplishes this. The intensified image is then transferred to a TV monitor for viewing. Cine and spot film cameras record fluoroscopic events. (*Bushong, pp 356–357*)

139. (A) Because using the master control's minus-density adjustment involves decreasing the exposure time (and this is not possible), the adjustment will be ineffective. Decreasing the kVp will effect a change in radiographic contrast. Because too long an exposure time results in excessive density, the best way to compensate is to decrease the mA. *(Saia, PREP, p 319)*

140. (C) The cassette is used to support the intensifying screens and x-ray film. It should be strong and provide good screen/film contact. The cassette front should be made of a sturdy material with a low atomic number, because attenuation of the remnant beam is undesirable. Bakelite (the forerunner of today's plastics) and magnesium (the lightest structural metal) are the most commonly used materials for cassette fronts. The high atomic number of tungsten makes it inappropriate as a cassette front material. *(Selman, p 274)*

141. (D) The AEC automatically adjusts the exposure required for body parts having different *thicknesses and densities*. Proper functioning of the phototimer depends on accurate positioning by the radiographer. The correct *photocell*(s) must be selected and the anatomic part of interest must completely cover the photocell in order to achieve the desired density. If *collimation* is inadequate, and a field size larger than the part is used, excessive scatter radiation from the body or tabletop can cause the AEC to terminate the exposure prematurely, resulting in an underexposed radiograph. Backup time should always be selected on the manual timer to prevent patient overexposure and to protect the x-ray tube from excessive heat production, should the AEC malfunction. Selection of the optimum kV for the part being radiographed is essential—no practical amount of mAs can make up for inadequate penetration (kV), and excessive kV can cause the AEC to terminate the exposure prematurely. *A technique chart is therefore strongly recommended for use with AEC; it should indicate the optimum kV for the part, the photocells that should be selected, and the backup time that should be set. (Carroll, p 337)*

142. (B) Subtraction technique is used most frequently with angiographic films. Tiny vessels needing evaluation are frequently superimposed on bony details. The blood vessels may be "subtracted out" by using either a digital or manual subtraction technique. Xeroradiography involves imaging a structure on an electrically charged plate. The electrostatic image is converted to a visible image in a special conditioner/processor. *(Carroll, p 498)*

143. (D) As the filament ages, vaporized tungsten may be deposited on the port window and act as an additional filter. Tungsten may also vaporize as a result of anode abuse. Exposures in excess of safe values deliver sufficient heat to cause surface melts, or pits, on the focal track. This results in roughening of the anode surface and decreased tube output. Delivery of a large amount of heat to a cold anode can cause cracking if the anode does not have sufficient time to disperse the heat. Loss of anode rotation would cause one large melt on the focal track as the electrons would bombard only one small area. If the anode was not heard rotating, the radiographer should not make an exposure. *(Selman, p 208)*

144. (D) The automatic exposure control (AEC) automatically terminates the exposure when the proper density has been recorded on the film. The important advantage of the phototimer, then, is that it can accurately duplicate radiographic densities. It is very useful in providing accurate comparison in follow-up examinations, and in decreasing patient exposure dose by decreasing the number of "retakes" due to improper exposure. The AEC automatically adjusts the exposure required for body parts having different *thicknesses and densities*. Remember that proper functioning of the phototimer depends on accurate positioning by the radiographer. The correct *photocell*(s) must be selected and the anatomic part of interest must completely cover the photocell in order to achieve the desired density. If *collimation* is inadequate, and a field size larger than the part is used, excessive scatter radiation from the body or tabletop can cause the AEC to terminate the exposure prematurely, resulting in an underexposed radiograph. *(Carroll, p 335)*

145. (A) The image intensifier's input phosphor receives the remnant beam from the patient and converts it to a fluorescent light image. In order to maintain resolution, the input phosphor is made of cesium iodide crystals. Cesium iodide is much more efficient in this conversion process than was the phosphor previously used, zinc cadmium sulfide. Calcium tungstate was the phosphor used in cassette intensifying screens for many years prior to the development of rare earth phosphors such as gadolinium oxysulfide. *(Bushong, p 356)*

146. (C) As filtration is added to the x-ray beam, the lower-energy photons are removed and the overall energy or wavelength of the beam is greater. As kilovoltage is increased, more high-energy photons are produced and again the overall, or average, energy of the beam is greater. An increase in mA serves to increase the number of photons produced at the target, but is unrelated to their energy. *(Selman, p 171)*

147. (A) A *phototimer* is one type of automatic exposure device (AED); it actually measures *light*. As x-ray photons penetrate and emerge from a part, a fluorescent screen beneath the cassette glows, and the fluorescent light charges a photomultiplier tube. Once a predetermined charge has been reached, the exposure automatically terminates. A parallel plate ionization chamber is another type of AED. A radiolucent chamber is beneath the patient (between the patient and film). As photons emerge from the patient, they enter the chamber and ionize the air within. Once a predetermined charge has been reached, the exposure is automatically terminated. *(Wolbarst, pp 204–205)*

148. (D) Vaporized tungsten may be deposited on the inner surface of the glass envelope at the tube (port) window. It acts as an *additional filter*, thereby *reducing tube output*. The tungsten deposit may also attract electrons from the filament, *creating sparking* and *causing puncture* of the glass envelope and subsequent tube failure. *(Selman, p 208)*

149. (B) The image intensifier's input phosphor receives the remnant radiation emerging from the patient and converts it into a fluorescent light image. Very close to the input phosphor, separated by a thin transparent layer, is the photocathode. The photocathode is made of a photoemissive alloy, usually an antimony and cesium compound. The fluorescent light image strikes the photocathode and is converted to an electron image that is focused by the electrostatic lenses to the output phosphor. *(Carlton, pp 356–357)*

150. (A) A synchronous spinning top test is used to test timer accuracy or rectifier function in three-phase equipment. Because three-phase, full-wave rectified current would expose a 360° arc per second, a $\frac{1}{12}$-sec exposure should expose a 30° arc. Anything more or less indicates timer inaccuracy. If exactly one half the expected arc appears, one should suspect rectifier failure. *(Saia, PREP, p 385)*

Subspecialty List

66. Components of basic radiographic unit
67. Types of units
68. Components of basic radiographic unit
69. Components of basic radiographic unit
70. Fluoroscopic unit
71. Types of units
72. Equipment calibration
73. Components of basic radiographic unit
74. Fluoroscopic unit
75. Equipment calibration
76. Components of basic radiographic unit
77. Types of units
78. Fluoroscopic unit
79. Components of basic radiographic unit
80. Equipment calibration
81. Components of basic radiographic unit
82. Equipment calibration
83. Types of units
84. X-ray generator, transformers, and rectification system
85. X-ray generator, transformers, and rectification system
86. Fluoroscopic unit
87. Fluoroscopic unit
88. Components of basic radiographic unit
89. Components of basic radiographic unit
90. Components of basic radiographic unit
91. Components of basic radiographic unit
92. Components of basic radiographic unit
93. X-ray generator, transformers, and rectification system
94. X-ray generator, transformers, and rectification system
95. Components of basic radiographic unit
96. Components of basic radiographic unit
97. Fluoroscopic unit
98. X-ray generator, transformers, and rectification system
99. Components of basic radiographic unit
100. Screens and cassettes
101. Fluoroscopic unit
102. X-ray generator, transformers, and rectification system
103. Screens and cassettes
104. Components of basic radiographic unit
105. Components of basic radiographic unit

106. Equipment calibration
107. Components of basic radiographic unit
108. Screens and cassettes
109. Components of basic radiographic unit
110. Components of basic radiographic unit
111. Components of basic radiographic unit
112. Fluoroscopic unit
113. Components of basic radiographic unit
114. Types of units
115. Recognition of malfunctions
116. Components of basic radiographic unit
117. Fluoroscopic unit
118. Recognition of malfunctions
119. Components of basic radiographic unit
120. Components of basic radiographic unit
121. Recognition of malfunctions
122. Fluoroscopic unit
123. Recognition of malfunctions
124. Components of basic radiographic unit
125. Components of basic radiographic unit
126. Components of basic radiographic unit
127. Components of basic radiographic unit
128. Components of basic radiographic unit
129. Components of basic radiographic unit
130. X-ray generator, transformers, and rectification system
131. Components of basic radiographic unit
132. Recognition of malfunctions
133. Components of basic radiographic unit
134. Components of basic radiographic unit
135. Components of basic radiographic unit
136. Components of basic radiographic unit
137. Recognition of malfunctions
138. Fluoroscopic unit
139. Recognition of malfunctions
140. Screens and cassettes
141. Components of basic radiographic unit
142. Types of units
143. Recognition of malfunctions
144. Components of basic radiographic unit
145. Fluoroscopic unit
146. Components of basic radiographic unit
147. Components of basic radiographic unit
148. Recognition of malfunctions
149. Fluoroscopic unit
150. Recognition of malfunctions

CHAPTER 6

Practice Test
Questions

DIRECTIONS (Questions 1 through 200): Each of the numbered items or incomplete statements in this section is followed by answers or by completions of the statement. Select the ONE lettered answer or completion that is BEST in each case.

1. Which of the following statements is true regarding the control badge that accompanies each shipment of personnel monitors?

    1. it should be stored away from all radiation sources
    2. it should be stored in the main work area
    3. it should be used to replace an employee's lost monitor

    (A) 1 only
    (B) 2 only
    (C) 1 and 3 only
    (D) 2 and 3 only

2. Which of the following is the approximate skin dose for 5 minutes of fluoroscopy performed at 1.5 mA?

    (A) 3.7 rad
    (B) 7.5 rad
    (C) 15 rad
    (D) 21 rad

3. Which of the following is an acceptable approximate entrance skin exposure (ESE) for a PA chest?

    (A) 6 mR
    (B) 20 mR
    (C) 38 mR
    (D) 0.6 R

4. Which of the dose–response curves pictured in Figure 6–1 illustrate(s) a linear threshold dose effect?

    1. curve number 1
    2. curve number 2
    3. curve number 3

    (A) 1 only
    (B) 3 only
    (C) 2 and 3 only
    (D) 1, 2, and 3

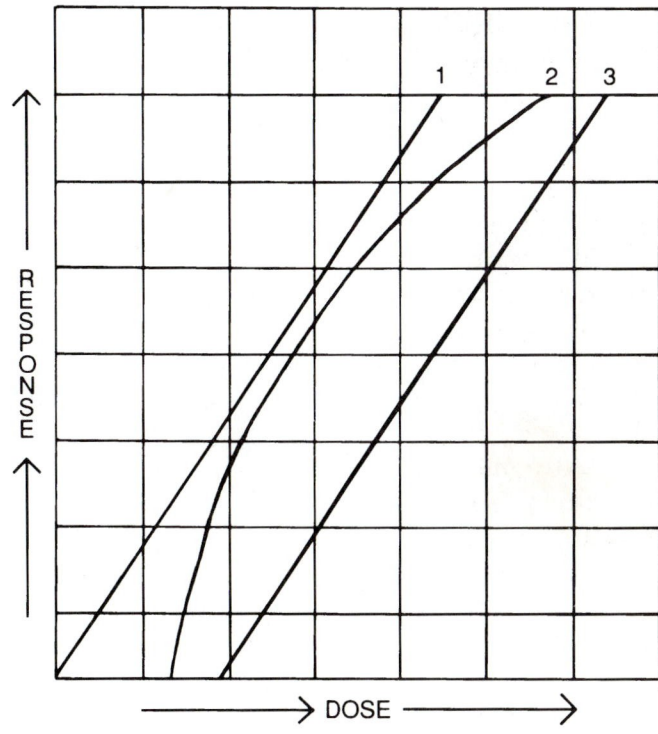

Figure 6–1. Courtesy of David Perri.

5. Which of the following is MOST likely to result in the greatest increase in patient exposure?

   (A) changing from a 400 film/screen combination to a 200 film/screen combination
   (B) increasing kVp 15 percent and cutting mAs in half
   (C) two tomographic cuts instead of two plain films
   (D) from nongrid technique to 8:1 grid

6. In 1906, Bergonié and Tribondeau theorized that undifferentiated cells are highly radiosensitive. Which of the following is (are) characteristics of undifferentiated cells?

   1. young cells
   2. highly mitotic cells
   3. precursor cells

   (A) 1 only
   (B) 1 and 2 only
   (C) 1 and 3 only
   (D) 1, 2, and 3

7. It is usually recommended that a TLD or film badge be worn

   (A) under the lead apron at waist level
   (B) outside the lead apron at waist level
   (C) under the lead apron at collar level
   (D) outside the lead apron at collar level

8. Which of the following types of adult tissues is (are) relatively insensitive to radiation exposure?

   1. muscle tissue
   2. nerve tissue
   3. epithelial tissue

   (A) 1 only
   (B) 1 and 2 only
   (C) 2 and 3 only
   (D) 1, 2, and 3

9. The late effects of radiation are considered to

   1. have no threshold dose
   2. be directly related to dose
   3. occur within hours of exposure

   (A) 1 only
   (B) 1 and 2 only
   (C) 2 and 3 only
   (D) 1, 2, and 3

10. The annual dose limit for occupationally exposed individuals is valid for

   (A) alpha, beta, and x-radiations
   (B) x- and gamma radiations only
   (C) beta, x-, and gamma radiations
   (D) all ionizing radiations

11. If a quantity of radiation is delivered to a body in a short period of time its effect

   (A) will be greater than if it were delivered over a long period of time
   (B) will be less than if it were delivered over a long period of time
   (C) has no relation to how it is delivered in time
   (D) is solely dependent on the radiation quality

12. Linear energy transfer may be BEST described as

   (A) the amount of energy delivered per distance traveled in tissue
   (B) the unit of absorbed dose
   (C) radiation equivalent man
   (D) radiation absorbed dose

13. If the exposure rate to a body standing 3 feet from a radiation source is 12 mR/min, what will be the exposure rate to that body at a distance of 7 feet from the source?

   (A) 2.2 mR/min
   (B) 5.1 mR/min
   (C) 28 mR/min
   (D) 36 mR/min

14. The NCRP recommends an annual effective (stochastic) dose equivalent limit of

(A) 50 mSv (5 rem)
(B) 100 mSv (10 rem)
(C) 25 mSv (2.5 rem)
(D) 200 mSv (20 rem)

15. How are LET and biologic response related?

(A) they are inversely proportional
(B) they are directly proportional
(C) in a reciprocal fashion
(D) they are unrelated

16. Which of the following is (are) considered long-term somatic effect(s) of exposure to ionizing radiation?

1. life-span shortening
2. carcinogenesis
3. cataractogenesis

(A) 1 only
(B) 1 and 2 only
(C) 2 and 3 only
(D) 1, 2, and 3

17. If a patient received 4500 mrad during a 6-min fluoroscopic examination, what was the dose rate?

(A) 0.75 rad/min
(B) 2.7 rad/min
(C) 7.5 rad/min
(D) 27 rad/hr

18. The radiation dose to an individual is dependent on which of the following?

1. type of tissue interaction(s)
2. quantity of radiation
3. biologic differences

(A) 1 only
(B) 1 and 2 only
(C) 1 and 3 only
(D) 1, 2, and 3

19. If the entrance dose for a particular radiograph is 320 mR, the radiation exposure at 1 meter from the patient will be approximately

(A) 32 mR
(B) 3.2 mR
(C) 0.32 mR
(D) 0.032 mR

20. What is the established fetal dose-limit guideline for pregnant radiographers during the entire gestation period?

(A) 0.1 rem
(B) 0.5 rem
(C) 5.0 rem
(D) 10 rem

21. All of the following will have an effect on patient dose, EXCEPT

(A) grid ratio
(B) kVp
(C) focal spot size
(D) SID

22. Which of the following statements regarding film badges is (are) correct?

1. film badges should be read quarterly
2. film badges must not leave the workplace
3. film badges measure quantity and type of radiation exposure

(A) 1 only
(B) 1 and 2 only
(C) 2 and 3 only
(D) 1, 2, and 3

23. Which of the following is (are) included in whole-body dose?

1. gonads
2. lens
3. extremities

(A) 1 only
(B) 1 and 2 only
(C) 2 and 3 only
(D) 1, 2, and 3

24. Which of the following defines the gonadal dose that, if received by every member of the population, would be expected to produce the same total genetic effect on that population as the actual doses received by each of the individuals?

(A) genetically significant dose
(B) somatically significant dose
(C) maximum permissible dose
(D) lethal dose

25. To radiograph an infant with suspected free air within the abdominal cavity, which of the following projections of the abdomen will demonstrate the condition with the LEAST exposure?

(A) PA erect with grid
(B) right lateral decubitus with grid
(C) left lateral decubitus without grid
(D) recumbent AP without grid

26. Which of the following is considered MOST radiosensitive?

(A) lymphocytes
(B) ova
(C) neurons
(D) myocytes

27. What percentage of x-ray attenuation is provided by a 0.5 mm lead equivalent apron at 100 kVp?

(A) 51 percent
(B) 66 percent
(C) 75 percent
(D) 94 percent

28. The primary function of filtration is to reduce

(A) patient skin dose
(B) operator dose
(C) film noise
(D) scattered radiation

29. Which of the groups of exposure factors below would deliver the lowest patient dose?

(A) 2.5 mAs, 100 kVp, 400-speed screens
(B) 10 mAs, 90 kVp, 200-speed screens
(C) 10 mAs, 70 kVp, 800-speed screens
(D) 10 mAs, 80 kVp, 400-speed screens

30. Which of the following would be the safest interval of time for a fertile woman to undergo abdominal radiography without significant concern for irradiating a recently fertilized ovum?

(A) the first 10 days following the cessation of menstruation
(B) the first 10 days following the onset of menstruation
(C) the 10 days preceding the onset of menstruation
(D) about 14 days before menstruation

31. X-ray photon energy is directly related to

1. applied kilovoltage
2. applied milliamperage
3. photon wavelength

(A) 1 only
(B) 1 and 2 only
(C) 1 and 3 only
(D) 1, 2, and 3

32. Which of the following x-ray circuit devices operate(s) on the principle of mutual induction?

1. high-voltage transformer
2. filament transformer
3. autotransformer

(A) 1 only
(B) 1 and 2 only
(C) 1 and 3 only
(D) 1, 2, and 3

33. If the primary coil of the high-voltage transformer is supplied by 110 volts and has 100 turns, and the secondary coil has 80,000 turns, what is the voltage induced in the secondary coil?

(A) 135 kilovolts
(B) 88 kilovolts
(C) 135 volts
(D) 88 volts

34. The source of electrons within the x-ray tube is a result of which process?

(A) electrolysis
(B) thermionic emission
(C) rectification
(D) induction

35. If 85 kVp, 400 mA, and ⅛ sec were used for a particular exposure using single-phase equipment, which of the following mA or time values would be required, all other factors being constant, to produce a similar density using three-phase twelve-pulse equipment?

(A) 200 mA
(B) 600 mA
(C) 0.125 sec
(D) 0.25 sec

36. Double focus x-ray tubes have two

1. focal spots
2. filaments
3. anodes

(A) 1 only
(B) 1 and 2 only
(C) 1 and 3 only
(D) 2 and 3 only

37. Which of the x-ray circuit devices seen in Figure 6–2 operates on the principle of self-induction?

(A) no. 1
(B) no. 2
(C) no. 3
(D) no. 7

38. Referring to the simplified x-ray circuit seen in Figure 6–2, what is indicated by the 5?

(A) step-up transformer
(B) autotransformer
(C) filament circuit
(D) rectification system

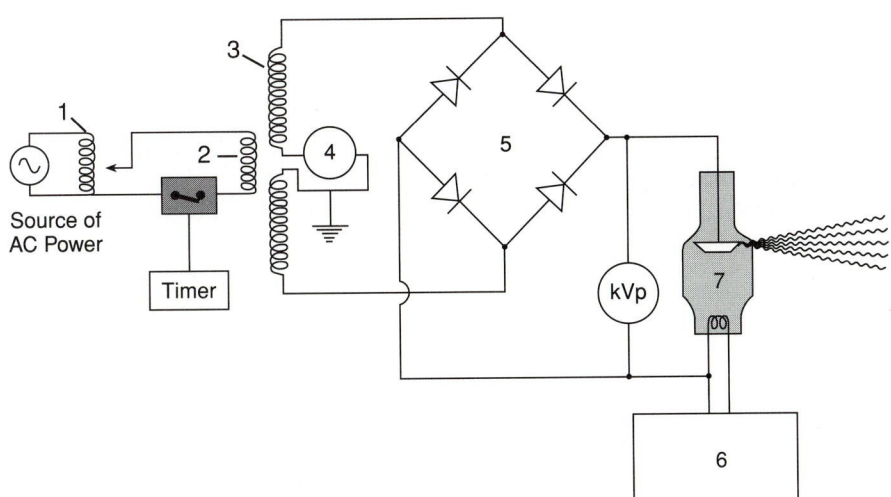

Figure 6–2. Reproduced with permission from Wolbarst AB. *Physics of Radiology*. East Norwalk, CT: Appleton & Lange, 1993.

39. TV camera tubes used in image intensification, such as the plumbicon and vidicon, function to

    (A) increase the brightness of the input phosphor image
    (B) transfer the output phosphor image to the TV monitor
    (C) focus and accelerate electrons toward the output phosphor
    (D) record the output phosphor image on 16 or 35-mm film

40. A light-absorbing dye is frequently incorporated during the manufacture of screens to

    (A) reduce the diffusion of fluorescent light
    (B) increase film contrast
    (C) increase screen speed
    (D) increase the useful life of the screen

41. Which of the following is (are) correct regarding care of protective leaded apparel?

    1. lead aprons should be fluoroscoped yearly to check for cracks
    2. lead gloves should be fluoroscoped yearly to check for cracks
    3. lead aprons should be hung on appropriate racks when not in use

    (A) 1 only
    (B) 1 and 2 only
    (C) 1 and 3 only
    (D) 1, 2, and 3

42. X-ray tube life may be extended by

    1. using low-mAs, high-kVp exposure factors
    2. avoiding lengthy anode rotation
    3. avoiding exposures to a cold anode

    (A) 1 only
    (B) 1 and 2 only
    (C) 1 and 3 only
    (D) 1, 2, and 3

43. 105-mm spot film images are recorded from the

    (A) spot film cassettes
    (B) output phosphor of the image intensifier
    (C) input phosphor of the image intensifier
    (D) TV monitor screen

44. Several types of exposure timers may be found on x-ray equipment. Which of the following types of timers functions to accurately duplicate radiographic densities?

    (A) synchronous
    (B) impulse
    (C) electronic
    (D) phototimer

45. According to the line focus principle, an anode with a small angle provides the following advantages.

    1. improved recorded detail
    2. improved heat capacity
    3. less heel effect

    (A) 1 and 2 only
    (B) 1 and 3 only
    (C) 2 and 3 only
    (D) 1, 2, and 3

46. Which of the following will serve to increase the effective energy of the x-ray beam?

    1. increase in added filtration
    2. increase in kilovoltage
    3. increase in milliamperage

    (A) 1 only
    (B) 2 only
    (C) 1 and 2 only
    (D) 1, 2, and 3

47. In order to eject a K shell electron from a tungsten atom, the incoming electron must have an energy of at least

    (A) 60 keV
    (B) 70 keV
    (C) 80 keV
    (D) 90 keV

48. The radiographic image in Figure 6–3 was obtained while testing

    (A) valve tube operation in a single-phase machine
    (B) valve tube operation in a three-phase machine
    (C) timer accuracy in a single-phase machine
    (D) timer accuracy in a three-phase machine

49. All of the following statements regarding three-phase power are true, EXCEPT

    (A) three-phase radiation is pulsed rather than constant
    (B) three-phase equipment produces more x-rays per mAs
    (C) three-phase produces higher average energy x-rays than single phase
    (D) the three-phase waveform has less ripple than the single phase

50. An automatic exposure device (AED) can operate on which of the following principles?

    1. a photomultiplier tube charged by fluorescent screen
    2. a parallel plate ionization chamber charged by x-ray photons
    3. motion of magnetic fields inducing current in a conductor

    (A) 1 only
    (B) 2 only
    (C) 1 and 2 only
    (D) 1, 2, and 3

51. The fact that x-ray intensity across the primary beam can vary as much as 45 percent describes the

    (A) line focus principle
    (B) transformer law
    (C) anode heel effect
    (D) Inverse Square Law

52. Which of the following constitutes the largest portion of the x-ray beam emerging from the target?

    (A) characteristic radiation
    (B) Bremsstrahlung radiation
    (C) Compton scatter
    (D) photoelectric

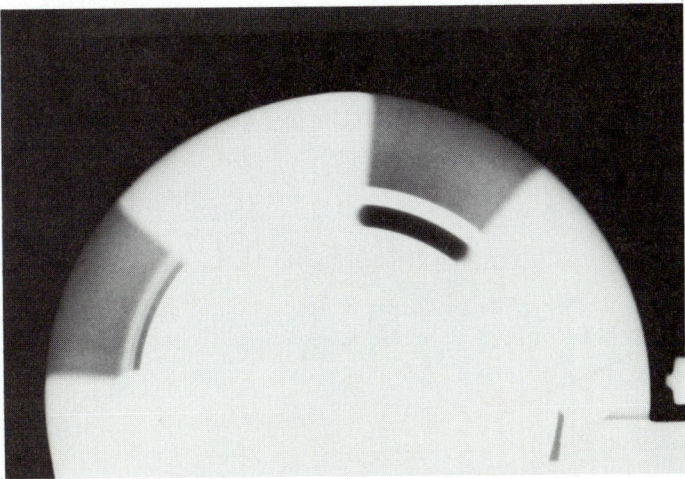

Figure 6–3. Courtesy of The Stamford Hospital, Department of Radiology.

53. The collimator light and actual irradiated area must be accurate to within

    (A) 2 percent of the SID
    (B) 5 percent of the SID
    (C) 10 percent of the SID
    (D) 15 percent of the SID

54. The purpose of inherent and added filtration in the x-ray tube is to

    (A) reduce patient skin dose
    (B) shorten the scale of contrast
    (C) reduce scatter radiation
    (D) soften the x-ray beam

55. Compared to the first half-value layer (HVL), the second half-value layer of a particular x-ray beam is generally

    (A) less than the first HVL
    (B) greater than the first HVL
    (C) equal to the first HVL
    (D) equal to or less than the first HVL

56. In which of the following examinations would a cassette front with very low absorption properties be especially desirable?

    (A) extremity radiography
    (B) abdominal radiography
    (C) mammography
    (D) angiography

57. The brightness level of the fluoroscopic image is dependent on

    1. milliamperage
    2. kilovoltage
    3. patient thickness

    (A) 1 only
    (B) 1 and 2 only
    (C) 1 and 3 only
    (D) 1, 2, and 3

58. The housing surrounding an x-ray tube functions to

    1. retain heat within the glass envelope
    2. protect from electric shock
    3. keep leakage radiation to a minimum

    (A) 1 and 2 only
    (B) 1 and 3 only
    (C) 2 and 3 only
    (D) 1, 2, and 3

59. Which of the following functions to protect the x-ray tube and patient from overexposure in the event the phototimer fails to terminate an exposure?

    (A) circuit breaker
    (B) fuse
    (C) back-up timer
    (D) rheostat

60. Which of the following combinations would deliver the least amount of heat to the anode of a three-phase, twelve-pulse x-ray unit?

    (A) 400 mA, 0.12 sec, 90 kVp
    (B) 300 mA, ½ sec, 70 kVp
    (C) 500 mA, ⅟₃₀ sec, 85 kVp
    (D) 700 mA, 0.06 sec, 120 kVp

61. The functions of the automatic processor's recirculation system include(s)

    1. keeping solution in contact with film emulsion
    2. maintaining constant temperatures
    3. mixing and agitating solutions

    (A) 1 only
    (B) 1 and 3 only
    (C) 2 and 3 only
    (D) 1, 2, and 3

62. When the collimated field must extend past the edge of the body, allowing primary radiation to strike the tabletop, as in a lateral lumbar spine, what may be done to prevent excessive radiographic density due to undercutting?

(A) reduce the mAs

(B) reduce the kVp

(C) use a shorter SID

(D) use lead rubber to absorb tabletop primary radiation

63. It would be necessary to repeat the radiograph of the pelvis seen in Figure 6–4 for the following reason.

(A) motion

(B) inadequate penetration

(C) scattered radiation fog

(D) excessive density

64. Magnification will increase with a(n)

1. decrease in SID

2. decrease in SOD

3. increase in OID

(A) 1 only

(B) 2 only

(C) 2 and 3 only

(D) 1, 2, and 3

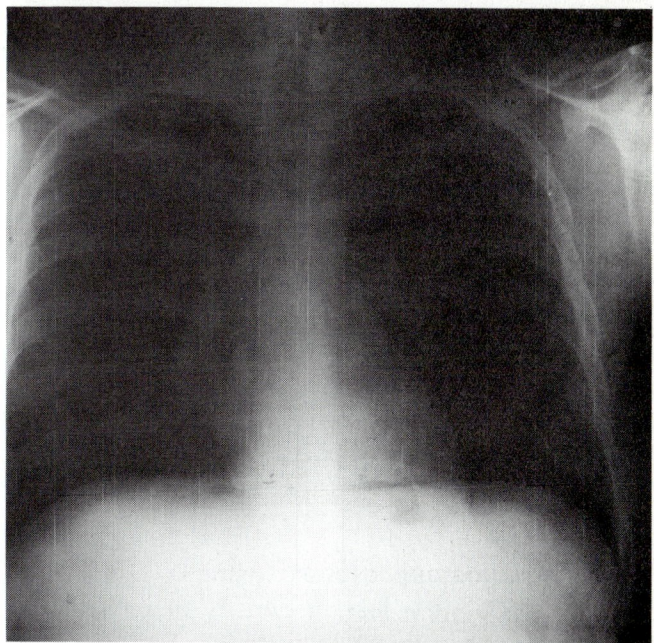

Figure 6–5. Courtesy of Fuji Medical Systems U.S.A., Inc.

65. The vertical lines present on the radiograph in Figure 6–5 represent which of the following?

(A) improperly placed guide shoes

(B) sediment on processor rollers

(C) pi lines

(D) grid lines

66. Which of the following groups of exposure factors would be MOST appropriate for an adult intravenous pyelogram?

(A) 300 mA, 0.02 sec, 72 kVp

(B) 300 mA, 0.01 sec, 82 kVp

(C) 150 mA, 0.01 sec, 94 kVp

(D) 100 mA, 0.03 sec, 82 kVp

67. Which of the following may be used to control the production of scatter radiation?

1. restricted field size

2. use of optimum kVp

3. use of grids

(A) 1 only

(B) 1 and 2 only

(C) 2 and 3 only

(D) 1, 2, and 3

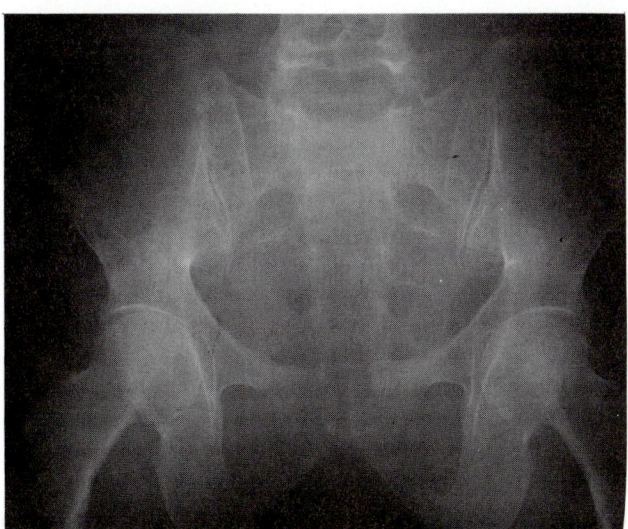

Figure 6–4. From the American College of Radiology Learning File. Courtesy of the ACR.

68. An exposure was made at a 36 inch SID using 12 mAs and 75 kVp with a 400 speed imaging system and an 8:1 grid. A second radiograph is requested with improved recorded detail. Which of the following groups of technical factors will BEST accomplish this task?

    (A) 15 mAs, 12:1 grid, 75 kVp, 400 speed system, 36 inch SID
    (B) 15 mAs, 12:1 grid, 75 kVp, 400 speed system, 40 inch SID
    (C) 30 mAs, 12:1 grid, 75 kVp, 200 speed system, 40 inch SID
    (D) 12 mAs, 8:1 grid, 86 kVp, 200 speed system, 36 inch SID

69. If a radiograph exposed using a 12:1 ratio grid exhibits a loss of density at its lateral edges, it is probably because the

    (A) SID was too great
    (B) grid failed to move during the exposure
    (C) x-ray tube was angled in the direction of the lead strips
    (D) central ray was off center

70. Which of the following will produce the greatest distortion?

    (A) AP projection of the skull
    (B) PA projection of the skull
    (C) 37° AP axial of the skull
    (D) 20° PA axial of the skull

71. Very low humidity in the darkroom can lead to which of the following?

    (A) crinkle marks
    (B) static electrical discharge
    (C) excessive emulsion swelling
    (D) chemical fog

72. Which of the following is (are) essential to high-quality mammographic examinations?

    1. small focal spot x-ray tube
    2. high radiographic contrast
    3. use of a compression device

    (A) 1 only
    (B) 1 and 2 only
    (C) 1 and 3 only
    (D) 1, 2, and 3

73. A technique chart should include which of the following information?

    1. recommended SID
    2. grid ratio
    3. screen/film combination

    (A) 1 only
    (B) 1 and 2 only
    (C) 1 and 3 only
    (D) 1, 2, and 3

74. The device that is used for the direct measurement of optical density is the

    (A) sensitometer
    (B) densitometer
    (C) penetrometer
    (D) H&D curve

75. The effect that differential absorption has on radiographic contrast can be minimized by

    1. using a compensating filter
    2. using high kVp exposure factors
    3. increased collimation

    (A) 1 only
    (B) 1 and 2 only
    (C) 2 and 3 only
    (D) 1, 2, and 3

76. Which of the following radiographic accessories functions to produce uniform density on a radiograph?

    (A) grid
    (B) intensifying screens
    (C) compensating filter
    (D) penetrometer

77. Improper spectral matching between rare earth intensifying screens and film emulsion results in

    (A) longer scale contrast
    (B) insufficient density
    (C) decreased recorded detail
    (D) excessive density

78. An exposure was made using 600 mA and 18 msec. If the mA is changed to 400, which of the following exposure times would MOST closely approximate the original radiographic density?

    (A) 16 msec
    (B) 0.16 sec
    (C) 27 msec
    (D) 0.27 sec

79. Silver reclamation may be accomplished in which of the following ways?

    1. metallic replacement cartridge
    2. electrolytic plating unit
    3. removal from used film

    (A) 1 only
    (B) 2 only
    (C) 1 and 2 only
    (D) 1, 2, and 3

80. Acceptable method(s) of minimizing motion unsharpness is (are)

    1. suspended respiration
    2. short exposure time
    3. patient instruction

    (A) 1 only
    (B) 1 and 2 only
    (C) 1 and 3 only
    (D) 1, 2, and 3

81. Which portion of the characteristic curve would MOST likely represent a density of 1.0?

    (A) toe
    (B) straight-line portion
    (C) shoulder
    (D) D-max

82. Shape distortion is influenced by the relationship between the

    1. x-ray tube and part to be imaged
    2. body part to be imaged and the film
    3. film and the x-ray tube

    (A) 1 and 2 only
    (B) 1 and 3 only
    (C) 2 and 3 only
    (D) 1, 2, and 3

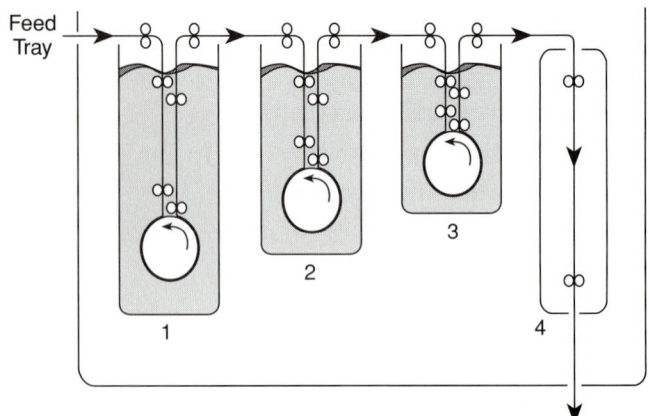

Figure 6–6. Reproduced with permission from Wolbarst AB. *Physics of Radiology.* East Norwalk, CT: Appleton & Lange, 1993.

83. In which section of the automatic processor seen in Figure 6–6 are the unexposed silver halide crystals removed from the emulsion?

(A) section 1
(B) section 2
(C) section 3
(D) section 4

84. Which of the following is MOST likely to occur as a result of using a 30-inch SID with a 14×17 inch film to radiograph a fairly homogeneous structure?

(A) production of quantum mottle
(B) density variation between opposite ends of the film
(C) production of scatter radiation fog
(D) excessively short-scale contrast

85. Which of the following is (are) characteristics of a 16:1 grid?

1. absorbs more primary radiation than an 8:1 grid
2. has more centering latitude than an 8:1 grid
3. used with higher kVp exposures than an 8:1 grid

(A) 1 only
(B) 1 and 3 only
(C) 2 and 3 only
(D) 1, 2, and 3

86. If a radiograph were made of an average size knee using automatic exposure control and all three photocells were selected, the resulting radiograph would demonstrate

(A) excessive density
(B) insufficient density
(C) poor detail
(D) adequate exposure

87. The radiograph in Figure 6–7 exhibits a loss of radiographic density due to

(A) x-ray tube angulation across grid lines
(B) exceeding the focusing distance
(C) a focused grid placed upside down
(D) insufficient SID

88. Which of the following is (are) classified as a rare earth phosphor?

1. lanthanum oxybromide
2. gadolinium oxysulfide
3. cesium iodide

(A) 1 only
(B) 1 and 2 only
(C) 2 and 3 only
(D) 1, 2, and 3

89. The reduction in x-ray photon intensity as it passes through material is termed

(A) absorption
(B) scattering
(C) attenuation
(D) divergence

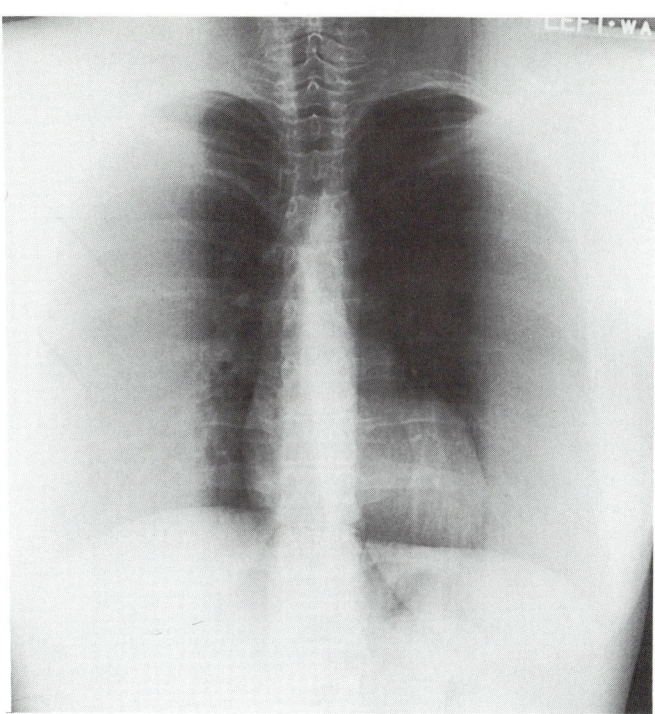

Figure 6–7. Courtesy of The Stamford Hospital, Department of Radiology.

90. An increase in the kilovoltage applied to the x-ray tube increases the

    1. percentage of high energy photons produced
    2. exposure rate
    3. patient absorption

    (A) 1 only
    (B) 1 and 2 only
    (C) 2 and 3 only
    (D) 1, 2, and 3

91. Which of the following groups of exposure factors will produce the longest scale of contrast?

    (A) 200 mA, 0.25 sec, 70 kVp, 12:1 grid
    (B) 500 mA, 0.10 sec, 90 kVp, 8:1 grid
    (C) 400 mA, 0.125 sec, 80 kVp, 12:1 grid
    (D) 300 mA, 0.16 sec, 70 kVp, 8:1 grid

92. Geometric unsharpness is most likely to be greater

    (A) at long SIDs
    (B) at the anode end of the image
    (C) with small focal spots
    (D) at the cathode end of the image

93. Greater latitude is available to the radiographer in which of the following circumstances?

    1. using high kVp techniques
    2. using slow film/screen combination
    3. using high-ratio grid

    (A) 1 only
    (B) 1 and 2 only
    (C) 2 and 3 only
    (D) 1, 2, and 3

94. An increase in kVp with appropriate compensation of mAs will result in

    1. increased exposure latitude
    2. higher contrast
    3. increased density

    (A) 1 only
    (B) 1 and 2 only
    (C) 2 and 3 only
    (D) 1, 2, and 3

95. If the quantity of black metallic silver on a particular radiograph is such that it allows 1 percent of the illuminator light to pass through the film, that film has a density of

    (A) 0.01
    (B) 0.1
    (C) 1.0
    (D) 2.0

96. Substituting intensifying screens having a speed of 200 in place of a 100 speed system will

   1. require 50 percent less exposure than 100 speed screens
   2. increase the production of scattered radiation
   3. enable the radiographer to decrease the exposure time

   (A) 1 only
   (B) 1 and 2 only
   (C) 1 and 3 only
   (D) 1, 2, and 3

97. Exposure factors of 80 kVp and 8 mAs are used for a particular nongrid exposure. What should be the new mAs if a 8:1 grid is added?

   (A) 16 mAs
   (B) 24 mAs
   (C) 32 mAs
   (D) 40 mAs

98. Figure 6–8 is an example of a

   (A) bar pattern test
   (B) Wisconsin test tool
   (C) star resolution test
   (D) screen/film contact test

99. Which of the following contribute to the radiographic contrast present on the finished radiograph?

   1. tissue density
   2. pathology
   3. muscle development

   (A) 1 and 2 only
   (B) 1 and 3 only
   (C) 2 and 3 only
   (D) 1, 2, and 3

Figure 6–8. From the American College of Radiology Learning File. Courtesy of the ACR.

100. Which of the following conditions require(s) a decrease in technical factors?

   1. emphysema
   2. osteomalacia
   3. atelectasis

   (A) 1 only
   (B) 1 and 2 only
   (C) 2 and 3 only
   (D) 1, 2, and 3

101. The amount of replenisher solution added as a film enters the automatic processor is related to the

   1. size of the film
   2. temperature of the solution
   3. thickness of the emulsion

   (A) 1 only
   (B) 1 and 2 only
   (C) 2 and 3 only
   (D) 1, 2, and 3

102. An overall film density arising from factors other than the light or radiation used to expose the film is called

    (A) fog
    (B) log relative exposure
    (C) optical density
    (D) artifact

103. An exposure was made at 40 inches SID using 5 mAs and 105 kVp with an 8:1 grid. In an effort to improve radiographic contrast, the film is repeated using a 12:1 grid and 90 kVp. Which of the following exposure times will be MOST appropriate, using 400 mA, in order to maintain the original density?

    (A) 0.01 sec
    (B) 0.03 sec
    (C) 0.1 sec
    (D) 0.3 sec

104. A satisfactory radiograph was made without a grid using a 72 inch source-image distance and 8 mAs. If the distance is changed to 40 inches and an 8:1 ratio grid is added, what should be the new mAs?

    (A) 10 mAs
    (B) 18 mAs
    (C) 20 mAs
    (D) 32 mAs

105. Recorded detail is directly proportional to

    1. focal-film distance
    2. tube current
    3. focal spot size

    (A) 1 only
    (B) 1 and 2 only
    (C) 2 and 3 only
    (D) 1, 2, and 3

106. The functions of the automatic processor's roller system include which of the following?

    1. film transport
    2. agitation
    3. squeegee action

    (A) 1 only
    (B) 1 and 2 only
    (C) 1 and 3 only
    (D) 1, 2, and 3

107. In order that a phosphor be suitable for use in intensifying screens, it should have which of the following characteristics?

    1. high conversion efficiency
    2. high x-ray absorption
    3. afterglow

    (A) 1 only
    (B) 1 and 2 only
    (C) 3 only
    (D) 1, 2, and 3

108. If the mAs is adjusted to produce equalized exposure, which of the following statements is FALSE?

    (A) a single-phase exam done at 10 mAs can be duplicated with three-phase, twelve-pulse at 5 mAs
    (B) there is greater patient dose with three-phase equipment than single-phase equipment
    (C) three-phase equipment can produce comparable radiographs with less heat unit buildup
    (D) three-phase equipment produces lower contrast radiographs than single-phase equipment

109. The instrument frequently used in quality control programs that measures varying degrees of film exposure is the

    (A) aluminum step wedge
    (B) spinning top
    (C) densitometer
    (D) sensitometer

110. Which of the following function(s) to reduce the amount of scatter radiation reaching the film?

1. grid devices
2. restricted focal spot size
3. beam restrictors

(A) 1 only
(B) 1 and 2 only
(C) 1 and 3 only
(D) 1, 2, and 3

111. Which of the following statements is true with regard to the two radiographs in Figure 6–9?

(A) film 1 was exposed on expiration
(B) film 2 was exposed on expiration
(C) film 1 is a lordotic position
(D) film 2 is a ventral decubitus position

112. Quiet, shallow breathing may be used during the exposure to obliterate prominent pulmonary vascular markings in which of the following?

1. RAO sternum
2. lateral thoracic spine
3. AP scapula

(A) 1 only
(B) 1 and 2 only
(C) 2 and 3 only
(D) 1, 2, and 3

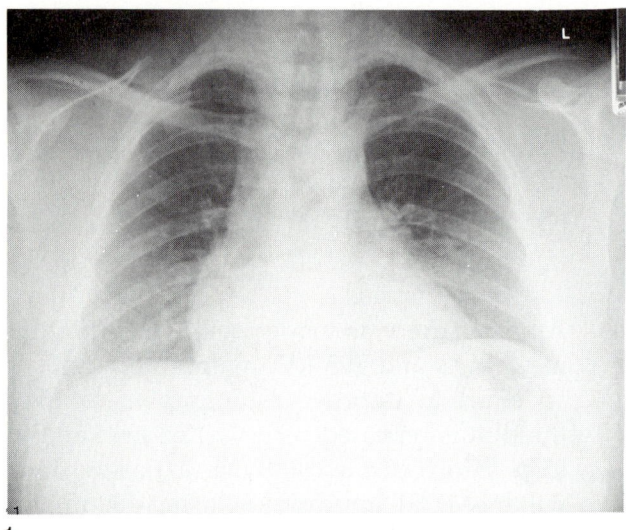

1

2

Figure 6–9. Courtesy of The Stamford Hospital, Department of Radiology.

113. With the patient positioned as illustrated in Figure 6–10, how should the CR be directed in order to best demonstrate the intercondyloid fossa?

(A) perpendicular to the popliteal depression
(B) 40° caudad to the popliteal depression
(C) perpendicular to the long axis of the femur
(D) 40° cephalad to the popliteal depression

114. Mammographic compression is applied until the breast tissue is a uniform thickness. This technique results in

1. increased radiation dose
2. increased radiographic detail
3. decreased motion unsharpness

(A) 1 only
(B) 1 and 2 only
(C) 2 and 3 only
(D) 1, 2, and 3

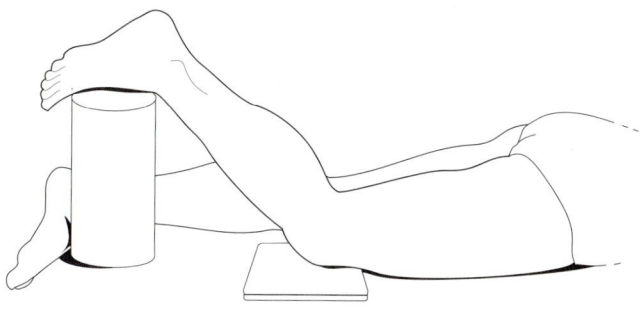

Figure 6–10.

115. The thoracic cavity is lined with

    (A) parietal pleura
    (B) visceral pleura
    (C) parietal peritoneum
    (D) visceral peritoneum

116. Which of the following is a functional study used to demonstrate the degree of AP motion present in the cervical spine?

    (A) moving mandible position
    (B) AP open-mouth projection
    (C) flexion and extension laterals
    (D) AP right and left bending

117. A lateral projection of the elbow should demonstrate which of the following?

    1. the olecranon process in profile
    2. superimposed humeral epicondyles
    3. radial tuberosity facing anteriorly

    (A) 1 only
    (B) 1 and 2 only
    (C) 2 and 3 only
    (D) 1, 2, and 3

118. Which of the labeled bones in Figure 6–11 identifies the tarsal navicular?

    (A) no. 2
    (B) no. 3
    (C) no. 6
    (D) no. 7

119. What does the 8 in Figure 6–11 identify?

    (A) medial malleolus
    (B) lateral malleolus
    (C) medial cuneiform
    (D) talus

120. The carpal navicular (scaphoid) may be demonstrated in which of the following projection(s) of the wrist?

    1. PA oblique
    2. PA with radial flexion
    3. PA with forearm elevated 20°

    (A) 1 only
    (B) 1 and 2 only
    (C) 1 and 3 only
    (D) 1, 2, and 3

121. Free air in the abdominal cavity may be demonstrated in which of the following?

    1. lateral recumbent abdomen
    2. erect AP abdomen
    3. left lateral decubitus abdomen

    (A) 1 only
    (B) 2 only
    (C) 1 and 2 only
    (D) 2 and 3 only

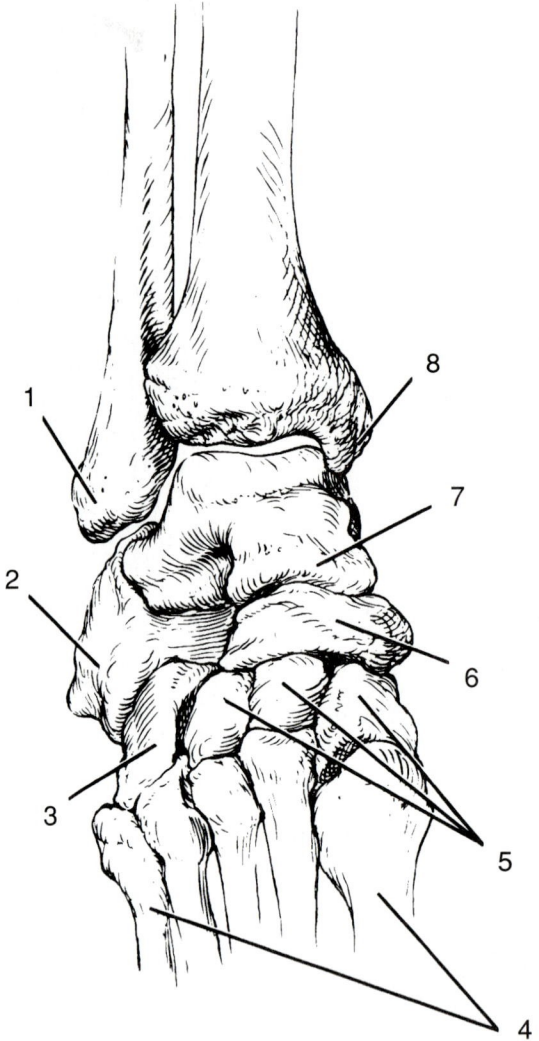

Figure 6–11. Reproduced with permission from Lindner HH. *Clinical Anatomy*. East Norwalk, CT: Appleton & Lange, 1989.

122. The AP oblique projection (medial rotation) of the elbow demonstrates which of the following?

1. radial head free of superimposition
2. olecranon process within the olecranon fossa
3. coronoid process free of superimposition

(A) 1 only
(B) 1 and 2 only
(C) 2 and 3 only
(D) 1, 2, and 3

123. Which of the following is the most likely site for a lumbar puncture?

(A) S 1-2
(B) L 3-4
(C) L 1-2
(D) C 6-7

124. Which of the four baselines illustrated in Figure 6–12 represents the IOML?

(A) baseline 1
(B) baseline 2
(C) baseline 3
(D) baseline 4

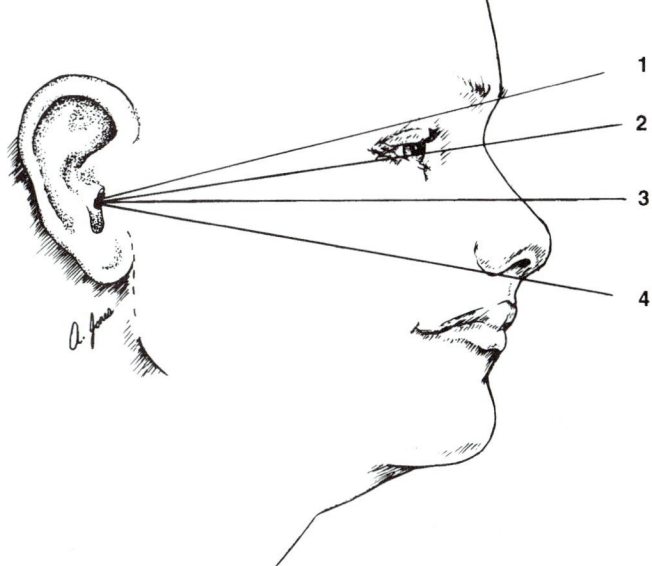

Figure 6–12. Reproduced with permission from Saia DA. *Radiography: Program Review and Exam Preparation.* Stamford, CT: Appleton & Lange, 1996.

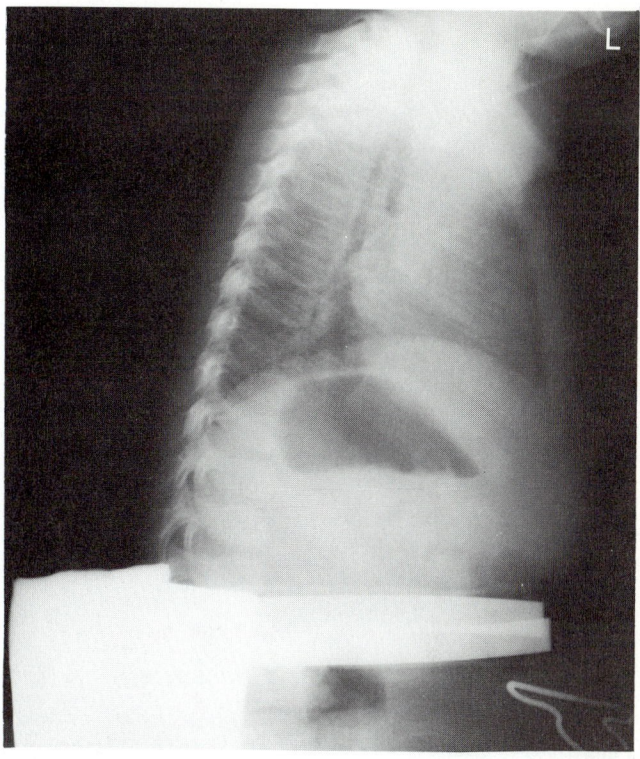

Figure 6–13. Courtesy of The Stamford Hospital, Department of Radiology.

125. Which of the following is true regarding the radiograph in Figure 6–13?

1. the part is rotated
2. the patient is not shielded correctly
3. there is excessive density

(A) 1 only
(B) 2 only
(C) 1 and 2 only
(D) 1, 2, and 3

126. The greater tubercle should be visualized in profile in which of the following?

(A) AP shoulder, external rotation
(B) AP shoulder, internal rotation
(C) AP elbow
(D) lateral elbow

127. Which cholangiographic procedure uses an indwelling drainage tube for contrast medium administration?

(A) endoscopic retrograde cholangiographic pancreatography
(B) operative cholangiography
(C) T-tube cholangiography
(D) percutaneous transhepatic cholangiography

128. The radiograph in Figure 6–14 could be improved in which of the following ways?

(A) the MSP should be placed 45° with the plane of the film
(B) the MSP should be placed 90° to the plane of the film
(C) the chin should be elevated slightly
(D) the head should be flexed slightly

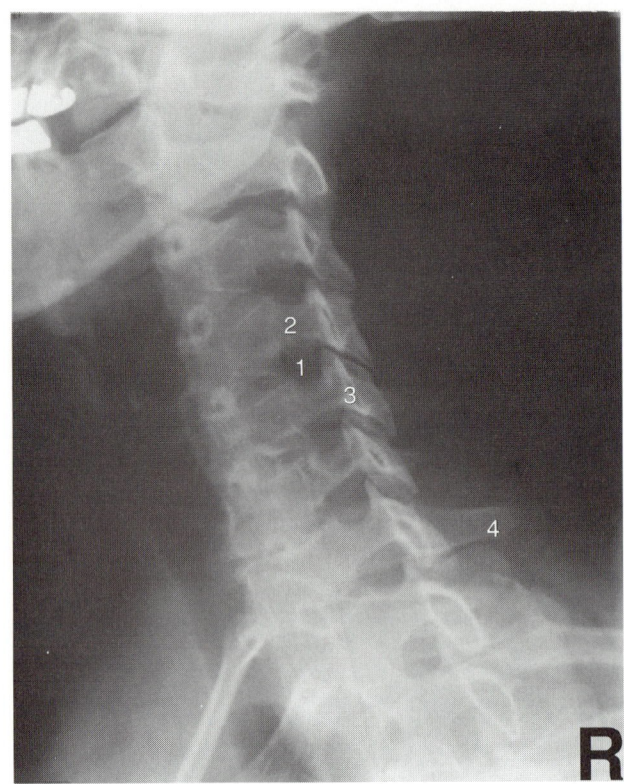

Figure 6–14. Courtesy of The Stamford Hospital, Department of Radiology.

129. What is the anatomic structure indicated by the number 2 in the radiograph in Figure 6–14?

    (A) spinous process
    (B) transverse process
    (C) pedicle
    (D) intervertebral foramen

130. What is the name of the plane indicated by the number 1 in Figure 6–15?

    (A) midcoronal plane
    (B) midsagittal plane
    (C) transverse plane
    (D) horizontal plane

131. The BEST projection to demonstrate the articular surfaces of the femoropatellar articulation is the

    (A) anteroposterior knee
    (B) posteroanterior knee
    (C) tangential ("sunrise") projection
    (D) "tunnel" view

132. Routine excretory urography usually includes a postmicturition radiograph of the bladder. This is done to demonstrate

    1. tumor masses
    2. residual urine
    3. prostatic enlargement

    (A) 2 only
    (B) 1 and 3 only
    (C) 2 and 3 only
    (D) 1, 2, and 3

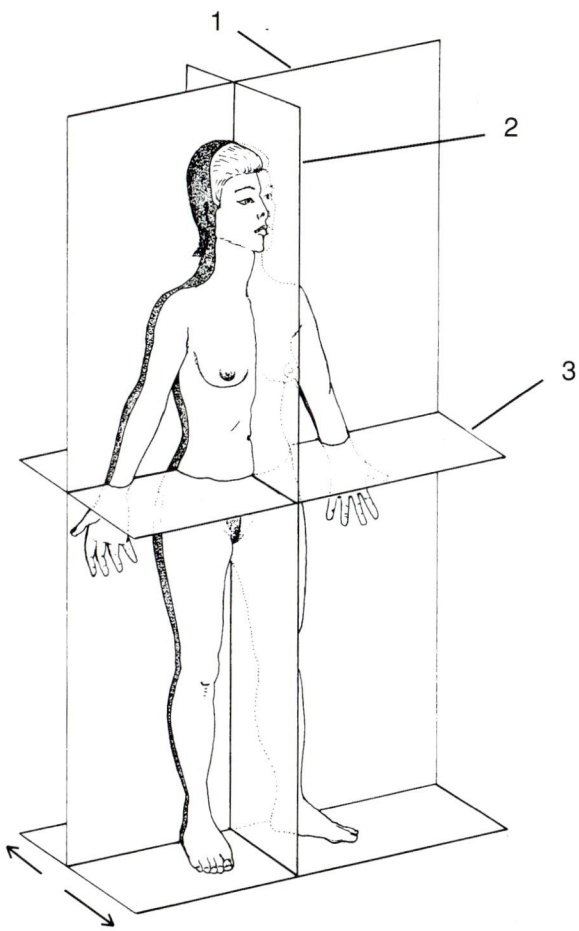

Figure 6–15. Reproduced with permission from Saia DA. *Radiography: Program Review and Exam Preparation.* Stamford, CT: Appleton & Lange, 1996.

133. Which surface of the forearm must be adjacent to the film in order to obtain a lateral projection of the fourth finger with optimal recorded detail?

(A) anterior

(B) posterior

(C) medial

(D) lateral

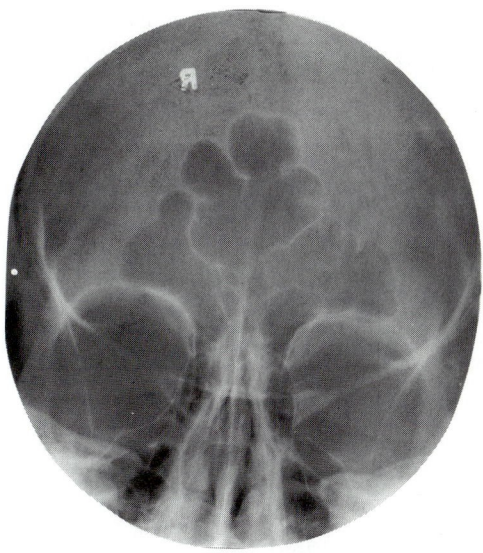

Figure 6–16. Courtesy of The Stamford Hospital, Department of Radiology.

134. Which of the following is (are) true regarding radiographic examination of the acromioclavicular joints?

1. the procedure is performed in the erect position

2. the use of weights enhances demonstration of the joint

3. the procedure is to be avoided if dislocation or separation is suspected

(A) 1 only

(B) 1 and 2 only

(C) 1 and 3 only

(D) 2 and 3 only

135. The radiograph illustrated in Figure 6–16 was made with the CR directed about

(A) 45° caudad

(B) 45° cephalad

(C) 20° caudad

(D) 10° cephalad

136. Which of the following positions may be used to effectively demonstrate the right axillary ribs?

    1. LAO
    2. RPO
    3. RAO

    (A) 1 and 2 only
    (B) 1 and 3 only
    (C) 2 and 3 only
    (D) 1, 2, and 3

137. The AP axial projection of the pulmonary apices requires the central ray to be directed

    (A) 15° cephalad
    (B) 15° caudad
    (C) 30° cephalad
    (D) 30° caudad

138. Proper and accurate film identification is essential in

    1. medicolegal cases
    2. follow-up examinations
    3. reducing patient exposure

    (A) 1 only
    (B) 1 and 2 only
    (C) 2 and 3 only
    (D) 1, 2, and 3

139. The patient with the MOST homogeneous glandular breast tissue would be the

    (A) postpubertal adolescent
    (B) 20 year old with one previous pregnancy
    (C) menopausal
    (D) post–menopausal 65 year old

140. The sternoclavicular joints will be best demonstrated in which of the following positions?

    (A) apical lordotic
    (B) anterior oblique
    (C) lateral
    (D) weight-bearing

141. Which of the following is true regarding the PA axial projection of the paranasal sinuses?

    1. central ray is directed caudally to the orbitomeatal line
    2. petrous pyramids are projected into the lower third of the orbits
    3. frontal sinuses are visualized

    (A) 1 only
    (B) 1 and 2 only
    (C) 1 and 3 only
    (D) 1, 2, and 3

142. Which of the following statements is (are) correct regarding the chest radiograph in Figure 6–17?

    1. rotation of the chest is demonstrated
    2. the pulmonary apices are not visualized
    3. the costophrenic angles are demonstrated

    (A) 1 only
    (B) 1 and 3 only
    (C) 2 and 3 only
    (D) 1, 2, and 3

143. Which of the following procedures does NOT demonstrate renal function?

    (A) intravenous pyelography
    (B) descending urography
    (C) retrograde urography
    (D) infusion nephrotomography

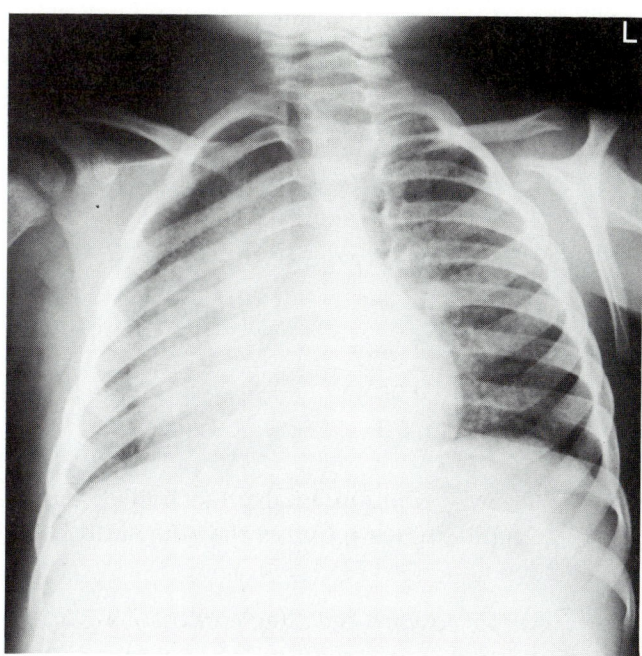

Figure 6–17. Courtesy of The Stamford Hospital, Department of Radiology.

144. Which of the following positions may be used to effectively demonstrate the hepatic flexure during radiographic examination of the large bowel?

1. RAO
2. LAO
3. LPO

(A) 1 only
(B) 1 and 2 only
(C) 1 and 3 only
(D) 2 and 3 only

145. Differences between body habitus types can affect all of the following, EXCEPT

(A) size and shape of an organ
(B) position of an organ
(C) position of the diaphragm
(D) degree of bone porosity

146. Which of the following is (are) demonstrated in the lateral projection of the cervical spine?

1. intervertebral joints
2. apophyseal joints
3. intervertebral foramina

(A) 1 only
(B) 1 and 2 only
(C) 2 and 3 only
(D) 1, 2, and 3

147. Which of the following statements is (are) correct with respect to postoperative cholangiography?

1. a T-tube is in place in the common bile duct
2. water-soluble contrast material is injected
3. the patency of biliary ducts is evaluated

(A) 1 only
(B) 1 and 2 only
(C) 2 and 3 only
(D) 1, 2, and 3

148. Which of the following is a contraindication for performing a barium enema examination?

(A) tumor
(B) polyp
(C) tarry stools
(D) recent biopsy of the colon

149. A dorsal decubitus projection of the chest may be used to evaluate small amounts of

1. fluid in the posterior chest
2. air in the posterior chest
3. fluid in the anterior chest

(A) 1 only
(B) 1 and 2 only
(C) 2 and 3 only
(D) 1, 2, and 3

150. Which of the following statements regarding the RAO position of the sternum is FALSE?

(A) the sternum is generally projected to the left of the vertebral column

(B) prominent pulmonary markings may be obliterated by shallow breathing during the exposure

(C) it is helpful to project the sternum over the heart

(D) a thin thorax requires a lesser degree of obliquity than a thicker thorax

151. What is the atomic structure indicated by the number 3 in the radiograph in Figure 6–18?

(A) mandibular angle

(B) coronoid process

(C) zygomatic arch

(D) mastoid air cells

152. What is the anatomic structure indicated by the number 2 in the radiograph in Figure 6–18?

(A) mandibular angle

(B) coronoid process

(C) zygomatic arch

(D) condyloid process

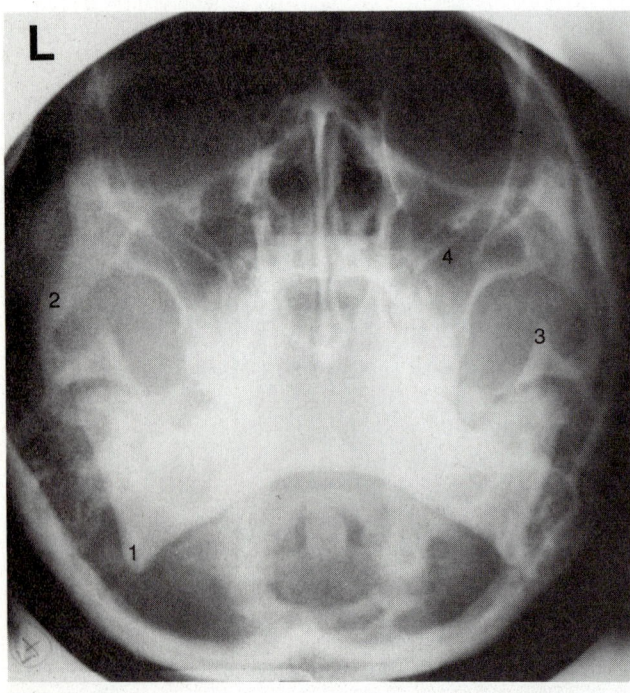

Figure 6–18. Courtesy of The Stamford Hospital, Department of Radiology.

153. Which of the following statements is (are) correct with respect to performing oblique projections of the chest for evaluation of the heart and great vessels?

1. 45° RAO is required
2. 45° LAO is required
3. 60° LAO is required

(A) 1 only

(B) 1 and 2 only

(C) 1 and 3 only

(D) 1, 2, and 3

154. The SMV oblique axial projection of the zygomatic arches requires that the skull be rotated

(A) 15° toward the affected side

(B) 15° away from the affected side

(C) 45° toward the affected side

(D) 45° away from the affected side

155. Which of the following is (are) evaluation criteria for a PA chest for heart and lungs?

1. ten posterior ribs should be seen above the diaphragm
2. the medial ends of the clavicles should be equidistant from the vertebral column
3. the scapulae should be seen through the upper lung fields

(A) 1 only

(B) 1 and 2 only

(C) 2 and 3 only

(D) 1, 2, and 3

156. With the patient positioned for an AP oblique projection, left side of the pelvis elevated 25°, and the central ray entering 1 inch medial to the ASIS, which of the following is demonstrated?

(A) left sacroiliac joint

(B) left ilium

(C) right sacroiliac joint

(D) right ilium

157. The AP axial projection of the chest for pulmonary apices

   1. requires 15 to 20° cephalad angulation
   2. projects the apices above the clavicles
   3. should demonstrate the medial ends of the clavicles equidistant from vertebra column

(A) 1 only
(B) 1 and 2 only
(C) 1 and 3 only
(D) 1, 2, and 3

158. What is the anatomic structure indicated by the number 2 in the radiograph in Figure 6–19?

(A) superior articular process
(B) inferior articular process
(C) pedicle
(D) lamina

159. What is the anatomic structure indicated by the number 3 in the radiograph in Figure 6–19?

(A) superior articular process
(B) inferior articular process
(C) pedicle
(D) lamina

160. Which of the following mammographic positions would BEST identify a tumor in the medial aspect of the breast?

(A) mediolateral
(B) lateromedial
(C) craniocaudad
(D) axillary

161. The pyloric canal and duodenal bulb are BEST demonstrated during a GI series in which of the following positions?

(A) RAO
(B) left lateral
(C) recumbent PA
(D) recumbent AP

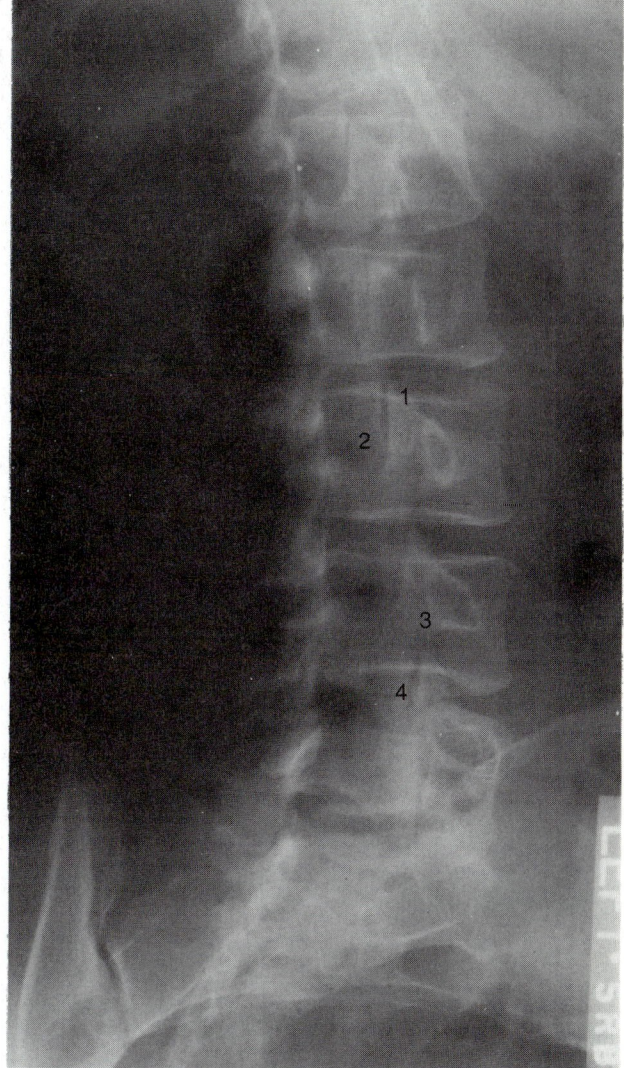

Figure 6–19. Courtesy of The Stamford Hospital, Department of Radiology.

162. Which of the following statements regarding the bony thorax is FALSE?

(A) the first seven pair of ribs are referred to as vertebrosternal, or true ribs
(B) the only articulation between the thorax and upper extremity is the sternoclavicular joint
(C) the gladiolus is the upper part of the sternum and is quadrilateral in shape
(D) the anterior ends of the ribs are about 4 inches below the level of the vertebral ends

163. Which of the following may be used to evaluate the glenohumeral joint dislocation?

1. inferosuperior axial
2. transthoracic lateral
3. scapular Y projection

(A) 1 only
(B) 1 and 2 only
(C) 3 only
(D) 1, 2, and 3

164. What is the anatomic structure identified as the number 5 in the illustration in Figure 6–20?

(A) coracoid process
(B) coronoid process
(C) trochlea
(D) capitulum

165. Which of the following correctly identifies the coronoid process in the illustration in Figure 6–20?

(A) no. 2
(B) no. 4
(C) no. 5
(D) no. 6

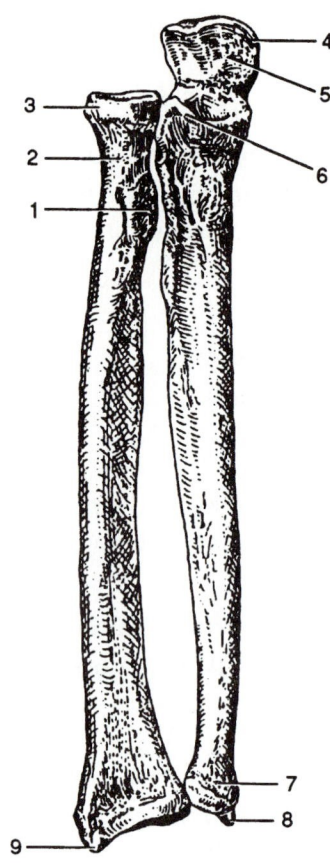

Figure 6–20. Reproduced with permission from Pansky B. *Review of Gross Anatomy* (5th ed). New York: McGraw-Hill, 1984, p 257.

166. In the parieto-orbital projection (Rhese method) of the optic canal, the median sagittal plane and central ray form what angle?

(A) 90°
(B) 37°
(C) 53°
(D) 45°

167. In myelography, the contrast medium is generally injected into the

(A) cisterna magna
(B) individual intervertebral discs
(C) subarachnoid space between the first and second vertebrae
(D) subarachnoid space between the third and fourth lumbar vertebrae

168. Which of the following is (are) valid criteria for a lateral projection of the forearm?

1. radius and ulna should be superimposed distally
2. coronoid process and radial head should be partially superimposed
3. humeral epicondyles should be superimposed

(A) 1 only
(B) 1 and 2 only
(C) 2 and 3 only
(D) 1, 2, and 3

169. To better demonstrate the interphalangeal joints of the toes, which of the following procedures may be employed?

   1. angle the CR 15° caudad

   2. angle the CR 15° cephalad

   3. place a sponge wedge under the foot with toes elevated 15°

(A) 1 only

(B) 1 and 2 only

(C) 1 and 3 only

(D) 2 and 3 only

170. Double-contrast examinations of the stomach or large bowel are performed to better visualize

(A) position of the organ

(B) size and shape of the organ

(C) diverticula

(D) gastric or bowel mucosa

171. The term *parenteral* refers to which of the following medication routes?

(A) oral

(B) sublingual

(C) mucosal

(D) any, other than alimentary

172. A signed consent form is NOT necessary prior to performing a(n)

(A) myelogram

(B) cardiac catheterization

(C) upper gastrointestinal series

(D) interventional vascular procedure

173. Which of the following is another name for an intermittent injection port?

(A) a hypodermic needle

(B) a butterfly needle

(C) a heparin lock

(D) an IV infusion

174. The radiographer must perform which of the following procedures prior to entering a strict isolation room with a mobile x-ray unit?

   1. wear a gown and gloves only

   2. wear a gown, gloves, mask, and a cap

   3. clean the mobile x-ray unit

(A) 1 only

(B) 2 only

(C) 1 and 3 only

(D) 1, 2, and 3

175. Patients' rights include which of the following?

   1. the right to refuse treatment

   2. the right to confidentiality

   3. the right to possess his or her radiographs

(A) 1 only

(B) 1 and 2 only

(C) 1 and 3 only

(D) 1, 2, and 3

176. The risk of inoculation with HIV is considered high for which of the following entry sites?

   1. broken skin

   2. shared needles

   3. conjunctiva

(A) 1 only

(B) 1 and 2 only

(C) 2 and 3 only

(D) 1, 2, and 3

177. Which of the following would be a violation of the ARRT Code of Ethics for the Profession of Radiologic Technology?

(A) a radiographer failing to wear a lead apron when performing portable radiography

(B) participating in continuing education

(C) revealing information regarding a suspected case of child abuse to the referring MD

(D) investigating a new and innovative technique

178. A patient's pulse may be accurately counted in which of the following locations?

    1. neck
    2. wrist
    3. foot

  (A) 1 only
  (B) 1 and 2 only
  (C) 2 and 3 only
  (D) 1, 2, and 3

179. The MOST frequent site of nosocomial infection is the

  (A) urinary tract
  (B) blood
  (C) respiratory tract
  (D) digestive tract

180. The infection streptococcal pharyngitis (strep throat) is caused by a

  (A) virus
  (B) fungi
  (C) protozoa
  (D) bacteria

181. Which of the following radiographic examinations require the patient to be NPO 8 to 10 hours prior to examination for proper patient preparation?

    1. esophagram
    2. upper gastrointestinal series
    3. barium enema

  (A) 1 and 2 only
  (B) 1 and 3 only
  (C) 2 and 3 only
  (D) 1, 2, and 3

182. While measuring blood pressure, the first pulse that is heard is recorded as the

  (A) diastolic pressure
  (B) systolic pressure
  (C) venous pressure
  (D) valvular pressure

183. The AIDS virus is transmitted

    1. by sharing contaminated needles
    2. mother to child during birth
    3. by intimate contact with body fluids

  (A) 1 only
  (B) 1 and 2 only
  (C) 1 and 3 only
  (D) 1, 2, and 3

184. Ipecac is a medication used to induce vomiting and is classified as a(n)

  (A) diuretic
  (B) antipyretic
  (C) antihistamine
  (D) emetic

185. Sterile technique is required when contrast agents are administered

  (A) through a nasogastric tube
  (B) intrathecally
  (C) rectally
  (D) orally

186. Proper body mechanics includes a wide base of support. The base of support is the portion of the body

  (A) in contact with the floor or other horizontal surface
  (B) in the mid portion of the pelvis or lower abdomen
  (C) passing through the center of gravity
  (D) none of the above

187. A radiolucent contrast agent

    1. absorbs a high number of x-ray photons
    2. permits anatomy to appear dark on the radiograph
    3. is composed of elements of low atomic numbers

  (A) 1 and 2 only
  (B) 1 and 3 only
  (C) 2 and 3 only
  (D) 1, 2 and 3

188. The most commonly used method of low-flow oxygen delivery is the

(A) oxygen mask
(B) nasal cannula
(C) respirator
(D) oxyhood

189. A patient who has been recumbent for some time and gets up quickly may suffer from light-headedness or feel faint. This is referred to as

(A) dyspnea
(B) orthopnea
(C) hypertension
(D) orthostatic hypotension

190. When disposing of contaminated needles, they are placed in special containers in which of the following ways?

(A) recap the needle, remove the syringe, and dispose
(B) do not recap needle, remove from syringe, and dispose
(C) recap the needle, dispose entire syringe
(D) do not recap needle, dispose entire syringe

191. Which of the following is a fast-acting vasodilator used to lower blood pressure and relieve the pain of angina pectoris?

(A) digitalis
(B) dilantin
(C) nitroglycerin
(D) Tagamet

192. A patient with an upper respiratory infection is transported to the radiology department for a chest examination. Who of the following should be masked?

1. technologist
2. transporter
3. patient

(A) 1 only
(B) 1 and 2 only
(C) 3 only
(D) 1, 2, and 3

193. The decision to deliver ionic or non-ionic contrast media should include a preliminary patient history including, but not limited to

1. patient age
2. history of respiratory disease
3. history of cardiac disease

(A) 1 and 2
(B) 1 and 3
(C) 2 and 3
(D) 1, 2 and 3

194. While performing radiography, improper support of a patient's fractured lower leg (tibia/fibula) could result in

1. movement of fracture fragments
2. tearing of soft tissue, nerves, and blood vessels
3. initiation of muscle spasm

(A) 1 and 2 only
(B) 1 and 3 only
(C) 2 and 3 only
(D) 1, 2 and 3

195. Which of the following involve(s) intentional misconduct?

1. invasion of privacy
2. false imprisonment
3. patient sustains injury from fall while left unattended

(A) 1 only
(B) 3 only
(C) 1 and 2 only
(D) 2 and 3 only

196. In what order should the following exams be performed?

1. upper GI
2. IVP
3. barium enema

(A) 3, 1, and 2
(B) 1, 3, and 2
(C) 2, 1, and 3
(D) 2, 3, and 1

197. The Centers for Disease Control (CDC) suggests that health care workers protect themselves and their patients from blood and body fluid contamination by using

(A) strict isolation precautions

(B) universal precautions

(C) respiratory precautions

(D) sterilization

198. Characteristics of the patient with pulmonary emphysema include

1. clubbed fingers
2. increased AP diameter of chest
3. hyperventilation

(A) 1 only

(B) 1 and 2 only

(C) 2 and 3 only

(D) 1, 2, and 3

199. The Hemovac or Penrose drains are used for

(A) bile duct drainage

(B) tissue drainage of wounds or in postoperative drainage

(C) decompression of the gastrointestinal tract

(D) feeding patients who are unable to swallow food

200. Drugs that may be used to prolong blood clotting time include

1. heparin
2. Benadryl
3. lidocaine

(A) 1 only

(B) 1 and 2 only

(C) 1 and 3 only

(D) 1, 2, and 3

Answers and Explanations

1. **(A)** The *control badge* is an important part of the monitoring system. It should be stored somewhere *away from radiation sources*. At the end of the monitoring period, when the badges are returned to the dosimetry service, the exposure to the control badge (which should be zero) is compared to the exposure received by the rest of the personnel monitors. If the control badge is stored near radiation, or used to replace someone's lost badge, there is no standard for comparison for the rest of the group of monitors. *(Ballinger, vol 1, p 35)*

2. **(C)** Fluoroscopic skin dose is greater than radiographic skin dose because the x-ray source is much closer to the patient. The generally accepted rule is that the skin receives 2 rad/min/mA. Therefore, 2 rad/min for 5 min equals 10 rad/mA. At 1.5 mA, the patient dose is 15 rad (2 rad/5 min/1.5 mA). *(Bushong, p 653)*

3. **(B)** If it is desired to determine entrance skin exposure, a small ionization chamber (pocket dosimeter) can be placed on the skin and the approximate ESE read immediately. These devices are readily imaged, however, and are awkward to position. For these reasons, TLDs are more easily used; they are precise and will not interfere with the radiographic image. The acceptable ESE for a PA chest is approximately 20 mR (12 to 26 mR is the acceptable range). An ESE of 6 mR would be underexposed and require repeating. Similarly, ESEs of 38 mR and 0.6 R (600 mR) would be overexposed and need to be repeated. *(Ballinger, vol 1, p 30)*

4. **(B)** Three dose–response (dose effect) curves are illustrated, representing the body's response to irradiation. *Dose* is indicated by the horizontal axis (increasing to the right); *response* is indicated by the vertical axis (increasing upward). Two of the curves (numbers 1 and 3) are *linear*, that is, a straight line. Curve 2 is not a straight line, and is therefore *nonlinear*. Curves 2 and 3 show that a particular dose (*threshold* quantity) of radiation is required before any effect will occur; therefore curve 2 is *nonlinear threshold* and 3 is *linear threshold*. Curve 1, however, shows that any dose of radiation (theoretically, even a single x-ray photon, that is, no threshold) can result in a particular biologic effect—*linear non-threshold*. *(Ballinger, vol 1, p 26)*

5. **(D)** Converting from nongrid to 8:1 grid requires about a fourfold increase in mAs. Increasing the kVp by 15 percent and cutting the mAs in half would reduce patient dose. When changing from plain film technical factors to tomographic factors, it is generally recommended that the kVp remain the same and that mAs is increased by 50 percent. Changing from 400 film/screen combination to 200 film/screen will require the mAs to be doubled. Therefore, the largest increase would be required by the implementation of a grid. *(Cullinan, p 228)*

6. **(D)** Cells that are termed undifferentiated are *immature* or young. They have no specific function and/or structure. They are usually precursor cells—their most important function is to divide. Mitosis is the most radiosensitive portion of the cell cycle. *(Travis, p 68)*

7. **(D)** Most of the occupational exposure received by radiographers is received during fluoroscopy and mobile radiography, and the use of lead aprons is required during both of these procedures. The position of the personnel monitor relative to that of the lead apron, then, becomes important. It is recommended that the badge be worn outside the lead apron at collar level. In this position, the badge will record the maximum possible exposure received by the radiographer, and provide a realistic estimate of thyroid and lens exposure. *(Ballinger, vol 1, p 36)*

8. **(B)** Because muscle and nerve tissues perform specific functions and do not divide, they are relatively insensitive to radiation exposure. Epithelial cells cover the outer surface of the body, and line body cavities as well as tubes and passageways leading to the exterior. They contain very little intercellular substance and are devoid of blood vessels. Because epithelial cells constantly regenerate through mitosis, they are very radiosensitive. *(Statkiewicz, p 123)*

9. **(B)** High doses of radiation exposure result in *early* effects. (Examples of early effects are blood changes and erythema.) If the exposed individual survives, then *late* or long-term effects must be considered. Individuals who receive *small* amounts of *low-level* radiation (such as the occupationally exposed) are concerned with the late effects of radiation exposure—those effects that can occur many years after the initial exposure. Late effects of radiation exposure such as carcinogenesis are considered to be related to the *linear nonthreshold* dose–response curve. That is, theoretically, there is *no safe dose*; even one photon can induce a later response. *(Ballinger, vol 1, p 26)*

10. **(C)** The occupational dose limit is valid for beta, x-, and gamma radiations. Because alpha radiation is so rapidly ionizing, traditional personnel monitors will not record alpha radiation. Because alpha particles are capable of penetrating only a few cm of air, they are practically harmless as an external source. *(Selman, p 504)*

11. **(A)** The effects of a quantity of radiation delivered to a body are dependent on a few factors: the amount of radiation received, the size of the irradiated area, and how the radiation is delivered in time. If the radiation is delivered in portions over a period of time, it is said to be fractionated, and has a less harmful effect than if the radiation was delivered all at once. Cells have an opportunity to repair; some recovery occurs between doses. *(Bushong, pp 533–537)*

12. **(A)** The velocity and charge of particulate radiation determines the amount of energy transferred (and, therefore, number of ionizations) to the tissue traversed. A greater LET (number of ionizations) is delivered by particles with a slower velocity and greater charge. The greater the LET and number of ionizations, the greater the biologic effect. The unit of absorbed dose is the rad (*radiation absorbed dose*). Rem is the acronym for *radiation equivalent man*—the unit of dose equivalent. *(Bushong, p 534)*

13. **(A)** The relationship between x-ray intensity and distance from the source is expressed in the Inverse Square Law of Radiation. The formula is:

$$\frac{I_1}{I_2} = \frac{D_2^2}{D_1^2}$$

Substituting known values:

$$\frac{12\ mR/min}{x\ mR/min} = \frac{7^2\ ft}{3^2\ ft} \frac{(=49\ ft.)}{(=9\ ft.)}$$

$$49\ x = 108$$
$$x = 2.20\ mR/min\ at\ 7\ ft$$

Note the inverse relationship between distance and dose. As distance from the source of radiation increases, dose rate decreases significantly. *(Selman, p 336)*

14. **(A)** A 1984 review of radiation exposure data revealed that the average annual dose equivalent for monitored radiation workers was approximately 0.23 rem (2.3 mSv). The fact that this is approximately one tenth the recommended limit indicates that the limit is

adequate for radiation protection purposes. Therefore the NCRP reiterates its 1971 recommended annual limit of 5 rem (50 mSv). *(NCRP Report no. 105, p 13)*

15. **(B)** LET expresses the rate at which photon or particulate energy is transferred to (absorbed by) biologic material (through ionization processes), and is dependent on the type of radiation and absorber characteristics. RBE describes the degree of response or amount of biologic change one can expect of the irradiated material. As the amount of transferred energy (LET) increases (from interactions occurring between radiation and biologic material) the amount of biologic effect/damage will also increase; that is, they are directly proportional. *(Bushong, p 534)*

16. **(D)** Follow-up studies have been done on individuals receiving accidental exposure of radiation (medical personnel, uranium miners, children irradiated in vivo). Pioneer radiation workers developed leukemia and other cancers, their vision was clouded by formation of cataracts, and their lives were shorter than those of their colleagues. Certainly today, with our sophisticated equipment and our knowledge of radiation protection, none of these situations should occur. *(Bushong, pp 580, 584, 588)*

17. **(A)** 4500 mrad is equal to 4.5 rad. If 4.5 rad were delivered in 6 min, then the dose rate must be 0.75 rad/min:

$$\frac{4.5 \text{ rad}}{6 \text{ min}} = \frac{x}{1 \text{ min}}$$
$$6x = 4.5$$
$$x = 0.75 \text{ rad/min}$$

(Selman, p 528)

18. **(D)** Photoelectric interaction in tissue involves complete absorption of the incident photon, whereas Compton interactions involve only partial transfer of energy. The larger the quantity of radiation, and the greater the number of photoelectric interactions, the greater the patient dose. Radiation

dose to more radiosensitive tissues, such as gonadal tissue or blood-forming organs, is more harmful than the same dose to muscle tissue. *(Selman, pp 179–181)*

19. **(C)** During radiography and fluoroscopy, radiation scatters in all directions from the patient. In fact, the patient is the single most important scattering object in both radiographic and fluoroscopic procedures. The approximate intensity (quantity) of scatter radiation at 1 meter from the patient is 0.1 percent of the entrance dose. Therefore, if the entrance dose for this film is 320 mR, the intensity of radiation at 1 meter from the patient is 0.1 percent of that, or 0.32 mR (0.001 × 320 = 0.32). *(Bushong, p 633)*

20. **(B)** The pregnant radiographer poses a special radiation protection consideration, as the safety of the unborn individual must be considered. It must be remembered that the developing fetus is particularly sensitive to radiation exposure. Established guidelines state that the occupational radiation exposure to the fetus must not exceed 0.5 rem (500 mrem, or 5 mSv) during the entire gestation period. *(Bushong, p 606)*

21. **(C)** As grid ratio increases, more scatter radiation is removed from the x-ray beam and exposure factors must be increased to maintain sufficient density. An appropriate decrease in/exposure factors is required when grid ratio is decreased. This results in an increase or decrease, respectively, of patient dose. Because an increase or decrease in kVp does affect the number of x-ray photons produced at the target, it will impact patient dose as well. As SID decreases, patient dose increases, and an appropriate mAs adjustment is required. Changing the size of the focal spot affects and/or requires no change in exposure factors; therefore, although it has impact on radiographic quality (recorded detail), it has no effect on patient dose. *(Bushong, pp 235, 298, 300)*

22. **(C)** Film badges are supplied by a dosimetry service. They contain pieces of dental film

held within a holder containing filters. When used properly, film badges measure quantity and quality of radiation exposure. Film within the badges is usually changed *monthly*. The sensitive film emulsion is susceptible to deterioration and false readings if worn for longer periods, or damaged by water, heat, light, and so on. In order to avoid the possibility of damage or exposure, film badges should not leave the workplace. *(Ballinger, vol 1, p 35)*

23. **(B)** Whole-body dose is calculated to include all the especially radiosensitive organs. The gonads, lens of eye, and blood-forming organs are particularly radiosensitive. The forearms may safely receive 30 rem per year and the hands have a dose limit of 75 rem per year. *(Bushong, p 606)*

24. **(A)** Genetically significant dose (GSD) illustrates that large exposures to a few people are cause for little concern when diluted by the total population. On the other hand, we all share the burden of that radiation received by the total population; and especially as the use of medical radiation increases, each individual's share of the total exposure increases. *(Curry, p 402)*

25. **(C)** Air/fluid levels are demonstrated in the erect or decubitus position. Grid radiography requires about three to four times greater dose than nongrid radiography. A left lateral decubitus without a grid, then, would demonstrate fluid levels with a considerably smaller dose to the infant. A recumbent AP would not demonstrate air/fluid levels. *(Bontrager, p 76)*

26. **(A)** Mature white blood cells (lymphocytes) are considered the most radiosensitive cells. Ova (female germ cells) are very radiosensitive but not to the degree of lymphocytes. Myocytes (muscle cells) and especially neurons (nerve cells) are actually radioresistant. *(Gurley, p 149)*

27. **(C)** Lead aprons are worn by occupationally exposed individuals during fluoroscopic and mobile x-ray procedures. Lead aprons are available with various lead equivalents; 0.25, 0.5, and 1.0 mm are the most common. The

1.0-mm lead equivalent apron will provide close to 100 percent protection at most kVp levels, but is rarely used because it weighs anywhere from 12 to 24 pounds! A *0.25 mm* lead equivalent apron will attenuate about 97 percent of a 50 kVp x-ray beam, 66 percent of a 75 kVp beam, 88 percent of a 75 kVp beam, and 51 percent of a 100-kVp beam. A *0.5 mm* lead equivalent apron will attenuate about 99 percent of a 50 kVp beam, 88 percent of a 75 kVp beam, and 75 percent of a 100-kVp beam. *(Ballinger, vol 1, p 34)*

28. **(A)** It is our ethical responsibility to minimize radiation dose to patients. X-rays produced at the target comprise a heterogeneous primary beam. There are many "soft" (low-energy) photons that, if not removed, would only contribute to greater patient dose. They are too weak to penetrate the patient and expose the film. These soft x-rays penetrate only a small thickness of tissue before being absorbed. *(Bushong, p 165)*

29. **(A)** Because patient dose is regulated by the quantity of x-ray photons delivered to the patient, mAs regulates patient dose. Highly energetic x-ray photons (high kVp) are more likely to penetrate the patient, rather than be absorbed by biologic tissue. Consequently, the use of high kVp and low mAs exposure factors is preferred in an effort to reduce patient dose. The use of high-speed screens can assist in the reduction of exposure factors. *(Statkiewicz, pp 169, 181, 184)*

30. **(B)** The most hazardous time for abdominal irradiation is in the earliest stages of pregnancy when many women are unaware that they are pregnant. For this reason *it is recommended that elective radiologic procedures be performed within the first 10 days following the onset of the menses.* It is during this time that the danger of irradiating a recently fertilized ovum is most unlikely. About 14 days before the onset of menses is when the ovarian follicle ruptures and liberates an ovum. *(Bushong, pp 612–613)*

31. **(A)** As kilovoltage is increased, more high-energy photons are produced and the overall

average energy of the beam is increased. Photon energy is inversely related to wavelength; as photon energy increases, wavelength decreases. An increase in mA serves to increase the number of photons produced at the target, but is unrelated to energy. *(Bushong, p 165)*

32. (B) In mutual induction, two coils are in close proximity and a current is supplied to one of the coils. As the magnetic field associated with every electric current expands and "grows up" around the first coil, it interacts with and "cuts" the coils of the second. This interaction, *motion* between magnetic field and coil (conductor), *induces* an emf in the second coil. This is mutual induction; the production of a current in a neighboring circuit. Transformers, such as the high-voltage transformer and filament (step-down) transformer, operate on the principle of mutual induction. The autotransformer operates on the principle of self-induction. Both the transformer and autotransformer require the use of alternating current. *(Bushong, pp 98–99)*

33. (B) The high-voltage, or step-up, transformer functions to *increase voltage* to the necessary *kilovoltage*. It *decreases* the *amperage to milliamperage*. The amount of increase or decrease depends on the transformer ratio, that is, the number of turns in the primary coil to the number of turns in the secondary coil. The transformer law is as follows.

To determine secondary V:

$$\frac{V_s}{V_p} = \frac{N_s}{N_p}$$

To determine secondary I:

$$\frac{I_s}{I_p} = \frac{V_p}{V_s}$$

Substituting known values:

$$\frac{x}{110} = \frac{80,000}{100}$$

$$100x = 8,800,000$$
$$x = 88,000 \text{ volts } (= 88 \text{ kV})$$

(Bushong, p 99)

34. (B) The thoriated tungsten filament of the cathode assembly is heated by its own filament circuit. The circuit provides current and voltage to heat the filament to incandescence, at which time it undergoes thermionic emission (the liberation of valence electrons from filament atoms). Electrolysis describes the chemical ionization effects of an electric current. Rectification is the process of changing AC to unidirectional current. Induction is a method of electrification. *(Bushong, p 118)*

35. (A) With three-phase equipment, the voltage never drops to zero and x-ray intensity is significantly greater. When changing *from single-phase to three-phase, six-pulse equipment, two thirds of the original mAs is required* to produce a radiograph with similar density (from 3ϕ 6P to 1ϕ, add ⅓ more mAs). When changing *from single-phase to three-phase, twelve-pulse equipment, only one-half the original mAs is required* (3ϕ 12P to 1ϕ requires 2× mAs). In this instance, we are changing from single-phase to three-phase, twelve-pulse equipment; therefore the new mAs should be twice the original 50 mAs, or *100 mAs*. The only selection that will provide 100 mAs is (D) 0.25 sec; (A) will produce 25 mAs: 200 mA × ⅛ sec = 25 mAs; (B) will produce 75 mAs (600 mA × 0.125 sec); (C) will produce 50 mAs (0.125 sec × 400 mA); (D) will produce 100 mAs (400 mA × 0.25 sec). *(Cullinan, pp 146–147)*

COMPARISON OF TECHNICAL FACTORS REQUIRED

Single ϕ	3ϕ 6p	3 ϕ 12p
x mAs	⅔x mAs	½x mAs

36. (B) A double-focus tube has two *focal spot* sizes available. These focal spots are actually two available paths on the focal track. There are also two *filaments*. As the small focal spot is selected, the small filament is heated, and electrons are driven across to the smaller portion of the focal track. As the large focal spot is selected, the large filament is heated, and electrons are driven across to the larger portion of the focal track. *(Bushong, p 119)*

37. **(A)** The *autotransformer* controls/selects the amount of voltage sent to the primary winding of the high-voltage transformer and operates on the principle of *self*-induction. The *step-up* (high voltage) *transformer* (primary coil is no. 2, secondary coil is no. 3) operates on the principle of *mutual* induction. The x-ray tube is identified as no. 7. *(Wolbarst, pp 79–80)*

38. **(D)** The *rectification system*, used to change alternating current to unidirectional current, is indicated by the number 5. The rectification system is located between the secondary coil of the high voltage transformer (3) and the x-ray tube (7). The autotransformer is labeled 1, the primary coil of the high voltage transformer is 2, the grounded mA meter is 4, and the filament circuit is 6. *(Wolbarst, pp 79–80)*

39. **(B)** Image intensification is a process that converts the dim fluoroscopic image into a much brighter image, much like normal daylight. As x-ray photons emerge from the patient and enter the image intensifier, they first encounter the *input phosphor*, which is generally composed of cesium iodide phosphors. At the input phosphor, x-ray photons are converted to light photons, which in turn strike the photocathode. The *photocathode* is a photoemissive metal (usually antimony and cesium compounds); when struck by light, it emits electrons in proportion to the intensity of light striking it. The electrons are then directed to the *output phosphor* via the *electrostatic focusing lenses*, speeded up in the neck of the tube by the *accelerating anode*, and directed to the output phosphor for further amplification. Most image intensifiers offer brightness gains of 5000 to 20,000. *From the output phosphor the image is taken by the TV camera, most often a plumbicon or vidicon tube, and transferred to the TV monitor.* A cine cam-

era is required to record images on 16 or 35 mm film. *(Thompson, p 370)*

40. **(A)** Remnant radiation emerging from the patient causes fluorescence of the cassette's intensifying screens. When activated by x-ray photons, the individual phosphors emit light isotropically. Fluorescent light that is not perpendicular to the film emulsion produces a blur resulting in geometric unsharpness and decreased resolution. During manufacture special dyes can be added to the active layer that will absorb much of the diffused fluorescence and allow a greater percentage of film density to be created by more perpendicular fluorescent light. These dyes improve resolution but a small loss of screen speed. *(Cullinan, p 98)*

41. **(D)** Proper care of leaded protective apparel is required in order to ensure its continued usefulness. If lead aprons and gloves are folded (or just left in a heap) cracks will develop and decrease their effectiveness. Both items should be fluoroscoped annually to check for the formation of cracks. *(Bushong, p 665)*

42. **(D)** X-ray tube life may be extended by using exposure factors that produce a minimum of heat (a lower mAs and higher kVp combination) whenever possible. When the rotor is activated, the filament current is increased to produce the required electron source (thermionic emission). Prolonged rotor time, then, can lead to shortened filament life due to early vaporization. Large exposures to a cold anode will heat the anode surface, and the temperature difference between surface and interior can cause cracking of the anode. This can be avoided by proper warming of the anode prior to use, thereby allowing sufficient dispersion of heat through the anode. *(Selman, pp 222–223)*

43. **(B)** Spot film cameras are most frequently used for 100 and 105 mm film, though 70 and 90 mm are also available. They operate somewhat like a movie camera, with the image *from the output phosphor* exposing the film one frame at a time. This differs considerably from

spot films that are exposed by intensifying screens (which also require a greater patient exposure). *(Thompson, p 373)*

44. **(D)** The synchronous timer is an older-type timer that does not permit very precise, short exposures. The impulse timer permits a shorter, more precise exposure, and the electronic timer may be used for exposures as short as 0.001 sec. The phototimer, however, automatically terminates the exposure when the proper density has been recorded on the film. The important advantage of the photo-timer, then, is that it can accurately duplicate radiographic densities. It is therefore very useful for providing accurate comparison in follow-up examinations, and in decreasing patient dose by decreasing the number of "retakes" due to improper exposure. Remember that proper functioning of the phototimer depends on accurate positioning (and centering) by the radiographer. *(Selman, pp 239–243)*

45. **(A)** The line focus principle design illustrates that as the target angle decreases, the effective focal spot decreases (providing improved recorded detail), but the actual area of electron interaction remains much larger (allowing for greater heat capacity). *It must be remembered, however, that a steep (small) target angle increases the heel effect, and film coverage may be compromised. (Bushong, p 124)*

46. **(C)** As *filtration* is added to the x-ray beam, the lower-energy photons are removed and the overall energy or wavelength of the beam is greater. As *kilovoltage* is increased, more high-energy photons are produced, and again, the overall or average energy of the beam is greater. *An increase in mA serves to increase the number of photons produced at the target, but is unrelated to their energy. (Selman, pp 161, 164)*

47. **(B)** X-ray photons are produced in two ways as high-speed electrons interact with target tungsten atoms. First, if the high-speed electron is attracted by the nucleus of a tungsten atom and changes its course, as the electron is "braked," energy is given up *in the form of*

an x-ray photon. This is called Bremsstrahlung (braking) radiation, and is responsible for the majority of x-ray photons produced at the conventional tungsten target. Second, a high-speed electron, *having an energy of at least 70 keV,* may eject a tungsten K shell electron, leaving a vacancy in the shell. An electron from the next energy level, the L shell, drops down to fill the vacancy, *emitting the difference in energy as a K characteristic ray.* Characteristic radiation comprises only about 15 percent of the primary beam. *(Thompson, p 224)*

48. **(D)** A spinning top test may be performed to evaluate timer accuracy or valve tube efficiency in single-phase equipment. The number of dots or dashes imaged on the film is counted and should equal the number of radiation "pulses" occurring during that exposure time. *Because three-phase equipment does not emit pulsed radiation, but rather, almost constant potential, a synchronous spinning top must be used to evaluate timer accuracy.* The resulting image is a solid black arc. The angle of the arc is measured and should correspond to the known correct angle. *(Saia, PREP, pp 384–385)*

49. **(A)** Three-phase current is obtained from three individual alternating currents superimposed on, but out of step with, each other by 180°. The result is an almost constant potential current, having a very small voltage ripple, and producing more x-rays/mAs. *(Bushong, pp 302–303)*

50. **(C)** A phototimer is one type of AED that actually measures *light.* As x-ray photons penetrate and emerge from a part, a fluorescent screen beneath the cassette glows and the fluorescent light charges a photomultiplier tube. Once a predetermined charge has been reached, the exposure automatically terminates. A parallel plate *ionization* chamber is another type of AED. A radiolucent *chamber* is beneath the patient (between the patient and film). As photons emerge from the patient, they enter the chamber and ionize the air within. Once a predetermined charge has been reached, the exposure is automatically terminated. Motion of magnetic fields induc-

ing a current in a conductor refers to the principle of mutual induction. *(Selman, pp 241, 243)*

51. **(C)** A beveled focal track extends around the periphery of the anode disc; when a small angle is used, the beveled edge allows for a smaller effective focal spot and better detail. The disadvantage, however, is that photons are noticeably absorbed by the "heel" of the anode, resulting in a smaller percentage of x-ray photons at the anode end of the x-ray beam and a concentration of x-ray photons at the cathode end of the beam. This is known as the anode heel effect and can cause a primary beam variation of up to 45 percent. The anode heel effect becomes more pronounced as the SID decreases, as film size increases, and as target angle decreases. *(Bushong, p 125)*

52. **(B)** X-ray photons are produced in two ways as high-speed electrons interact with target atoms. First, if the high-speed electron is attracted by the nucleus of a tungsten atom and changes its course, as the electron is "braked," energy is given up *in the form of an x-ray photon*. This is called Bremsstrahlung (braking) radiation, and is responsible for the majority of x-ray photons produced at the conventional tungsten target. Second, a high-speed electron may eject a tungsten K shell electron, leaving a vacancy in the shell. An electron from the next energy level, the L shell, drops down to fill the vacancy, *emitting the difference in energy as a K characteristic ray*. Characteristic radiation comprises only about 15 percent of the primary beam. *(Selman, pp 156–158)*

53. **(A)** Restriction of field size is one important method of patient protection. However, the accuracy of the light field must be evaluated periodically as part of a QA program. Guidelines set forth for patient protection state that the collimator light and actual irradiated area must be accurate to within 2 percent of the SID employed. *(Bushong, p 630)*

54. **(A)** The x-ray tube's glass envelope and oil coolant are considered *inherent* filtration. Thin sheets of aluminum are *added* to make a

total of 2.5 mm Al equivalent filtration in equipment operated above 70 kVp. The function of aluminum filtration is to remove from the x-ray beam the soft (long-wavelength) x-ray photons that do not contribute to patient dose. These soft x-rays penetrate only a small thickness of tissue before being absorbed. *(Selman, p 199)*

55. **(B)** The HVL is that thickness of material that decreases the exposure rate of a particular x-ray beam to one half its initial value. The material absorbs the lower-energy photons, and consequently, the emerging beam has a greater overall energy. Therefore a greater thickness of material will be needed to cut the intensity of the filtered beam in half. *(Curry, p 410)*

56. **(C)** Because mammographic techniques operate at very low kVp levels, cassette front material becomes especially important. The use of soft, low-energy x-ray photons is the underlying principle of mammography; any attenuation of the beam would be most undesirable. Special plastics that resist impact and heat softening are frequently used as cassette front material, such as polystyrene and polycarbonate. *(Cullinan, p 103)*

57. **(D)** The thicker and more dense the anatomic part being studied, the less bright will be the fluoroscopic image. Both mA and kVp affect the fluoroscopic image similar to the way they affect the radiographic image. For optimum contrast, and especially in consideration of patient dose, higher kVp and lower mA are generally preferred. *(Bushong, p 353)*

58. **(C)** When high-speed electrons strike surfaces other than the tungsten target, x-rays may be produced and emitted in all directions. X-ray tubes therefore have a lead-lined metal protective housing to absorb much of this "leakage radiation." Leakage radiation must not exceed 100 mR/hr at a distance of one meter from the tube. Because the production of x-radiation requires the use of exceedingly high voltage, the tube housing also serves to protect from electric shock. The pro-

duction of x-rays involves the production of large quantities of heat, which can be damaging to the x-ray tube. Therefore an oil coolant surrounds the x-ray tube to further insulate and to absorb heat from the x-ray tube structures. (*Bushong, p 633*)

59. **(C)** A phototimer is one type of automatic exposure device (AED). When it is installed in the x-ray unit, it is calibrated to produce radiographic densities as required by the radiologist. Once the part being radiographed has been exposed to produce the proper film density, the phototimer automatically terminates the exposure. *The manual timer should be used as a backup timer, should the phototimer fail to terminate the exposure, thus protecting the patient from overexposure and the x-ray tube from excessive heat load.* Circuit breakers and fuses are circuit devices used to protect circuit elements from overload. In case of current surge, the circuit will be broken, thus preventing equipment damage. A rheostat is a type of variable resistor. (*Cullinan, p 25*)

60. **(A)** Radiographic rating charts enable the operator to determine the maximum safe mA, exposure time, and kVp for a particular exposure using a particular x-ray tube. An exposure made using the large focal spot may not be safe when using the small focal spot of the same x-ray tube. The total number of heat units (HU) an exposure generates also influences the amount of stress (in the form of heat) imparted to the anode. *Single-phase heat units* are determined by the product of mAs and kVp. *Three-phase, six-pulse heat units* are determined from the product of mA × sec × kVp × 1.35. *Three-phase, twelve-pulse heat units* are determined from the product of mA × sec × kVp × 1.41. In the examples given, then, group A produces 6091 HU, group B produces 14805 HU, group C produces 1997 HU, and D produces 7106 HU. Therefore, group A exposure factors will deliver the least amount of heat to the anode. (*Carlton, p 127*)

61. **(D)** The processor's pumping mechanisms transport solution through heating devices in order to maintain proper temperature. Solu-

tion is then returned under pressure for recirculation. The added pressure functions to agitate solution and keep it in close contact with the film emulsion. (*Bushong, p 212*)

62. **(D)** Sometimes, while restricting the primary beam to an area near the periphery of the body, part of the illuminated area overhangs the edge of the body. If the exposure is then made, scatter radiation from the tabletop (where there is no absorber) will undercut the part, causing excessive film density. If, however, a lead rubber mat is placed on the overhanging illuminated area, most of this scatter will be absorbed. This is frequently helpful in lateral lumbar spines and AP shoulders. (*Cullinan, p 73*)

63. **(C)** Radiographic contrast is greatly affected by changes in kilovoltage (Fig. 6–21). As kVp increases, a greater number of high-energy photons are produced at the target. These photons are more penetrating, but also produce more scatter radiation, contributing to *lower* radiographic contrast as a result of *scattered radiation fog*. The radiograph in question, radiograph B was made using 100 kVp and 18 mAs. Radiograph A was made of the same part using 80 kVp and 75 mAs, all other factors constant. The image details in radiograph A are far more perceptible due to the production of less scattered radiation. (*Saia, PREP, p 307*)

64. **(D)** *Size distortion is magnification* and is controlled by distance. As the part is moved away from the film (increased OID), magnification is produced. Similarly, as the focus is moved toward the object (decreased SOD) or film (decreased SID), magnification distortion is produced. (*Cullinan, p 124*)

65. **(A)** A misaligned guide shoe in the turnaround assembly will create evenly spaced minus density lines due to emulsion scratching. Dirty rollers cause multiple black specks on the finished radiograph—these are present as well on the pictured radiograph. A pi line is a plus-density artifact found ¾ inch from the leading edge of the film. Grid lines appear as minus density lines resulting from

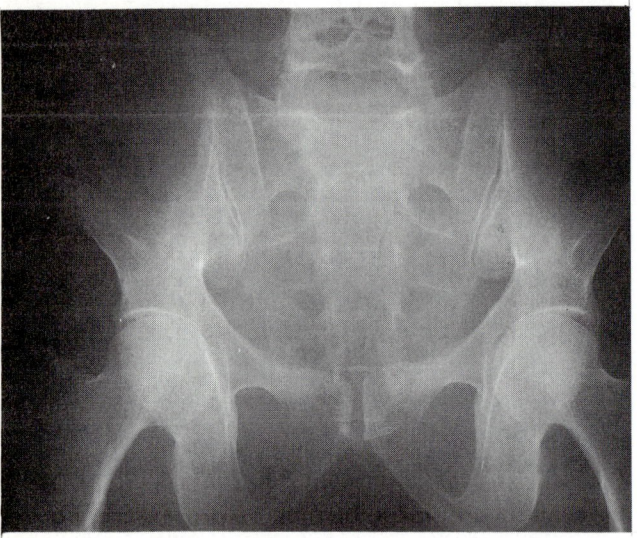

A

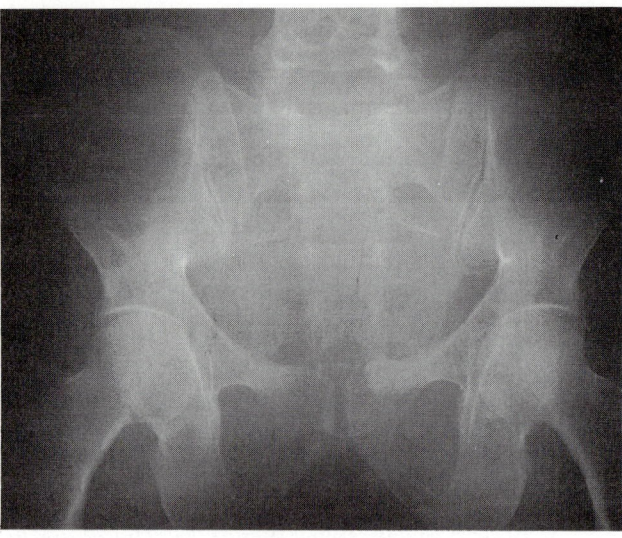

B

Figure 6–21. From the American College of Radiology Learning File. Courtesy of the ACR.

improper positioning or centering of the grid or x-ray tube. *(Sweeney, p 291)*

66. **(A)** Intravenous urography requires the use of iodinated contrast media. Low kilovoltage (about 70) is usually employed to enhance the photoelectric effect, and in turn to better visualize the renal collecting system. High kilovoltage will produce excessive scatter radiation and obviate the effect of the contrast agent. A higher mA with a short exposure time is generally preferable. *(Cullinan, p 144)*

67. **(B)** As kVp is increased, x-ray photons begin to interact with atoms of tissue via the Compton scatter interaction. Scattered x-ray photons result, which serve only to add unwanted, undiagnostic densities (scatter radiation fog) to the radiologic image. (Whereas Compton scatter reduces patient dose, compared to photoelectric interactions, it can pose a significant radiation hazard to personnel during fluoroscopic procedures.) Therefore, the use of *optimum* kVp is recommended to reduce the production of scatter radiation fog. Scatter radiation is also a function of the size and content of the irradiated field. The greater the volume and atomic number of the tissue, the greater the production of scatter radiation. So although there is little we can do

about the atomic number of the structure to be radiographed, we can make every effort to keep the field size restricted to only the essential area of interest in an effort to decrease production of scatter radiation. Grids have no effect on the *production* of scatter radiation, but they are very effective in removing scatter radiation from the beam before it strikes the film. *(Cullinan, pp 67–69)*

68. **(C)** Look over the choices again, keeping in mind the factors that affect recorded detail. Looking first to SID, the options may be reduced to B and C because the increase to a 40-inch SID will certainly improve recorded detail. There is one other factor that will impact detail: the speed of the system (intensifying screens). Because a slower system will render better recorded detail, the best answer is C. *(Cullinan, p 300)*

69. **(A)** If the source-image distance is above or below the recommended focusing distance, the primary beam at the lateral edges will not coincide with the angled lead strips. Consequently, there will be absorption of the primary beam, termed grid cutoff. If the grid failed to move during the exposure, there would be grid lines throughout. Central ray

angulation in the direction of lead strips is appropriate and will not cause grid cutoff. If the central ray was off center, there would be uniform loss of density. *(Selman, pp 372–373)*

70. **(C)** Distortion is the result of misalignment of the x-ray tube, the anatomic part, and the film. If these three parts are not parallel with each other, shape distortion occurs. The greater the misalignment, the greater the distortion. In the example cited, the image made with the most tube angle will produce the greatest distortion. Distortion is often introduced intentionally in order to visualize some structure to better advantage. The 37° (caudad) AP axial projection of the skull, for example, projects the facial bones inferiorly so that the occipital bone can be visualized to better advantage. *Cullinan, p. 126, 157)*

71. **(B)** X-ray film emulsion is sensitive and requires proper handling and storage. Several kinds of artifacts can be produced by careless handling. Tree-like black branching marks are usually due to static electrical discharge especially prevalent during cold, dry weather. Crinkle marks are produced by acute bending of the film. Excessive emulsion swelling and chemical fog can be caused by increased solution temperature or insufficient replenishment. *(Selman, pp 309–310)*

72. **(D)** Breast tissue has very low subject contrast, but microcalcifications and subtle density differences are imperative to visualize. Fine detail is necessary to visualize any microcalcifications; therefore, a small focal spot tube is essential. High contrast (therefore, low kilovoltage) is needed to accentuate any differences in tissue density. A compression device serves to even out differences in tissue thickness (thicker at chest wall, thinner at nipple), decrease object-film distance, and helps decrease the production of scatter radiation. *(Selman, pp 344–347)*

73. **(D)** Technique charts are exposure factor *guides* that help technologists produce radiographs with consistent density and contrast. They suggest a group of exposure factors to be used at a particular SID, with a particular

grid ratio, and screen/film combination, a particular focal spot size and central ray angulation. *Technique charts do not take into account the nature of the part* (disease, atrophy, and so on). *(Cullinan, pp 134–137)*

74. **(B)** A densitometer indicates optical density by providing a digital readout of the quantity of light transmitted through a film. A sensitometer is a device used to produce a consistent gray scale exposure on film, usually in conjunction with processor sensitometry. A penetrometer is an aluminum step wedge that may be radiographed to produce a gray scale. H&D curves are sensitometric curves of film emulsion's response to light and radiation. *(Pizzutiello & Cullinan, p 139)*

75. **(B)** Differential absorption refers to the different attenuation, or absorption, properties of adjacent body tissues. Two parts of widely differing absorption characteristics will produce a high radiographic contrast. Frequently, exposure factors that would properly expose one part will severely overexpose or underexpose the neighboring part (as with lungs versus thoracic spine). This effect can be minimized with the use of a compensating filter, or by the use of high kilovoltage (for more uniform penetration). Increased collimation is important in the control of patient dose and scatter radiation. *(Cullinan, p 118)*

76. **(C)** When the anatomic part is comprised of greatly differing densities, a compensating filter is frequently helpful. Compensating filters can be accommodated for by tracks in the tube head. They can be wedge shaped (the thicker part of the wedge paralleling the thinner body part), thus compensating for greater or lesser tissue densities (as in a large decubitus abdomen). A grid is used to absorb scatter radiation before it reaches the film; intensifying screens amplify the action of x-rays; and a penetrometer (aluminum step wedge) is used to illustrate the effect of kVp on contrast. *(Cullinan, pp 37–42)*

77. **(B)** Calcium tungstate intensifying screens had a broad range of emitted light, and it was more likely that somewhat different film

emulsions could still be compatible with them. However, rare earth phosphors emit light over a relatively short range, usually in the green portion of the spectrum. Film emulsion must be sensitive and responsive to that particular color, or the expected results will not occur. If, for example, a blue-sensitive emulsion were matched with green-emitting screens, the resulting radiograph would be underexposed because the blue-sensitive film emulsion was not responsive to the green-emitting phosphors. (*Cullinan, p 95*)

78. **(C)** 18 msec (milliseconds) is equal to 0.018 sec, and because mA × sec = mAs, then 10.8 was the original mAs. Now it is only necessary to determine what exposure time must be used with 400 mA to provide the same 10.8 mAs (and thus, the same radiographic density). Because mA × sec = mAs, then:

$$400x = 10.8$$
$$x = 0.027 \text{ sec} (= 27 \text{ msec})$$

(*Selman, p 332*)

79. **(D)** Unexposed, undeveloped silver is removed from film emulsion in the fixer solution. It is recovered (reclaimed) from the solution and sold. The silver may be reclaimed electrolytically or with a metallic replacement cartridge. Silver may also be reclaimed from processed radiographs or unexposed film. (*Cullinan, pp 112–113*)

80. **(D)** Motion causes unsharpness that destroys detail. Careful and accurate patient instruction is essential for minimizing voluntary motion. Suspended respiration eliminates respiratory motion. Using the shortest possible exposure time is essential to decrease involuntary motion. Immobilization also can be very useful in eliminating motion unsharpness. (*Selman, p 327*)

81. **(B)** A *characteristic curve* is used to predict the speed, contrast, and exposure latitude of a particular film emulsion. It compares the exposure given the film with the resultant density. It has *three portions*: the toe, the

straight-line portion (region of correct exposure), and the shoulder. The *toe* occurs immediately after base-plus fog, whose density must not exceed 0.2. The ascending straight-line portion follows the toe; the *straight-line portion is the portion of correct exposure* and extends from about 0.25 to 2.5. The curve then bends and levels off at the *shoulder* (D-max) portion of the curve. (*Saia, PREP, p 315*)

82. **(D)** Shape distortion is caused by misalignment of the x-ray tube, body part to be radiographed, and film. An object can be falsely imaged (foreshortened, elongated) by incorrect placement of tube, part, or film. Only one of the three need be misaligned for distortion to occur. (*Selman, pp 351–352*)

83. **(B)** As the exposed film enters the processor from the feed tray, it first enters the *developer* section (1), where *exposed* silver bromide crystals are reduced to black metallic silver. The film then enters the *fixer* (2), where the *unexposed* silver grains are removed from the film by the clearing agent. The film then enters the wash section (3), where chemicals are removed from the film to preserve the image. From the wash the film enters the dryer section (4). (*Saia, PREP, p 335*)

84. **(B)** As x-ray photons are produced at the tungsten target, they more readily diverge toward the cathode end of the x-ray tube. As they try to diverge toward the anode, they interact with and are absorbed by the anode "heel." Consequently, there is a greater intensity of x-ray photons at the cathode end of the x-ray beam. This phenomenon is known as the anode heel effect. Because shorter SIDs and larger film sizes require greater divergence of the x-ray beam in order to provide coverage, the anode heel effect will be accentuated. (*Selman, pp 394–396*)

85. **(B)** High-kilovoltage exposures produce large amounts of scattered radiation, and high-ratio grids are often used with high kV techniques in an effort to absorb more of this scatter radiation. However, as more scattered radiation is absorbed, more primary radiation is absorbed as well. This accounts for the increase in mAs

required when changing from 8:1 to 16:1 grid. Additionally, precise centering and positioning become more critical; a small degree of inaccuracy is more likely to cause grid cutoff in a high ratio grid. *(Selman, pp 362–363)*

86. **(B)** Proper functioning of the phototimer depends on accurate positioning by the radiographer. The correct *photocell*(s) must be selected and the anatomic part of interest must completely cover the photocell in order to achieve the desired density. If a photocell is left uncovered, scattered radiation from the part being examined will cause premature termination of exposure and an underexposed radiograph. *(Carroll, p 337)*

87. **(C)** If the x-ray tube is angled significantly across the lead strips of a focused grid, there is uniform loss of density (grid cutoff). Insufficient or excessive distance with focused grids causes loss of density (grid cutoff) along the periphery of the image. Figure 6–7 demonstrates grid cutoff everywhere except a central vertical strip of the image. This density loss is due to the focused grid being placed upside down. Thus, the middle vertical lead strips allow x-rays to pass, but because the lead strips cant laterally, they are directly opposite the direction of the x-ray photons (rather than parallel to them), and severe grid cutoff results. *(Selman, pp 376–377)*

88. **(B)** Rare earth phosphors have a greater conversion efficiency than do other phosphors. Lanthanum oxybromide is a blue-emitting phosphor, and gadolinium oxysulfide is a green-emitting phosphor. Cesium iodide is the phosphor used on the input screen of image intensifiers; it is not rare earth. *(Selman, p 283)*

89. **(C)** Absorption occurs when an x-ray photon interacts with matter and disappears, as in the photoelectric effect. Scattering occurs when there is *partial* transfer of energy to matter, as in the *Compton effect*. The *reduction in the intensity* of an x-ray beam as it passes through matter is called attenuation. *(Selman, pp 179–181)*

90. **(B)** As the kilovoltage is increased, a greater number of electrons are driven across to the anode with greater force. Therefore, as energy conversion takes place at the anode, more high-energy photons are produced. However, because they are higher-energy photons, there will be less patient absorption. *(Selman, p 164)*

91. **(B)** Of the given factors, kilovoltage and grid ratio will have a significant effect on the scale of radiographic contrast. The mAs values are almost identical. Because an increased kilovoltage and low-ratio grid combination would allow the greatest amount of scatter radiation to reach the film, thereby producing more gray tones, B is the best answer. Group D also uses a low-ratio grid, but the kV is too low to produce as much gray as C. *(Cullinan, p 300)*

92. **(D)** The x-ray tube anode is designed according to the line focus principle, that is, with the focal track beveled (Fig. 6–22). This allows a larger *actual* focal spot to project a smaller *effective* focal spot, resulting in improved recorded detail with less penumbral blur. However, due to the target angle, penumbral *blur varies along the longitudinal tube axis*, being greater at the cathode end of the image and less at the anode end of the image. *(Saia, PREP, p 260)*

93. **(B)** In the low kilovoltage ranges, a difference of just a few kVp makes a very noticeable radiographic difference. High kVp techniques offer much greater margin for error, as do slow film/screen combinations. Grid ratio is unrelated to exposure latitude, but higher-ratio grids offer less tube centering latitude (leeway, margin for error) than low-ratio grids. *(Cullinan, p 138)*

94. **(A)** As the kilovoltage is increased, more penetration will occur and a greater variety of densities (grays) will be apparent on the film. This is termed *long scale or low contrast*. Additionally, as the kVp and scale of grays increase, the *exposure latitude increases*; the "margin for error" in technical factors becomes greater. As the mAs is decreased to

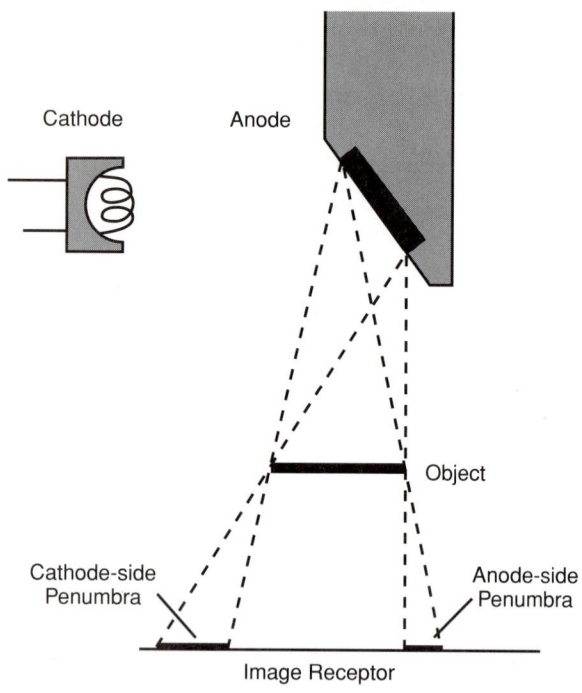

Cathode   Anode

Object

Cathode-side
Penumbra

Anode-side
Penumbra

Image Receptor

Figure 6–22. Reproduced with permission from Wolbarst AB. *Physics of Radiology.* East Norwalk, CT: Appleton & Lange, 1993.

compensate for the increased kVp, density should remain the same. *(Cullinan, p 121)*

95. **(D)** If a film was placed on an illuminator and 100 percent of the illuminator's light was transmitted through the film, that film must have a density of 0. According to the equation

$$\text{density} = \log \frac{\text{incident light intensity}}{\text{transmitted light intensity}}$$

if 10 percent of the illuminator's light passes through the film, that film has a density of 1. If 1 percent of the light passes through the film, that film has a density of 2. *(Bushong, p 272)*

96. **(C)** At a given exposure, higher speed intensifying screens will emit more fluorescent light, thereby increasing radiographic density. Faster speed intensifying screens allow a considerable reduction in mAs, and therefore in patient dose and motion unsharpness. Intensifying screen speed is unrelated to scattered radiation. *(Selman, p 380–385)*

97. **(C)** In order to change nongrid to grid exposure, or to adjust exposure when changing from one grid ratio to another, remember the factor for each grid ratio:

no grid = 1 original mAs
5:1 grid = 2 original mAs
6:1 grid = 3 original mAs
8:1 grid = 4 original mAs
12:1 grid = 5 original mAs
16:1 grid = 6 original mAs

Therefore, to change from nongrid to an 8:1 grid, multiply the original mAs by a factor of four. A new mAs of 32 is required. *(Cullinan, p 292)*

98. **(D)** The figure illustrates a *wire mesh* test for *screen film contact.* If intensifying screens and film do not make perfect contact, recorded detail can be seriously compromised. In this test, a wire mesh supported between two rigid pieces of clear plastic is used to evaluate screen/film contact. The mesh is placed on a cassette and radiographed. Upon viewing, any areas that appear unsharp or blurry are indicative of poor screen/film contact. A bar pattern is used to evaluate screen resolution, a star pattern is used to evaluate the focal spot, and a Wisconsin cassette can be used to evaluate kVp calibration. *(Saia, PREP, p 270)*

99. **(D)** The radiographic subject (the patient) is composed of many different tissue types of varying densities resulting in varying degrees of photon attenuation and absorption. This differential absorption contributes to the various shades of gray (scale of radiographic contrast) on the finished radiograph. Normal tissue density may be significantly altered in the presence of pathology. For example, destructive bone disease can cause a dramatic decrease in tissue density. Abnormal accumulation of fluid (as in ascites) will cause a significant increase in tissue density. Muscle atrophy, or highly developed muscles, will similarly decrease or increase tissue density. *(Cullinan, pp 93, 121)*

100. (B) Subcutaneous emphysema is a pathologic distention of tissues with air; pulmonary emphysema is a chronic disease characterized by overdistention of the alveoli with air. Osteomalacia is a softening of bone so that it becomes flexible, brittle, and deformed. Both of these conditions involve a *decrease in tissue density*, and therefore require a *decrease in exposure factors*. Atelectasis is a collapsed or airless lung, and requires an *increase* in exposure factors. *(Carlton, p 257)*

101. (A) Each film passing through the processor solutions takes with it a certain amount of solution. Replenishment is also essential for maintaining each solution's level of concentration, in order to maintain solution activity and avoid chemical fog from exhausted solutions. One way to determine the quantity of replenisher solution to be added is by the length of the film entering the processor. A microswitch initiates and terminates replenishment as it senses the beginning and end of each film. *(McKinney, p 106)*

102. (A) This is the definition of *fog*. Log relative exposure is the amount of exposure required to produce a given density as measured on the sensitometric graph. Optical density is normal radiographic density. An artifact is anything foreign to the image; it could include fog, but also many physical interferences. *(Pizzutiello & Cullinan, p 109)*

103. (B) The use of high kVp with a fairly low-ratio grid will be ineffective in ridding the remnant beam of scatter radiation. In order to improve contrast in this example, it has been decided to decrease the kilovoltage by 15 percent, thus making it necessary to increase the mAs from 5 to 10 mAs. Because it is also desired to increase the grid ratio to 12:1, another change in mAs will be required (remember, 10 mAs is now the *old* mAs):

$$\frac{\text{old grid factor } (8:1)}{\text{new grid factor } (12:1)} = \frac{\text{old mAs}}{\text{new mAs}}$$

$$\frac{4}{5} = \frac{10}{x}$$

$$4x = 50$$

$$x = 12.5 \text{ mAs required with } 12:1 \text{ grid}$$

Now, determine the exposure time required with 400 mA to produce 12.5 mAs:

$$400x = 12.5$$

$$x = 0.03 \text{ second exposure}$$

(Cullinan, p 292)

104. (A) According to the Inverse Square Law of Radiation, as the distance between the radiation source and film decreases, the exposure rate increases. Therefore, a decrease in technical factors is first indicated to compensate for the distance change. The following formula is used to determine new mAs values, when changing distance:

$$\frac{\text{mAs}_1}{\text{mAs}_2} = \frac{D_1^2}{D_2^2}$$

Substituting known values:

$$\frac{8}{x} = \frac{5184}{1600}$$

$$5184\,x = 12800$$

$$x = 2.4 \text{ mAs at a 40 inch SI}$$

To then compensate for adding an 8:1 grid, you must multiply the 2.4 mAs by a factor of 4. Thus, 9.6 mAs is required to produce a film density similar to the original radiograph. The following are the factors used for mAs conversion from nongrid to grid:

no grid = 1 original mAs
5:1 grid = 2 original mAs
6:1 grid = 3 original mAs
8:1 grid = 4 original mAs
12:1 grid = 5 original mAs
16:1 grid = 6 original mAs

(Cullinan, p 292)

105. (A) As SID increases, so does recorded detail, because magnification is decreased. As focal spot size increases, recorded detail decreases because more penumbra is produced. Therefore, SID is directly proportional to recorded detail. Focal spot size is inversely proportional to radiographic sharpness or recorded detail. Tube current impacts radiographic density, and is unrelated to recorded detail. (*Cullinan, p 124*)

106. (D) The processor's roller system consists of a series of rollers, propelled by gears that function to transport the film from one solution to another and through the dryer section. As the film is being transported, solution is agitated along the film surface. Rollers coming in very close contact with the film provide a squeegee action to remove excess solution from the emulsion surface. (*Carlton, p 302*)

107. (B) Phosphors used in intensifying screens must *absorb* a high percentage of the incident x-ray photons and convert x-ray photon energy to fluorescent light energy. Afterglow is an undesirable characteristic of phosphors, as continued fluorescence causes unpredictable and increased density. (*Cullinan, p 147*)

108. (B) If the same kV is used with single-phase and three-phase equipment, the three-phase unit will require about 50 percent less mAs to produce similar radiographs. Because three-phase equipment has much higher effective voltage than single-phase equipment, the three-phase radiograph will possess lower contrast. Lower mAs can be used with three-phase equipment; heat units are not built up as quickly. *When technical factors are adjusted to obtain the same density and contrast there is no difference in patient dose.* (*Selman, p 256*)

109. (C) Every radiographic image is composed of a number of different densities. These densities may be measured and given a numeric value with a device called a densitometer. A sensitometer is another device used in quality control programs; it is used to give a precise exposure to a film emulsion. An aluminum step wedge (penetrometer) may be used to show the effect of kVp on contrast. A spinning top is used to test the accuracy of the x-ray machine's timer or rectifiers. (*Selman, p 331*)

110. (C) There are several ways to reduce the amount of scatter radiation reaching the film. First, use of optimum kVp is essential; excessive kVp will increase the production of scatter radiation. Second, conscientious use of the beam restrictor (collimator); the smaller the volume of irradiated tissue, the less scatter radiation is produced. The use of grids helps clean up scatter radiation before it reaches the film. The size of the tube focus has an impact on image geometry and recorded detail, but no effect on scatter radiation. (*Cullinan, pp 67–78*)

111. (A) Inspiration and expiration films are frequently requested when examining patients for pneumothorax, or to demonstrate degree of diaphragm excursion or the presence of a foreign body. The expiration radiograph (film 1) demonstrates fewer ribs projected above the diaphragm. Because a smaller volume of air is contained within the lungs, the expiration radiograph requires an increase in exposure of approximately 6 to 8 kVp. A lordotic position projects the clavicles above the pulmonary apices. A ventral decubitus position may be used to demonstrate air/fluid levels and is made using a horizontal x-ray beam with the patient in the prone position. (*Ballinger, vol 1, p 444*)

112. (D) Pulmonary vascular markings are often prominent in the elderly and in smokers. Quiet, shallow breathing may be used during a long exposure (and a compensating low mA) to blur them out. Oblique sternum, AP scapula, ribs, and the lateral thoracic spine are examinations where this technique is useful. (*Ballinger, vol 1, pp 164, 360, 407, 422*)

113. (B) In order to demonstrate the intercondyloid fossa the CR must be directed perpendicular to the long axis of the tibia (Fig. 6–23). Because the knee is flexed so that the tibia forms a 40° angle with the film, the CR must be directed 40° caudad in order to place the CR perpendicular to the long axis of the tibia. Directing the CR to the popliteal depression

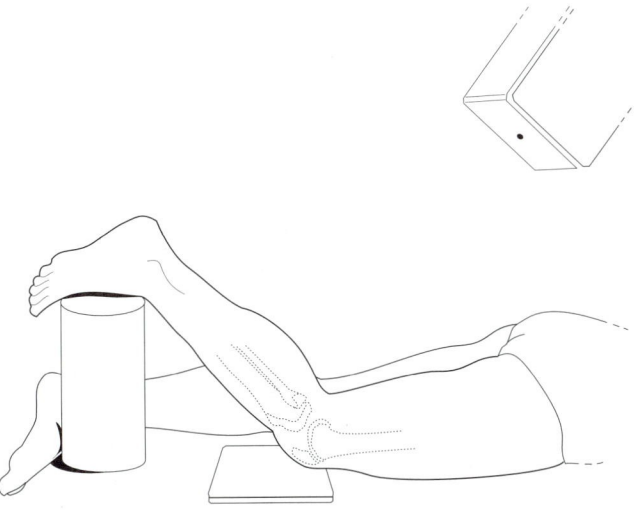

Figure 6–23.

aligns the CR parallel with the knee joint space. *(Ballinger, vol 1, p 252)*

114. **(C)** Breast tissue compression accomplishes several things: it decreases the tissue thickness to be irradiated, brings tissue closer to the film, draws more tissue onto the film, and immobilizes the tissue for the lengthy exposure. Thinner breast tissue is easier to penetrate, and therefore requires less radiation. When an object (breast tissue) is closer to the film, the OID is decreased and the recorded detail is increased. *(Ballinger, vol 2, p 467)*

115. **(A)** The thoracic and abdominal cavities are associated with serous membranes: the thoracic cavity with the *pleura* and the abdominal cavity with the *peritoneum*. The pleura and peritoneum each have two walls, a parietal or (outer) wall and a visceral (inner) wall. The parietal pleura lines the thoracic cavity while the visceral pleura is reflected over the surface of the lungs and projects between the fissures. The parietal peritoneum lines the abdominal cavity and the visceral peritoneum invests the abdominal viscera. *(Ballinger, vol 1, p 439)*

116. **(C)** The degree of anterior and posterior motion is occasionally diminished with a "whiplash" type injury. Anterior (forward,

flexion) and posterior (backward, extension) motion is evaluated in the lateral position with the patient assuming flexion and extension as best as he or she can. Left and right bending films are frequently obtained of the thoracic and lumbar vertebrae when evaluating scoliosis. The AP open-mouth projection is used to evaluate the first two cervical vertebrae. The moving mandible AP is used to demonstrate the entire cervical spine while blurring out the superimposed mandible. *(Ballinger, vol 1, p 338)*

117. **(D)** For the lateral projection of the elbow, the patient is positioned with the elbow forming a 90° angle, and the forearm and hand in the lateral position. This superimposes the humeral epicondyles and projects the olecranon process in profile. The proximal radius and ulna are only slightly superimposed, and the radial head faces anteriorly. *(Ballinger, vol 1, p 85)*

118. **(C)** and 119. **(A)** An anterior view of the foot and ankle is illustrated. The ankle joint is formed by the articulation of the tibia, fibula, and talus (7). The tibial (medial) malleolus is labeled 8; the fibular (lateral) malleolus is labeled 1. The talus articulates with the calcaneus (2) inferiorly and the navicular (6) anteriorly. The cuboid (3) is seen anterior to the calcaneus and the three cuneiforms (5) are anterior to the navicular. *(Saia, PREP, p 92)*

120. **(C)** Lateral carpals, especially the navicular, are demonstrated in the PA oblique projection and the ulnar flexion maneuver. The navicular may also be demonstrated with the wrist PA and elevated 20°. The CR is directed perpendicular to the carpal navicular. The medial carpals, especially the pisiform, are well demonstrated in the AP oblique projection and with the radial flexion maneuver. *(Ballinger, vol 1, pp 89–91)*

121. **(D)** When air/fluid levels are to be demonstrated, it is important to direct the central ray horizontally. If the CR is angled or directed vertically, the air or fluid level will be distorted or entirely obliterated. Free air in the abdominal cavity is best visualized when

the patient is left lateral decubitus. This allows the air to accumulate around the homogeneous liver. *(Ballinger, vol 2, p 41)*

122. **(C)** The AP oblique projection (medial rotation) of the elbow superimposes the radial head and neck on the proximal ulna. It demonstrates the olecranon process within the olecranon fossa, and projects the coronoid process free of superimposition. The radial head is projected free of superimposition in the AP oblique projection (lateral rotation) of the elbow. *(Ballinger, vol 1, p 104)*

123. **(B)** The spinal cord is a column of nervous tissue about 17 inches (44 cm) in length. It is somewhat flattened anteroposteriorly and extends from the medulla oblongata of the brain to the level of L2 within the spinal canal. Because the adult spinal cord ends at the level of L2, a lumbar puncture is usually performed below that level—generally at the level of L 3-4. A lumbar puncture may be performed for the removal of spinal fluid for diagnostic purposes, or for the injection of medications. *(Ballinger, vol 2, p 500)*

124. **(C)** The IOML (infraorbitomeatal line) is an imaginary line extending from the infraorbital margin to the EAM (external auditory meatus) and is represented by no. 3. Number 1 is the glabellomeatal line, 2 is the OML (orbitomeatal line), and 4 is the acanthomeatal line. These baselines are used to obtain accurate positioning in skull radiography. *(Ballinger, vol 2, p 233)*

125. **(B)** The patient is well positioned; the spinous processes and sternum are clearly seen without superimposition. Adequate penetration and long-scale contrast are present without excessive radiographic density. The patient had been properly shielded for the PA projection; however, the shield was not moved to the correct location prior to the lateral exposure. *(Ballinger, vol 3, p 35)*

126. **(A)** The greater and lesser tubercles are prominences on the proximal humerus, separation by the bicipital groove. The AP projection of the humerus in *external rotation* demon-

strates the *greater tubercle* in profile. With the arm placed in *internal rotation*, the humerus is placed in a true lateral position and the *lesser tubercle* is demonstrated. *(Saia, PREP, p 86)*

127. **(C)** Contrast media may be administered in a variety of manners in cholangiography, including

1. An endoscope with a cannula placed in the hepatopancreatic ampulla (of Vater) for an ERCP
2. A needle or small catheter placed directly in the common bile duct for an operative cholangiogram
3. A very fine needle through the patient's side and into the liver for a percutaneous transhepatic cholangiogram
4. Via an indwelling T-tube for postoperative or T-tube cholangiogram *(Ballinger, vol 2, p 54)*

128. **(C)** An oblique projection of the cervical spine is pictured. The first two cervical vertebrae are not well visualized due to superimposition of the mandible. The chin should be elevated somewhat to avoid this problem. Otherwise, the positioning is satisfactory, with good demonstration of the remainder of the cervical intervertebral foramina. The patient has been accurately rotated 45° with a 15 to 20° cephalic tube angle. *(Ballinger, vol 1, p 342)*

129. **(C)** An oblique projection of the cervical spine is pictured. The patient has been accurately positioned RAO with the MSP 45° to the film and the CR angled 15 to 20° caudad. The chin should be elevated to better visualize the first two cervical vertebrae. This position offers excellent delineation of the intervertebral foramina (no. 1) formed by the adjacent vertebral notches of pedicles (2). This projection gives an "on-end" view of the transverse processes (3). A portion of the spinous processes (4) may be seen, especially in the lower cervical vertebrae. *(Ballinger, vol 1, p 342)*

130. **(A)** The *midcoronal* plane (1) divides the body into anterior and posterior halves. A *coronal* plane is any plane parallel to the midcoronal plane. The *midsagittal* plane (2) divides the body into left and right halves. A *sagittal* plane is any plane parallel to the midsagittal plane. A *transverse* or *horizontal* plane (3) is perpendicular to the midsagittal plane and midcoronal plane, dividing the body into superior and inferior portions. *(Saia, PREP, p 61)*

131. **(C)** The tangential ("sunrise") projection is used to demonstrate the articular surfaces of the femur and patella. It is also used to demonstrate vertical fractures of the patella. The AP, PA, and oblique projections of the knee are used primarily to evaluate the joint space and articulating structures. The "tunnel" view is used to demonstrate the intercondyloid fossa. *(Ballinger, vol 1, p 264)*

132. **(D)** Variance from the normal bladder contour will be noted while the bladder is full of contrast medium. However, a postmicturition (postvoiding) radiograph is also an essential part of an IVP. The presence of residual urine may be an indication of small tumor masses, or in male patients, enlargement of the prostate gland. *(Ballinger, vol 2, p 169)*

133. **(C)** A lateral projection of the fourth finger is best obtained if the finger is positioned so that there is as little OID as possible. Therefore, with only the fourth finger extended in the lateral position, the arm is positioned on the ulnar (medial) surface. This places the finger closer to the film than if it were positioned radial side down. Excessive magnification distortion is avoided and better recorded detail is obtained. *(Bontrager, p 94)*

134. **(B)** Evaluation of the acromioclavicular joints requires bilateral AP or PA erect projections with and without the use of weights. Weights are used to emphasize the minute changes within a joint caused by separation or dislocation. The use of weights should be avoided if a fracture of the affected area is suspected. *(Ballinger, vol 1, pp 152–156)*

135. **(C)** A PA axial Caldwell position is illustrated, demonstrating the frontal and ethmoid sinuses. The Caldwell position requires an an-

gle of 15° caudad, exiting the nasion. The petrous ridges should be projected in the lower third of the orbits. The illustrated radiograph demonstrates somewhat excessive angulation because the petrous pyramids are projected at the bottom of the orbits. *(Ballinger, vol 2, pp 380–381)*

136. **(A)** In order to place the right axillary ribs parallel to the film, an LAO or RPO position is required. The RAO position will demonstrate the left axillary ribs. *(Ballinger, vol 1, p 420)*

137. **(A)** It is occasionally necessary to view the lung apices free of superimposition with clavicles. This objective can be achieved in the AP axial projection. The patient is positioned AP erect with the CR directed 15° cephalad, entering the manubrium. An AP axial projection can also be obtained with the patient in the lordotic position. If sufficient lordosis can be assumed, the CR is directed perpendicular to the film. *(Ballinger, vol 1, pp 468, 472–473)*

138. **(D)** Correct and complete patient information on each radiograph is of paramount importance. Correct ID is required on follow-up examinations that need to be accurately compared. Incorrect or delayed diagnosis can result from careless errors. Repeat exams may be required, resulting in needless additional radiation exposure—hence, poor radiation protection practice. *(Ballinger, vol 1, p 16)*

139. **(A)** Breast tissue is the most dense, glandular, and radiographically homogeneous in appearance in the postpubertal adolescent. Pregnancy causes changes within the breast that reduce the glandular tissue and replace it with fatty tissue (a process called fatty infiltration). Menopause causes glandular tissue to atrophy further. *(Ballinger, vol 2, pp 460–461)*

140. **(B)** The (diarthrotic) sternoclavicular joints are formed by the medial (sternal) extremities of the clavicles with the clavicular notches of the manubrium (of the sternum).

They can be demonstrated in the LAO and RAO positions. The LAO demonstrates the left sternoclavicular joint while the RAO demonstrates the joint on the right. The patient is obliqued about 15° with the side of interest adjacent to the film. (*Ballinger, vol 1, p 416*)

141. **(D)** The PA axial (Caldwell) projection of the paranasal sinuses is used to demonstrate the frontal and ethmoid sinuses. The central ray is angled caudally 15° to the OML. This projects the petrous pyramids into the lower one third of the orbits, thus permitting optimum visualization of the frontal and ethmoid sinuses. (*Ballinger, vol 2, pp 380–381*)

142. **(B)** Rotation of the chest is evidenced in the following ways: the distance between the medial aspect of the clavicles and lateral portion of the vertebral column is asymmetrical, the air-filled trachea is off midline, and the scapulae and air-filled lungs are asymmetric. The exposure was made during reasonably good inspiration as evidenced by visualization of eight ribs above the diaphragm. Upper and lateral aspects of the lungs, including pulmonary apices and costophrenic angles, are demonstrated. Even minimal rotation of the chest introduces significant distortion of the heart. (*Ballinger, vol 1, p 442*)

143. **(C)** Retrograde urography is not considered a functional study of the urinary system. Intravenous pyelography (IVP), descending urography, and infusion nephrotomography are all considered functional urinary tract studies because the contrast medium is introduced intravenously and *excreted* by the kidneys. Retrograde urography involves introduction of contrast medium into the kidneys via catheter, thereby demonstrating *structure*, but not *function*. (*Ballinger, vol 2, p 158*)

144. **(C)** The hepatic and splenic flexures are not generally well demonstrated in the AP and PA projections. In order to "open" the flexures, oblique projections are required. The hepatic flexure is usually well demonstrated in the RAO (right PA oblique) or LPO (left AP oblique) positions. The LAO or RPO positions are used to demonstrate the splenic flexure. (*Ballinger, vol 2, pp 135–136, 140–141*)

145. **(D)** The four types of body habitus are (from upper extreme to lower extreme) hypersthenic, sthenic, hyposthenic, and asthenic. The gallbladder and stomach are higher and more lateral and the large bowel more peripheral in the hypersthenic. The diaphragm is in a higher position in the hypersthenic individual. Recognition of a patient's body habitus, and its characteristics, is an important part of accurate radiography. Bone porosity is generally unrelated to body habitus type. (*Ballinger, vol 1, p 41*)

146. **(B)** Intervertebral joints are well visualized in the lateral projection of all the vertebral groups. Cervical articular facets (forming apophyseal joints) are 90° to the midsagittal plane and are therefore well demonstrated in the lateral projection. The cervical intervertebral foramina lie 45° to the midsagittal plane (and 15 to 20° to a transverse plane), and are therefore demonstrated in the oblique position. (*Ballinger, vol 1, pp 337, 358, 369*)

147. **(D)** Postoperative, or T-tube, cholangiography is frequently performed to evaluate patency of the biliary ducts and to identify any previously undetected stones. A T-tube is left in place within the common bile duct, following surgery, with the vertical portion of the *T* extending outside the body. Water-soluble iodinated medium is injected and fluoroscopic examination is carried out. (*Ballinger, vol 2, p 78*)

148. **(D)** Recent biopsy of the colon is a contraindication to performing a barium enema examination. A biopsy of the colon introduces a perforation in the colon wall, which must be allowed to heal before introducing barium. Extravasation, or leakage of barium into the abdominal cavity can result in peritonitis. If radiographic examination of the colon cannot wait, a water-soluble radiographic contrast agent that is absorbed, such as gastrographin, should be selected. Barium enema examination is not usually contraindicated, but rather indicated for eval-

uation of suspected tumor or polyp and in the case of tarry stools. *(Adler & Carlton, p 306)*

149. **(A)** The dorsal decubitus position is obtained with the patient supine and the x-ray beam directed horizontally. The finished radiograph looks similar to a routine lateral projection of the chest. However, small amounts of fluid will gravitate posteriorly, and small amounts of air will rise anteriorly. *(Ballinger, vol 1, p 55)*

150. **(D)** A thin chest would require a greater obliquity to separate the vertebrae and sternum from superimposition than would a thick chest. With the patient in the RAO position, the sternum is projected to the left of the vertebral column and is superimposed on the heart. This superimposition promotes more uniform tissue density, and, therefore more uniform radiographic density. Prominent pulmonary vascular markings may be obliterated by allowing the patient to breathe (shallow breaths only) during a long exposure (with a very low mA). *(Ballinger, vol 1, p 406)*

151. **(B) and 152. (C)** A parietoacanthial projection (Water's position) of the maxillary sinuses and facial bones is pictured. The chin is elevated sufficiently to project the petrous ridges below the maxillary sinuses (number *4*). Note foramen rotundum seen near the upper margin of the maxillary sinuses. Other sinus groups are not well visualized in this position, although a modification with the mouth open may be taken to demonstrate the sphenoid sinuses. This is also the single best projection to demonstrate the facial bones. The zygomatic arch (number *2*) is well demonstrated; the mandible, its angle (number *1*), and coronoid process (number *3*) are also well demonstrated. The odontoid process is seen projected through the foramen magnum. The mastoid air cells are seen adjacent to the mandibular angle as multiple small air-filled bony spaces. *(Ballinger, vol 2, pp 382–383)*

153. **(C)** A 45° RAO is adequate for evaluation of the left side of the chest. (The side furthest from the film is the side of interest.) However, to separate the aorta and spine, a 55 to 60° left oblique is necessary. Left and right 45° obliques are generally performed to demonstrate the lungs. *(Ballinger, vol 1, p 463)*

154. **(A)** The oblique axial projection is valuable when the zygomatic arches cannot be demonstrated bilaterally with the submentovertical projection, because they are not prominent enough or because of a depressed fracture. The patient may still be positioned as for an SMV, but the head is obliqued 15° toward the side being examined. This serves to move the zygomatic arch away from superimposed structures and provides a slightly oblique axial projection of the arch. *(Ballinger, vol 2, p 320)*

155. **(B)** Sufficient inspiration is demonstrated by the visualization of ten posterior ribs projected above the diaphragm. Rotation of the chest is detected by asymmetry of the distance between the medial ends of the clavicles and vertebral column. The scapulae should be free of superimposition with the lung fields; this is accomplished by rolling the shoulders forward while positioning for the PA. *(Ballinger, vol 1, p 455)*

156. **(A)** The sacroiliac joints angle posteriorly and medially 25° to the MSP. Therefore, in order to demonstrate them with an AP oblique projection, the *affected side* must be elevated 25°. This places the joint space perpendicular to the film and parallel to the central ray. When performed with the PA oblique projection the *unaffected side* will be elevated 25°. *(Ballinger, vol 1, pp 380–381)*

157. **(C)** The AP axial projection is used to project the clavicles from superimposition on the pulmonary apices. A 15 to 20° cephalad angle projects the clavicles above the apices. The radiograph is evaluated for rotation by checking the distance between the medial ends of the clavicles and lateral border of vertebral column. *(Ballinger, vol 1, p 468)*

158. **(C) and 159. (D)** An LPO of the lumbar spine is pictured. The patient is positioned so that the lumbar spine forms a 45° angle with the film. The apophyseal joints (those closest to the film) are well demonstrated in this position. The typical "scotty dog" image is depicted. The "ear" of the scotty is the superior articular process (no. 1) and the front foot is the inferior articular process (4). The scotty's eye is the pedicle (2) and its "body" is the lamina (3). *(Ballinger, vol 1, p 372)*

160. **(C)** The craniocaudal projection is performed in the seated position with the cassette under the breast, snugly against the chest wall. The breast is drawn across the cassette to place the nipple in profile. Using the nipple as the midline, a tumor can be identified as medial or lateral in location. *(Ballinger, vol 2, p 470)*

161. **(A)** The RAO position affords a good view of the pyloric canal and duodenal bulb. It is also a good position for the barium-filled esophagus, projecting it between the vertebrae and heart. The left lateral projection of the stomach demonstrates the left retrogastric space; the recumbent PA is used as a general survey of the gastric surfaces; and the recumbent AP with slight left oblique affords a double contrast of the pylorus and duodenum. *(Ballinger, vol 2, p 114)*

162. **(C)** The sternum has three parts: the uppermost portion is the manubrium (and is quadrilateral in shape), the midportion is the body or gladiolus, and the distal portion is the ensiform or xiphoid process. The sternum supports the clavicles superiorly and provides attachment for the ribs laterally. The first seven pair of ribs are true or vertebrosternal ribs, as they attach directly to the sternum. The ribs angle obliquely anteriorly and inferiorly so that their anterior portions are 3 to 5 inches inferior to their posterior attachment. The sternoclavicular joints afford the only bony attachment between the thorax and upper extremity. *(Ballinger, vol 1, pp 400–403)*

163. **(D)** The inferosuperior axial projection and transthoracic lateral projection are used to evaluate the glenohumeral joint and upper humerus when the patient is unable to abduct the arm (as in dislocation). The scapular Y projection is an oblique projection of the shoulder and is used in demonstrating anterior or posterior dislocation. *(Ballinger, vol 1, pp 132, 142)*

164. **(C) and 165. (D)** An AP projection of the elbow is pictured. The proximal anterior surface of the ulna presents a rather large pointed process at the anterior margin of the semilunar (trochlear) notch called the *coronoid process* (6). The distal anterior humerus presents a depression, the coronoid fossa (1), and several prominences, the capitulum (2), the medial epicondyle (4), and the trochlea (5). *(Saia, PREP, p 78)*

166. **(B)** In the parieto-orbital projection (Rhese method), the patient is prone with the acanthomeatal line perpendicular to the film. The head rests on the forehead, nose, and chin, and the MSP should form 53° with the film (37° with the CR). Radiographically, the optic canal should appear in the *lower outer quadrant* of the orbit. Incorrect *rotation* of the MSP results in *lateral displacement*, and incorrect positioning of the *baseline* results in *longitudinal displacement of the optic canal*. *(Ballinger, vol 2, p 272)*

167. **(D)** Generally, contrast medium is injected into the subarachnoid space between the third and fourth lumbar vertebrae. Because the spinal cord ends at the level of the first or second lumbar vertebrae, this is considered to be a relatively safe injection site. The cisterna magna can be used, but the risk of contrast entering the ventricles and causing side effects increases. Discography requires injection of contrast medium into the individual intervertebral discs. *(Ballinger, vol 2, p 500)*

168. **(D)** To accurately position a lateral forearm, the elbow must form a 90° angle with the humeral epicondyles superimposed. The radius and ulna are superimposed distally. Proximally, the coronoid process and radial head are partially superimposed. Failure of the elbow to form a 90° angle, or the hand to be lateral, results in a less than satisfactory lateral projection of the forearm. *(Ballinger, vol 1, p 102)*

169. **(D)** Because the toes curve naturally downward, the interphalangeal joints are not well demonstrated in the AP (dorsoplantar) projection. In order to "open" the interphalangeal joints the CR should be directed 15° cephalad. Another method is to place a 15° foam sponge wedge under the foot, elevating the toes 15° from the film; the CR would then be directed perpendicularly. *(Ballinger, vol 1, p 186)*

170. **(D)** Double-contrast studies of the stomach or large intestine involve coating the organ with a thin layer of barium sulfate, and then introducing air. This permits seeing through the organ to structures behind it, and most especially allows visualization of the mucosal lining of the organ. A barium-filled stomach or large bowel demonstrates position, size, and shape of the organ, and any lesion that projects out from its walls, such as diverticula. Polypoid lesions, which project inward from the wall of an organ, may go unnoticed unless a double-contrast exam is performed. *(Ehrlich & McCloskey, p 186)*

171. **(D)** Medication is administered parenterally when it cannot be given by mouth. Examples of parenteral routes are subcutaneous, intravenous, intramuscular, or intracardiac. The speed of absorption varies with the route used. *(Torres, p 131)*

172. **(C)** A signed consent form (informed consent) is not necessary prior to performing an upper gastrointestinal series. Informed consent is necessary before performing any procedure that is considered invasive or carries considerable risk. A myelogram, a cardiac catheterization, and an interventional vascular procedure are all invasive procedures, and all carry some degree of risk. A physician should explain to the patient what those risks are, as well as the risk of not having the procedure. Additionally, the patient should be made aware of alternative procedures and the risks associated with the alternatives. Only after the patient has been made aware and all questions answered appropriately, should the informed consent be signed. A radiographer is not responsible for obtaining informed consent. In some institutions, it may be departmental procedure for the radiographer to check the chart and see if there is a signed consent form in place. *(Adler & Carlton, p 350)*

173. **(C)** Another name for an intermittent injection port is a heparin lock. As the name suggests, heparin locks are used for patients who will require intermittent injections. An intravenous catheter is placed in the vein, and an external adapter with a diaphragm allows for repeated injections. Heparin locks provide more freedom than an IV infusion, which also allow for repeated access. Hypodermic needles are usually used for drawing blood, or drawing up fluids, whereas a butterfly needle is usually used for venipuncture. *(Ehrlich & McCloskey, pp 147–149)*

174. **(B)** When performing bedside radiography in a strict isolation room, the radiographer should wear a gown, gloves, mask, and cap. The cassette should be prepared for the examination by placing a pillowcase over the cassette to protect it from contamination. Whenever possible, one radiographer should manipulate the mobile unit and remain "clean," while the other handles the patient. The mobile unit should be cleaned with a disinfectant before exiting the patient's room. *(Ballinger, vol 1, p 10)*

175. **(B)** The American Hospital Association identifies 12 important areas in their "Patients' Bill of Rights." These include the right to refuse treatment (to the extent allowed by law), the right to confidentiality of records and communication, and the right to continuing care. Other patient rights identified are the right to informed consent, privacy, respectful care, access to the records, refusal to participate in research projects, and an explanation of the hospital bill. *(Torres, p 5)*

176. **(B)** The overall chance that a person will become infected with HIV is high with entry sites such as the anus, broken skin, shared needles, infected blood products, and perinatal exposure. Low-risk entry sites include oral and nasal, conjunctiva, and accidental needle stick. *(Hopp, p 68)*

177. **(A)** A radiographer who fails to wear a lead apron when performing portable radiography is in direct violation the American Registry of Radiologic Technologists (ARRT) Code of Ethics for the Profession of Radiologic Technology. Although this may seem to some to be a "personal" decision, the fact is that our profession demands that we protect not only others, but ourselves as well, from unnecessary radiation exposure. Participating in continuing education is every radiographer's duty, and is in keeping with not only the ARRT code of ethics, but is now also mandatory for renewal of ARRT certification. Under normal circumstances, patient confidentiality is of the utmost importance, and radiographers must always respect a patient's right to privacy. There are special circumstances, however, where a radiographer is negligent for not revealing confidential information to the proper individuals. These cases include suspected cases of child abuse, or any instance where the welfare of an individual or community is at risk. The professional radiographer is encouraged to investigate new and innovative techniques. As technology continues to grow, we must grow with it. There are often new and better ways to perform procedures, especially as equipment changes come about. *(Saia, PREP, p 4)*

178. **(D)** There are five commonly used pulse points. Perhaps the most familiar is the radial pulse in the wrist. The femoral pulse in the inguinal region and the carotid pulse in the neck are also commonly used to count pulse rate. Other pulse points are the temporal and the dorsalis pedis of the foot. *(Ehrlich & McCloskey, p 179)*

179. **(A)** Nosocomial infections are infections acquired in hospitals. Despite the efforts of Infectious Disease Departments, nosocomial infections continue to be a problem in hospitals today. This is at least partly due to a greater number of older, more vulnerable patients and an increase in the number of invasive procedures performed today (needles, catheters, and so on). The most frequent site of nosocomial infection is the urinary tract, followed by wounds, respiratory tract, and blood. *(Torres, p 12)*

180. **(D)** Streptococcal pharyngitis (strep throat) is a bacteria. To know this, you have to remember bacteria are classified according to their morphology (i.e. size and shape). The three classifications are spirals, rods (bacilli), and spherical (cocci). A virus, unlike bacteria, cannot live outside of a human cell. Viruses attach themselves to a host cell and invade the cell with its genetic information. Various fungal infections may grow on the skin, cutaneously, or they may enter the skin. Fungal infections that enter the circulatory or lymphatic system can be deadly. Protozoa are one-celled organisms classified by their motility. Ameboids move by locomotion, flagella use their protein tail, cilia posses numerous short protein tails, and sporozoans are actually not mobile. *(Adler & Carlton, pp 196–199)*

181. **(C)** There is no preparation required for an esophagram, unless the upper gastrointestinal tract is routinely examined with the esophagram, as is the routine protocol in some institutions. For an upper gastrointestinal series (UGI) and lower gastrointestinal series or barium enema (BE), the patient should be NPO, or have nothing by mouth for 8 to 10 hours prior to the examination. Additionally, a low-residue diet may be imposed, fluid intake may be increased, and cleansing enemas and laxatives may be prescribed to rid the colon of fecal matter. *(Adler & Carlton, p 307)*

182. **(B)** With the blood pressure cuff wrapped snugly around the patient's brachial artery, and the pump inflated to approximately 180 mm Hg, the valve is opened only slightly to release pressure very slowly. With the stethoscope over the brachial artery, listen for the pulse while watching the Hg column (gauge). Note the point at which the first pulse is heard as the systolic pressure. As the valve is opened further, the sound is louder; the point at which it suddenly becomes softer is recorded as the diastolic pressure. *(Ehrlich & McCloskey, p 81)*

183. **(D)** Epidemiologic studies indicate that AIDS can be transmitted only by intimate contact with body fluids of an infected individual. This can occur through the sharing of contaminated needles, through sexual contact, and from mother to baby at childbirth (perinatal). AIDS can also be transmitted by transfusion of contaminated blood. *(Hopp, p 66)*

184. **(D)** Ipecac is a medication used to induce vomiting and is classified as an emetic. This is easy to remember if you think of what an emesis basin is for. A diuretic is a medication that stimulates the production of urine. Lasix (furosemide) is an example of a diuretic. An antipyretic is used to reduce fever. Tylenol (acetaminophen) is an example of an antipyretic. An antihistamine is used to relieve allergic effects. Benadryl (diphenhydramine hydrochloride) is an example of an antihistamine often on hand in radiology departments, in the event of a minor reaction to contrast media. *(Ehrlich & McCloskey, p 136)*

185. **(B)** Sterile technique is required for administration of contrast media by the intravenous and intrathecal (intraspinal) methods. Sterile technique is also required for injection of contrast media during arthrography. Aseptic technique is used for administration of contrast media by means of the oral and rectal routes, as well as through the nasogastric tube. *(Torres, pp 178–179)*

186. **(A)** Proper body mechanics includes a wide base of support. The base of support is the part of the body in touch with the floor or other horizontal plane. The center of gravity is the midpoint of the pelvis or lower abdomen, depending on body build. The line of gravity is the abstract line passing through the center of gravity, vertically. Proper body mechanics can help prevent painful back injuries by making proficient use of the muscles in the arms and legs. *(Ehrlich & McCloskey, p 61)*

187. **(C)** A radiolucent contrast agent is composed of elements of low atomic number. An example of a radiolucent contrast agent is air. Air permits anatomy to appear dark on a radiograph, because it absorbs a low number of x-ray photons, so the photons pass through

the anatomy more easily. The opposite is true of positive contrast agents, such as barium or iodine, where an element with a high atomic number causes the outlined anatomy to appear light on the resulting radiograph because many photons are absorbed, and a smaller number pass entirely through the anatomy. *(Adler & Carlton, p 301)*

188. **(B)** The most commonly used method of low-flow oxygen delivery is the nasal cannula. It can be used to deliver oxygen from 1 to 4 mL/min at concentrations of 24 to 36 percent. The nasal cannula also provides increased patient freedom to eat and talk, that a mask does not. Masks are used for higher-flow concentrations of oxygen, over 5 mL/min, and depending on the type of mask, can deliver anywhere from 35 to 60 percent oxygen. Respirators or ventilators are high-flow delivery mechanisms and are used for patients who are in severe respiratory distress or unable to breathe on their own. Oxyhoods or tents are generally used for pediatric patients who may not tolerate a mask or cannula. The amount of oxygen delivered is somewhat unpredictable, especially if the opening is frequently accessed. Oxygen delivery may be between 20 and 100 percent. *(Adler & Carlton, pp 188–190)*

189. **(D)** A patient who has been recumbent for some period of time and quickly gets up may suffer from lightheadedness or feel faint. This is referred to as orthostatic hypertension. It is best to have patients sit up and dangle their feet from the table, while being supported for a moment, and then assist them off the table. Patients will also feel better emotionally if they are not rushed or treated like they are on an assembly line. Always assist patients on and off of the radiographic table. Even healthy, young outpatients can injure themselves. Patients with dyspnea or orthopnea are unable to lie supine. Dyspnea and orthopnea refer to difficulty breathing and may be due to a heart condition, asthma, strenuous exercise, or excessive anxiety. Hypertension refers to the condition of elevated blood pressure. *(Adler & Carlton, pp 148–149)*

190. **(D)** Most needle sticks occur while attempting to recap a needle. Several diseases, including hepatitis and HIV, can be transmitted via a needle stick. Therefore, do not attempt to recap a needle, but rather, dispose of the entire syringe with needle attached in the special container available. (*Ehrlich & McCloskey, p 97*)

191. **(C)** Angina pectoris is a spasmodic chest pain frequently due to oxygen deficiency in the myocardium. The pain often radiates down the left arm and up to the left jaw. Angina pectoris attacks are frequently associated with exertion or emotional stress in individuals with coronary artery disease. Pain may be relieved with a vasodilator such as nitroglycerine given sublingually or transdermally. Digitalis is used to treat congestive heart failure. Dilantin is used in the control of seizure disorders, and Tagamet is used to treat duodenal ulcers. (*Torres, p 174*)

192. **(C)** A patient with a respiratory disease can transmit infectious organisms via airborne contamination (if the patient sneezes or coughs). Therefore, patients with upper respiratory infection should be transported wearing a mask, to prevent the possibility of airborne contamination. It is not necessary for the radiographer to be masked. (*Ehrlich & McCloskey, p 103*)

193. **(D)** All of the choices listed in the question should be part of a preliminary patient history before deciding to inject ionic or nonionic contrast media. As patients age, their general health decreases, and they are therefore more likely to suffer from adverse reactions. Patients with a history of respiratory disease such as asthma or emphysema, and COPD (chronic obstructive pulmonary disease), are more likely to have a reaction, and suffer greater distress in the event of a reaction. Patients with cardiac disease run an increased risk of changes in heart rate, and myocardial infarction. Patients should also be screened for decreased renal or hepatic function, sickle-cell disease, diabetes, and pregnancy. (*Adler & Carlton, pp 314–317*)

194. **(D)** While performing radiography, improper support of a patient's fractured lower leg (tibia/fibula) could result in movement of the fracture fragments, which can cause tearing of the soft tissue, nerves, and blood vessels. Additionally, lack of support may cause muscle spasm, which can make closed reduction of some fractures difficult. (*Ehrlich & McCloskey, p 180*)

195. **(C)** Invasion of privacy, that is, public discussion of privileged and confidential information, is intentional misconduct. False imprisonment, such as unnecessarily restraining a patient, is also intentional misconduct. However, if a radiographer left a weak patient standing while leaving the room to check films or get supplies, and the patient fell and sustained injury, that would be considered unintentional misconduct or negligence. (*Gurley, pp 123–124*)

196. **(D)** When scheduling patient examinations, it is important to avoid the possibility of residual contrast medium covering areas of interest on later examinations. The IVP should be scheduled first, because the contrast medium used is excreted rapidly. The BE should be scheduled next. Lastly, the GI is scheduled. Any barium remaining from the previous BE should not be enough to interfere with the stomach or duodenum (a preliminary scout film should be taken in each case). (*Ehrlich & McCloskey, p 170*)

197. **(B)** Universal blood and body fluid precautions serve to protect health care workers and patients from the spread of diseases such as AIDS and AIDS-related complex. Although the precautions are indicated for *all* patients, special care must be emphasized when working with patients whose infectious status is unknown (for example, the emergency trauma patient). Gloves must be worn if the radiographer may come in contact with blood or body fluids. A gown should be worn if the clothing may become contaminated. Blood spills should be cleaned with a solution of one part bleach to ten parts water. (*Torres, p 30*)

198. **(B)** Emphysema is a COPD characterized by pathologic distention of the pulmonary alveoli with (destructive) changes in their walls, resulting in a loss of elasticity. Emphysema is occasionally seen following asthma or TB, but most frequently is caused by cigarette smoking. Because the emphysematous patient's greatest difficulty is exhalation, it becomes a conscious, forced effort. Breathing is shallow and rapid. Forced and ineffective breathing results in expansion of the AP diameter of the chest and clubbed fingertips in established emphysema. Hyperventilation results from too frequent deep breaths in the anxious or tense individual. This results in a feeling of dizziness and tingling of the extremities. *(Ehrlich & McCloskey, p 80)*

199. **(B)** The Hemovac or Penrose drains are used for tissue drainage of wounds or in postoperative drainage. Drainage tubes help prevent the formation of infection or fistulas in wound and postoperative sites with large amounts of drainage. Bile duct drainage, when necessary, is performed with a T-tube, and radiographers often perform radiographic examinations of the T-tube to verify patency. Nasogastric and nasoenteric tubes may be used for either decompression of the gastrointestinal tract or to feed patients who are unable to swallow food normally. Additionally, radiographic examination of the gastrointestinal tract may be performed by introducing a contrast agent into a nasogastric or nasoenteric tube. *(Torres, p 158)*

200. **(A)** Heparin is produced by the body (especially in the liver) and functions to prevent intravascular clotting. Heparin is also produced artificially and used to treat thromboembolytic disorders. Lidocaine and Benedryl are drugs usually available on "crash carts" for emergency use. Lidocaine is used to treat ventricular arrythmias and Benedryl is used to treat allergic reactions and acute anaphylaxis. *(Torres, p 174)*

Subspecialty List

52. Equipment operation and maintenance (Chapter 5)
53. Equipment operation and maintenance (Chapter 5)
54. Equipment operation and maintenance (Chapter 5)
55. Equipment operation and maintenance (Chapter 5)
56. Equipment operation and maintenance (Chapter 5)
57. Equipment operation and maintenance (Chapter 5)
58. Equipment operation and maintenance (Chapter 5)
59. Equipment operation and maintenance (Chapter 5)
60. Equipment operation and maintenance (Chapter 5)
61. Image production and evaluation (Chapter 4)
62. Image production and evaluation (Chapter 4)
63. Image production and evaluation (Chapter 4)
64. Image production and evaluation (Chapter 4)
65. Image production and evaluation (Chapter 4)
66. Image production and evaluation (Chapter 4)
67. Image production and evaluation (Chapter 4)
68. Image production and evaluation (Chapter 4)
69. Image production and evaluation (Chapter 4)
70. Image production and evaluation (Chapter 4)
71. Image production and evaluation (Chapter 4)
72. Image production and evaluation (Chapter 4)
73. Image production and evaluation (Chapter 4)
74. Image production and evaluation (Chapter 4)
75. Image production and evaluation (Chapter 4)
76. Image production and evaluation (Chapter 4)
77. Image production and evaluation (Chapter 4)
78. Image production and evaluation (Chapter 4)
79. Image production and evaluation (Chapter 4)
80. Image production and evaluation (Chapter 4)
81. Image production and evaluation (Chapter 4)
82. Image production and evaluation (Chapter 4)
83. Image production and evaluation (Chapter 4)
84. Image production and evaluation (Chapter 4)
85. Image production and evaluation (Chapter 4)
86. Image production and evaluation (Chapter 4)
87. Image production and evaluation (Chapter 4)
88. Image production and evaluation (Chapter 4)
89. Image production and evaluation (Chapter 4)
90. Image production and evaluation (Chapter 4)
91. Image production and evaluation (Chapter 4)
92. Image production and evaluation (Chapter 4)
93. Image production and evaluation (Chapter 4)
94. Image production and evaluation (Chapter 4)
95. Image production and evaluation (Chapter 4)
96. Image production and evaluation (Chapter 4)
97. Image production and evaluation (Chapter 4)
98. Image production and evaluation (Chapter 4)
99. Image production and evaluation (Chapter 4)
100. Image production and evaluation (Chapter 4)
101. Image production and evaluation (Chapter 4)
102. Image production and evaluation (Chapter 4)
103. Image production and evaluation (Chapter 4)
104. Image production and evaluation (Chapter 4)
105. Image production and evaluation (Chapter 4)
106. Image production and evaluation (Chapter 4)
107. Image production and evaluation (Chapter 4)
108. Image production and evaluation (Chapter 4)
109. Image production and evaluation (Chapter 4)
110. Image production and evaluation (Chapter 4)
111. Radiographic procedures (Chapter 2)
112. Radiographic procedures (Chapter 2)
113. Radiographic procedures (Chapter 2)
114. Radiographic procedures (Chapter 2)
115. Radiographic procedures (Chapter 2)
116. Radiographic procedures (Chapter 2)
117. Radiographic procedures (Chapter 2)
118. Radiographic procedures (Chapter 2)
119. Radiographic procedures (Chapter 2)
120. Radiographic procedures (Chapter 2)
121. Radiographic procedures (Chapter 2)
122. Radiographic procedures (Chapter 2)
123. Radiographic procedures (Chapter 2)
124. Radiographic procedures (Chapter 2)
125. Radiographic procedures (Chapter 2)
126. Radiographic procedures (Chapter 2)
127. Radiographic procedures (Chapter 2)
128. Radiographic procedures (Chapter 2)
129. Radiographic procedures (Chapter 2)
130. Radiographic procedures (Chapter 2)
131. Radiographic procedures (Chapter 2)
132. Radiographic procedures (Chapter 2)
133. Radiographic procedures (Chapter 2)
134. Radiographic procedures (Chapter 2)
135. Radiographic procedures (Chapter 2)
136. Radiographic procedures (Chapter 2)
137. Radiographic procedures (Chapter 2)
138. Radiographic procedures (Chapter 2)
139. Radiographic procedures (Chapter 2)
140. Radiographic procedures (Chapter 2)
141. Radiographic procedures (Chapter 2)
142. Radiographic procedures (Chapter 2)

143. Radiographic procedures (Chapter 2)
144. Radiographic procedures (Chapter 2)
145. Radiographic procedures (Chapter 2)
146. Radiographic procedures (Chapter 2)
147. Radiographic procedures (Chapter 2)
148. Radiographic procedures (Chapter 2)
149. Radiographic procedures (Chapter 2)
150. Radiographic procedures (Chapter 2)
151. Radiographic procedures (Chapter 2)
152. Radiographic procedures (Chapter 2)
153. Radiographic procedures (Chapter 2)
154. Radiographic procedures (Chapter 2)
155. Radiographic procedures (Chapter 2)
156. Radiographic procedures (Chapter 2)
157. Radiographic procedures (Chapter 2)
158. Radiographic procedures (Chapter 2)
159. Radiographic procedures (Chapter 2)
160. Radiographic procedures (Chapter 2)
161. Radiographic procedures (Chapter 2)
162. Radiographic procedures (Chapter 2)
163. Radiographic procedures (Chapter 2)
164. Radiographic procedures (Chapter 2)
165. Radiographic procedures (Chapter 2)
166. Radiographic procedures (Chapter 2)
167. Radiographic procedures (Chapter 2)
168. Radiographic procedures (Chapter 2)
169. Radiographic procedures (Chapter 2)
170. Radiographic procedures (Chapter 2)
171. Patient care and management (Chapter 1)
172. Patient care and management (Chapter 1)
173. Patient care and management (Chapter 1)
174. Patient care and management (Chapter 1)
175. Patient care and management (Chapter 1)
176. Patient care and management (Chapter 1)
177. Patient care and management (Chapter 1)
178. Patient care and management (Chapter 1)
179. Patient care and management (Chapter 1)
180. Patient care and management (Chapter 1)
181. Patient care and management (Chapter 1)
182. Patient care and management (Chapter 1)
183. Patient care and management (Chapter 1)
184. Patient care and management (Chapter 1)
185. Patient care and management (Chapter 1)
186. Patient care and management (Chapter 1)
187. Patient care and management (Chapter 1)
188. Patient care and management (Chapter 1)
189. Patient care and management (Chapter 1)
190. Patient care and management (Chapter 1)
191. Patient care and management (Chapter 1)
192. Patient care and management (Chapter 1)
193. Patient care and management (Chapter 1)
194. Patient care and management (Chapter 1)
195. Patient care and management (Chapter 1)
196. Patient care and management (Chapter 1)
197. Patient care and management (Chapter 1)
198. Patient care and management (Chapter 1)
199. Patient care and management (Chapter 1)
200. Patient care and management (Chapter 1)

NAME _____

ADDRESS _____

Street

City State Zip

DIRECTIONS Mark your social security number from top to bottom in the appropriate boxes on the right.

MAKE
ERASURES
COMPLETE

PLEASE USE NO.2 PENCIL ONLY.

SOC SEC NUMBER

	S O C

↓BEGIN HERE

01 Ⓐ Ⓑ Ⓒ Ⓓ	21 Ⓐ Ⓑ Ⓒ Ⓓ	41 Ⓐ Ⓑ Ⓒ Ⓓ	61 Ⓐ Ⓑ Ⓒ Ⓓ
02 Ⓐ Ⓑ Ⓒ Ⓓ	22 Ⓐ Ⓑ Ⓒ Ⓓ	42 Ⓐ Ⓑ Ⓒ Ⓓ	62 Ⓐ Ⓑ Ⓒ Ⓓ
03 Ⓐ Ⓑ Ⓒ Ⓓ	23 Ⓐ Ⓑ Ⓒ Ⓓ	43 Ⓐ Ⓑ Ⓒ Ⓓ	63 Ⓐ Ⓑ Ⓒ Ⓓ
04 Ⓐ Ⓑ Ⓒ Ⓓ	24 Ⓐ Ⓑ Ⓒ Ⓓ	44 Ⓐ Ⓑ Ⓒ Ⓓ	64 Ⓐ Ⓑ Ⓒ Ⓓ
05 Ⓐ Ⓑ Ⓒ Ⓓ	25 Ⓐ Ⓑ Ⓒ Ⓓ	45 Ⓐ Ⓑ Ⓒ Ⓓ	65 Ⓐ Ⓑ Ⓒ Ⓓ
06 Ⓐ Ⓑ Ⓒ Ⓓ	26 Ⓐ Ⓑ Ⓒ Ⓓ	46 Ⓐ Ⓑ Ⓒ Ⓓ	66 Ⓐ Ⓑ Ⓒ Ⓓ
07 Ⓐ Ⓑ Ⓒ Ⓓ	27 Ⓐ Ⓑ Ⓒ Ⓓ	47 Ⓐ Ⓑ Ⓒ Ⓓ	67 Ⓐ Ⓑ Ⓒ Ⓓ
08 Ⓐ Ⓑ Ⓒ Ⓓ	28 Ⓐ Ⓑ Ⓒ Ⓓ	48 Ⓐ Ⓑ Ⓒ Ⓓ	68 Ⓐ Ⓑ Ⓒ Ⓓ
09 Ⓐ Ⓑ Ⓒ Ⓓ	29 Ⓐ Ⓑ Ⓒ Ⓓ	49 Ⓐ Ⓑ Ⓒ Ⓓ	69 Ⓐ Ⓑ Ⓒ Ⓓ
10 Ⓐ Ⓑ Ⓒ Ⓓ	30 Ⓐ Ⓑ Ⓒ Ⓓ	50 Ⓐ Ⓑ Ⓒ Ⓓ	70 Ⓐ Ⓑ Ⓒ Ⓓ
11 Ⓐ Ⓑ Ⓒ Ⓓ	31 Ⓐ Ⓑ Ⓒ Ⓓ	51 Ⓐ Ⓑ Ⓒ Ⓓ	71 Ⓐ Ⓑ Ⓒ Ⓓ
12 Ⓐ Ⓑ Ⓒ Ⓓ	32 Ⓐ Ⓑ Ⓒ Ⓓ	52 Ⓐ Ⓑ Ⓒ Ⓓ	72 Ⓐ Ⓑ Ⓒ Ⓓ
13 Ⓐ Ⓑ Ⓒ Ⓓ	33 Ⓐ Ⓑ Ⓒ Ⓓ	53 Ⓐ Ⓑ Ⓒ Ⓓ	73 Ⓐ Ⓑ Ⓒ Ⓓ
14 Ⓐ Ⓑ Ⓒ Ⓓ	34 Ⓐ Ⓑ Ⓒ Ⓓ	54 Ⓐ Ⓑ Ⓒ Ⓓ	74 Ⓐ Ⓑ Ⓒ Ⓓ
15 Ⓐ Ⓑ Ⓒ Ⓓ	35 Ⓐ Ⓑ Ⓒ Ⓓ	55 Ⓐ Ⓑ Ⓒ Ⓓ	75 Ⓐ Ⓑ Ⓒ Ⓓ
16 Ⓐ Ⓑ Ⓒ Ⓓ	36 Ⓐ Ⓑ Ⓒ Ⓓ	56 Ⓐ Ⓑ Ⓒ Ⓓ	76 Ⓐ Ⓑ Ⓒ Ⓓ
17 Ⓐ Ⓑ Ⓒ Ⓓ	37 Ⓐ Ⓑ Ⓒ Ⓓ	57 Ⓐ Ⓑ Ⓒ Ⓓ	77 Ⓐ Ⓑ Ⓒ Ⓓ
18 Ⓐ Ⓑ Ⓒ Ⓓ	38 Ⓐ Ⓑ Ⓒ Ⓓ	58 Ⓐ Ⓑ Ⓒ Ⓓ	78 Ⓐ Ⓑ Ⓒ Ⓓ
19 Ⓐ Ⓑ Ⓒ Ⓓ	39 Ⓐ Ⓑ Ⓒ Ⓓ	59 Ⓐ Ⓑ Ⓒ Ⓓ	79 Ⓐ Ⓑ Ⓒ Ⓓ
20 Ⓐ Ⓑ Ⓒ Ⓓ	40 Ⓐ Ⓑ Ⓒ Ⓓ	50 Ⓐ Ⓑ Ⓒ Ⓓ	80 Ⓐ Ⓑ Ⓒ Ⓓ

081 Ⓐ Ⓑ Ⓒ Ⓓ   111 Ⓐ Ⓑ Ⓒ Ⓓ   141 Ⓐ Ⓑ Ⓒ Ⓓ   171 Ⓐ Ⓑ Ⓒ Ⓓ

082 Ⓐ Ⓑ Ⓒ Ⓓ   112 Ⓐ Ⓑ Ⓒ Ⓓ   142 Ⓐ Ⓑ Ⓒ Ⓓ   172 Ⓐ Ⓑ Ⓒ Ⓓ

083 Ⓐ Ⓑ Ⓒ Ⓓ   113 Ⓐ Ⓑ Ⓒ Ⓓ   143 Ⓐ Ⓑ Ⓒ Ⓓ   173 Ⓐ Ⓑ Ⓒ Ⓓ

084 Ⓐ Ⓑ Ⓒ Ⓓ   114 Ⓐ Ⓑ Ⓒ Ⓓ   144 Ⓐ Ⓑ Ⓒ Ⓓ   174 Ⓐ Ⓑ Ⓒ Ⓓ

085 Ⓐ Ⓑ Ⓒ Ⓓ   115 Ⓐ Ⓑ Ⓒ Ⓓ   145 Ⓐ Ⓑ Ⓒ Ⓓ   175 Ⓐ Ⓑ Ⓒ Ⓓ

086 Ⓐ Ⓑ Ⓒ Ⓓ   116 Ⓐ Ⓑ Ⓒ Ⓓ   146 Ⓐ Ⓑ Ⓒ Ⓓ   176 Ⓐ Ⓑ Ⓒ Ⓓ

087 Ⓐ Ⓑ Ⓒ Ⓓ   117 Ⓐ Ⓑ Ⓒ Ⓓ   147 Ⓐ Ⓑ Ⓒ Ⓓ   177 Ⓐ Ⓑ Ⓒ Ⓓ

088 Ⓐ Ⓑ Ⓒ Ⓓ   118 Ⓐ Ⓑ Ⓒ Ⓓ   148 Ⓐ Ⓑ Ⓒ Ⓓ   178 Ⓐ Ⓑ Ⓒ Ⓓ

089 Ⓐ Ⓑ Ⓒ Ⓓ   119 Ⓐ Ⓑ Ⓒ Ⓓ   149 Ⓐ Ⓑ Ⓒ Ⓓ   179 Ⓐ Ⓑ Ⓒ Ⓓ

090 Ⓐ Ⓑ Ⓒ Ⓓ   120 Ⓐ Ⓑ Ⓒ Ⓓ   150 Ⓐ Ⓑ Ⓒ Ⓓ   180 Ⓐ Ⓑ Ⓒ Ⓓ

091 Ⓐ Ⓑ Ⓒ Ⓓ   121 Ⓐ Ⓑ Ⓒ Ⓓ   151 Ⓐ Ⓑ Ⓒ Ⓓ   181 Ⓐ Ⓑ Ⓒ Ⓓ

092 Ⓐ Ⓑ Ⓒ Ⓓ   122 Ⓐ Ⓑ Ⓒ Ⓓ   152 Ⓐ Ⓑ Ⓒ Ⓓ   182 Ⓐ Ⓑ Ⓒ Ⓓ

093 Ⓐ Ⓑ Ⓒ Ⓓ   123 Ⓐ Ⓑ Ⓒ Ⓓ   153 Ⓐ Ⓑ Ⓒ Ⓓ   183 Ⓐ Ⓑ Ⓒ Ⓓ

094 Ⓐ Ⓑ Ⓒ Ⓓ   124 Ⓐ Ⓑ Ⓒ Ⓓ   154 Ⓐ Ⓑ Ⓒ Ⓓ   184 Ⓐ Ⓑ Ⓒ Ⓓ

095 Ⓐ Ⓑ Ⓒ Ⓓ   125 Ⓐ Ⓑ Ⓒ Ⓓ   155 Ⓐ Ⓑ Ⓒ Ⓓ   185 Ⓐ Ⓑ Ⓒ Ⓓ

096 Ⓐ Ⓑ Ⓒ Ⓓ   126 Ⓐ Ⓑ Ⓒ Ⓓ   156 Ⓐ Ⓑ Ⓒ Ⓓ   186 Ⓐ Ⓑ Ⓒ Ⓓ

097 Ⓐ Ⓑ Ⓒ Ⓓ   127 Ⓐ Ⓑ Ⓒ Ⓓ   157 Ⓐ Ⓑ Ⓒ Ⓓ   187 Ⓐ Ⓑ Ⓒ Ⓓ

098 Ⓐ Ⓑ Ⓒ Ⓓ   128 Ⓐ Ⓑ Ⓒ Ⓓ   158 Ⓐ Ⓑ Ⓒ Ⓓ   188 Ⓐ Ⓑ Ⓒ Ⓓ

099 Ⓐ Ⓑ Ⓒ Ⓓ   129 Ⓐ Ⓑ Ⓒ Ⓓ   159 Ⓐ Ⓑ Ⓒ Ⓓ   189 Ⓐ Ⓑ Ⓒ Ⓓ

100 Ⓐ Ⓑ Ⓒ Ⓓ   130 Ⓐ Ⓑ Ⓒ Ⓓ   160 Ⓐ Ⓑ Ⓒ Ⓓ   190 Ⓐ Ⓑ Ⓒ Ⓓ

101 Ⓐ Ⓑ Ⓒ Ⓓ   131 Ⓐ Ⓑ Ⓒ Ⓓ   161 Ⓐ Ⓑ Ⓒ Ⓓ   191 Ⓐ Ⓑ Ⓒ Ⓓ

102 Ⓐ Ⓑ Ⓒ Ⓓ   132 Ⓐ Ⓑ Ⓒ Ⓓ   162 Ⓐ Ⓑ Ⓒ Ⓓ   192 Ⓐ Ⓑ Ⓒ Ⓓ

103 Ⓐ Ⓑ Ⓒ Ⓓ   133 Ⓐ Ⓑ Ⓒ Ⓓ   163 Ⓐ Ⓑ Ⓒ Ⓓ   193 Ⓐ Ⓑ Ⓒ Ⓓ

104 Ⓐ Ⓑ Ⓒ Ⓓ   134 Ⓐ Ⓑ Ⓒ Ⓓ   164 Ⓐ Ⓑ Ⓒ Ⓓ   194 Ⓐ Ⓑ Ⓒ Ⓓ

105 Ⓐ Ⓑ Ⓒ Ⓓ   135 Ⓐ Ⓑ Ⓒ Ⓓ   165 Ⓐ Ⓑ Ⓒ Ⓓ   195 Ⓐ Ⓑ Ⓒ Ⓓ

106 Ⓐ Ⓑ Ⓒ Ⓓ   136 Ⓐ Ⓑ Ⓒ Ⓓ   166 Ⓐ Ⓑ Ⓒ Ⓓ   196 Ⓐ Ⓑ Ⓒ Ⓓ

107 Ⓐ Ⓑ Ⓒ Ⓓ   137 Ⓐ Ⓑ Ⓒ Ⓓ   167 Ⓐ Ⓑ Ⓒ Ⓓ   197 Ⓐ Ⓑ Ⓒ Ⓓ

108 Ⓐ Ⓑ Ⓒ Ⓓ   138 Ⓐ Ⓑ Ⓒ Ⓓ   168 Ⓐ Ⓑ Ⓒ Ⓓ   198 Ⓐ Ⓑ Ⓒ Ⓓ

109 Ⓐ Ⓑ Ⓒ Ⓓ   139 Ⓐ Ⓑ Ⓒ Ⓓ   169 Ⓐ Ⓑ Ⓒ Ⓓ   199 Ⓐ Ⓑ Ⓒ Ⓓ

110 Ⓐ Ⓑ Ⓒ Ⓓ   140 Ⓐ Ⓑ Ⓒ Ⓓ   170 Ⓐ Ⓑ Ⓒ Ⓓ   200 Ⓐ Ⓑ Ⓒ Ⓓ

APPLETON & LANGE REVIEW SERIES

Health Related

Appleton & Lange's Review of Cardio-vascular-Interventional Technology
Vitanza
1995, ISBN 0-8385-0248-2

Appleton & Lange's Review for the Chiropractic National Boards, Part I
Shanks
1992, ISBN 0-8385-0224-5

Appleton & Lange's Review for the Dental Assistant, 3/e
Andujo
1992, ISBN 0-8385-0135-4

Appleton & Lange's Review for the Dental Hygiene National Board Review, 4/e
Barnes and Waring
1995, ISBN 0-8385-0230-X

Appleton & Lange's Review for the Medical Assistant, 4/e
Palko and Palko
1994, ISBN 0-8385-0197-4

Medical Technology Examination Review, 2/e
Hossaini
1984, ISBN 0-8385-6283-3

Appleton & Lange's Review of Pharmacy, 5/e
Hall and Reiss
1993, ISBN 0-8385-0162-1
Appleton & Lange's Review for the

Appleton & Lange's Review for the Radiography Examination, 2/e
Saia
1993, ISBN 0-8385-0058-7

Radiography: Program Review & Exam Preparation (PREP)
Saia
1996, ISBN 0-8385-8244-3

Appleton & Lange's Review for the Surgical Technology Examination, 4/e
Allmers and Verderame
June 1996, ISBN 0-8385-0270-9

Appleton & Lange's Review for the Ultrasonography Examination, 2/e
Odwin
1993, ISBN 0-8385-9073-X

First Aid

1996 First Aid for the USMLE Step 1
A Student-to-Student Guide
Bhushan, Le, and Amin
1996, ISBN 0-8385-2597-0

First Aid for the USMLE Step 2
A Student-to-Student Guide
Go, Curet-Salim, and Fullerton
1996, ISBN 0-8385-2591-1

First Aid for the Wards
A Student-to-Student Guide
Le, Bhushan, and Amin
1996, ISBN 0-8385-2596-2

First Aid for the Match
A Student-to-Student Guide
Le, Bhushan, and Amin
1996, ISBN 0-8385-2596-2

Instant Exam

The Instant Exam Review for the USMLE Step 2, 2/e
Goldberg
1996, ISBN 0-8385-4328-6

The Instant Exam Review for the USMLE Step 3
Goldberg
1994, ISBN 0-8385-4334-0

Comprehensive A&L Reviews

Appleton & Lange's Review for the USMLE Step 1, 2/e
Barton
1996, ISBN 0-8385-0265-2

Appleton & Lange's Review for the USMLE Step 2, 2/e
Catlin
1996, ISBN 0-8385-0266-0

Appleton & Lange's Review for the USMLE Step 3
Schultz
1994, ISBN 0-8385-0227-X

Basic Science

Appleton & Lange's Review of Anatomy for the USMLE Step 1, 5/e
Montgomery
1995, ISBN 0-8385-0246-6

Appleton & Lange's Review of Epidemiology & Biostatistics for the USMLE
Hanrahan and Madupu
1994, ISBN 0-8385-0244-X

Appleton & Lange's Review of Microbiology and Immunology for the USMLE Step 1, 3/e
Yotis
October 1996, ISBN 0-8385-0273-3

Appleton & Lange's Review of General Pathology, 3/e
Lewis and Barton
1993, ISBN 0-8385-0161-3

Clinical Science

Appleton & Lange's Review of Internal Medicine
Goldlist
1996, ISBN 0-8385-0251-2

Appleton & Lange's Review of Obstetrics and Gynecology, 5/e
Julian, et al.
1995, ISBN 0-8385-0231-8

Appleton & Lange's Review of Pediatrics, 5/e
Lorin
1993, ISBN 0-8385-0057-9

Appleton & Lange's Review of Psychiatry, 5/e
Easson
1994, ISBN 0-8385-0247-4

Public Health and Preventive Medicine Review, 2/e
Penalver
1984, ISBN 0-8385-5936-2

Specialty Board Reviews

The MGH Board Review of Anesthesiology, 4/e
Dershwitz
1994, ISBN 0-8385-8611-4

(More on Reverse)

APPLETON & LANGE QUICK REVIEW SERIES

Health Related

MEPC: Medical Assistant
Examination Review, 4/e
Dreizen and Audet
1989, ISBN 0-8385-5772-4

MEPC: Medical Record,
Examination Review, 6/e
Bailey
1994, ISBN 0-8385-6192-6

MEPC: Occupational Therapy
Examination Review, 5/e
Dundon
1988, ISBN 0-8385-7204-9

MEPC: Optometry
Examination Review, 4/e
Casser et al.
1994, ISBN 0-8385-7449-1

MEPC: Physician Assistant
Examination Review, 3/e
Rahr and Niebuhr
1996, ISBN 0-8385-8094-7

Comprehensive A & L Reviews

MEPC: USMLE Step 1 Review
Fayemi
1995, ISBN 0-8385-6269-8

MEPC: USMLE Step 2 Review
Jacobs
1996, ISBN 0-8385-6270-1, A6270-1

MEPC: USMLE Step 3 Review
Chan
1996, ISBN 0-8385-6339-2, A6339-4

Basic Science

MEPC: Anatomy, 10/e
A USMLE STEP 1 Review
Wilson
1995, ISBN 0-8385-6218-3

MEPC: Biochemistry, 11/e
A USMLE STEP 1 Review
Glick
1995, ISBN 0-8385-5779-1

MEPC: Microbiology, 11/e
A USMLE Step 1 Review
Kim
1995, ISBN 0-8385-6308-2

MEPC: Pathology, 10/e
A USMLE STEP 1 Review
Fayemi
1994, ISBN 0-8385-8441-1

MEPC: Pharmacology, 8/e
A USMLE STEP 1 Review
Krzanowski et al.
1995, ISBN 0-8385-6227-2

MEPC: Physiology, 9/e
A USMLE STEP 1 Review
Penney
1995, ISBN 0-8385-6222-1

Clinical Science

MEPC: Neurology, 10/e
A USMLE Step 2 Review
Slosberg
1993, ISBN 0-8385-5778-3

MEPC: Pediatrics, 9/e
A USMLE Step 2 Review
Hansbarger1995, ISBN 0-8385-6223-X

MEPC: Preventive Medicine and Public Health, 10/e
A USMLE Step 2 Review
Hart
June 1996, ISBN 0-8385-6319-8

MEPC: Psychiatry, 10/e
A USMLE Step 2 Review
Chan and Prosen
1995, ISBN 0-8385-5780-5

MEPC: Surgery, 11/e
A USMLE Step 2 Review
Metzler
1995, ISBN 0-8385-6195-0

Specialty Board Reviews

MEPC: Anesthesiology, 9/e
Specialty Board Review
Dekornfeld and Sanford
1995, ISBN 0-8385-0256-3

MEPC: Otolaryngology
Specialty Board Review
Head & Neck Surgery
Willett and Lee
1995, ISBN 0-8385-7580-3

MEPC: Neurology, 4/e
Specialty Board Review
Giesser and Kanof
1995, ISBN 0-8385-8650-3

To order or for more information,
visit your local health science bookstore
or call Appleton & Lange toll free at
1-800-423-1359.